EPIZOOTIOLOGY
OF INSECT DISEASES

EPIZOOTIOLOGY
OF INSECT DISEASES

Edited by

JAMES R. FUXA

Louisiana State University

YOSHINORI TANADA

University of California at Berkeley

A Wiley-Interscience Publication

JOHN WILEY & SONS

New York / Chichester / Brisbane / Toronto / Singapore

Library of Congress Cataloging-in-Publication Data:

Epizootiology of insect diseases.

    "A Wiley-Interscience publication."
    Includes indexes.
    1. Insects—Diseases—Epidemiology. I. Fuxa,
James R. II. Tanada, Yoshinori.
SB942.E65 1987   595.7′023   86-26795
ISBN 0-471-87812-X

Printed in the United States of America

10 9 8 7 6 5 4 3 2 1

CONTRIBUTORS

Theodore G. Andreadis, Department of Entomology, Connecticut Agricultural Experiment Station, New Haven, Connecticut

Georg Benz, Department of Entomology, Swiss Federal Institute of Technology, Zurich, Switzerland

Grayson C. Brown, Department of Entomology, University of Kentucky, Lexington, Kentucky

Raymond I. Carruthers, United States Department of Agriculture, Agricultural Research Service, Plant Protection Research Unit, Boyce Thompson Institute, Cornell University, Ithaca, New York

Philip F. Entwistle, Natural Environment Research Council, Institute of Virology, Oxford, England

Hugh F. Evans, Research and Development Division, Forestry Commission, Farnham, Surrey, England

James R. Fuxa, Department of Entomology, Louisiana State University Agricultural Center, Louisiana State University, Baton Rouge, Louisiana

Randy R. Gaugler, Department of Entomology and Economic Zoology, Rutgers University, New Brunswick, New Jersey

James D. Harper, Department of Zoology and Entomology, Alabama Agricultural Experiment Station, Auburn University, Auburn, Alabama

Tosihiko Hukuhara, Faculty of Agriculture, Tokyo University of Agriculture and Technology, Fuchu, Tokyo, Japan

Harry K. Kaya, Division of Nematology, Department of Entomology, University of California, Davis, California

Aloysius Krieg, Biologische Bundesanstalt für Land-und Forstwirtschaft, Darmstadt, Federal Republic of Germany

Joseph V. Maddox, Illinois Natural History Survey and Illinois Agricultural Experiment Station, Champaign, Illinois

Richard S. Soper, United States Department of Agriculture, Agricultural Research Service, Plant Protection Research Unit, Boyce Thompson Institute, Cornell University, Ithaca, New York

Yoshinori Tanada, Department of Entomological Sciences, University of California, Berkeley, California

Hitoshi Watanabe, Faculty of Agriculture, University of Tokyo, Tokyo, Japan

Jaroslav Weiser, Institute of Entomology, Academy of Sciences, Prague, Czechoslovakia

PREFACE

The epizootiology of insect diseases has its roots at the beginnings of the history of the science of insect pathology, although its progress and development have been slow and undisciplined. If Agostino Bassi laid the foundations for insect pathology in the early 1880s, then, with his conclusions about transmission, the involvement of certain environmental factors, and the control or prevention of muscardine caused by *Beauveria bassiana* in silkworms (*Bombyx mori*), he also did the same for the epizootiology of insect diseases. If Louis Pasteur gave insect pathology its first moment of glory and major practical success in controlling pebrine in the silkworm, then he also did the same for insect epizootiology since his recommendations for impeding the transmission of *Nosema bombycis* resulted from epizootiology. And if Elie Metchnikoff was the first to experiment with microbial control of insects, then he also was the first to demonstrate experimentally that pathogen population density is important to disease epizootics in insects. Yet there has been only little coalescing of this science with underlying theories to stimulate research, concepts, and practical results. Its terminology has been haphazard with little attention to or strict usage of definitions and concepts. It has many gaps as a quantitative science, and there has not even been a comprehensive review since 1963.

This book has several purposes or goals. Perhaps the most important is to organize insect epizootiology into a well-defined discipline and to call scientists' attention to the standardized definitions and principles of this science. A second purpose is to review at least the landmark literature, though a complete review of everything pertaining to insect epizootiology is beyond the scope of this book. The third is to explore the interrelationships among the factors pertinent to epizootiology and thereby evaluate and suggest research trends. It is hoped these goals will provoke thought, controversy, and the testing of underlying theories.

This is a particularly appropriate time for a text in insect epizootiology. The more general fields of insect pathology, entomology, biological control, integrated pest management, and, indeed, general biology are at the beginning of a new era due to advances in genetic engineering and other biotechnology. In the late 1970s the field of microbial control of insects was foundering due partly to inattention to epizootiology. Just as the microbial control specialists began to realize this and correct it, the genetic engineering revolution arrived. Now epizootiological research, which is the foundation for practical applications of entomopathogens such as microbial control, may again be threatened with the advent of engineered microorganisms. A text at this critical time would emphasize epizootiology and the involvement of biotechnology in this science. Moreover, the text should help to stimulate cooperation between laboratory and field specialists, plus accumulate crucial ecological information for integrated pest management (IPM) specialists and ecologists so that insect diseases can be considered more frequently, as they should be, in evolving IPM research and control programs.

This text is intended for a range of readers. It should be useful to students and lecturers requiring information about advanced topics as well as researchers in insect pathology, epizootiology, and microbial control. It also is aimed toward readers interested in the ecology of insect–entomopathogen interactions, including insect ecologists, integrated pest management specialists, and those who rear insects.

The text has four parts. In Part I, "Concepts and Methodology," the authors introduce general terminology and methodology from other disciplines, specific methodology for quantification, and modeling. The "Key Factors" are discussed in Part II. Major factors—the host population, pathogen population, environment, and transmission—are reviewed and evaluated. They cause certain patterns of disease over place and time; a crucial strategy in ecology or epizootiology is to recognize these patterns, try to explain them, and to develop hypotheses. It would be difficult to have a text about epizootiology of insect diseases without consideration of the "Disease Groups" (Part III). These range from the epizootics of the seldom-researched subject of noninfectious diseases to the spectacular viral and fungal diseases, the inapparent but widespread protozoa, the enigmatic bacteria, and the diverse and biologically interesting nematodes. Finally, insect epizootiology has its "Practical Aspects" (Part IV). These are usually directed toward increasing disease in host populations (control of pest insects), but also are of great importance in the prevention and control of diseases in beneficial insects or rearing facilities.

The 16 chapters of this book cover the present status of insect epizootiology, though obviously not all pertinent references could be included without greatly increasing the number of pages. Even so, it will quickly become evident to the reader that certain topics have been the subjects of much research, others have not, and all require much additional work, especially to further develop the basic principles of the epizootiology of insect diseases.

The editors thank Wayne M. Brooks for his advice and assistance in the early stages of this project, all the colleagues who reviewed chapters, and the publisher for providing the opportunity to expound on the discipline of epizootiology of insect diseases.

JAMES R. FUXA
YOSHINORI TANADA

Baton Rouge, Louisiana
Berkeley, California
February, 1987

CONTENTS

EPIZOOTIOLOGY
OF INSECT DISEASES

CONCEPTS AND METHODOLOGY

1

EPIDEMIOLOGICAL CONCEPTS APPLIED TO INSECT EPIZOOTIOLOGY

JAMES R. FUXA

Department of Entomology
Louisiana State University Agricultural Center
Louisiana State University
Baton Rouge, Louisiana

YOSHINORI TANADA

Department of Entomological Sciences
University of California
Berkeley, California

1. INTRODUCTION

Epizootiology of insect diseases, or "insect epizootiology," is a relatively young science. In 1949 Steinhaus (1) was the first to present the principles of insect epizootiology. Since then the discipline has made progress but has never coalesced with consistent terminology, methodology, or purpose.

Plant epiphytology and especially medical epidemiology are well-established disciplines that have potentially useful concepts and techniques for insect epizootiologists. Although not all of these concepts and methods are likely to prove adequate or applicable, their standardization in insect epizootiology is essential and overdue.

The purpose of this chapter is to review, define, and analyze epidemiological concepts applicable to insect epizootiology. These concepts should provide a foundation upon which insect epizootiology can become more organized as a science and more consistent in terminology. Ideally, this in turn should stimulate more critical evaluation of research, concepts, and objectives.

2. OBJECTIVES OF EPIZOOTIOLOGY

Epizootiology is the science of causes and forms of the mass phenomena of disease at all levels of intensity in a host population (modified slightly from Sinnecker (2)). It includes noninfectious as well as infectious diseases—

medical epidemiology is now concerned primarily with noninfectious diseases (3). The objectives of epizootiology, like those of epidemiology and epiphytology, can be theoretical or practical.

The major theoretical objective of insect epizootiology is the explanation of disease patterns (or lack thereof) in host populations, though the description of the patterns is a usual prerequisite to that objective. Description of patterns builds general knowledge from which hypotheses can be formulated. The explanations of patterns would further man's general knowledge of disease and parasite dynamics.

Description and explanation of disease patterns can do more than just satisfy man's curiosity; they can lead to more practical reasons for insect epizootiology. Seldom, if ever, will entomologists be satisfied with natural levels of disease in insect populations; they wish either to increase disease levels in pestiferous insects (see Ch. 15) or to decrease levels in beneficial insects or in insects in rearing facilities (see Ch. 16). Just as in epidemiology, where more thorough understanding of disease has usually led to better prevention (4), better microbial control of insects can be expected as knowledge of epizootiology increases.

Within the discipline of insect pathology, epizootiology also can contribute to the identification of causes of disease. This of course is one of the primary purposes, along with prevention, of epidemiology. Though not considered directly in this book, epidemiology has potentially useful techniques for establishing the etiology of insect disease (5).

It is necessary at this point to consider two points that are important to this chapter and the entire book: the causes of disease and the definition of an epizootic.

3. DISEASE CAUSATION

Epizootiology is concerned with infectious and noninfectious diseases and the principle of multiple etiology. Insect epizootiology now concentrates on infectious diseases just as epidemiology did in its early history. Only time will tell whether noninfectious diseases will ever predominate in insect studies as they do now in studies of humans.

Disease is a condition in which a state of physiological equilibrium of the host with its environment (i.e., health) becomes unbalanced (4, 6). The many factors that contribute to the balance or the health of an insect can be upset by noninfectious (abiotic) or infectious (biotic) causes. Though a disease usually has a primary cause, for example an infectious agent, there may also be a complex web of secondary causes and contributing or predisposing factors (multiple etiology), including environmental conditions (6). In medical epidemiology, if the presence of some factor increases the occurrence of a disease, then that factor is a "cause" of the disease (7). Thus the insect epizootiologist must concern himself with everything related to the disease: etiological agents (see Chs. 5, 9–14), the host population (see Ch. 4), and the environment (see Ch. 7). It may be just wishful thinking to hope for success in practical epizootiology (e.g., microbial control) by concentrating effort on just one casual aspect such as the inoculum of a pathogen in the host's environment.

4. DEFINITION OF EPIZOOTIC

The insect epizootiologist must have a clear idea of what constitutes an epizootic if he is to learn the causes of epizootic vs. enzootic disease. The definition of "epidemic" used in medical epidemiology and that of "epizootic" in this book is "an unusually large number of cases of disease" in a host population (4, 8, 9). The main point is that there must be an "unusually" large number. Thus, for example, relatively few cases of a disease constitute an epizootic if none are expected (8).

One question raised by this definition is, what is an unusually large number? To find an answer, it is necessary to have a past history (previous experience) of the disease for a number of years (4). The past history provides data which can be subjected to statistical or other mathematical analyses. For example, human pneumonia and influenza data have been subjected to time series modeling to calculate expected prevalence and an epidemic threshold (Fig. 1.1). If the ob-

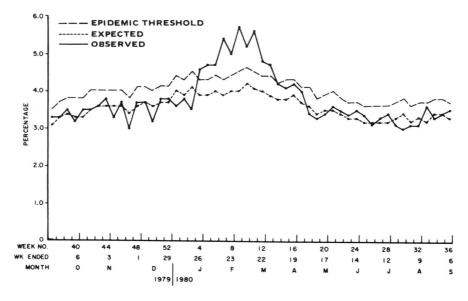

Figure 1.1. Observed and expected percentage of deaths attributed to pneumonia and influenza in 121 US cities, 1979–1980 (10).

served prevalence surpasses the epidemic threshold, then influenza is epidemic (10).

Establishment of a space-time framework is important in the study of epizootics. In the influenza example (Fig. 1), 4% of deaths due to pneumonia and influenza would constitute an epidemic in July but not in January. The epidemic thresholds also might be different 20 years from now or outside the USA due to differences in treatment, detection, or natural prevalence of the disease. Similarly, 50% prevalence of nuclear polyhedrosis virus (NPV) disease in fall armyworms, *Spodoptera frugiperda,* might constitute an epizootic in Georgia sorghum (11) but not in Louisiana pastures in early August (12). On the other hand, 3% prevalence of the fungus *Entomophaga aulicae,* which is rarely present in Louisiana *S. frugiperda* (13), would be an epizootic in that state at any time of year.

It should be noted that medical epidemiologists do not always agree about what is an "unusually large number" of cases, and perhaps for this reason they often avoid the term "epidemic" (7). It would not be surprising if insect epizootiologists have the same problem. The concept is nevertheless important.

In addition to an operational viewpoint, the meaning of epizootic should be considered from a biological viewpoint. A convenient way to do this is to contrast the epizootic vs. the enzootic state of disease in a population.

Enzootic diseases are those that are usually low in prevalence and constantly present in a population. There is no clear-cut distinction between enzootic and epizootic; rather they are opposite ends of a range.

There is some agreement among plant, animal, and human "epidemiologists" that time is critical to the distinction of enzootic from epizootic (2, 14, 15). Time is relatively unimportant in enzootic disease but is the distinguishing feature in an epizootic. Epizootics are sporadic and limited in duration and are characterized by a sudden change in prevalence and incidence. Enzootic disease levels are of long duration. Since time is important in epizootics, the pathogens involved are usually characterized by a short generation time (15). This generalization seems to hold true for entomopathogens, for example, the commonly epizootic NPVs as opposed to the usually enzootic protozoan diseases.

Balance and coexistence between pathogen and host are characteristic of enzootic disease. Enzootic disease is in a balanced (15) or climax (6) state with the host population. The disease agent and host coexist, implying that the disease is always present (15). This in turn implies that there is very little spread, if any, of endemic disease in the area where it is endemic (15), though a disease may have been introduced and spread before it became endemic. Other characteristics of endemic disease include low virulence (14) and a ratio of daughter to parent infections that approximately equals one (15).

Epizootics are characterized by imbalance over a short time period, with rapid change in prevalence, and a ratio of daughter to parent infections that substantially exceeds one (15). This change in prevalence depends first on massive reproduction of the causal organism (infectious diseases only!) reinforced by environmental and host factors (2). At the beginning of the epizootic the host population must be largely susceptible to the causal agent, possibly due to an increase in virulence after passage through some susceptible hosts. Finally, there must be efficient transmission or "adequate contact" (16) between the agent and host. One or more factors trigger the epizootic, such as stress or an increase in the susceptible:resistant host ratio (2). Later, excessive multiplication of the disease agent, which initiated the epizootic, continues but eventually becomes unimportant because the proportion of susceptible hosts has declined (2).

The above discussion might lead one to believe that we are discussing specific enzootic host-pathogen systems vs. epizootic ones. This is not necessarily true. An enzootic system can become epizootic through some change in one or more of the major components: host, pathogen, or environment. For example, migration might introduce susceptible hosts, a pathogen might change to a more virulent strain, or an environmental change might stress hosts or favor the pathogen.

The epizootic wave can be composed of smaller successive waves and is only part of a larger curve which includes the enzootic and possibly the panzootic states. Steinhaus (1) divided the epizootic (wave) into preepizootic, epizootic and postepizootic phases. Sinnecker (2) discussed these phases more thoroughly and with different terminology. Fuxa (12) proposed the term "enzootic wave" for situations in which there is a relatively high, seasonal prevalence of a disease, which is therefore not "unusually" high.

The term "panzootic" simply refers to an epizootic with an unusually wide spatial distribution (2). Thus space *and* time are critical to consideration of panzootics (2). Due to the geographically limited nature of insect populations of a given species, panzootics in insects are unlikely to reach worldwide extent as they can, for example, for influenza in humans.

5. BASIC MEASUREMENTS AND OBSERVATIONS IN EPIZOOTIOLOGY

Insect epizootiologists have not been very critical in their terminology. This is unfortunate because measurements and observations in epidemiology have evolved with precise and valuable definitions, including operational terminology. Not all epidemiological measures and terms are likely to be useful in insect epizootiology; there are many differences between the two fields, especially in collection of data. Nevertheless it would be beneficial to learn which concepts are useful and which are not and to standardize terminology in our discipline.

In this section we shall consider terminology from certain disciplines, in particular medical epidemiology, and briefly discuss applications of some of them to insect epizootiology.

5.1. Rates, Ratios, and the Population at Risk

Rates in epidemiology are essentially proportions with a time specification. The denominators in these proportions are of critical importance in order to describe and compare groups (7, 17). Denominators in these proportions represent the "reference population" (18) or the population among which affected hosts are observed. If the denominator is restricted to the number of hosts to which the event (disease related), represented by the numerator, *could* have happened, whether it did or not, then this denominator is called the "population at risk" (7).

The term "rate" implies that the proportion involves a time factor. This time factor also must be specified and will be discussed in subsequent sections, particularly 5.3 and 5.4 about the two most important rates.

A "ratio" expresses the relationship of one number, for example a rate, to another rate, or of affected hosts to unaffected but not to the total population (18).

We shall follow conventional epidemiological terminology in this chapter, though, for example, incidence and prevalence "rates" are actually probability estimates (19). The reader is referred to basic epidemiology texts cited in this chapter for the discussion of other ratios and types of risks and to Morgenstern et al. (19) for a more theoretical discussion of rate and risk.

5.2. Mortality Rates

There are probably almost as many types of mortality rates in epidemiology as there are reasons for determining them. One can almost design a mortality rate to suit his purpose as long as all conditions are specified. For example, in addition to rates mentioned below, other host characters (e.g., sex, race) or time periods can be specified (16). Some of the more common rates might prove useful in insect epizootiology. A time factor is always specified when these rates are used (16).

5.2.1. ANNUAL DEATH RATE.   Annual death rate is simply the total number of deaths from all causes during a specified year (or perhaps shorter time span for insects) divided by the number of hosts in the population (usually at midyear). This fraction is usually multiplied by some number (e.g., 1000) to give a rate/number in the population (i.e., deaths/1000 hosts) (7, 17).

5.2.2. CAUSE-SPECIFIC MORTALITY RATE.   The cause-specific mortality rate is the same as the annual death rate but for one cause of death only. It is the number of hosts dying due to a particular cause per unit time divided by the total number in the host population during that time (7, 17). This rate, in one form or another, is used often in insect epizootiology.

5.2.3. AGE-SPECIFIC MORTALITY RATE.   This is the number of hosts dying in a particular age group (e.g., an instar or stage in insects) per unit time divided by the total number in that age group in the host population during that time (17).

5.2.4. CASE-FATALITY RATE.   This is a measure that should probably be investigated more often in insect pathology. Case-fatality rate is the number of hosts dying due to a particular disease divided by the total number of hosts with that disease (7, 17). The use of the word rate in this case is a misnomer because often a time factor is not specified. Only when the case-fatality rate is 100% is mortality rather than infection a good measure of incidence or prevalence (see below). Also, the case-fatality rate is more useful with acute than with chronic diseases (4).

5.3. Prevalence

Prevalence is an important measure and is often confused with incidence. It is the most commonly used rate in insect epizootiology. Prevalence of a disease can be defined as the number of hosts (or proportion of hosts at risk) afflicted with that disease *at a given point in time* (e.g., 8). As such, prevalence depends on two things: the proportion of hosts afflicted *and* the duration of the disease (7, 18). If one counts diseased insects on a given day, he counts new cases of the disease as well as old ones. This is the critical distinction between prevalence and incidence, which is discussed in section 5.4. If the duration of the disease is relatively short, then prevalence and incidence are similar (see 5.4).

Theoretically, prevalence is an inferior measure to incidence (see 5.4), but for practical purposes prevalence has certain advantages in entomology. Prevalence measures the total effect on a host population at a given time, whereas incidence measures only a proportion of the cases at that time. Thus prevalence is more valuable for the determination of how an insect population is being affected in terms of pest control and damage reduction. Also, as far as the measurement process itself is concerned, prevalence is relatively easy to determine in insects, whereas the measurement of incidence poses problems in insect field populations.

5.4. Incidence

"Incidence" is probably the most frequently misused term in insect pathology and general entomology. It has a very specific meaning in medical epidemiology. Incidence of a disease is defined as "the number of new cases of disease in a defined population in a given interval of time" (8). The reader will recall that prevalence includes *all* the cases, new and old, at a given time. Use of the words "incidence" and "prevalence" synonymously should be avoided.

Since prevalence includes new and old cases of disease whereas incidence includes only new, prevalence is always larger except for rapidly fatal or curable diseases (20). There is an often-cited relationship, where prevalence (P) varies as the product of incidence (I) and duration (D) of the disease, or, if the disease is stable (incidence and duration are constant over time), then $P = ID$ (18). Thus if two of the factors are known, the third can be determined.

Theoretically, incidence should be more useful than prevalence for determination of disease causation (18). For example, if a researcher relates incidence to some environmental factor, he can relate the number of new cases of disease to the occurrence or level of that factor at one point in time. But if he uses prevalence, he is relating new *and* old cases to the factor even though the onset of old cases actually occurred at some prior level.

While incidence is potentially more useful than prevalence, there are some problems in its estimation in entomological research. It is difficult or even impossible to keep records of individual insects in field research, in contrast to studies of humans, domestic animals, or plants, which, for various reasons, are generally more amenable to individual record-keeping. Also, there are few diagnostic techniques for recognizing an insect in the early stages of disease without killing it. Nevertheless, incidence measurement should be possible with relatively sedentary insects and with marking or caging techniques (see 21). A life-table study of a cohort over short time intervals also should provide a good estimate of incidence.

Incidence also might be estimated by other somewhat less accurate methods. Collection of prevalence data at time intervals substantially shorter than D could provide a reasonable estimate of incidence (i.e., increase in % infection or number infected per day), but such a study that did not keep records of individuals would have to assume dangerously that the ratio of infected vs. noninfected insects removed from the population is the same as that remaining in the total population. However, this assumption might be less dangerous in the early stages of an epizootic when the causal factors would be of greatest importance. Fuxa and Geaghan (22) attempted a multiple regression analysis of incidence collected as prevalence data, but it failed probably because the time intervals were too long.

Mortality rates, substituted for infection rates, are most reliable in the estimation of incidence when D is short and the disease has a high case-fatality rate (18). Fortunately both appear to be true of many diseases of greatest interest in insect epizootiology. Morgenstern et al. (19) further considered the estimation of incidence.

5.5. Infection Rate

Infection rate R and apparent infection rate r are important measurements in epiphytology and epizootiology for the description or explanation of disease patterns in populations (e.g., 15, 23, 24).

The infection rate R is defined as the average infection rate per unit of inoculum per unit of time, or the rate of disease increase per unit of infectious tissue per unit of time (15). Thus R concentrates on units of the pathogen and, unlike incidence or prevalence, is used only with infectious diseases. R depends on the ability of the infectious units to produce disease in the host population. Thus R is a substantially different concept from incidence or prevalence.

The apparent infection rate r is the rate of disease increase per unit of infected host tissue (15). The relationship between r and R is analogous to that between prevalence and incidence. Environmental or other factors can affect R in two ways: they can affect infectiousness of pathogenic units or the production of new units. On the other hand, r can be affected in the same two ways and also by latent and noninfectious periods (15). Thus, for example, a change in environment is quickly reflected in R but more slowly in r, which is a more complex parameter just as prevalence is more complex than incidence. Unfortunately, whereas r is not difficult to measure (15, 24), R can only rarely be estimated in the field (15). Van der Plank (15) discussed infection rates and their usefulness more extensively.

Anderson and May (25) defined a parameter similar to r called the "infection reproductive rate." This rate is defined as the expected number of secondary infections per infected host during the infectious period. The disease is maintained in the host population if this rate is greater than one.

5.6. Transmission Rate

"Transmission" is a somewhat broader term than "infection" since the latter process is only one component of the former. There are different operational definitions in the literature for rates of different kinds of transmission. For horizontal transmission by contact between individual hosts it is assumed that the transmission rate is a proportionality constant for the number of encounters between susceptible and infected hosts (25, 26). Brown and Nordin (27) demonstrated a regression method for measuring horizontal transmission. However, different procedures are required if the causal agent has free-living infective stages, sexual horizontal transmission (26), or vertical transmission (28). Transmission and transmission rates will be discussed further in Chapter 6.

6. STRATEGIES FOR EPIZOOTIOLOGY

In the remainder of Chapter 1 we shall discuss types of studies and general procedures for epizootiology. This material is again based largely on medical epidemiology and to a lesser degree on epiphytology. Consideration of general strategies is important because few other sciences rely so heavily on a combination of observational studies, complex factors, and confusing or embryonic mathematical techniques. As with other sciences, epizootiology relies on the empirical cycle: observation, induction or formulation of hypotheses, deduction, testing of hypotheses, and evaluation.

Section 6 will briefly consider types of studies in epidemiology, quantification of studies, and a few common pitfalls.

6.1. Types of Studies

The major types of studies in medical epidemiology are applicable in various degrees to insect epizootiology. It is fundamental for epidemiologists to be well acquainted with the procedures, advantages, and disadvantages of each type. We shall now review the types of studies with their advantages and disadvantages and discuss their applicability to insect epizootiology.

The two main types of studies are *experimental* and *observational.*

6.1.1. EXPERIMENTAL STUDIES. Experimental testing of cause-effect hypotheses involves deliberate manipulation (elimination, addition, or change in level) of the proposed cause, followed by observation of the effect on the disease (18). Opportunities are much greater for experimental studies in insects than in humans due to ethical problems, numbers, life spans, etc. However, even with insects, experimental studies have certain difficulties and disadvantages.

In experimental or interventional studies the scientist simplifies nature so that a hypothesis can be tested under controlled conditions (29). The investigator changes or controls an independent variable and observes changes in the dependent variable. The strength of the method depends on the investigator's ability to control the situation and rule out the effects of extraneous variables (16, 17). Occasionally natural experiments occur in which the investigator can take advantage of natural changes in the independent variable. The reader is probably familiar with procedures including design, randomization and assigning of control and treatment groups, sampling, and statistical tests, all of which are beyond the scope of the present text. Further discussion can be found in Susser (5) and elsewhere.

The major advantage of a rigorously designed experimental study is that it provides the closest thing to proof of a hypothesis. Experimental studies usually provide more precise tests of hypotheses than observational ones (5).

There are also disadvantages to experimental studies (5). One is that nature is so complex it is difficult to sort out the numerous important factors, individual effects, and interactions among causes. Also, it is difficult to control variables without altering the "natural setting" of the host and pathogen populations. The more manipulation there is, the more difficult it is to relate results to the natural setting. Another disadvantage is that an unbiased representation of a population can be difficult to achieve, though this is not as difficult a problem in entomology as in epidemiology. Finally, experimental studies are not usually as practical as observational ones. They are more difficult to set up and do not permit the variety of tests of hypotheses that can be achieved in observational research.

In insect epizootiology there are numerous examples of laboratory and field experimental studies, but few comprehensive ones examining many facets of a host-pathogen system. Even short-term microbial control experiments, in which pathogens are sprayed or otherwise introduced into field populations, constitute epizootiological experiments. There are a few examples of systems that have been experimentally studied from many aspects, such as the entomo-pathogenic fungus *Nomuraea rileyi* in soybean fields. Experimenters have re-searched temperature, humidity, host susceptibility parameters, fungal isolates, conidial stability, transmission, field inoculum, and certain aspects of the eco-system (see 30).

6.1.2. OBSERVATIONAL STUDIES.   Observational epizootiology involves the observation of disease occurrence, analysis, and inference about causes of dis-ease distribution. Investigators do not manipulate variables; they simply collect data as nature takes its course, and relate changes or differences in one variable to those in another. Principles of observational studies are similar for infectious and noninfectious diseases except that there is no underlying germ theory for the latter (16). Just as in experimental studies, observational studies can include proposal of a hypothesis and prediction of events based on that hypothesis. However the testing of those predictions is not deliberately controlled by the investigator.

Medical epidemiology has demonstrated the value of observational studies, which comprise much of its history (3, 16). Insect epizootiologists should not hesitate to do observational studies, but they should not rely on them as heavily as epidemiologists since insects are more amenable than humans to the experi-mental approach.

The advantages and disadvantages of observational studies are somewhat the reverse from experimental. Observational studies usually provide a less precise test of hypothesis (5). Relationships and specificity between variables are less precise or well known or documented; the impossibility of random assignment to control or treatment groups in some study types can cause bias; and statistical analysis is less efficient or powerful. On the other hand, observa-tional studies are more practical (lower cost and easier to arrange) and do not affect the "natural setting" as drastically.

There are two categories of observational studies: descriptive and analytic (17). As the name implies, descriptive studies provide a relatively superficial, general view of patterns of disease. Analytic studies attempt an explanation of those patterns, or in other words, associate causes with disease occurrence. Analytic studies usually address a more specific question directed toward test-ing a hypothesis with more rigorous study design and analysis. Analytic studies in medical epidemiology often attempt to answer questions raised by descrip-tive studies.

Both descriptive and analytic studies can be divided into prevalence or incidence studies. A third type, the case-control study, is usually analytic (17).

6.1.3. CASE-CONTROL STUDIES.   In case-control studies, hosts with or with-out the disease (i.e., some already *have* the disease) are identified, and "case histories" of both groups (which are broadly comparable, e.g., in age) are collected and analyzed for the presence or absence or different levels of causa-tive factors (17). Thus the dependent variable (the disease) is used to pick the study groups. Case-control studies are generally observational and analytic. Such studies have a long history of successes in medical epidemiology (3), but they also have serious disadvantages and have only very limited potential usefulness in insect epizootiology.

Case-control studies have their advantages and disadvantages. They are good for initial research because they are relatively short and inexpensive (20). They are more appropriate than incidence or cohort studies (see below) when the disease is relatively rare. On the other hand, the time-sequence of causal relationships is difficult to establish in case-control studies, and they yield only prevalence data, not incidence (20).

Case-control studies are possible with insects but only in certain situations. Insects do not have memories or "medical" records from which the case histories can be obtained. But case histories are occasionally kept for small groups of insects or, rarely, individuals. Examples might include careful records kept in rearing regimes, or experiments in which insects unexpectedly become diseased in some treatments (i.e., not designed initially as pathology experiments). Steinhaus (31) discussed the important points in the collection of a history of a group of diseased insects.

6.1.4. PREVALENCE OR CROSS-SECTIONAL STUDIES. Prevalence studies, which can be either descriptive or analytic, are similar to case-control studies except that the causative factors and the effect (independent and dependent variables) are not only *measured* at the same time in a defined population, they also *occur* at the same time (18). The proportion of hosts affected by disease is related to one or more quantitative attributes. Hosts have or are exposed to those attributes (or causes or independent variables) at the time of the study. Whereas in case-control studies the dependent variable is used to choose the study population, in prevalence studies both variables can be used (17, 20).

One sample date yields a disease *point* prevalence (prevalence at one point in time), and more than one sample (i.e., repeated point-prevalence studies) yield an estimation of *period* prevalence (prevalence during some time period) if the spacing is appropriate to the duration of the disease (20). However, prevalence studies do not provide incidence rates.

Ease and versatility are among the advantages of prevalence studies. Hypotheses can be formulated and tested relatively inexpensively and quickly (7). Results of prevalence studies are more likely than case-control studies to yield generalizations for a whole population because, in the latter type, groups are based on already-diseased cases. Prevalence studies are often the most practical with insects because it is not necessary to keep records of individual hosts.

Like case-control studies, a disadvantage of prevalence studies is that only associations can be inferred, and not necessarily cause-effect time sequences (20). In fact, causal inference is even weaker in prevalence than in case-control studies because there is no time sequence involved at all (16). Physiological factors in particular may have been caused by the disease rather than contributed to its presence (16). Also, conclusions may not be meaningful if the independent variables are not permanent characteristics of the host or change too extensively (18). However, a time-sequence element can be introduced to partially counteract these disadvantages by doing a sequence of prevalence studies rather than just one.

Most field studies in insect epizootiology have been prevalence studies. Descriptive, period-prevalence studies comprise the majority of these and are too numerous to be listed here, though a few examples are cited (32–35). There are also examples of descriptive point prevalence studies (36, 37), but far too few of analytic (period) prevalence (22, 38-40).

6.1.5. INCIDENCE AND COHORT STUDIES.   Incidence and cohort studies start with a host population at risk for the disease of interest and follow that population for a certain time period to relate disease development to various independent variables (7). The words "incidence" and "cohort" are often used synonymously, but cohort refers to an age group of individuals (with no immigration) whereas incidence implies that records are kept on individual hosts (16). A cohort study yields period prevalence data, and an incidence study yields incidence data (20). Both types are better than prevalence or case-control studies for learning about factors that contribute to the *development* of disease in a population (17).

In incidence and cohort studies the independent variables are used to pick the population to be studied (20). Unlike prevalence or case-control studies, they actually *measure* all variables at different points in time (18). The studies are arranged so that all hosts are free of the disease at the beginning, and at the end, all hosts with the disease will have developed it during the study (20). Then hosts with disease vs. those without disease, or subgroups with different levels of disease can be analyzed for differences in the independent variables (17). Data from these types of studies lend themselves well to life table analysis (16).

The major advantage of incidence and cohort studies is that *change* in the dependent variable can be related to change in the independent variables (16); there is greater variety and depth of information (17). Other advantages over prevalence or case-control studies include less chance of bias, the possibility of including several diseases in the same study, and less chance of missing diseased hosts when the disease kills rapidly (16). They are also better when there is a precise hypothesis to be tested and when some sort of intervention is intended (20). Like prevalence studies, incidence and cohort studies can be readily generalized to the general population (20). One final major advantage of incidence studies is that they yield true incidence rates, which are the best for the determination of causative factors (Section 5.4).

The disadvantages of incidence and cohort studies are mainly concerned with the difficulty of keeping records of individuals or cohorts through time (16). In that respect, entomology has a major advantage (shorter host life cycles) and disadvantage (identification of individuals) compared to medical epidemiology. Incidence and cohort studies usually involve greater difficulty and expense than prevalence studies. They are at a disadvantage for studying relatively rare diseases (18), for searching for hypotheses, or for evaluation of a large number of factors with doubtful significance (16).

The closest example to a cohort study in insect epizootiology is a life table which includes disease, such as that by Harcourt (41). There are no good examples of a true incidence study in a field population.

6.1.6. RETROSPECTIVE AND PROSPECTIVE STUDIES. "Retrospective" and "prospective" studies are commonly used terms in medical epidemiology. Basically, retrospective refers to case-control studies whereas prospective is roughly synonymous with incidence or cohort studies (17). It has been suggested that these terms be dropped (3) due to confusion in their use (17).

6.2. Quantification of Studies

Medical epidemiology and epiphytology became quantitative, a necessity for a respectable science, in the early 1900s and 1960s, respectively (23, 29). Insect epizootiology has meandered in that direction, but in-depth quantitative studies are rare and mathematical techniques are even rarer and not well developed.

Rules for the quantification of epidemiological or epiphytological studies have been proposed. Zadoks (29) set forth five rules for testing the validity of quantitative evidence on epiphytotics:

1. The source(s) of inoculum at the onset of the epidemic must be known, and the amount of the pathogen in the source(s) must be expressed quantitatively.
2. The effect of environmental conditions on the development of an epidemic of the pathogen must be known in terms of quantitative relations between independent (usually abiotic) variables and dependent (biotic) variables.
3. The rate of development of the epidemic under the prevailing conditions must be calculated.
4. The successive levels of the epidemic must be calculated from the known amount of pathogen in the source(s) and the calculated rates of development of the epidemic.
5. The calculated terminal level and the calculated intermediate levels of the epidemic must be equal to the observed terminal and intermediate levels.

Similarly, parallels of Koch's postulates have been proposed for medical epidemiology (16). At first glance these appear to relate only to the primary etiological agent, but actually "cause" is to be interpreted in the broad sense discussed earlier:

1. Prevalence of the disease should be significantly higher in those exposed to the hypothesized cause than in controls not so exposed.

2. Exposure to the hypothesized cause should be more frequent among those with the disease than in controls without the disease—when all other risk factors are held constant.

3. Incidence of the disease should be significantly higher in those exposed to the cause than in those not so exposed, as shown by prospective studies.

4. Temporally, the disease should follow exposure to the hypothesized causative agent with a distribution of incubation periods on a bell-shaped curve.

5. A spectrum of host responses should follow exposure to the hypothesized agent along a logical biological gradient from mild to severe.

6. A measurable host response following exposure to the hypothesized cause should have a high probability of appearing in those lacking this before exposure (e.g., antibody, cancer cells), or should increase in magnitude if present before exposure; this response pattern should occur infrequently in persons not so exposed.

7. Experimental reproduction of the disease should occur (more) frequently in animals or man appropriately exposed to the hypothesized cause than in those not so exposed; this exposure may be deliberate in volunteers, experimentally induced in the laboratory, or may represent a regulation of natural exposure.

8. Elimination or modification of the hypothesized cause should decrease the incidence of the disease (i.e., attentuation (sic) of a virus, removal of tar from cigarettes).

9. Prevention or modification of the host's response on exposure to the hypothesized cause should decrease or eliminate the disease (i.e., immunization, drugs to lower cholesterol, specific lymphocyte transfer factor in cancer).

10. All of the relationships and finding should make biological and epidemiological sense.

A few insect epizootiology studies (22, 38-40) provide estimates or attempts at four or five of Zadoks' rules (29), but none has actually completed all five. A worthwhile goal in insect epizootiology would be to complete initially at least a few such quantitative studies, including one or more for each pathogen group and type of habitat. In doing so, insect epizootiologists should take advantage of the advent of the computer. These masses of data that need to be collected have discouraged attempts not only at quantifying an epizootic, but also at attempting a study in this area. Suitable and practical programs need to be developed for insect epizootiology.

6.3. Common Pitfalls

Insect epizootiologists should be aware of possible pitfalls in study design and interpretation common in epidemiology. Fox et al. (4), Susser (5), Austin and

Werner (7), and Fisher (20) provide detailed discussion. Common problems include: faulty sampling leading to unintentional bias or artificially spurious associations, improper extrapolation of results, imprecise definition of variables, improper conclusions of cause-effect (e.g., relating two factors to one another instead of a third, unsuspected factor which is the true cause), subjectivity in evaluation, errors in disease classification, and ecological fallacy (applying results from a main group to a subgroup without verification).

7. CONCLUSION

Some of the concepts borrowed from other disciplines and discussed in this chapter might have limited application to insect epizootiology. However, we would not be able to determine which are useful unless we standardize terminology and attempt some of the various approaches in insect epizootiology. Even if one does not agree with our concept of "epizootic," he at least must define how he uses the term in his work. Attempts should be made to measure incidence, as distinct from prevalence, for one or more insect diseases to determine its usefulness. Similarly, the measurement of infection rates in the field is potentially beneficial. Observational studies, particularly prevalence, cohort, and incidence (if possible), are more important in epizootiology than in many sciences and should not be scorned. However, there is greater opportunity for experimental studies in insect epizootiology than in medical epidemiology or even veterinary epizootiology. Finally, there is already a certain degree of quantification in insect epizootiology, but much remains to be done, particularly in field situations.

ACKNOWLEDGMENTS

We thank Wayne M. Brooks, H. Denis Burges, and Harry V. Hagstad for reviewing the manuscript.

REFERENCES

1. E. A. Steinhaus, *Principles of Insect Pathology,* McGraw-Hill, New York, 1949.

2. H. Sinnecker, *General Epidemiology,* Wiley, New York, 1976.

3. P. E. Sartwell, "Trends in Epidemiology," in G. T. Stewart, Ed., *Trends in Epidemiology. Application to Health Service Research and Training,* Charles C. Thomas, Springfield, Illinois, 1972, p. 3.

4. J. P. Fox, C. E. Hall, and L. R. Elveback, *Epidemiology: Man and Disease,* Macmillan, New York, 1970.

5. M. Susser, "Procedures for Establishing Causal Associations," in G. T. Stewart, Ed., *Trends in*

Epidemiology. Application to Health Service Research and Training, Charles C. Thomas, Springfield, Illinois, 1972, p. 23.

6. W. H. LeRiche and J. Milner, *Epidemiology as Medical Ecology*, Churchill Livingstone, London, 1971.

7. D. F. Austin and S. B. Werner, *Epidemiology for the Health Services. A Primer on Epidemiologic Concepts and Their Uses*, Charles C. Thomas, Springfield, Illinois, 1974.

8. A. R. Barr, Epidemiological concepts for entomologists, *Bull. Entomol. Soc. Am*, **25**, 129 (1979).

9. E. A. Steinhaus and M. E. Martignoni, *An Abridged Glossary of Terms Used in Invertebrate Pathology*. 2nd ed., Pac. N. W. For. Range Exp. Stn., USDA For. Ser., 1970.

10. Anonymous, Influenza mortality surveillance—United States, *Morbid. Mortal. Weekly Rep.* **29**, 578 (1980).

11. R. D. Schwehr and W. A. Gardner, Natural occurrence of entomogenous pathogens in *Spodoptera frugiperda* and *Heliothis zea* larvae collected from sorghum, *Sorghum Newslett.*, **23**, 87 (1980).

12. J. R. Fuxa, Prevalence of viral infections in populations of fall armyworm, *Spodoptera frugiperda*, in southeastern Louisiana, *Environ. Entomol.*, **11**, 239 (1982).

13. J. R. Fuxa, unpublished data.

14. R. M. Anderson and R. M. May, Population biology of infectious diseases: part I, *Nature*, **280**, 361 (1979).

15. J. E. Van der Plank, *Principles of Plant Infection*, Academic Press, New York, 1975.

16. A. M. Lilienfeld, *Foundations of Epidemiology*, Oxford Univ. Press, New York, 1976.

17. G. D. Friedman, *Primer of Epidemiology*, McGraw-Hill, New York, 1974.

18. B. MacMahon and T. F. Pugh, *Epidemiology. Principles and Methods*, Little, Brown, and Company, Boston, 1970.

19. H. Morgenstern, D. G. Kleinbaum, and L. L. Kupper, Measures of disease incidence used in epidemiologic research, *Int. J. Epidemiol.*, **9**, 97 (1980).

20. F. D. Fisher, *An Introduction to Epidemiology. A Programmed Text*, Appleton-Century-Crofts, New York, 1975.

21. T. R. E. Southwood, *Ecological Methods with Particular Reference to the Study of Insect Populations*, 2nd ed., Halsted Press, Wiley, New York, 1978.

22. J. R. Fuxa and J. P. Geaghan, Multiple-regression analysis of factors affecting prevalence of nuclear polyhedrosis virus in *Spodoptera frugiperda* (Lepidoptera: Noctuidae) populations, *Environ. Entomol.*, **12**, 311 (1983).

23. N. T. J. Bailey, *The Mathematical Theory of Epidemics*, Hafner, New York, 1957.

24. J. C. Cunningham and P. F. Entwistle, "Control of Sawflies by Baculovirus," in H. D. Burges, Ed., *Microbial Control of Pests and Plant Diseases 1970–1980*, Academic Press, New York, 1981, p. 379.

25. R. M. Anderson and R. M. May, Infectious diseases and population cycles of forest insects, *Science*, **210**, 658 (1980).

26. R. M. May and R. M. Anderson, Population biology of infectious diseases: part II, *Nature*, **280**, 455 (1979).

27. G. C. Brown and G. L. Nordin, An epizootic model of an insect-fungal pathogen system, *Bull. Math. Biol*, **44**, 731 (1982).

28. P. E. M. Fine and E. S. Sylvester, Calculation of vertical transmission rates of infection, illustrated with data on an aphid-borne virus, *Am. Nat.*, **112**, 781 (1978).

29. J. C. Zadoks, Methodology of epidemiological research, *Annu. Rev. Phytopathol.*, **10**, 253 (1972).

30. C. M. Ignoffo, "The Fungus *Nomuraea rileyi* as a Microbial Insecticide," in H. D. Burges, Ed.,

Microbial Control of Pests and Plant Diseases 1970–1980, Academic Press, New York, 1981, p. 513.

31. E. A. Steinhaus, "Background for the Diagnosis of Insect Diseases," in E. A. Steinhaus, Ed., *Insect Pathology. An Advanced Treatise,* Vol. 2, Academic Press, New York, 1963, p. 549.

32. Y. Tanada and E. M. Omi, Epizootiology of virus diseases in three lepidopterous insect species of alfalfa, *Res. Popul. Ecol.,* **16,** 59 (1974).

33. S. Keller and H. Suter, Epizootiologische untersuchungen uber das *Entomophthora*--auftreten bei feldbaulich wichtigen blattlausarten, *Acta Oecologia Oecol. Applic.,* **1,** 63 (1980).

34. G. Remaudière, J.-P. Latgé, and M.-F Michel, Écologie comparée des Entomophthoracées pathogènes de pucerons en France littorale et continentale, *Entomophaga,* **26,** 157 (1981).

35. T. E. T. Trought, T. A. Jackson, and R. A. French, Incidence and transmission of a disease of grass grub *(Costelytra zealandica)* in Canterbury, *N. Z. J. Exp. Agric.,* **10,** 79 (1982).

36. R. J. Barney, P. L. Watson, K. Black, J. V. Maddox, and E. J. Armbrust, Illinois distribution of the fungus *Entomophthora phytonomi* (Zygomycetes: Entomophthoraceae) in larvae of the alfalfa weevil, *Great Lakes Entomol.,* **13,** 149 (1980).

37. B. A. Federici, Baculovirus epizootic in a larval population of the clover cutworm, *Scotogramma trifolii,* in southern California, *Environ. Entomol.,* **7,** 423 (1978).

38. L. P. Kish and G. E. Allen, *The Biology and Ecology of* Nomuraea rileyi *and a Program for Predicting its Incidence on* Anticarsia gemmatalis *in Soybean,* Univ. Fla. Agric. Exp. Stn. Bull. (Tech.) 795, 1978.

39. R. S. Soper and D. M. MacLeod, *Descriptive Epizootiology of an Aphid Mycosis,* USDA Tech. Bull. No. 1632, 1981.

40. G. L. Nordin, G. C. Brown, and J. A. Millstein, Epizootic phenology of *Erynia* disease of the alfalfa weevil, *Hypera postica* (Gyllenhal) (Coleoptera: Curculionidae), in central Kentucky, *Environ. Entomol.,* **12,** 1350 (1983).

41. D. G. Harcourt, Major factors in survival of the immature stages of *Pieris rapae* (L.), *Can. Entomol.,* **98,** 653 (1966).

2

ECOLOGICAL METHODS

JAMES R. FUXA

Department of Entomology
Louisiana State University Agricultural Center
Louisiana State University
Baton Rouge, Louisiana

---

1. INTRODUCTION

A wide range of ecological techniques is commonly used in insect epizootiology. Both ecology and epizootiology are concerned with responses of individuals, populations, and communities to the abiotic and biotic environment. Insect epizootiology depends heavily on techniques from both microbial and insect ecology.

The purpose of this chapter is to briefly review and discuss sampling methods and other ecological techniques relating to the study of insect diseases. Chapter 1 presented types of studies for achieving the respectability of being a quantitative science, and now Chapter 2 reviews some of the operational methods for doing so. Some techniques are presented more for their potential than their realized usefulness in epizootiology.

This chapter has two major sections, one about sampling and the other about ecological techniques and analyses. There is also a brief discussion about pathogens as tools in insect ecology.

2. SAMPLING AND MEASUREMENT

The purpose of this section is to review sampling methods useful or potentially useful for insect epizootiology and to discuss some of the problems peculiar to this field. It is not intended to be a comprehensive review of entomological or statistical methods, or design of sampling programs. Green (1) discussed sampling design for environmental studies.

Sampling in insect epizootiology involves three subjects: host insects, pathogen units, and environmental parameters.

2.1. Host Insects

The basic problem in sampling hosts in insect epizootiology that makes it different from general entomology is that infected and noninfected hosts must be counted separately. This creates possibilities for error and bias in addition to those (2, 3) usually encountered in entomology.

2.1.1. BASIC ENTOMOLOGICAL TECHNIQUES. It is beyond the scope of this chapter to discuss the extensive literature about insect sampling. The insect epizootiologist should familiarize himself with the text by Southwood (2), an excellent introduction to methods and problems likely to be encountered. Sampling programs, absolute methods, relative methods, mark-release-recapture techniques, trapping, and estimates based on products and effects of insects are some of the areas he discussed.

This section is concerned with sampling bias inherent to insect epizootiology. Pathogens can affect sampling results in two ways: by affecting behavior, or by the infection of insects during or after the sampling procedure.

2.1.2. SAMPLING BIAS DUE TO ALTERED BEHAVIOR OF INFECTED INSECTS. Sampling bias can be introduced if the behavior of infected insects is affected in such a way that they become more or less likely than usual to be counted or captured by a given procedure. Many such behavioral changes are known: movement to higher than usual positions on host plants (4–7); failure to burrow deeply into soil (7, 8); failure to return to nests (5); seclusion in litter and debris (5); change in flight habits or capability (6, 9); feeding during daylight rather than only at night (6); and reduced activity resulting in easier capture or less likelihood of being attracted to traps (6, 10). The possibility of these or other behavioral changes must be considered when sampling techniques and schemes are being chosen or designed.

The best way to counteract such sampling bias is with the use of preliminary research to identify possible aberrant behavior and then the design of sampling around that information. In one example, preliminary research aided the placement of light traps so that an unbiased estimate could be made of percent infection by a microsporidium (10). In another, two sampling methods were compared in order to select the better one for determining prevalence of a fungus in a soybean insect (4). Similarly, bolt samples were better than traps for determining prevalence of a nematode in pine beetles (11). If a biased sampling method must be used, then a mathematical transformation can be derived from preliminary research to correct for the bias (e.g., 12).

2.1.3. SAMPLING BIAS DUE TO HORIZONTAL TRANSMISSION (CONTAMINA-
TION) DURING OR AFTER SAMPLING. The sampling procedure or subsequent
handling of the insects can cause errors in estimation of disease prevalence. The
major concern is overestimation of prevalence: uninfected insects can come
into contact with pathogens due to agitation or by way of contaminated equipment or hands; insects sampled in groups can spread disease to one another
before they are separated; or the stress of being sampled can increase susceptibility or perhaps induce latent infections that would not otherwise be activated.
Underestimation, a less common occurrence, also is possible. For example,
small, infected insects can die and decompose before they are counted (10).
Cannibalism after sampling could conceivably cause underestimation or overestimation of prevalence.

There are countermeasures for overestimation of prevalence due to contamination from sampling. These depend to some degree on the method of diagnosis. If large numbers of insects can be diagnosed without laboratory rearing,
then it is sometimes possible to kill the insects and disease organisms for later
examination. For example, insects can be frozen immediately after sampling
(13), killed in a preservative (14), or squashed on microscope slides (15). Similarly, the widely used technique of chilling insects while returning them to the
laboratory probably reduces chances for contamination.

Complications arise if the disease cannot be diagnosed quickly, but countermeasures are still possible. Prior to sampling, the investigator can arrange that
no one disturb sampling sites and perhaps artificially spread a disease (16, 17). A
common procedure is to separate insects into individual rearing containers as
they are collected or quickly thereafter. For example, *Heliothis* spp. larvae that
were separated soon after sampling had a lower prevalence of viral disease than
those returned to the laboratory in groups (18). Another procedure, particularly
useful if diagnosis is based on symptomatology at the time of death, is to rear
collected insects for a shorter time than the disease's period of lethal infection
(18, 19); insects that contract the disease during or after sampling are counted as
healthy. This procedure can lead to underestimation of disease if the period of
lethal infection varies too greatly.

Disinfectant techniques comprise some of the simplest, most effective ways
to avoid contamination. Such techniques have been reviewed for insect pathology (20, 21). They have been used for sampling equipment, sampler's clothes
and hands (17), and the sampled insects (22).

Experimental evaluation of possible cross-infection is beneficial but done
only infrequently. Ignoffo et al. (23) assessed contamination during sampling
with "sentinel insects" (see Section 2.2.2). They added laboratory-reared *Heliothis zea* larvae to plastic bags containing plants that had been sampled to
determine prevalence of *Nomuraea rileyi* in *Trichoplusia ni;* the *H. zea* larvae
did not become infected, and thus the sampling procedure was not causing
cross-infections. Fuxa (17) sampled field plots in a specific pattern to help
determine whether samplers were spreading disease in a field. Harcourt (24), in
a preliminary experiment, determined that rearing of field-collected insects
provided a good estimate of disease prevalence in a *Pieris rapae* population.

2.2. Pathogen Units

Sampling, detection, and quantification of pathogen units have been neglected in insect epizootiology. Pathogen units seldom are sampled except in infected hosts, yet research in other fields, such as plant pathology (25), indicates that quantification of the pathogen in the host's habitat is one of the most important factors for understanding natural prevalence of disease. Dosage-response results in microbial control research certainly indicate that the amount of the pathogen present is important in epizootics, as do the relatively rare, more strictly epizootiological studies in which pathogen units have been quantified (e.g., 12, 19).

This section consists of a brief discussion of the ways in which pathogen units (e.g., spores, conidia, polyhedral inclusion bodies) have been or could be quantified in insect epizootiology.

2.2.1. SAMPLING INFECTED HOSTS. Sampling infected hosts can provide information ranging from simple presence or absence of a pathogen to an estimate of the number of pathogen units per unit area. In the former case, insects can be pooled so that there is only one sample from each locality, and then subsamples from each pool can be examined to determine at which localities the pathogen is present (26). This technique is useful for a survey of many localities or a large area. The next level of information obtainable from sampling hosts is very common in the literature: individual insects are diagnosed to determine disease prevalence, often at certain time intervals throughout a season. This provides a rough estimate of the amount of a pathogen in the habitat, but not in the form of pathogen units per unit area. Still more precise estimates of pathogen amounts can be made if the sampler calculates the number of hosts infected or killed by the disease in a certain area (e.g., 27). Pathogen units per mg of infected insect tissue also have been measured in field plots (28), but again this is not an absolute sample of pathogen units per area units. This has been termed the assessment of "pathogen load" per individual (29). Finally, when numbers of pathogen units produced in insects of certain sizes or stage categories can be determined and numbers of such infected hosts can be counted in the field, then estimates can be made of pathogen units per unit area (19, 30, 31).

2.2.2. BIOASSAY OF SAMPLES FROM THE HABITAT. Bioassay is a valuable method for detection or quantification of insect pathogens. Even when other methods can be used, they probably should be related to bioassay results at some point in a study because bioassay is the best indicator of biological activity. With preliminary standard curves bioassays can be used to estimate numbers of pathogen units per unit of habitat.

Bioassays for pathogen sampling have been limited to the use of live insects, but assays in cell cultures are also possible (32, 33). Live insects have been used in two ways: as detection "tools" in the laboratory, or as "sentinel insects" placed for a time in the host insect's habitat. With either method, however, the researcher must take care not to assume that the bioassay insects are fully comparable to field insects with regard to susceptibility, behavior, and other attributes. Bioassay has been used to detect and quantify insect pathogens both in biotic and in abiotic components of the habitat, and it has been used for all major groups of insect pathogens.

Laboratory insects have been used for quantification as well as detection of pathogens. Basically, the method involves the sampling of some portion of the host insect's habitat, the preparation of a suspension resulting in partial selection for the microorganism to be detected, and the exposure of bioassay insects to that suspension, usually on artificial diet or portions of foliage if feeding is involved. The insects can also be exposed to unprocessed material, for example soil (34). Bioassays have been used to determine relative amounts of pathogens in soil (e.g., 35), on foliage (e.g., 23, 36), in bird feces (37), and in water (38). They been used to estimate absolute numbers of pathogen units in soil (e.g., 39), on foliage (40), in avian and mammalian feces (41), and in avian and mammalian tissues (42). Bioassays for virus in soil have been related to subsequent disease prevalence in one study (12).

Vail (43) and Martignoni and Ignoffo (44) discussed the advantages, disadvantages, and necessity for standardization of bioassays in relation to microbial control. Portions of these discussions are pertinent to epizootiology.

The other way of using bioassay for sampling the host's habitat for pathogens is with *sentinel insects.* With this method appropriate host animals are placed for a time in the habitat, usually in cages, and then observed there or in the laboratory for the development of disease.

Sentinel insects have been used frequently in studies of aquatic insects but only rarely with terrestrial hosts. Examples with aquatic insects are numerous; the reader is referred to Chapman et al. (45) for the general methodology and to Westerdahl et al. (46), Jaronski and Axtell (47), and Mulligan et al. (38) for examples involving a nematode, fungus, and bacterium, respectively. In terrestrial situations, Ignoffo et al. (23) released and collected sentinel insects in soybeans without the use of cages, and Lynch et al. (48) released adult moths into cages and later collected their eggs to survey for egg diseases.

2.2.3. PATHOGEN COUNTS.   Direct counts of pathogen units have certain advantages: they are more accurate than pathogen numbers estimated from sampled hosts and usually quicker and cheaper than bioassays. However, since the presence of pathogen units does not necessarily mean they are infective, such counts should always be related to bioassay results, if possible, at some point in a study. In this section extraction of microorganisms from samples is discussed initially, then counting, though the two topics are sometimes inseparable.

Microorganisms have been separated from their substrates, or samples of habitat, by various filtration-related methods. Nematodes have been separated from samples of substrate, particularly soil, by *flotation, sieving,* and a *funnel technique* (6, 49). Sieving or *filtration* have not been used with other pathogen groups in insect pathology, but they have been used in other disciplines for bacteria, viruses, and fungi in soil or water (50–54). Similarly, spores of nonentomogenous fungi have been extracted by flotation (53, 54).

Another group of techniques takes advantage of *adhesion.* For example, in the *impression film or slide* technique, a sticky tape was pressed to a plant surface and then transferred to a microscope slide for examination for a nuclear polyhedrosis virus (NPV) (27, 55). In general microbiology, variations of this technique have been used to sample fungi from soil (54) and plant surfaces (56). Also, soil has been sampled for microorganisms by mixture of the soil with melted agar for subsequent staining and counting, by the Rossi-Cholodny *buried slide,* and by *microcapillaries* incubated in soil (57).

Aerial trapping for pathogens is potentially valuable, particularly for studies of fungi, but is seldom used with insect diseases. Entomopathogenic fungal conidia have been sampled with adhesives on glass slides (18, 19, 58) or a rotating dish (58). In general microbiology, various types of aerial traps have been used to sample bacteria, viruses, rickettsiae, and particularly fungi (50, 59–61).

The most generally useful method for sampling, isolation, and counting is the use of *selective media.* These have been used extensively in insect pathology, mostly in studies of pathogen persistence. Generally, the method consists of taking the sample (e.g., piece of foliage, soil, host insect), either washing it or suspending it in sterile water, and then using standard pour-plate or streak-plate isolation techniques. Air also can be sampled by placing petri plates with media in the habitat, though other methods are more effective (see above). In field studies with insect pathogens, selective media have been used to sample bacteria on foliage (36) and in soil from terrestrial (62) or aquatic (63) habitats, and they have been used to sample fungi from soil (64) and tree bark (65). Use of selective media has been reviewed for insect pathology (20, 21), for the bacterial (50) and fungal (66, 67) groups, and for sampling from water (68) and soil (52, 53). Similar methods also are possible for viruses (69) and mycoplasmas (70, 71).

Sampling considerations, precision, and disadvantages of sampling with selective media have been discussed previously (52, 53, 72). The major disadvantages are that the method favors microorganisms with vigorous growth, that the selection of methods and media heavily biases which microorganisms are sampled or counted, and that growth on media does not necessarily relate to pathogenicity. These disadvantages can be partially offset by the use of more than one isolation medium or more than one technique (e.g., microscope counts) for each sample (53). Alexander (57) discussed disadvantages in relation to soil sampling and pointed out that numerous subsamples are more valuable than plating replicates because there is more variability in the soil than in the counting procedure.

Microscopy is commonly used to count microorganisms in a sample. Usually the sample is washed or suspended in water, though techniques such as embedding and sectioning (73) or filtration and removal (see above) can be used, and then a count is made with electron, light, or fluorescence microscopy. In insect pathology, microscope counts have been used for bacteria on foliage (36) and in soil (74); for a protozoan in soil (75); and for virus in soil (73, 76, 77), on foliage (27), and in bird feces (37). Microscopy also can be used with nematodes (78) and fungi (54, 56). Techniques have been reviewed for fluorescence, luminescence, or electron microscopy of microorganisms in soil (53, 79). Microscopy is a good technique to use in conjunction with media: microscopy is better for microorganisms that do not grow readily on media or in cell cultures, but it cannot usually distinguish viable from nonviable organisms. Burges and Thomson (72) discussed methods and precision of counting by microscopy.

2.2.4. DETECTION OR QUANTIFICATION BY SEROLOGY.   Other than the direct counting of pathogen units or propagules, other, more indirect methods can be used to quantify a pathogen in a host's habitat provided these methods are appropriately calibrated. These include various serological methods, chromatography, electrophoresis, and DNA probe techniques, though only serology has been used in insect pathology. These methods are potentially quicker and cheaper than bioassay and more specific and possibly more sensitive than microscopy for detection and quantification of microorganisms.

In insect pathology, serological methods have been developed to the point where they can detect insect pathogens, but few studies have done so in the host's habitat. NPVs have been detected in bird feces by radioimmunoassay (80), in soil by enzyme-linked immunosorbent assay (ELISA) (39), and in the feces of predatory insects by immunofluorescent microscopy (81). Other techniques or types of pathogens have been investigated, but only in basic studies or for diagnosis. These systems, which have the potential to be used in more ecologically oriented studies, include: single or double immunodiffusion for NPVs (82, 83); latex agglutination for virus (84); ELISA for CPV (85), iridescent virus (86), nonoccluded baculovirus (87), and a microsporidium (88); and a radioimmunoassay for NPV (89).

2.2.5. COMPARISONS OF QUANTIFICATION METHODS IN INSECT PATHOLOGY.   Some of the methods for quantification of entomopathogens have been directly or indirectly compared experimentally. These include bioassay vs. microscope counts for NPV in soil (90), NPV in bird feces (37), and *Bacillus thuringiensis* (Bt) on foliage (91); bioassay vs. microscope count vs. selective media for Bt on foliage (36); bioassay vs. estimates made from sampled hosts infected with NPV from foliage (27); sentinel insects vs. laboratory bioassay for *Bacillus sphaericus* in water (38); and bioassay vs. ELISA for NPV in soil (39). In most of these studies a conclusion was reached about the relative usefulness of the methods: bioassay was always the most sensitive, the most biologically meaningful, and sometimes the most accurate method for detection or quantification. Flotation-sieving, centrifugation-flotation, and the Baermann funnel have been compared for sampling a nematode in soil (49).

2.3. Environmental Parameters

Environmental parameters are of critical importance in any epizootiological study (see Ch. 7). The measurement and recording of such factors have been reviewed comprehensively and are considered here only briefly.

This topic can be divided into a consideration of abiotic and biotic parameters.

2.3.1. ABIOTIC FACTORS. The most important abiotic factors in insect epizootiology for terrestrial insects are temperature, relative humidity, sunlight, precipitation, wind, soil pH and composition, and moisture in soil or on vegetation (see Ch. 7). Hydrostatic force, dew point, and oxygen concentration are also sometimes important. In aquatic habitats important factors include water temperature, depth, pH, oxygen content, current velocity, and type of substrate. The measurement of many of these factors has been reviewed and discussed previously (92–94), as have the methods for computation of degree days (95). It is possible to record environmental data in the field so that they can be transferred directly to a computer file (92, 94). This can be a convenient and relatively error-free way to collect and process data.

2.3.2. BIOTIC FACTORS. Measurement of biotic factors in the environment depends on the system being studied and a search for appropriate methods in the literature. Such factors might include identification, enumeration, or measurement of host plants; sampling of vectors or alternate hosts; and population parameters of the pathogen and primary host, which can, at times, be considered part of each other's biotic environment.

3. ECOLOGICAL TECHNIQUES AND ANALYSES

Methodology in insect ecology has been largely ignored by insect epizootiologists in spite of the close relationship between the two fields. The value of many methods in this section is unknown because examples in insect epizootiology are either rare or nonexistent. Thus many of these methods are mentioned more for their potential than their realized value.

This section is divided into two parts, ecological field techniques or measurements, and types of analyses of data.

3.1. Techniques

Methods to be considered in this section will be those used strictly in field situations or to calculate parameters before they are subjected to various data analyses.

3.1.1. INTRODUCTION, INCREASE, OR ELIMINATION OF THE PATHOGEN POP-
ULATION.   The artificial introduction or increase of a pathogen population, a
technique researched extensively in microbial control experiments, is a powerful tool for the study of the effects of pathogen numbers on epizootics. Pathogen
numbers have been increased in many ways including by the release of infected
insects or healthy carriers and by spraying or dusting methods normally used
for chemical insecticides. This subject has been reviewed previously (e.g.,
see 96).

   In certain situations, the pathogen population can be eliminated for an
ecological study. For example, a fungicide can be sprayed to eliminate an
entomopathogenic fungus from an insect population which can then be compared to an unsprayed population (97). Soil decontamination techniques (98)
provide similar opportunities for pathogens with a soil reservoir.

3.1.2. RELEASE AND RECOLLECTION OF THE PATHOGEN.   Insect pathologists usually have released and recollected entomopathogens to determine their
persistence in the host's habitat (e.g., 27, 28, 36, 75), though investigations of
germination (54) and other characteristics are possible. This technique involves
methods already discussed, usually the introduction of the pathogen (see 3.1.1)
followed after a time by sampling of the habitat and quantification (see 2.2)

3.1.3. MARKING TECHNIQUES.   Marking techniques have not been used in
insect epizootiology but have potential value for learning the fates of individual
hosts or a limited number of pathogens.

   The marking of individual *hosts* for later identification would be one way of
conducting a true incidence or cohort study (see Ch. 1, section 6.1.5). Such
work is laborious but well worth at least a few attempts by insect epizootiologists. Southwood (2) has reviewed methods for marking insects in field studies.

   The marking and release of *pathogens* is potentially valuable for the study of
biological "decay" (99), growth rates, biomass (67), and the spread of a pathogen population in a host population or over a certain geographical area. Radio-
isotopes have been used to mark viruses (99) and fungi (100). Methods for
identifying strains or isolates of microorganisms also could be used for "marking" as long as the strain released at a study site is not the same as one already
present. Strains or isolates of entomopathogenic viruses, fungi, protozoa, and
bacteria have been identified by restriction endonuclease analysis (101), allo-
zyme analysis (102), electrophoresis (103), and serotyping (104), respectively.

3.1.4. BIOMASS.   The measurement of biomass is better than counts of organisms for the study of trophic levels because the size of the organisms makes no
difference with the former (2, 105). Zadoks (25) discussed biomass in relation to
epiphytology of plant diseases. Biomass of insects is usually measured as dry
weight (2), but with microorganisms it can be measured by volume (2, 57, 67),
chemical analyses, autoradiography (67), and amount of ATP (106). Biomass
rarely has been measured in insect epizootiology (107).

3.2. Analyses

Methods in this section include data analyses or presentations that are potentially useful in insect epizootiology. Although methods in this section are considered "analyses," appropriate sampling and experimental designs are necessary for the collection of suitable data.

3.2.1. DISPERSION. Dispersion, "the description of the pattern of the distribution or disposition of the animals in space" (2), has considerable significance ecologically, both from practical and theoretical viewpoints. Southwood (2) discussed its significance and the mathematical distributions and indices used in dispersion analysis. Taylor (3) discussed the strengths and weaknesses of those indices. Dispersion analysis is important to insect epizootiology for the same reasons as it is to ecology: design of sampling programs for hosts, infected hosts, and the pathogen population; data analysis; and description of populations. Additionally, dispersion analysis has concentrated research on the "empty set," where host and parasite do not make contact (108), and dispersion of the parasite population has been proposed as an important factor determining whether that population will effectively regulate a host population (109) or result in stability (110). Simple description of populations has been the purpose of the few dispersion analyses in insect epizootiology (17, 30). Such research is necessary to lay a groundwork, to provide data from several systems as a basis for hypotheses, and to study "thresholds" of host and pathogen dispersion in relation to the development of epizootics.

3.2.2. EPICENTERS AND SPREAD. The dynamics and extent of a disease epizootic depend to some degree on its epicenters or foci and spread. Data about epicenters and spread of insect diseases have come from experimental as well as observational studies and have been analyzed by mathematical, graphing, and mapping techniques. Experimental approaches have involved the release of infected insects with subsequent observations of disease spread on host plants (111), in small plots (112), and on islands (16, 113). Epicenters of disease have been identified mathematically (30) and by mapping (17), and the spread of insect diseases has been evaluated by regression analyses (112, 113), graphing (113), and mapping (17, 112). Contour (114) and demographic mapping (115), both of which have been used in human epidemiology, could have applications in insect epizootiology.

3.2.3. LIFE TABLES AND RELATED ANALYSES. Life table analysis is a tool by which insect epizootiologists can study how disease fits into the entire life system of a particular insect host and how important a factor infectious disease is compared to other causes of death. Furthermore, the design of a life table study in entomology is similar to that of an incidence or cohort study (see Ch. 1, section 6.1.5). For this reason alone insect epizootiologists should become familiar with life tables. Finally, life table analysis, along with key factor analysis (below), is a tool for evaluation of density-dependence (2, 105), an important factor in insect epizootiology.

Southwood (2), Price (105), and van den Bosch et al. (116) discussed the usefulness of life tables, the different kinds with their advantages and disadvantages, and procedures for data collection and analyses.

Life tables including insect diseases are not common. Harcourt (24) identified a granulosis virus as the most important factor in survival of *Pieris rapae* in a life table of 18 generations. Fine and Sylvester (117) used life table techniques to assess the role of transmission in the maintenance of a plant-pathogenic virus which is also pathogenic to the insect vector, though other mortality causes were not examined.

Key factor analysis uses life table data to find the most important mortality agent(s) associated with fluctuations in an insect population. Southwood (2) and van den Bosch et al. (116) reviewed and evaluated the types of such analyses and the methods. Auer (18) included an insect disease in a key factor analysis.

3.2.4. MULTIPLE REGRESSION. At the beginning of a complex epizootiological study, survey sampling of a variety of abiotic and biotic independent variables followed by multiple regression analysis is a good way to determine which relationships are important and should be studied experimentally. Whereas life table analysis is similar to an incidence or cohort study, a survey subjected to multiple regression analysis constitutes a prevalence or cross-sectional study (Ch. 1, section 6.1.4). The reader should consult statistical texts for the methodology and pitfalls of such analysis. Multiple regression has been used in insect epizootiology in research of an entomopathogenic virus (12) and a fungus (30).

4. PATHOGENS AS TOOLS IN INSECT ECOLOGY

Entomopathogens occasionally can be tools in ecology rather than the subjects studied. Suzuki and Kunimi (119) studied survival and dispersal of moths with NPV as a marker; Fuxa (unpublished data) used susceptibility patterns to isolates of NPV to study the geographical origin of a population of the migratory *Spodoptera frugiperda;* and pathogens can be introduced to selectively eliminate populations of certain insects so that others can be studied with less complication, for example the reduction of numbers of lepidopterous larvae with *Bacillus thuringiensis.* Though widespread use of pathogens as tools should not be anticipated, fertile imaginations will undoubtedly discover other ways of using them in ecology.

CONCLUSION

Insect epizootiologists have not often used ecological techniques, yet insect ecology and epizootiology are closely related. Information from sampling or measurement can be worthless if errors or biases, including some peculiar to insect epizootiology, are not eliminated. There are myriad methods available for all three major types of factors sampled in epizootiology: the host population, the pathogen population, and environmental parameters. Insect epizootiologists are thoroughly familiar with some ecological techniques, such as the introduction of pathogen units into the host's habitat, but not with others, such as calculation of biomass and dispersion analysis. Finally, Chapter 2 provides some possible methodology for the quantification necessary in types of studies mentioned in Chapter 1, such as use of life tables in incidence studies or multiple regression in prevalence studies.

ACKNOWLEDGMENTS

I thank Leslie C. Lewis, K. P. Lim, and L. D. Newsom for reviewing the manuscript.

REFERENCES

1. R. H. Green, *Sampling Design and Statistical Methods for Environmental Biologists,* Wiley, New York, 1979.
2. T. R. E. Southwood, *Ecological Methods with Particular Reference to the Study of Insect Populations,* 2nd ed., Halsted Press, Wiley, New York, 1978.
3. L. R. Taylor, Assessing and interpreting the spatial distributions of insect populations, *Annu. Rev. Entomol.,* **29,** 321 (1984).
4. G. G. Newman and G. R. Carner, Disease incidence in soybean loopers collected by two sampling methods, *Environ. Entomol.,* **4,** 231 (1975).
5. H. C. Evans, Entomogenous fungi in tropical forest ecosystems: an appraisal, *Ecol. Entomol.,* **7,** 47 (1982).
6. G. O. Poinar, Jr., *Entomogenous Nematodes,* E. J. Brill, Leiden, The Netherlands, 1975.
7. G. Benz, "Physiopathology and Histochemistry," in E. A. Steinhaus, Ed., *Insect Pathology, An Advanced Treatise,* Vol. 1, Aacademic Press, New York, 1963, p. 299.
8. J. Kalmakoff and A. M. Crawford, "Enzootic Virus Control of *Wiseana* spp. in the Pasture Environment," in E. Kurstak, Ed., *Microbial and Viral Pesticides,* Marcel Dekker, New York, 1982, p. 435.
9. B. Zelazny, *Oryctes rhinoceros* populations and behavior influenced by a baculovirus, *J. Invertebr. Pathol.,* **29,** 210 (1977).
10. J. Vavra and J. V. Maddox, "Methods in Microsporidiology," in L. A. Bulla, Jr., and T. C. Cheng, Eds., *Comparative Pathobiology,* Vol. 1, *Biology of the Microsporidia,* Plenum Press, New York, 1976, p. 281.
11. T. H. Atkinson and R. C. Wilkinson, Microsporidan and nematode incidence in live-trapped and reared southern pine beetle adults, *Fla. Entomol.,* **62,** 169 (1979).
12. J. R. Fuxa and J. P. Geaghan, Multiple-regression analysis of factors affecting prevalence of nuclear polyhedrosis virus in *Spodoptera frugiperda* (Lepidoptera: Noctuidae) populations, *Environ. Entomol.,* **12,** 311 (1983).
13. J. E. Henry, K. Tiahrt, and E. A. Oma, Importance of timing, spore concentrations, and levels

of spore carrier in applications of *Nosema locustae* (Microsporida: Nosematidae) for control of grasshoppers, *J. Invertebr. Pathol.*, **21**, 263 (1973).

14. C. J. Mitchell, *Coelomomyces psorophorae*, an aquatic fungus parasitizing *Aedes vexans* mosquito larvae in Knox county, Nebraska, *Mosq. New*, **36**, 501 (1976).

15. J. Burke and J. Percy, Survey of pathogens in the large aspen tortrix, *Choristoneura conflictana* (Lepidoptera: Tortricidae), in Ontario and British Columbia with particular reference to granulosis virus, *Can. Entomol.*, **114**, 457 (1982).

16. E. C. Young, The epizootiology of two pathogens of the coconut palm rhinoceros beetle, *J. Invertebr. Pathol.*, **24**, 82 (1974).

17. J. R. Fuxa, Dispersion and spread of the entomopathogenic fungus *Nomuraea rileyi* (Moniliales: Moniliaceae) in a soybean field, *Environ. Entomol.*, **13**, 252 (1984).

18. G. R. Carner, "Sampling Pathogens of Soybean Pests," in M. Kogan and D. Herzog, Ed., *Sampling Methods in Soybean Entomology,* Springer-Verlag, New York, 1980, p. 559.

19. L. P. Kish and G. E. Allen, *The Biology and Ecology of* Nomuraea rileyi *and a Program for Predicting its Incidence on* Anticarsia gemmatalis *in Soybean,* Univ. Fla. Agric. Exp. Stn. Tech. Bull. 795, 1978.

20. G. Wittig, "Techniques in Insect Pathology," in E. A. Steinhaus, Ed., *Insect Pathology, An Advanced Treatise,* Vol. 2, Academic Press, New York, 1963, p. 591.

21. G. M. Thomas, "Diagnostic Techniques," in G. E. Cantwell, Ed., *Insect Pathology,* Vol. 1, Marcel Dekker, New York, 1974, p. 1.

22. R. A. Clark, R. A. Casagrande, and D. B. Wallace, Influence of pesticides on *Beauveria bassiana,* a pathogen of the Colorado potato beetle, *Environ. Entomol.*, **11**, 67 (1982).

23. C. M. Ignoffo, N. L. Marston, D. L. Hostetter, and B. Puttler, Natural and induced epizootics of *Nomuraea rileyi* in soybean caterpillars, *J. Invertebr. Pathol.*, **27**, 191 (1976).

24. D. G. Harcourt, Major factors in survival of the immature stages of *Pieris rapae* (L.), *Can. Entomol.*, **98**, 653 (1966).

25. J. C. Zadoks, Methodology of epidemiological research, *Annu. Rev. Phytopathol.*, **10**, 253 (1972).

26. C. Reinganum, S. J. Gagen, S. B. Sexton, and H. P. Vellacott, A survey for pathogens of the black field cricket, *Teleogryllus commodus,* in the western district of Victoria, Australia, *J. Invertebr. Pathol.*, **38**, 153 (1981).

27. H. F. Evans and P. F. Entwistle, "Epizootiology of the Nuclear Polyhedrosis Virus of European Spruce Sawfly with Emphasis on Persistence of Virus Outside the Host," in E. Kurstak, Ed., *Microbial and Viral Pesticides,* Marcel Dekker, New York, 1982, p. 449.

28. L. C. Lewis, Persistence of *Nosema pyrausta* and *Vairimorpha necatrix* measured by microsporidiosis in the European corn borer, *J. Econ. Entomol.*, **75**, 670 (1982).

29. P. J. Wigley, "Assessing the Influence of Disease on Insect Numbers," in J. Kalmakoff and J. F. Longworth, Eds., *Microbial Control of Insect Pests,* New Zealand Dep. Sci. and Ind. Res. Bull. 228, 1980, p. 10.

30. R. S. Soper and D. M. MacLeod, *Descriptive Epizootiology of an Aphid Mycosis,* U. S. Dep. Agric. Tech. Bull. No. 1632, 1981.

31. W. J. Kaupp, Estimation of nuclear polyhedrosis virus produced in field populations of the European pine sawfly, *Neodiprion sertifer* (Geoff.) (Hymenoptera: Diprionidae), *Can. J. Zool.* **61**, 1857 (1983).

32. P. J. Wigley and P. D. Scott, The seasonal incidence of cricket paralysis virus in a population of the New Zealand small field cricket, *Pteronemobius nigrovus* (Orthoptera: Gryllidae), *J. Invertebr. Pathol.*, **41**, 378 (1983).

33. E. M. Dougherty, "Baculovirus *in Vitro* Quantification," in C. M. Ignoffo, M. E. Martignoni, and J. L. Vaughn, Eds., *Characterization, Production and Utilization of Entomopathogenic Viruses,* Proc. 2nd Conf. US/USSR Joint Wkg. Grp. on the Production of Substances by Microbiological Means, National Tech. Serv., Springfield, VA, 1980, p. 119.

34. W. H. Ko, J. K. Fujii, and K. M. Kanegawa, The nature of soil pernicious to *Coptotermes formosanus, J. Invertebr. Pathol.,* **39,** 38 (1982).

35. Z. Mracek, The use of "*Galleria* traps" for obtaining nematode parasites of insects in Czechoslovakia (Lepidoptera: Nematoda, Steinernematidae), *Acta Entomol. Bohemoslovaca,* **77,** 378 (1980).

36. K. L. H. Leong, R. J. Cano, and A. M. Kubinski, Factors affecting *Bacillus thuringiensis* total field persistence, *Environ. Entomol.,* **9,** 593 (1980).

37. P. F. Entwistle, P. H. W. Adams, and H. F. Evans, Epizootiology of a nuclear polyhedrosis virus in European spruce sawfly *(Gilpinia hercyniae):* the rate of passage of infective virus through the gut of birds during cage tests, *J. Invertebr. Pathol.,* **31,** 307 (1978).

38. F. S. Mulligan III, C. H. Schaefer, and T. Miura, Laboratory and field evaluation of *Bacillus sphaericus* as a mosquito control agent, *J. Econ. Entomol.,* **71,** 774 (1978).

39. J. R. Fuxa, G. W. Warren, and C. Kawanishi, Comparison of bioassay and enzyme-linked immunosorbent assay for quantification of *Spodoptera frugiperda* nuclear polyhedrosis virus in soil, *J. Invertebr. Pathol.,* **46,** 133 (1985).

40. E. D. Thomas, A. M. Heimpel, and J. R. Adams, Determination of the active nuclear polyhedrosis virus content of untreated cabbages, *Environ. Entomol.,* **3,** 908 (1974).

41. R. A. Lautenschlager and J. D. Podgwaite, Passage of nucleopolyhedrosis virus by avian and mammalian predators of the gypsy moth, *Lymantria dispar, Environ. Entomol.,* **8,** 210 (1979).

42. R. A. Lautenschlager, J. D. Podgwaite, and D. E. Watson, Natural occurrence of the nucleopolyhedrosis virus of the gypsy moth, *Lymantria dispar* (Lep.: Lymantriidae) in wild birds and mammals, *Entomophaga,* **25,** 261 (1980).

43. P. V. Vail, "Standardization and Quantification: Insect Laboratory Studies," in M. Summers, R. Engler, L. A. Falcon, P. V. Vail, Eds., *Baculoviruses for Insect Pest Control: Safety Considerations,* Am. Soc. Microbiol., Washington, 1975, p. 44.

44. M. E. Martignoni and C. M. Ignoffo, "Biological Activity of Baculovirus Preparations: *in Vivo* Assay," in C. M. Ignoffo, M. E. Martignoni, and J. L. Vaughn, Eds., *Characterization, Production and Utilization of Entomopathogenic Viruses,* Proc. 2nd Conf. US/USSR Joint Wkg. Grp. on the Production of Substances by Microbiological Means, National Tech, Serv., Springfield, VA, 1980, p. 138.

45. H. C. Chapman, D. B. Woodard, T. B. Clark, and F. E. Glenn, Jr., A container for use in field studies of some pathogens and parasites of mosquitoes, *Mosq. New,* **30,** 90 (1970).

46. B. B. Westerdahl, R. K. Washino, and E. G. Platzer, Successful establishment and subsequent recycling of *Romanomermis culicivorax* (Mermithidae: Nematoda) in a California rice field following postparasite application, *J. Med. Entomol.,* **19,** 34 (1982).

47. S. Jaronski and R. C. Axtell, Persistence of the mosquito fungal pathogen *Lagenidium giganteum* (Oomycetes: Lagenidiales) after introduction into natural habitats, *Mosq. News,* **43,** 332 (1983).

48. R. E. Lynch, L. C. Lewis, and T. A. Brindley, Bacteria associated with eggs and first-instar larvae of the European corn borer: isolation techniques and pathogenicity, *J. Invertebr. Pathol.,* **27,** 325 (1976).

49. M. C. Saunders and J. N. All, Laboratory extraction methods and field detection of entomophilic rhabditoid nematodes from soil, *Environ. Entomol.,* **11,** 1164 (1982).

50. B. M. Mitruka, *Methods of Detection and Identification of Bacteria,* CRC Press, Cleveland, 1976.

51. C. P. Gerba and S. M. Goyal, Eds., *Methods in Environmental Virology,* Marcel Dekker, New York, 1982.

52. S. T. Williams and T. R. G. Gray, "General Principles and Problems of Soil Sampling," in R. G. Board and D. W. Lovelock, Eds, *Sampling—Microbiological Monitoring of Environments,* Academic Press, New York, 1973, p. 111.

53. L. F. Johnson and E. A. Curl, *Methods for Research on the Ecology of Soil-Borne Plant Pathogens,* Burges Publishing, Minneapolis, 1972.

54. G. L. Barron, "Soil Fungi," in C. Booth, Ed., *Methods in Microbiology,* Academic Press, New York, 1971, p. 405.

55. C. J. Elleman, P. F. Entwistle, and S. R. Hoyle, Application of the impression film technique to counting inclusion bodies of nuclear polyhedrosis viruses on plant surfaces, *J. Invertebr. Pathol.,* **36,** 129 (1980).

56. B. I. Lindsey, "A Survey of Methods Used in the Study of Microfungal Succession on Leaf Surfaces," in C. H. Dickinson and T. F. Preece, Eds., *Microbiology of Aerial Plant Surfaces,* Academic Press, New York, 1976, p. 217.

57. M. Alexander, Introduction to Soil Microbiology, 2nd ed., Wiley, New York, 1977.

58. J. A. Millstein, G. C. Brown, and G. L. Nordin, Microclimatic humidity influence on condidial discharge in *Erynia* sp. (Entomophthorales: Entomophthoraceae), an entomopathogenic fungus of the alfalfa weevil (Coleoptera: Curculionidae), *Environ. Entomol.,* **11,** 1166 (1982).

59. A. B. Akers and W. D. Won, "Assay of Living, Airborne Microorganisms," in R. L. Dimmick and A. B. Akers, Eds., *An Introduction to Experimental Aerobiology,* Wiley, New York, 1969, p. 59.

60. R. R. Davies, "Air Sampling for Fungi, Pollens and Bacteria," in C. Booth, Ed., *Methods in Microbiology,* Vol. 4, Academic Press, New York, 1971, p. 367.

61. J. Lacey, "The Aerobiology of Conidial Fungi," in G. T. Cole and B. Kendrick, Eds., *Biology of Conidial Fungi,* Vol. 1, Academic Press, New York, 1981, p. 373.

62. R. J. Milner, A method for isolating milky disease, *Bacillus popilliae* var. *rhopaea,* spores from the soil, *J. Invertebr. Pathol.,* **30,** 283 (1977).

63. B. C. Hertlein, R. Levy, and T. W. Miller, Jr., Recycling potential and selective retrieval of *Bacillus sphaericus* from soil in a mosquito habitat, *J. Invertebr. Pathol.,* **33,** 217 (1979).

64. J. C. R. Pereira, O. D. Dhingra, and G. M. Chaves, A selective medium for population estimations of *Metarhizium* in soil, *Trans. Br. Mycol. Soc.,* **72,** 495 (1979).

65. J. W. Doberski and H. T. Tribe, Isolation of entomogenous fungi from elm bark and soil with reference to ecology of *Beauveria bassiana* and *Metarhizium anisopliae, Trans. Br. Mycol. Soc.,* **74,** 95 (1980).

66. C. Booth, Ed., *Methods in Microbiology,* Academic Press, New York, 1971.

67. D. Parkinson, "Techniques for the Study of Soil Fungi," in T. Rosswall, Ed., *Modern Methods in the Study of Microbial Ecology,* Rotobeckman, Stockholm, 1973, p. 29.

68. V. G. Collins, J. G. Jones, M. S. Hendrie, J. M. Shewan, D. D. Wynn-Williams, and M. E. Rhodes, "Sampling and Estimation of Bacterial Populations in the Aquatic Environment," in R. G. Board and D. W. Lovelock, Eds., *Sampling—Microbiological Monitoring of Environments,* Academic Press, New York, 1973, p. 77.

69. E. M. Smith and C. P. Gerba, "Laboratory Methods for the Growth and Detection of Animal Viruses," in C. P. Gerba and S. M. Goyal, Eds., *Methods in Environmental Virology,* Marcel Dekker, New York, 1982, p. 15.

70. R. J. Fallon, "Isolation Methods for Mycoplasmas from Man and Rodents," in D. A. Shapton and G. W. Gould, Eds., *Isolation Methods for Microbiologists,* Academic Press, New York, 1969, p. 41.

71. C. H. Liao and T. A. Chen, "Media and Methods for Culture of Spiroplasmas," in M. J. Daniels and P. G. Markham, Eds., *Plant and Insect Mycoplasma Techniques,* Halsted Press, Wiley, New York, 1982, p. 174.

72. H. D. Burges and E. M. Thomson, "Standardization and Assay of Microbial Insecticides," in H. D. Burges and N. W. Hussey, Eds., *Microbial Control of Insects and Mites,* Academic Press, New York, 1971, p. 591.

73. T. Hukuhara and H. Namura, Microscopical demonstration of polyhedra in soil, *J. Invertebr. Pathol.,* **18,** 162 (1971).

74. A. W. West, N. E. Crook, and H. D. Burges, Detection of *Bacillus thuringiensis* in soil by immunofluorescence, *J. Invertebr. Pathol., 43*, 150 (1984).

75. J. J. Germida, Persistence of *Nosema locustae* spores as determined by fluorescence microscopy, *Appl. Environ. Microbiol., 47*, 313 (1984).

76. T. Hukuhara, Demonstration of polyhedra and capsules in soil with scanning electron microscope, *J. Invertebr. Pathol., 20*, 375 (1972).

77. S.-Y, Kuo, Persistence of the Nuclear Polyhedrosis Virus of the Soybean Looper, *Pseudoplusia includens* (Walker) (Lepidoptera: Noctuidae) in the Soybean Ecosystem, M. S. Thesis, Dep. Entomol., Louisiana St. Univ., Baton Rouge, 1974.

78. V. H. Dropkin, *Introduction to Plant Nematology,* Wiley, New York, 1980.

79. T. Rosswall, Ed., *Modern Methods in the Study of Microbial Ecology,* Rotobeckman, Stockholm, 1973.

80. A. M. Crawford and J. Kalmakoff, Transmission of *Wiseana* spp. nuclear polyhedrosis virus in the pasture habitat, *N. Z. J. Agric. Res., 21*, 521 (1978).

81. D. J. Cooper, The role of predatory Hemiptera in disseminating a nuclear polyhedrosis virus of *Heliothis punctiger, J. Aust. Entomol. Soc., 20*, 145 (1981).

82. S. Y. Young, W. C. Yearian, and H. A. Scott, Detection of nuclear polyhedrosis virus infection in *Heliothis* spp. by agar gel double diffusion, *J. Invertebr. Pathol., 26*, 309 (1975).

83. H. A. Scott, W. C. Yearian, and S. Y. Young, Evaluation of the single radial diffusion technique for detection of nuclear polyhedrosis virus (NPV) infection in *Heliothis zea, J. Invertebr. Pathol., 28*, 229 (1976).

84. S. Shimizu, M. Ohba, K. Kanda, and K. Aizawa, Latex agglutination test for the detection of the flacherie virus of the silkworm, *Bombyx mori, J. Invertebr. Pathol., 42*, 151 (1983).

85. P. Payment, D. J. S. Arora, and S. Belloncik, An enzyme-linked immunosorbent assay for the detection of cytoplasmic polyhedrosis virus, *J. Invertebr. Pathol., 40*, 55 (1982).

86. D. C. Kelly, M. L. Edwards, and J. S. Robertson, The use of enzyme-linked immunosorbent assay to detect, and discriminate between, small iridescent viruses, *Ann. Appl. Biol., 90*, 369 (1978).

87. J. F. Longworth and G. P. Carey, The use of an indirect enzyme-linked immunosorbent assay to detect baculovirus in larvae and adults of *Oryctes rhinoceros* from Tonga, *J. Gen. Virol., 47*, 431 (1980).

88. M. H. Greenstone, An enzyme-linked immunosorbent assay for the *Amblyospora* sp. of *Culex salinarius* (Microspora: Amblyosporidae), *J. Invertebr. Pathol. 41*, 250 (1983).

89. M. Ohba, M. D. Summers, P. Hoops, and G. E. Smith, Immunoradiometric assay for baculovirus enveloped nucleocapsids and polyhedrin, *J. Invertebr. Pathol., 30*, 362 (1977).

90. H. F. Evans, J. M. Bishop, and E. A. Page, Methods for the quantitative assessment of nuclear-polyhedrosis virus in soil, *J. Invertebr. Pathol., 35*, 1 (1980).

91. R. Engler and M. H. Rogoff, Entomopathogens: ecological manipulation of natural associations, *Environ. Health Perspectives, 14*, 153 (1976).

92. D. M. Unwin, *Microclimate Measurement for Ecologists,* Academic Press, New York, 1980.

93. P. C. Doraiswamy, "Instrumentation and Techniques for Microclimate Measurements," in J. L. Hatfield and I. J. Thomason, Eds., *Biometeorology in Integrated Pest Management,* Academic Press, New York, 1982, p. 43.

94. J. C. Sutton, T. J. Gillespie, and P. D. Hildebrand, Monitoring weather factors in relation to plant disease, *Plant Disease, 68*, 78 (1984).

95. K. P. Pruess, Day-degree methods for pest management, *Environ. Entomol., 12*, 613 (1982).

96. H. D. Burges, *Microbial Control of Pests and Plant Diseases 1970–1980,* Academic Press, New York, 1981.

97. R. L. Brandenberg, Impact of *Erynia phytonomi* (Zygomycetes: Entomophthorales), a fungal

pathogen, on alfalfa weevil, *Hypera postica* (Gyllenhall) (Coleoptera: Curculionidae) populations in Missouri, *J. Econ. Entomol.,* **78,** 460 (1985).

98. D. Mulder, Ed., *Soil Disinfestation,* Elsevier, New York, 1979.

99. J. C. Spendlove and K. F. Fannin, "Methods of Characterization of Virus Aerosols," in C. P. Gerba and S. M. Goyal, Eds., *Methods in Environmental Virology,* Marcel Dekker, New York, 1982, p. 261.

100. R. L. Lucas, "Autoradiographic Techniques in Mycology," in C. Booth, Ed., *Methods in Microbiology,* Vol. 4, Academic Press, New York, 1971, p. 501.

101. L. C. Loh, J. J. Hamm, C. Kawanishi, and E.-S. Huang, Analysis of the *Spodoptera frugiperda* nuclear polyhedrosis virus genome by restriction endonucleases and electron microscopy, *J. Virol.,* **44,** 747 (1982).

102. D. G. Boucias, C. W. McCoy, and D. J. Joslyn, Isozyme differentiation among 17 geographical isolates of *Hirsutella thompsonii, J. Invertebr. Pathol.,* **39,** 329 (1982).

103. D. A. Streett and J. D. Briggs, An evaluation of sodium dodecyl sulfate-polyacrylamide gel electrophoresis for the identification of microsporidia, *J. Invertebr. Pathol.,* **40,** 159 (1982).

104. H. de Barjac, "Identification of H-Serotypes of *Bacillus thuringiensis,"* in H. D. Burges, Ed., *Microbial Control of Pests and Plant Diseases 1970–1980,* Academic Press, New York, 1981, p. 35.

105. P. W. Price, *Insect Ecology,* Wiley, New York, 1975.

106. O. Holm-Hansen, "The Use of ATP Determinations in Ecological Studies," in T. Rosswall, Ed., *Modern Methods in the Study of Microbial Ecology,* Rotobeckman, Stockholm, 1973, p. 215.

107. J. Fargues, O. Reisinger, P. H. Robert, and C. Aubart, Biodegradation of entomopathogenic hyphomycetes: influence of clay coating on *Beauveria bassiana* blastospore survival in soil, *J. Invertebr. Pathol.,* **41,** 131 (1983).

108. J. H. Whitlock, *The Population Biology of Disease,* Cornell Univ. Press, Ithaca, New York, 1977.

109. R. M. Anderson and R. M. May, Regulation and stability of host-parasite population interactions. I. Regulatory processes, *J. Anim. Ecol.,* **47,** 219 (1978).

110. M. P. Hassel, Patterns of parasitism by insect parasitoids in patchy environments, *Ecol. Entomol.,* 7, 365 (1982).

111. C. M. Ignoffo, C. Garcia, D. L. Hostetter, and R. E. Pinnell, Laboratory studies of the entomopathogenic fungus *Nomuraea rileyi:* soil-borne contamination of soybean seedlings and disperal of diseased larvae of *Trichoplusia ni, J. Invertebr. Pathol.,* **29,** 147 (1977).

112. H. F. Evans and G. P. Allaway, Dynamics of baculovirus growth and dispersal in *Mamestra brassicae* L. (Lepidoptera: Noctuidae) larval populations introduced into small cabbage plots, *Appl. Environ. Microbiol.,* **45,** 493 (1983).

113. P. F. Entwistle, P. H. W. Adams, H. F. Evans, and C. F. Rivers, Epizootiology of a nuclear polyhedrosis virus (Baculoviridae) in European spruce sawfly, *(Gilpinia hercyniae):* spread of disease from small epicentres in comparison with spread of baculovirus diseases in other hosts, *J. Appl. Ecol.,* **20,** 473 (1983).

114. P. J. J. Angulo, C. A. A. Pederneiras, M. E. Sakuma, C. K. Takiguti, and P. Megale, Contour mapping of the temporal-spatial progression of a contagious disease, *Bull. Soc. Pathol. Exotique,* **72,** 374 (1979).

115. F. Forster, Use of a demographic base map for the presentation of areal data in epidemiology, *Br. J. Prev. Soc. Med.,* **20,** 165 (1966).

116. R. van den Bosch, P. S. Messenger, and A. P. Gutierrez, *An Introduction to Biological Control,* Plenum Press, New York, 1982.

117. P. E. M. Fine and E. S. Sylvester, Calculation of vertical transmission rates of infection, illustrated with data on an aphid-borne virus, *Am. Nat.,* **112,** 781 (1978).

118. C. Auer, "A Simple Mathematical Model for 'Key-Factor' Analysis and Comparison in Population Research Work," in G. P. Patil, E. C. Pielou, and W. E. Waters, Eds., *Statistical Ecology,* Vol. 2, *Sampling and Modeling Biological Populations and Population Dynamics,* The Pennsylvania State Univ. Press, University Park, 1971, p. 33.

119. N. Suzuki and Y. Kunimi, Dispersal and survival rate of adult females of the fall webworm, *Hyphantria cunea* Drury (Lepidoptera: Arctiidae), using the nuclear polyhedrosis virus as a marker, *Appl. Entomol. Zool.,* **16,** 374 (1981).

3

MODELING

GRAYSON C. BROWN

Department of Entomology
University of Kentucky
Lexington, Kentucky

1. INTRODUCTION

The field of insect epizootiology has lagged behind general crop protection entomology in its adoption of systems technology. A recent review of the mathematical literature (1) found fewer than five documented attempts at a systems approach describing epizootics in insect populations. These attempts varied widely in their theoretical foundations from purely empirical constructs to standard epidemiological approaches. The significance of this point is that, if insect epizootiology is to proceed as a scientific discipline, quantitative theory must develop coincident with the acquisition of empirical knowledge. This means that we should seek common principles driving epizootiological phenomena and apply them in a management system.

This is the central philosophy underlying this chapter. In order to pursue it, I shall first present some basic definitions and concepts concerning systems and the systems approach. I do not intend to review the sizeable literature on systems analysis and, instead, refer the interested reader to earlier reviews (2-5). I shall then present a tutorial overview of general epizootiology theory and models with emphasis on entomological applications. Finally, I shall attempt to synthesize these two areas with emphasis on incorporating epizootiological modeling activities into ongoing insect pest management efforts.

2. DEFINITIONS AND CONCEPTS

The concept of a system is central to this discussion. A system is defined here as a group of regularly interacting or interdependent items forming a unified whole. The term "unified whole" explicitly means that a system has a specified boundary across which inputs and outputs are exchanged. Furthermore, a system reacts to inputs but is unaffected by its own outputs (6). Thus, a system is defined by a specified boundary, its inputs and outputs, and by regular interaction of its components.

The systems analyst specifies the boundary of a system early in a particular investigation. Usually, this boundary is envisioned as one level in a hierarchy of potential levels. Potential levels of abstraction in finer resolution are referred to as *subsystems* unless they cannot be usefully further subdivided, in which case they are referred to as *system components*. Potential levels of abstraction in lower resolution are collectively referred to as the *environment*. Mechanisms that govern interactions or flow (inputs and outputs) between components are referred to as *processes* (when a process lacks detail, it is said to have low resolution and is sometimes derisively referred to as a "black box"). For example, an alfalfa production system might be specified to include the crop, its pests, one or more of the pests' natural enemies, and some management criteria. In this case, anything else would be considered as the environment. A subsystem might be the interaction between a pest, such as the alfalfa weevil, and its predominant fungal pathogen, *Erynia* sp. A component could be the plant itself. A process could be the larval aging routine.

The view of a system being defined in relatively artificial terms and consisting of pieces that are also arbitrarily defined confers a sense of abstraction that is characteristic of the concept of *systems approach*. This approach "permits the reduction of a complex problem into relatively simple component parts and . . . provides rules (via mathematics) for reassembling the parts without losing any essential behavior or properties of the original system" (7). Steps for applying the systems approach to pest management problems can be found elsewhere (5, 8).

An important aspect of the systems approach is the construction of a *systems model,* simplistically defined as an imitation and representation of a real system. A typical integrated pest management (IPM) systems model consists of three parts: a biological-conceptual framework formulated from previous traditional agricultural research; a mathematical representation of that framework; and a computer program that implements the mathematics.

Much has been written on the subject and classification of these models (e.g., 3, 5, 8–12). One classification is the distinction between management and research models. Management models, as their name implies, are used directly for pest management either in elucidating better management practices or in aiding management decisions such as forecasting biotic events or scheduling agricultural practices. Models used for management decision aids by individual farmers are often referred to as *on-line management models* (13).

Research models are used to organize and guide research of a particular system for two reasons. First, researchers can formally specify hypotheses, inferences, and data and thus objectively compare their perceptions about a system with what is actually known or observed. Second, such models precisely identify what additional information is needed to further understand a given system and thus help reduce nonessential empirical research. Research models often evolve into or give rise to management models, but the converse is infrequent.

An example of a research model that illustrates some of these concepts is shown in Figure 3.1, a simplified gypsy moth-nuclear polyhedrosis virus (NPV) epizootic compartmental model (14). The entire diagram is the "system," the interaction between healthy larvae and polyhedra on foliage to produce newly infected larvae is a "subsystem," and each box in Figure 3.1 is a "component." This is a research model because, although it was too crude to accurately predict field events (14), it was useful for investigating the general host density-NPV epizootic relationship for this system and how it might be modified by insecticide applications. The model's domain or boundary has since been expanded by incorporating additional components, principally those dealing with leaf production and availability, to derive additional results that were similarly useful at providing insight and organizing research. This general approach has been successfully used with other gypsy moth-NPV models (15, 16).

I shall now discuss how these definitions and concepts apply to insect epi-

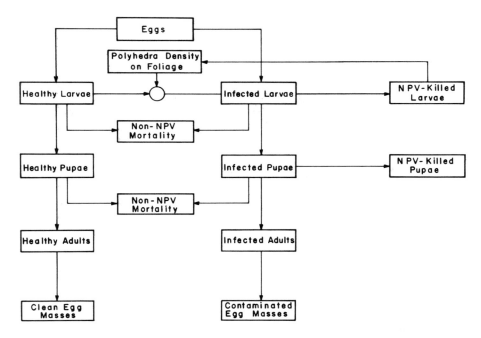

Figure 3.1. Gypsy moth-NPV model developed by Etter (14). Mortality of adults and eggs is not shown in this figure.

zootiology. It will be useful to divide this discussion into generalized and practical models.

3. EPIZOOTIC THEORY AND GENERAL MODELS

3.1. Theory and Epizootic Models

The mathematical theory of epizootics in insect populations is a complex albeit specialized field. However, the basic concepts are straightforward and necessary if one is to comprehend the more applied aspects of modeling insect pathogens. In this section, basic concepts will be described in an introductory fashion, but more advanced mathematical treatments are available (1, 17–19).

It is appropriate at the outset to define "epizootiological dynamics" as the study of the change in host and pathogen numbers over time and space along with the intrinsic and extrinsic factors responsible for that change. Changes over time comprise *temporal dynamics* and changes over space are *spatial dynamics*.

The key concepts in this definition are "intrinsic and extrinsic factors." Strictly speaking, intrinsic factors ultimately mean genetic effects, and extrinsic factors are nongenetic effects. However for practical uses, intrinsic factors refer to fixed or constant effects intrinsic to a specific population or system (for example, a host population's intrinsic growth rate), while extrinsic factors are variable or independent effects of the population's environment such as weather. The ability to distinguish between intrinsic and extrinsic factors is thus closely allied with the precision at which a given system's boundary is specified.

This implies that epizootiological phenomena are composites of the three basic and well-known building blocks of host, pathogen, and environment. The question is: what are the principles that govern the interaction among the various components in such a system? Let us approach this question by analyzing the temporal dynamics (a proper discussion of spatial dynamics is beyond the scope of this chapter) of a hypothetical insect host population in a constant environment and, for the moment, assume that all host stages, eggs to adult, are equally susceptible to a horizontally transmitted pathogen. Let us further decide that we can estimate the pathogen population simply by the number of infected hosts rather than by counting vegetative stages or infectious units directly.

With these simplifying assumptions, the dynamics of the "system" can be represented by the common approach of describing the susceptible and infected populations independently (c.f. Fig. 3.1):

$$\begin{pmatrix} \text{Number of Susceptible} \\ \text{Hosts Tomorrow} \end{pmatrix} = \begin{pmatrix} \text{Number of} \\ \text{Susceptibles Today} \end{pmatrix} - \begin{pmatrix} \text{Number of} \\ \text{Susceptibles that} \\ \text{are Infected Today} \end{pmatrix}$$

and,

$$\begin{pmatrix} \text{Number of Infected} \\ \text{Hosts Tomorrow} \end{pmatrix} = \begin{pmatrix} \text{Number of Infected} \\ \text{Hosts Alive Today} \end{pmatrix}$$
$$+ \begin{pmatrix} \text{New Infections} \\ \text{Today} \end{pmatrix} - \begin{pmatrix} \text{Number of Infected} \\ \text{that Die Today} \end{pmatrix}$$

The number of susceptible hosts that become infected on any day depends on the number of contacts between susceptible and infective hosts. If the number of susceptibles today is represented by S_t and infectives by I_t then the potential number of contacts is $S_t \cdot I_t$. For now, assume that some constant proportion p of these contacts actually results in transmission so that:

$$\begin{matrix} \text{Number of Susceptibles that} \\ \text{are Infected Today} \end{matrix} = \begin{matrix} \text{New Infections} \\ \text{Today} \end{matrix} = p\, S_t\, I_t$$

The parameter, p, is usually referred to as the *transmission efficiency* or pathogen *transmissibility*.

The term $p\,S_t\,I_t$ is the most common term in epizootiological or epidemiological models. Usually there is an allowance for exponential powers of one or both dependent variables. For example, the term might be more generally expressed as: $p\,S_t\,I_t^a$ where the exponent a is referred to as an "activity coefficient" that, in this case, determines the effective density of infective individuals. Although no insect epizootiological model of which I am aware of at this writing incorporates this exponent, it generally should be used because many insect hosts become moribund at about the same time that they become infective (i.e., releasing infectious units), and thus their potential contacts with susceptibles are restricted to those individuals in the immediate area.

Returning to the original word equation, the other term that must be estimated is the number of infected hosts that die on any given day. This value depends on the number of infecteds and their mortality rate. For simplicity, if we assume that infected hosts die only from the pathogen, then the average daily mortality rate m is inversely proportional to the pathogen's incubation or latent period. If the incubation period is 4 days, then an average of 25% of the infected individuals die each day, and $m = .25$. Therefore:

$$\text{Number of Infecteds} = m \cdot I_t$$
$$\text{that Die Today}$$

In the insect epizootic modeling literature, the parameter m is usually referred to as the pathogenicity of the disease, though it is actually more similar to insect pathology's definition for virulence.

Now the word equations can be written as algebraic expressions:

$$S_{t+1} = S_t - p\,S_t\,I_t$$
$$I_{t+1} = I_t + p\,S_t\,I_t - m\,I_t$$

$$(1)$$

This simple model, originally proposed in 1906 (20), reveals three important principles of epizootics, the first of which is density-dependence. Because the rate at which individuals become infected depends on the number of individuals, the growth of a natural epizootic is fundamentally a *density-dependent process.*

The next two principles can best be illustrated by examining the behavior of this system. If $p = .005$ (i.e., of all potential contacts, 1/2% of them result in disease transmission each day) and $m = .25$ (4-day incubation period) with 100 susceptibles and one infected individual initially, then we can use these equations to calculate the daily number of susceptibles and infecteds. The results are shown in Figure 3.2.

One feature of this figure is that there are still susceptible individuals remaining after the epizootic has died out (day 50). Stated another way, our second principle becomes: *an epizootic does not cease because of a lack of susceptible hosts* (19).

The third general principle is evidenced by the peak of the infected population. The infected population peaks when the rate of new infections is precisely equal to the death rate of infected individuals. With very little algebra we can see

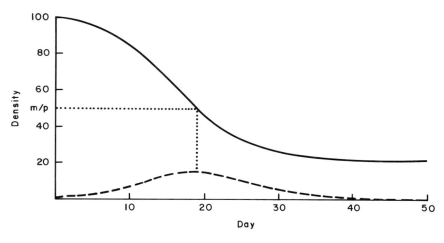

Figure 3.2. Solution of equation (1) when $p = 0.005$, $m = 0.25$, and the population initially consists of 100 susceptibles (solid line) and one infected host (dashed line).

that this occurs when $S_t = m/p$, that is, when the susceptible population equals the pathogenicity (or virulence) divided by the transmissibility. This ratio is called the *threshold density*. Thus, since all diseases have, by definition, some pathogenicity and some transmissibility, our third principle is: *all epizootics have a host threshold density conceptually defined by pathogenicity over transmissibility.* When the host (susceptible) population exceeds this threshold, disease prevalence can increase; when the host population is below this threshold, prevalence can only decrease.

The following terms must be defined at this point:

Disease Prevalence: the proportion of total host population
that is infected, $I_t/(I_t+S_t)$
Pathogen Burden: the mean number of available infectious units per host.

Note that in the simple case of equation (1), where it was assumed that pathogen densities could be estimated by the density of infected hosts that disease prevalence and pathogen burden are the same.

Prevalence is often used to characterize three stages in an epizootic cycle, depicted in Figure 3.3. There are various terminologies for these stages (c.f., Ch. 1), but I shall use the terms preepizootic, epizootic, and postepizootic phases, respectively. Factors responsible for these stages have been discussed in Chapter 1. Figure 3.3 is presented here because it represents a general behavior for an insect pathogen-host model.

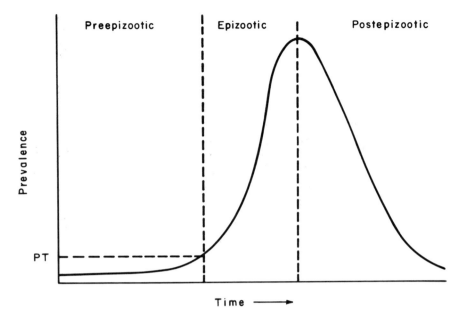

Figure 3.3. The three general stages of an epizootic cycle reflected in the rise and fall of prevalence. The preepizootic phase, which may have prolonged duration, exists as long as prevalence is below some perception threshold (PT). The epizootic phase is evidenced by a dramatic increase in prevalence and is normally short lived relative to the other two stages. The postepizootic stage occurs as prevalence recedes from a peak to below the perception threshold.

Equation (1) is normally written in the more traditional form of ordinary differential equations as:

$$\frac{dS}{dt} = rS - pSI$$

$$\frac{di}{dt} = pSI - mI$$

(2)

where the term dS/dt means the rate of change in the susceptible population over time (dI/dt is similar); r is the Malthusian "intrinsic rate of increase;" and other terms are as before. All of the assumptions for equation (1) hold for (2) except that reproduction of the susceptible host population is now incorporated.

There has been much biomathematical research of this equation, primarily as a base for predator-prey theory. This has led some workers to draw biological analogies between mathematical predator-prey theory and epizootic theory. However, it has been shown that the mathematical similarity between (2) and the classical Lotka-Volterra predator-prey model dissipates when the two systems are restated in biologically analogous terms (21).

One similarity between equation (2) and the Lotka-Volterra model that does hold is its neutral oscillatory behavior. An example is shown in Figure 3.4 which shows the familar sequence of very regular oscillations. The periodicity and frequency of these oscillations (along with the other theoretical results) are obtained from an analytic method generically referred to as stability analysis (22), the discussion of which is beyond the scope of this chapter. However, the important aspect of Figure 3.4 is that this system has no preepizootic behavior. That is, instead of looking like a string of asymmetrical three-phase cycles from Figure 3.3, it is a mechanical string of symmetrical two-phase cycles that bear little resemblance to real epizootic cycles.

The reason that this model fails to mimic actual cycles is that it lacks detail. Fortunately, several workers have enhanced this general framework by adding realistic aspects of insect and pathogen biology. The impact of age classes and seasonality on the ability of a pathogen to regulate its host have been analyzed (21), as well as the general effects of parasite-induced reduction of host reproduction, vertical transmission, latent periods, host stress, density-dependent constraints, and free-living infective stages (18).

As a general rule, the addition of these factors into equation (2) complicates

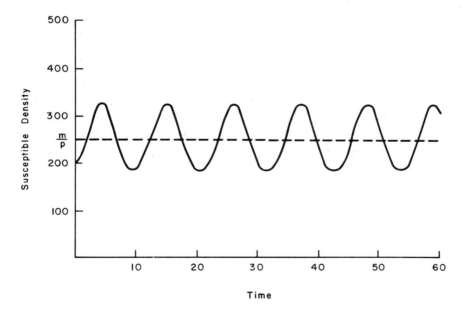

Figure 3.4. The solution to equation (2) when $r = 0.7$, $p = 0.002$, $m = 0.5$, and there are 200 initial susceptible hosts and 30 infected hosts. The oscillations are centered on $S = m/p$ but their amplitude depends on the initial susceptible and infected host densities.

the mathematical analysis somewhat; thus the details of these studies will not be presented here. However, the reader should note that these modifications to the basic model often have provided interesting insights into the dynamic behavior of these systems as well as a base for more applied efforts.

For example, the schematic model in Figure 3.5 represents the vertical transmission model (both transovarial and transovum) developed by Anderson and May (18). The letters on each arrow indicate rate constants or parameters similar to those presented in equation (2). This model is interesting for several reasons. First, it requires one to think in terms of the total host population (susceptibles + infecteds) when analyzing this system because both suscepti- bles and infecteds can reproduce. This impacts the mathematical representa- tion relative to equation (2).

Second, it has dynamic implications considerably different from those in Figures 3.3 or 3.4. The host threshold density for this system, H_T (where $H_t = S_t + I_t$), is

$$H_T = (m + b + \gamma - rq)/\beta$$

From Figure 3.5 we can see that r is the reproduction rate and q is the proportion of infected host's progeny that are born infected ($0 > q > 1$). Thus, if

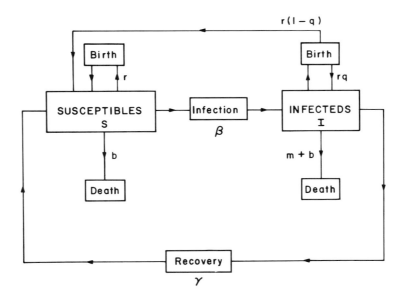

Figure 3.5. The vertical transmission model of Anderson and May (18). Susceptible and infected hosts give birth at rate r, but a proportion q of the infected host's progeny also are infected. The horizontal transmission efficiency in this model is β and the pathogenicity is m. Death due to sources other than the pathogen occurs at rate b. Finally, this model allows infected hosts to recover, thereby becoming susceptible again. This feature can be eliminated by setting $\gamma = 0$.

$q > 0$, vertical transmission reduces the host threshold density allowing the pathogen to maintain itself at lower host densities.

Another behavioral difference between this model and Figures 3.3 and 3.4 is that it can reach a steady state equilibrium where both infected and susceptible populations stabilize at a constant level. This will happen only if $m > (r\text{-}b)$, that is, if the pathogen can kill hosts faster than they can net reproduce (18). If this condition does not hold, both populations increase until suppressed by some external force not accounted for in this model.

A final interesting aspect of Figure 3.5 is its implications regarding the vertical transmission process itself. One can create a "vertical-transmission only" model out of Figure 3.5 by setting $\beta = 0$ and, thus, eliminate horizontal transmission. Note that this sends the host threshold density, as determined above, to infinity. Anderson and May examined this situation more formally (18) and concluded that a pathogen cannot generally maintain itself by vertical transmission alone. This implies that those pathogen systems in which little or no horizontal transmission is known (23–25) must incorporate it in an as yet unknown form or must have a general system structure significantly different from that in Figure 3.5. Such research guidance is useful, regardless of the outcome.

Incorporating one realistic factor into equation (2), such as the vertical transmission just discussed, significantly complicates mathematical analysis of the resultant model. Incorporating several factors usually yields a model that is mathematically intractable; simulation studies (response mapping, Monte Carlo methods, sensitivity testing, etc.) are the main tools used for analyzing such models. A discussion of these techniques is beyond the scope of this chapter, but the point is that when an epizootic model reaches this level of complexity it can properly be considered a simulation model. I shall therefore examine a generalized simulation model that is built on the above dynamic considerations.

3.2. A Generalized Simulation Model

It might seem unrealistic to attempt a generalized yet meaningful simulation model given the vast diversity of pathogen and host biologies. However, there are basic underlying mechanisms in these systems that can be generalized so as to represent broad classes of host-pathogen relationships. Moreover, if the resultant model has a modular design, then each module can be of arbitrary complexity and therefore detailed to any extent necessary.

A schematic representation of such a model is shown in Figure 3.6. This model incorporates age-dependent effects, inoculum loads, a host stress-susceptibility relationship, a latent period, an environmental "pool" of infectious units which are inactivated over time, and nonpathogen sources of host mortality. It thus has many biological phenomena commonly found in these systems. The actual model in its computer implementation (Microsoft™ BASIC) is given in the Appendix along with suggestions for modifying it to include effects

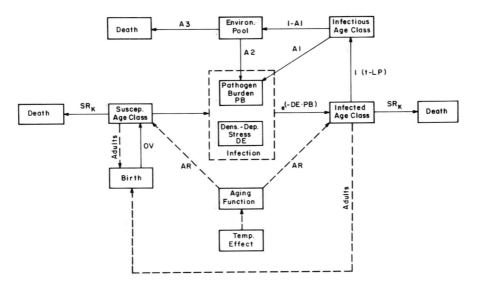

Figure 3.6. Schematic representation of the generalized simulation model. The notation of the rate constants follows that used in the computer implementation (c.f. Appendix). Solid arrows denote material flows while dashed arrows denote information (control) flow.

such as vertical transmission, temperature-dependent latent periods, seasonality, etc.

The insect host in Figure 3.6 is divided into three categories: susceptibles, infected but not yet infectious, and infectious. Infectious hosts are considered dead and exist only long enough to release the pathogen's infectious units which are tallied in third-instar larval equivalents (c.f., Appendix). Infectious units, once released, may either immediately contact a susceptible host or may accumulate in the environmental pool and be made available for infection later (if they are not inactivated first).

Both susceptible and infected, but not yet infectious, hosts produce susceptible eggs which then hatch and molt through a normal sequence of larval instars, according to some temperature-dependent aging function, to finally become adults. The effect of host stress on susceptibility is density-dependent in this model. There is assumed to be an "indifference density" which is merely a convenient standard at which an inoculum load-infection response is known. Presumably, then, when host densities are below this standard, susceptibles are less likely to become infected, but when the density exceeds this threshold, they are more likely to become infected. Susceptibility is also age-dependent, and infection is treated as a Poisson process in which individuals become infected randomly in each age class.

Figure 3.7 shows the behavior of this model over a 300 day simulation when the initial conditions are chosen to represent an early season situation (100 adults; 50 eggs; 20 first instars; two second instars; no other larval instars, infected, or infectious hosts; and 500 third instar equivalents of infectious units in the environmental pool). Although this model is still quite crude by simulation model standards, it is nevertheless clear that its general behavior has many realistic attributes. The host density profile, after initial stabilization, looks like the string of three phase cycles (c.f., Figure 3.3) with about four host generations elapsing between the peak of each cycle. If seasonality were incorporated into the model (c.f., Appendix), this periodicity would approximately double, depending on the specific assumptions. Furthermore, prevalence peaks about 8 to 10 days after peak host density (not unusual for naturally occurring epizootics), and the environmental pool logically peaks one latent period after prevalence.

Although this model is simplistic, its relatively realistic behavior suggest several general uses. First, it can serve as a preliminary framework which can easily be modified to account for most biological idiosyncrasies one is likely to encounter in a specific system. Second, it can be a useful guide to the information necessary for a model of a particular system (this application is demonstrated below). Many parametric values and functional forms can be found in the literature for the more widely studied host-pathogen species.

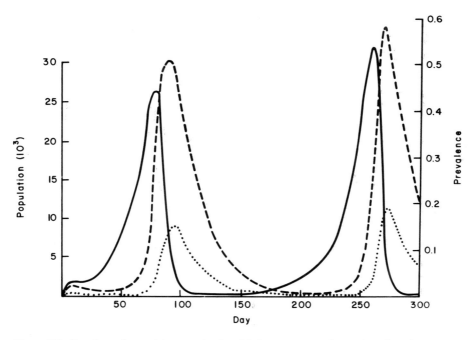

Figure 3.7. Sample simulation of the generalized model when parameter values and initial conditions are as described in the Appendix. The total host population (solid line) and environmental pool of infectious units (dotted line), in third instar equivalents, use the left ordinate while prevalence (dashed line) uses the right ordinate.

A third general use for this model is in evaluation of the potential impact of specific pest management practices on the host-pathogen dynamics. For example, the computer program in the Appendix can easily be modified so that whenever the host exceeds some critical level, an insecticide application is simulated by drastically reducing host population densities. Similarly, the effects of host plant resistance (e.g., reduced oviposition, survival, stress tolerance, etc.), innundative releases (microbial insecticides), density-dependent host mortalities, and natural enemies can all be studied either independently or in any combination. This educational value has resulted in models such as this being used in formal coursework in ecology and insect pathology.

The variety of potential studies and rich dynamic behavior of this "simple" model indicate that one must plan controlled experiments on this system in much the same manner as with a real system. If the objectives of such experiments are intended to elucidate a particular insect host-pathogen system, the system must be modeled in detail.

4. MODELING EPIZOOTICS IN PRACTICE

Generalizations about applied models are difficult because so few exist in the literature. However, published models can be divided into two categories: research models designed to guide epizootiological studies, and management models designed to evaluate or incorporate pathogens in a pest management program.

4.1. Epizootic Research Models

Most of the published models of pathogens have been added after the insect host (and usually its host plant) had already been modeled. Depending on the objectives of the pathogen model, this approach usually is beneficial because the basic utilities for crop management, climatic data processing, input and output routines, etc. are already available, and most of the host's bionomic processes such as oviposition, aging, dispersal, overwintering, etc. are easily adaptable. The general approach, then, has been to capitalize on existing model components to describe the infected host population as parallel but interacting with the uninfected host (e.g., Fig. 3.1). There are examples of this approach in the literature (1, 26, 27) as well as in the generalized epizootiology simulation model in the Appendix.

This approach tends to underemphasize the pathogen's biology. A lack of detail in the pathogen component was not necessarily a deficiency in the models cited above because they were primarily designed to evaluate management strategies for the host insect. Low resolution in pathogen biology also has been the case with those few models dealing with microbial insecticide testing (28, 29). However, where the objective of modeling a specific system is to aid in the study of a particular pathogen's epizootiology, a more traditional systems or component approach at the pathogen biology level is needed.

A notable exception to this trend was a model developed by Kish and Allen (30) to predict *Nomuraea rileyi* prevalence on velvetbean caterpillars in soybean fields. This model was of lower resolution than the generalized simulation model discussed above (in fact, it was presented essentially as a single equation), but the approach used by these authors embodied many of the principles discussed throughout this chapter. For example, the effect of weather was incorporated by separate analysis of various climatic factors (humidity, rain, wind, ultraviolet light, etc.) and their effects on the *N. rileyi* infectious process.

The *N. rileyi* model was concerned solely with predicting prevalence. Where a broader application is desired, the generalized simulation model (Appendix) can be used to identify potentially necessary model components and processes. Table 3.1 lists these components in three levels: pathogen life stages external to the host, those internal to the hosts, and the interactions that take place at the host-pathogen interface. Few, if any, pathogen models would combine all of the items in Table 3.1 in a single model. Others, depending on the specific pathogen's biology and life cycle, may expand or subdivide some of these items. Furthermore, the ease with which these items can be implemented is highly variable. Some factors, such as the influence of pathogens on the host's feeding and aging rates, would normally involve no more than reparameterizing existing host processes. On the other hand, those items in the first two columns would require completely new functional description (that is, new to the original host model).

Some of the items in the first two columns have been researched for models. A two-dimensional model, originally developed to describe the aerial dispersal of phytopathogenic fungal spores (31), was used to model the dispersal of *Erynia* sp. conidia (1). An exponential decay process has been used to account for virus particle survival (18). Conidial loss has been treated as linear functions

TABLE 3.1

Components and Processes that Must be Considered Initially in Modeling Pathogen Populations and the Interaction with Their Insect Hosts[a]

Pathogen Stages External to Host	Factors at the Host-Pathogen Interface	Pathogen Stages Internal to Host
Dispersal	Host stress	Incubation (latency) period
Survival or inactivation	Inoculum load	Host feeding rate
Transmission mode	Infection mode	Host aging rate
Alternative forms or hosts	Age-specific susceptibility	Host competitiveness (natural enemies)

[a] Not all of these items would necessarily be incorporated in any one model.

of wind and rain as well as inversely proportional to the leaf area index (30). The effect of host stress ("larval vigor") and age-specific susceptibility of the western tent caterpillar, *Malacosoma californicum,* to a pathogenic virus was modeled by Wellington et al. (26). Thus, documented modeling attempts at many of the items in Table 3.1 are sparse, and more work is needed if general conclusions are to be drawn about these components and processes.

The data base for the items in Table 3.1 varies between a great deal of information on some (such as incubation period) to very little on others (such as host competitiveness). Sources for the available data can be found elsewhere in this book. However, although data and models may be currently lacking in many areas, in any new pathogen modeling effort a list of major components and processes such as that in Table 3.1 should be developed early in the project. Ruesink wrote a step-by-step procedure detailing other considerations for new modeling efforts (8).

4.2. Epizootic Models in Insect Pest Management

If the published models of insect host-pathogen dynamics are sparse, those dealing with the use of pathogens in on-line pest management programs are almost nonexistent. The latter mainly have dealt with the use of pathogens as microbial insecticides either theoretically (18) or from the standpoint of efficacy evaluation (28, 29) with few dynamic properties. These are certainly valid studies, yet the variability among pathogen life cycles and the rich dynamic behavior possible with models of such life cycles is so great that other strategies for enhancement or augmentation of naturally occurring pathogen populations certainly must be possible. For example, an alfalfa weevil (*Hypera postica*)-*Erynia* sp. model (32) was used under various insecticide and harvesting strategies to determine if the Kentucky alfalfa weevil management recommendations then in use (and developed prior to the appearance of *Erynia* sp. in Kentucky) could be profitably modified to capitalize on this fungus. The results showed that a strategy incorporating early-season insecticide decision thresholds and early harvesting and relying on *Erynia* sp. to decimate the postharvest weevil population could increase net profits (increased yield and decreased management costs) by as much as 20%. A new set of alfalfa weevil recommendations was generated from this study and is summarized in Table 3.2 alongside the pre-*Erynia* recommendations. The complete scheme can be found in (33).

Field evaluation of this strategy was conducted in 1983 and 1984, and the results were convincing enough so that in 1984 a version of this model was added to the computer-based extension decision tools where it generates weevil management recommendations via computer and printed media (principally newsletters). The system can now be considered fully implemented, and its use is solely controlled by extension specialists.

Alfalfa is particularly well suited to such a study principally because it compensates for pest damage and because harvesting can be used as a pest control device. However, other types of pathogen enhancement or augmentation are possible in other crop systems. Nordin (34) reviewed the possibilities for fungi.

TABLE 3.2

Synoptic Comparison of pre-*Erynia* and *Erynia*-Enhanced Alfalfa Weevil Management
Recommendations When Weevil Densities Exceed Action Thresholds, Tabulated by
Degree-Day Accumulations

Degree Days (base 48°F)	Pre-*Erynia*	*Erynia*-Based Weevil Management
190–225	Long residual insecticide	Long residual
226–275	Long residual insecticide	Short residual
276–325	Short residual insecticide	Higher threshold
376–450	Short residual insecticide	No spray
451–525	Short residual insecticide	Harvest
526–600	Very short residual insecticide or early harvest	*Erynia* active
ca. 700	Harvest	—
100 DD after harvest	Short residual stubble spray	None required

Some of these strategies have been modeled in a preliminary manner. For example, a simulation model of a generalized insect host-NPV-parasitoid system incorporating light and pheromone traps for autodissemination has recently been constructed (G. C. Brown, unpublished). This model has five components: a pheromone trap attraction component, a mating submodel, an epizootic model, a parasitoid model, and a host population dynamics component. Simulations with this model demonstrated several features of the system, two of which are particularly important. First, a host-pathogen system with parasitoid-dependent phoretic transmission can sustain enzootics at very low host densities (the basis for this result is similar to that mentioned previously for vertical transmission). This means that the pathogen, at least theoretically, always has the potential to create an epizootic should the host population surge to high enough densities.

A second feature was that, if the early season growth rate of the host population is high relative to the NPV's transmission efficiency, an autodissemination strategy could indeed initiate epizootics significantly earlier than by natural means. However, because such a strategy would optimally be implemented just prior to a pest outbreak, it also would occur during a period of rapid increase in parasitoid abundance. The resultant decrease in parasitoids would have a destabilizing effect on the total system and could result in additional pest control in subsequent years. This adverse affect could be avoided by proper timing of the autodissemination strategy. However, the lesson is clear. That is, on paper, an autodissemination technique can prove beneficial, but its dynamic effects, particularly with respect to its net influence on the parasitoid population, should be clearly understood so that both populations can be manipulated in a desirable and predictable manner.

Most of the pathogen modeling of alternative management strategies is at a similarly preliminary stage. However, such models have potential applications rich with exciting, if not sometimes bizarre, behavior well worth the effort for both pathologists and modelers. The limiting factor is that this dynamic behavior needs to be more clearly understood. This understanding will come only after more modeling work is directed at specific systems.

The need for additional modeling efforts for specific systems that could eventually be implemented in pest management programs is made more urgent by the continuing advance of computer-based extension delivery systems technology. This technology has been reviewed relative to pest management (35) and it is apparent that there are already several advanced systems in the USA and that the technology continues to increase in sophistication. The concept of decision support systems in IPM has been outlined (36), and the alfalfa weevil system (33) demonstrated the applicability to IPM of an area of artificial intelligence known as "expert systems" (37).

Thus the technology of applied systems research in pest management is advancing at an accelerating rate. In some of these areas, such as decision support systems which are highly modular by design, pathogen models can be added after the total system is operational. In other areas, such as in the expert systems, significant modifications to the knowledge base can be major undertakings because of the single, total enterprise (crop production, storage, and marketing) plan of current designs. Because the systems engineer cannot afford frequent major upgrades, allowances for pathogen effects should be made as early as possible in the system's design stage. Consequently, the earlier pathogen modeling activities commence, the more likely they are to be incorporated into state-of-the-art pest management systems in a timely manner.

5. CONCLUSION

This chapter has dealt with modeling and systems analysis as applied to insect pathogens in a descending topical order according to the available information for each topic. A good deal is known of the theory of epizootics in insect populations, largely owing to the work of R. M. Anderson and R. M. May. Considerably less is known about the design of general epizootic simulation models, and what we do have is largely abstracted from other fields of pest management. Finally, the discussion of applied research was limited to only a few studies, and much work is only in a preliminary and unpublished stage due to inadequate understanding of pathogens as dynamic ecological forces. Fortunately this situation has been changing recently and the ecological role of insect pathogens is gaining visibility and attracting more modeling attention.

A continued emphasis on modeling insect pathogens and their epizootiology is necessary because many avenues of work remain untouched. General topics such as spatial considerations, particularly in the early stages of an epizootic when a disease is concentrated in separate epicenters, should be explored from both the theoretical and practical standpoints. Many more specific systems should be modeled so that comparative studies could elucidate general principles and techniques. Such studies could compare pathogen groups, ecotypes, processes, or entire models. The efficient research structure imposed by a true systems approach can help identify critical research needs and reduce work on tangential or marginally important components.

Correspondingly, the field of insect pathology has much to offer to other fields, and modeling, as a formal mathematical language, can act as the interdisciplinary medium. Most of our concepts concerning diseases in host populations have come from medical epidemiology or plant pathology. Workers in both fields have routinely used modeling as a tool to organize their empirical and theoretical research. However, the unique properties of insects and their pathogens as model systems, such as the relative ease of culture, great diversity, inexpensive subjects, etc., offer a great potential to the general field of disease dynamics in much the same way that the study of fruit fly genes can be studied in a general sense only because of the common language of DNA and RNA. For insect epizootiology to similarly contribute to its general, parent field, it must use the common language of mathematics.

This is the natural order of maturation for any scientific discipline. "The descriptive, natural history stage of science is eventually replaced by a deductive theoretical stage, basically mathematical in nature, which creates the abstractions and measurements necessary to deepen causal analysis" (38). Insect epizootiology will not be an exception.

REFERENCES

1. G. C. Brown, "A Modeling Approach to Epizootiological Dynamics of Insect Host-Pathogen Interactions and an Implementation Example," in C. A. Shoemaker and G. C. Brown, Eds., *Insect Pest Management Modeling,* John Wiley & Sons, New York, in press.

2. J. France and J. H. M. Thornley, *Mathematical Models in Agriculture,* Butterworths, London, 1984.

3. W. M. Getz and A. P. Gutierrez, A perspective on systems analysis in crop production and insect pest management, *Annu. Rev. Entomol.,* **27,** 447 (1982).

4. J. B. Dent and M. J. Blackie, *Systems Simulation in Agriculture,* Applied Science Publ., London, 1979.

5. W. G. Ruesink, Status of the systems approach to pest management, *Annu. Rev. Entomol.,* **21,** 27 (1976).

6. C. R. W. Spedding, *An Introduction to Agricultural Systems,* Applied Science Publ., London, 1979.

7. Anonymous, Integrated Pest Management, A Program of Research for the State Agricultural Experiment Stations and the Colleges of 1890, a study conducted by the Intersociety Consortium for Plant Protection (J. L. Apple, Chair), 1979, p. 57.

8. W. G. Ruesink, "Analysis and Modeling in Pest Management." in R. L. Metcalf and W. H. Luckman, Eds. *Introduction to Insect Pest Management,* Wiley-Interscience, New York, 1982, p. 353.

9. C. A. Shoemaker, "The Role of Systems Analysis in Integrated Pest Management," in C. B. Huffaker, Ed., *New Technology of Pest Control,* Wiley, New York, 1980, p. 25.

10. H. J. Gold, *Mathematical Modeling of Biological Systems—An Introductory Guidebook,* Wiley, New York, 1977.

11. B. C. Patten, "A Primer for Ecological Modeling and Simulation with Analog and Digital Computers," in B. C. Patten, Ed., *Systems Analysis and Simulation in Ecology,* Vol. 1, Academic Press, New York, 1971, p. 3.

12. K. E. F. Watt, Use of mathematics in population ecology, *Annu. Rev. Entomol.,* 7, 243 (1962).

13. R. L. Tummala, "Concept of On-Line Pest Management," in R. L. Tummala, D. L. Haynes, and B. A. Croft, Eds., *Modeling for Pest Management: Concepts, Applications and Techniques,* Michigan State Univ., 1976, p. 28.

14. D. O. Etter, "Pest Management Systems Development," in C. Doane and M. L. McManers, Eds., *The Gypsy Moth: Research Toward Integrated Pest Management,* USDA-APHIS Tech. Bull. 1584, 1981, p. 697.

15. H. T. Valentine, "A Model of Oak Forest Growth Under Gypsy Moth Influence," in C. C. Doane and M. L. McManers, Eds., *The Gypsy Moth: Research Toward Integrated Pest Management,* USDA-APHIS Tech. Bull. 1584, 1981, p. 50.

16. H. T. Valentine and J. D. Podgwaite, Modeling the role of NPV in gypsy moth population dynamics, *Proc. 15th Annu. Meeting Soc. Invertebr. Pathol. Brighton.,* p. 353 (1982).

17. R. M. Anderson and R. M. May, Infectious diseases and population cycles of forest insects, *Science,* 210, 658 (1980).

18. R. M. Anderson and R. M. May, The population dynamics of microparasites and their invertebrate hosts, *J. Anim. Ecol.,* 291, 451 (1981).

19. P. Waltman, *Deterministic Threshold Models in the Theory of Epidemics,* in S. Levin, Ed., *Lecture Notes in Biomathematics,* Vol. 6, Springer-Verlag, New York, 1978.

20. W. H. Hammer, Epidemic disease in England—the evidence of variability and the persistence of type, *The Lancet,* 11, 733 (1906).

21. G. C. Brown, Stability in an epizootiological model with age-dependent immunity and seasonal host reproduction, *Bull. Math. Biol.,* 46, 139 (1984).

22. R. M. May, *Stability and Complexity in Model Ecosystems,* Princeton Univ. Press, Princeton, New Jersey, 1974.

23. T. G. Andreadis and D. W. Hall, Significance of transovarial infections of *Amblyospora* sp. (Microspora: Thelohaniidae) in relation to parasite maintenance in the mosquito *Culex salinarius* Coquillett, *J. Invertebr. Pathol.,* 34, 152 (1979).

24. W. R. Kellen, H. C. Chapman, T. B. Clark, and J. E. Lindegren, Transovarian transmission of some *Thelohania* (Nosematidae: Microsporidian) in mosquitoes of California and Louisiana, *J. Invertebr. Pathol.,* 8, 355 (1966).

25. H. C. Chapman, D. B. Woodard, W. R. Kellen, and T. B. Clark, Host-parasite relationship of *Thelohania* associated with mosquitoes in Louisiana (*Nosematida, Microsporidia*), *J. Invertebr. Pathol.,* 8, 452 (1966).

26. W. G. Wellington, P. J. Cameron, W. A. Thompson, I. B. Vertinsky, and A. S. Landsberg, A stochastic model for assessing the effect of external and internal heterogenecity on an insect population, *Res. Popul. Ecol.,* 7, 1 (1975).

27. Y. Ito, M. Shiga, N. Oho, and H. Nakazawa, A granulosis virus, possible biological control agent for control of *Adoxophes orana* (Lepidoptera: Tortricidae) in apple orchards. III. A preliminary model for population management, *Res. Popul. Ecol.,* 19, 33 (1978).

28. R. J. Brand and D. E. Pinnock, "Application of Biostatistical Modeling to Forecasting the Results of Microbial Control Trials," in H. D. Burges, Ed., *Microbial Control of Pests and Plant Diseases 1970-1980,* Academic Press, New York, 1981, p. 667.

29. D. E. Pinnock, R. J. Brand, J. E. Milstead, M. E. Kirby, and N. F. Coe, Development of a model for prediction of target insect mortality following field application of a *Bacillus thuringiensis* formulation, *J. Invertebr. Pathol.,* 31, 31 (1978).

30. L. P. Kish and G. E. Allen, The biology and ecology of *Nomuraea rileyi* and a program for predicting its incidence on *Anticarsia gemmatalis* in soybean, Bull. Fla. Agric. Exp. Stn. 795, 1978.

31. B. J. Legg and F. A. Powell, Spore dispersal in a barley crop: a mathematical model, *Agric. Meterol.,* **20,** 47 (1979).

32. G. C. Brown and G. L. Nordin, An epizootic model of an insect-fungal pathogen system, *Bull. Math. Biol.,* **44,** 731 (1982).

33. G. C. Brown, W. Marek, V. Yeh, S. Mehs, and N. Freedman, Expert systems in agricultural pest management: an implementation example, *N. Cent. Compu. Inst. Software J.,* **1**(2), 1 (1985).

34. G. L. Nordin, Enhancement strategies for entomogenous fungi in integrated pest management, *U.S. Nat. Acad. Sci. — China Acad. Sci. Jt. Symp. Biol. Control Insects,* September, 1981, Beijing, China, in press.

35. S. M. Welch, Developments in computer-based IPM extension delivery systems, *Annu. Rev. Entomol.,* **29,** 359 (1983).

36. E. J. Rykiel, M. C. Saunders, T. L. Wagner, D. K. Loh, R. H. Turnbow, L. C. Hu, P. E. Pulley, and R. N. Coulson, Computer-aided decision making and information accessing in pest management systems, with emphasis on the southern pine beetle (Coleoptera: Scolytidae), *J. Econ. Entomol.,* **77,** 1073 (1984).

37. F. Hayes-Roth, D. A. Waterman, and D. B. Lenat, *Building Expert Systems,* Addison-Wesley, London, 1983.

38. E. O. Wilson and W. H. Bossert, *A Primer of Population Biology,* Sinauer Assoc., Inc. Sunderland, Mass., 1971.

39. T. J. Manetsch and G. L. Park, *System Analysis and Simulation with Application to Economic and Social Systems,* Mich. St. Univ. Press, East Lansing, 1974.

40. W. G. Rudd, Population modeling for pest management studies, *Math. Biosci.,* **26,** 283 (1975).

41. R. M. Feldman and G. L. Curry, Mathematical foundations for modeling poikilotherm mortality, *Math. Biosci.,* **71,** 81 (1984).

APPENDIX

This appendix describes and lists the generalized simulation model discussed in the main body of Chapter 3. The purpose of a detailed documentation is to provide enough information so that interested readers can perform their own simulation studies. Toward this end, the program was written in Microsoft™ BASIC 2.0 which is available on most microcomputers and easily adaptable to computer systems that have other, similar languages (FORTRAN, PASCAL, etc.). Furthermore, the program's design as well as the programming and modeling techniques were chosen so that novice programmers and modelers will have minimum difficulty in adapting the model to their needs.

The computer program is written in two basic sections. The first section sets the parameter values and initial biological components and appears on lines 90–220. Table 3.3 alphabetically lists the initialized variables and parameters; most of the descriptions are straightforward. The array PR, which is initialized in lines 100–110, represents the instar-specific production rate of infectious units relative to that of the third-instar. For example, PR(5)—the fourth-instar

TABLE 3.3

List of Variables, Their Initial Values, and A Description of How They Are Used in the Simulation Model

Symbol	Type[a]	Initial Value	Description
A1	R, f	0.1	Proportion of infectious units that can actually reach hosts without first passing into the environmental pool
A2	R, f	0.01	Daily proportion of the environmental pool's infectious units that can inoculate hosts
A3	R, f	0.95	Daily survival rate of infectious units (infectious unit half-life $= 2$ weeks)
AE	R, v	0	Exposed ovipositing adults
AR	R, v	None	Stage-dependent aging rate after correction for the daily temperature
CC	R, f	500	Standard host stress density
DE	R, v	None	Density-dependent host stress effect
E	R, a	0	A 7×5 matrix holding the exposed population, rows are instars (1-7) and columns are days after exposure (1-5)
EP	R, v	500	Environmental infectious unit density in units equivalent to 3rd-instar release rate
I	R, a	0	Stage-dependent infectious hosts
J	I, v	None	Days-after-exposure index (1-5)
JD	I, v	1	Julian date
K	I, v	None	Stage index (1-7) with 1 being the egg stage and 7 being the adult stage
LP	I, f	5	Latent period in days
MP	R, a	Stage dependent	Half the instar period (in days) at constant, median temperature
OV	R, f	5	Eggs laid/day/adult
PA	R, v	None	Total infectious units actually reaching hosts
PB	R, v	0	Pathogen burden
PP	R, v	None	Poisson probability of infection accounting for the pathogen burden and host stress
PR	R, a	Stage dependent	The per host production rate of infectious units for each stage relative to the 3rd-instar
S	R, a	Stage dependent	Susceptible population array
TD	R, v	None	Total dead hosts
TE	R, v	None	Random temperature effect varies between 0 and 1
TH	R, v	172	Total hosts (includes the infectious hosts on any given day)

TABLE 3.3

(continued)

Symbol	Type[a]	Initial Value	Description
TS	R, v	172	Total susceptible hosts
TX	R, v	0	Total exposed hosts

[a]R = real, I = index or integer, a = array or matrix, f = fixed parameter, v = simple variable.

— has a value of 1.5 which means that a fourth-instar releases 1.5 times as many infectious units as does a third-instar. The choice of a third-instar standard is purely arbitrary, but the use of a consistent larval equivalent provides a convenient unit base for enumerating pathogen stages in the environmental pool and for determining pathogen burden. If the standard production rates were determined at optimum temperature, they could be made temperature dependent by multiplying each rate by the daily fractional temperature effect (see below). Finally, the daily host survival rate from nonpathogen hazards is initially set for all stages on line 120. If exposed hosts (infected but not yet infectious) have a different survival rate, another array of the same size would be needed here.

The second basic section of the computer program is the model itself and is enclosed in the daily loop between lines 250 and 700. It is subdivided into six modules: (1) pathogen burden and infectious unit population (lines 270-310); (2) host stress effect on susceptibility and the infection process (lines 330-360); (3) density-independent mortality from nonpathogen hazards (lines 380-400); (4) oviposition (lines 410-450); (5) aging process (lines 470-610); and (6) an output routine which prints the results at the end of each day for all components (lines 630-700).

In the oviposition module, exposed adults lay susceptible (uninfected) eggs at the same rate as uninfected adults (line 450). If line 450 is changed to read:

450 S(1)=S(1)+OV*(S(7)+AE*(1−Q)):E(1,1)=E(1,1)+OV*Q*AE

then the model will have a vertical transmission pathway identical to that in Figure 3.5 where Q is the proportion of infected eggs laid by infected adults. Different oviposition rates for susceptible and infected adults also could be used.

The aging process used here is simplistic. The "temperature effect," TE (line 480), is a uniform random number between 0 and 1 with an expected value of 0.5 (this is the rationale behind using half periods for stadial duration in array MP). If seasonal daily mean temperatures are desired, they can be generated by inserting

485 DT=MN+(MX−MN)*SIN((2*PI/365)*(JD−121)+2*MD*(TE−0.5)

where DT is the daily temperature, MX is the maximum normal summer temperature, MN is the average between MX and the minimum normal winter temperature, PI is 3.1416, and MD is the maximum allowable random deviation from normal. This function will generate a minimum temperature around January 30 and a maximum temperature around July 31, and the temperature scale used (C or F) will be the same as that for MX, MN, and MD (all must use the same scale).

Once temperatures can be generated, the daily aging rate array, AR (line 520), can be determined for a specific system by interpolating between data points on development time (or latent period)-temperature data for a given species to simulate daily percent development more realistically than is done in line 520. However, use of calculated daily development rates in aging insect life stages requires a knowledge of distributed delays which are beyond the scope of this chapter (see 39-41).

As indicated in the main text, management strategies can be incorporated relatively easily into this model. For example, a microbial insecticide application can be simulated with the statement,

315 IF TH>EL THEN PB=A4*IR+PB:EP=EP+A5*IR

where EL is the economic level of the host population, IR the infectious units applied (in third instar larval equivalents), A4 the proportion of those units that immediately contribute to the pathogen burden, and A5 the proportion that goes directly into the environmental pool. Note that it is necessary that A4+A5≤1 and that the values for EL, A4, A5, and IR be determined beforehand as in the initialization section. If the microbial species is different from the natural pathogen, it will, of course, then be necessary to keep track of both pathogens separately.

Similarly, an insecticide application can be simulated with

395 IF TH>EL THEN S(K)=S(K)*SI(K)
405 IF TH>EL THEN E(K,J)=E(K,J)*SI(K)

where SI(K) is the age-dependent host survival from the insecticide (between 0 and 1) and must be predefined in the initialization section. Differential survival of susceptibles vs. infected hosts could be simulated with two different survival arrays at this point in the program.

These are but a few of many possible modifications of this model. Simple simulation models such as this provide a convenient laboratory in which to test ideas and conceptions concerning the dynamic interaction between populations of insect pathogens and those of their hosts.

```
10 '----------------------------------------------------------
20 '
30 '          Demonstration Simulation Model for an
40 '               Entomopathogen System
50 '
60 '----------------------------------------------------------
70 RANDOM:DIM S(7),E(7,5),I(7),AR(7),PR(7),SR(7),MP(7)
80 '
90 '          Initialize and set parameter constants
100 FOR K=1 TO 7:READ PR(K):NEXT K 'per capita production rate
110 DATA 0.0, 0.1, 0.5, 1.0, 1.5, 2.0, 1.0
120 FOR K=1 TO 7:SR(K)=.95             '5% daily mortality rate
130 READ MP(K):NEXT K
140 DATA 3., 3., 4., 4., 4., 4., 6.
150 A1=.1:A2=.01:A3=.95:CC=500:OV=5:LP=5 'See text for definitions.
160 '
170 '          Initialize biological components
180 EP=500      '500 3rd instar larval equivalents
190 FOR K=1 TO 7:I(K)=0 'No initial infectious or exposed hosts.
200 READ S(K)
210 DATA 50., 20., 2., 0., 0., 0., 100.
220 FOR J=1 TO LP:E(K,J)=0:NEXT J:NEXT K:TH=172
230 '
240 '  Actual simulation begins with the following daily loop.
250 FOR JD=1 TO 1000 'Julian date.
260 '
270 'Determine the pathogen burden (PB).
280 PB=0:FOR K=1 TO 7:PB=PB+I(K)*PR(K):NEXT K
290 PA=A1*PB+A2*EP   'Infectious units actually reaching hosts.
300 EP=A3*EP+PB*(1-A1) 'Adjust remaining IU's.
310 PB=PA/TH       'Pathogen burden on a per host basis
320 '
330 'Determine density-dependent stress effect on susceptibility.
340 DE=TH/CC:PP=1.-EXP(-DE*PB) 'Poisson infection probability
350 FOR K=1 TO 7: E(K,1)=PP*S(K)  'Exposed and newly infected hosts.
360 S(K)=S(K)-E(K,1):NEXT K   'Adjust susceptible for loss.
370 '
380 'Determine non-pathogen-induced daily mortality
390 FOR K=1 TO 7:S(K)=S(K)*SR(K)   'susceptibles
400 AE=0:FOR J=1 TO LP:E(K,J)=E(K,J)*SR(K) 'Exposed
410 IF K=7 THEN AE=AE+E(K,J)   'Adult exposed needed for oviposition
420 NEXT J:NEXT K
430 '
440 'Susceptible and exposed adults lay eggs
450 S(1)=S(1)+OV*(S(7)+AE)
460 '
470 'Age all components
480 TD=0:TS=0:AR(0)=0:TE=RND(0)   'TE is a random temperature effect
490 FOR K=1 TO 7
500 I(K)=E(K,LP)   'Exposed become infectious on last day of latent period.
510 TD=TD + I(K)  'Infectious hosts are dead.
520 AR(K)=TE/MP(K) 'Stage-dependent aging rate.
530 NEXT K
540 TH=0:TX=0:TS=0 'Zero totals for output needs.
550 FOR K=7 TO 1 STEP -1
560 S(K)=S(K)*(1-AR(K))+S(K-1)*AR(K-1):IF S(K)<1E-5 THEN S(K)=0
570 TS=TS+S(K)  'Total susceptibles across instar
580 FOR J=LP TO 1 STEP -1
590 E(K,J)=E(K,J-1)*(1-AR(K))+E(K-1,J-1)*AR(K-1)
595 IF E(K,J)<1E-5 THEN E(K,J)=0
600 TH=TH+E(K,J):TX=TX+E(K,J) 'Total exposed
610 NEXT J:TH=TH+I(K)+S(K):NEXT K 'Total hosts
620 '
630 '          Output daily results
```

```
640 LPRINT "DAY NO. ";JD;"    TEMP. EFFECT";TE
650 LPRINT "PATH. BURDEN ";PB;"    PREVALENCE ";TX/TH
660 LPRINT "TOTAL HOSTS ";TH;"    EXPOSED   ";TX;"   SUSCEP. ";TS
670 LPRINT:LPRINT "STAGE","SUSCEP","E1","E2","E3","E4","E5","INFECT."
680 FOR K=1 TO 7
690 LPRINT K,S(K),E(K,1),E(K,2),E(K,3),E(K,4),E(K,5),I(K):NEXT K
700 LPRINT:LPRINT:NEXT JD
710 END
```

II

KEY FACTORS

4

THE HOST POPULATION

HITOSHI WATANABE

Faculty of Agriculture
University of Tokyo
Tokyo, Japan

1. INTRODUCTION

The primary factors that are involved in the cause, initiation, and development of epizootics of infectious diseases in insects are the pathogen population with its variable virulence and infectivity (see Ch. 5), an efficient means of transmission (see Ch. 6), and the susceptibility of the host population to the pathogen. All of these factors are affected by the abiotic and biotic environments (1,2) (see Ch. 7).

The present chapter treats the properties of the host population that are most significant in the epizootics of infectious diseases: (1) susceptibility, including genetic resistance, and (2) population parameters such as density, behavior, and association with other insect and animal populations, in relation to the spread of diseases. There are reviews dealing with insect resistance to microbial pathogens. (3–5).

2. POPULATION SUSCEPTIBILITY TO MICROBIAL PATHOGENS

2.1. Variability in the Susceptibility of Populations

Many reports point out differences between two or more isolated populations of an insect species in responding to microbial pathogens. Ossowski (6) observed that the nuclear polyhedrosis virus (NPV) in a population of the wattle bagworm *Kotochalia junodi* 200 miles away was more infectious than the native virus to the local bagworm population. David and Gardiner (7) reported significant differences in the resistance of cultures of the large white butterfly *Pieris brassicae* to a granulosis virus (GV). Hunter and Hoffmann (8), in their tests with a GV, showed a greater than seven fold difference in the LC_{50} between two isolates of the Indian meal moth *Plodia interpunctella* from Georgia and

California. Considerable variability in response to a GV was also found in field populations of the potato tuberworm, *Phthorimaea operculella* (9).

In the silkworm *Bombyx mori* there are large differences among various strains in their susceptibility to viral infections. The same interstrain difference is also true among their progeny, indicating that the susceptibility is inherited (10–12).

Rinderer et al. (13) tested honeybees *Apis mellifera* from three different stocks (hairless-black resistant, hairless-black susceptible, and commercial) by feeding them, over a 3-day period, a total dose per bee of 4×10^9 virus particles in sucrose syrup. By the last day of the experiment, the mortality in the control groups of all three stocks averaged less than 4%, whereas samples of susceptible, commercial, and resistant stocks receiving the virus particles averaged 94, 60, and 40%, respectively. Differences in the response of honeybee colonies to sacbrood virus (SBV) (14), and to chronic bee paralysis virus (CBPV) (15, 16) have also been reported.

Rothenbuhler and Thompson (17) fed to honeybee larvae the spores of *Bacillus larvae,* the causative agent of American foulbrood, and found significant differences in the larval mortalities of different colonies. Hoage and Rothenbuhler (18) confirmed this difference among strains. However, the slope of the dosage-mortality regression was lower in the more resistant strain, indicating variability due to the mixing of resistant and susceptible insects in the population. Similarly, Cherepov (19) reported differences among five Russian honeybee races in the survival rates of worker and drone larvae diseased with European foulbrood.

Papierok and Widing (20) reported differences in two clones of the aphid *Acyrthosiphon pisum* to the infection of the fungus *Entomophthora obscura*. A laboratory population of the wireworm, *Conoderus falli,* treated with the fungus *Metarhizium anisopliae* had a significantly longer LT_{50} than a population originating in the field (23).

Differences in susceptibility occur among honeybee colonies to the fungus *Ascosphaera apis,* which is a causative agent to chalkbrood (24, 25). DeJong (26) administered standard doses of *Ascosphaera* spores to 3-day old larvae of the honeybee reared under standard conditions. He confirmed that the infection rates varied from 0.5% to 7% among colonies. According to Milne (21, 22), the honeybee colony resistant to the chalkbrood had more efficient hygienic behavior, especially brood removal, than a susceptible colony.

Certain races of the moths *Antheraea pernyi* and *Platysamia cecropia* are more resistant to microsporidia than others (27). Such resistant races also occur in the silkworm and honeybee. Moreover, some honeybee colonies inoculated with *Nosema* spores displayed a significant resistance, apparently caused by genetic differences (28). Burnside and Revell (29) noted that some colonies of the honeybee heavily-infected with *Nosema apis* were able to tolerate the disease and build up rapidly, while others weakened or died. Some inoculated queens survived and were found free of spores, suggesting that selection for resistance to *Nosema* disease was possible (30). Kartman (31) found differences in susceptibility to infection by the nematode *Dirofilaria immitis* between populations of the mosquito *Aedes aegypti* obtained from various geographic zones. The differences in susceptibility were inversely correlated to the percentage encapsulation of microfiliaria. He was able to select for increased resistance in eight generations. Thus differences in susceptibility of insect colonies, populations, races, or strains to diverse pathogens may be due to their degree of exposure to a pathogen over many generations.

2.2. Genetical Resistance

2.2.1. MODE OF INHERITANCE.   There is a need to compare the dosage-infection responses of resistant and susceptible parent strains and their F_1 and F_2 offsprings for a better understanding of the mode of inheritance of susceptibility to microbial pathogens (32). As shown in Figure 4.1, if an insect population is almost homozygous in response to the infection of a microbial pathogen, the regression of the log-dosage-probit infection forms a straight line as in A. The slope of the line indicates the degree of homogeneity; that is, the steeper the slope the more homozygous is the response. On the other hand, if individuals with distinctly different responses to infection are mixed in one population, the dosage-infection regression is not a straight line, but a curve with plateaus in certain parts of the curve. For example, when individuals with two distinctly different responses are involved in a population, the regression line forms a curve with one plateau as in B. When three members with distinctly different responses are involved in a population, the regression line forms a curve with two plateaus as in C. The more different minor responses there are in a population, the more the regression curve becomes linear as in D with a low slope.

The inheritance of susceptibility to viral infection has been studied more intensively in the silkworm *B. mori* than in other species primarily to gain basic

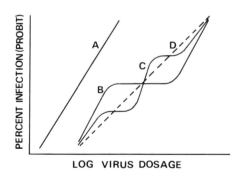

Figure 4.1. Dosage-infection regression lines of silkworm populations with members responding differently to virus infection. (A) Homozygous, susceptible population; (B) population with two members having distinctly different susceptibilities; (C) populations with three members having distinctly different susceptibilities; and (D) population with many members with minor differences in susceptibility.

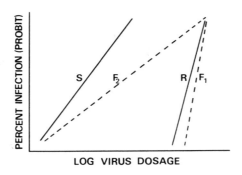

Figure 4.2. Concentration infection regressions lines for CPV in resistant (R) and susceptible (S) parent silkworm strains, and their F_1 and F_2 offspring.

data on silkworm varieties resistant to diseases and thus reduce economic loss in sericulture. There are many silkworm varieties or strains whose genetic constitutions are almost homozygous, and they could be studied intensively to establish the basis for genetic resistance to diseases.

Figure 4.2 is an example of the dosage-infection response to a cytoplasmic polyhedrosis virus (CPV) in resistant (R) and susceptible (S) parent strains and their F_1 and F_2 offsprings. The regression line of F_1 hybrid is linear and much closer to that of the resistant parent strain than that of the susceptible strain. The F_1 hybrid larvae show heterosis in their resistance to infection by CPV, i.e., they are more resistant than the larvae of their parent strains. The regression line of F_2 hybrid is also linear, but it has a fairly low slope. These results indicated that the silkworm resistance to CPV infection is controlled by a multifactorial genetic system, that is, by polygenes. Similarly, the silkworm resistance to a NPV and to an infectious flacherie virus (IFV) is controlled, in general, by polygenes (10, 11).

Although silkworm resistance to CPV infection is generally controlled by polygenes, there is a special case in a strain called Daizo which has a major gene for CPV resistance (33). This was demonstrated by exposing to CPV larvae of F_1, F_2, and of backcrossed hybrids obtained from crosses between the two inbred strains, Daizo and Okusa.Okusa is one of the most highly susceptible strains to CPV. The larval resistance of F_1 hybrids and of backcrossed hybrid to Daizo were nearly the same as that of Daizo, while the resistance of backcrossed hybrid to Okusa was approximately intermediate between the two inbred strains. This indicated that the resistance in the Daizo strain was inherited as a complete dominance. Furthermore, the dosage-infection regression lines of the F_1 hybrid between Daizo and Okusa, and of the backcrossed hybrid to Daizo, were quite similar to that of Daizo, while the dosage-infection regression lines of the F_2 hybrid and the backcrossed hybrid to Okusa formed curves with plateaus, in part, at the 25% and 50% infection levels, respectively (Fig. 4.3). This suggested that resistant and susceptible larvae occurred at a 3:1 ratio in the F_2

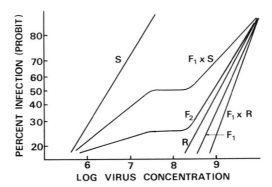

Figure 4.3. Concentration infection regression lines for CPV in Daizo (R) and Okusa (S) silkworm strains, their F_1, F_2 offspring, and backcrossed hybrids ($F_1 \times S$, $F_1 \times R$) to Daizo and Okusa.

hybrid and at a 1 : 1 ratio in the background hybrid to Okusa. Thus, although the resistance to CPV infection is generally controlled by polygenes, a strain like Daizo, which is highly resistant, possesses a dominant major gene enhancing the resistance to CPV.

Most of the silkworm strains are nonsusceptible to infection by densonucleosis virus (DNV) even after intrahemocoelic inoculation of a high dosage. Tests with susceptible and nonsusceptible parent strains, their reciprocal F_1 hybrids, the F_2 hybrids, and the backcrossed hybrids to either of the parents demonstrated that the susceptibility to DNV infection was inherited as a completely recessive character. The genetical segregation for nonsusceptibility took place in the F_2 hybrid and in the backcrossed hybrid to the nonsusceptible parent (34). This was confirmed as follows. The dosage-infection regression lines of reciprocal F_1 hybrids were the same as that of the susceptible parent, while those of the F_2 hybrid and of the backcrossed hybrids to the nonsusceptible parent formed lines with plateaus at the 75% and 50% infection levels, respectively (Fig. 4.4). These results indicated that the nonsusceptibility to DNV infection was controlled by a recessive gene, and the susceptible and nonsusceptible larvae segregated in a Mendelian manner at a 3 : 1 ratio in the F_2 hybrid, and at a 1 : 1 ratio in the backcrossed hybrid to the nonsusceptible parent. The recessive gene controlling nonsusceptibility was not sex-linked because the same degree of susceptibility occurred between the two reciprocal F_1 hybrids.

In *Spodoptera frugiperda*, Reichelderfer and Benton (35) made a probit analysis of the dosage-response of two insect isolates to a NPV, and found a greater than fivefold difference in the median lethal dose (LD_{50}). The response of the F_1 progeny of reciprocal crosses between the two parental isolates demonstrated that resistance to the virus was not sex linked. The LD_{50} values of F_1 and F_2 backcrosses suggested that the major variation in resistance was due to one or more genes that lacked dominance.

A laboratory larval population of the potato tuberworm *P. operculella* was more resistant to infection by GV than larvae collected from field populations. The resistance shown by the laboratory population was controlled by a single dominant autosomal gene that segregated in Mendelian ratios (36).

There are relatively few studies of the genetic basis for resistance to fungi, protozoa, and nematodes. The pathogenicity of many fungi to insects are apparently due to their toxic metabolites. *Aspergillus* spp. produce an insecticidal toxin Aflatoxin B_1 (AFB_1). Llewellyn and Chinnici (37) have found variability in the response of five wild-type laboratory strains of the fruit fly *Drosophila melanogaster* to AFB_1 administered perorally. Studies of hybrid crosses indicate that the variable resistance to AFB_1 in *D. melanogaster* is due to a complex polygenic system in which each of the two major autosomes carries genes exerting some influence on resistance (38). A wide range is found in the susceptibility of the mosquito *A. aegypti* to infection with the protozoan *Plasmodium gallinaceum* (39). The resistance can be rapidly selected and is controlled by a single autosomal recessive gene. The F_1 hybrid honeybee lines are more resistant than their parental lines to the infection by *N. apis* (40). This may be due to a heterosis suggesting that the differences in susceptibility are controlled by a polygenic system. The susceptibility of the mosquito *A. aegypti* to several nematode species is controlled by a sex-linked recessive gene (41). Variations in susceptibility shown by field population can be correlated with differences in the frequency of this gene.

2.2.2. DEVELOPMENT OF RESISTANT POPULATIONS. There is evidence that an insect population develops resistance after prolonged association with disease, simply by a process of selection. Bergold (42) speculated that a population of the spruce budworm *Choristoneura fumiferana*, which was resistant to NPV, developed resistance from a previous sublethal infection of the virus which occurred as an enzootic in the population. Bird and Elgee (43) reported that there might be increased host resistance in a population of *Diprion hercyniae* that had been kept in check by a NPV over a period of 20 years or more (about

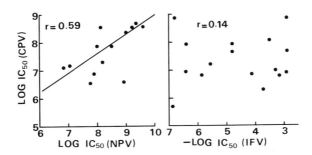

Figure 4.4. Concentration infection regression lines for DNV in susceptible strains (S), F_1 and F_2 hybrids between susceptible and nonsusceptible silkworm strains (N), and backcrossed hybrids ($F_1 \times S, F_1 \times N$).

40 generations). When Bird and Burk (44) tested the same virus against a formerly disease-free population, they could not show any decrease in its effectiveness. It seems unlikely that there has been any significant change in the virus in this case.

The increase in resistance of insect stocks to virus diseases has been reported in *P. brassicae.* Sidor (45) and Rivers (46) have reported that resistance to GV infection in *P. brassicae* increases after breeding the survivors of a GV outbreak for several generations. This stock is also more resistant to CPV than other stocks of the cabbageworm (45). Subsequently, David and Gardiner (47) discovered a stock of *P. brassicae* resistant to GV after a severe outbreak of the virus.

Martignoni and Schmidt (48) reported differences in resistance to virus diseases among populations of *Phryganidia californica* and *Pieris rapae.* They found that the two populations of *P. californica,* a native insect, differed in their susceptibility to NPV; but two populations of *P. rapae,* a recent immigrant, showed no difference in their susceptibility to a GV. This indicated that some populations of native species that had long association with their viruses might eventually develop some resistance to them.

In the larch bud moth *Eucosma griseana,* Martignoni (49) found about a 38 times increase in LD_{50} for GV following a natural epizootic in the field. The dosage-mortality regression slope was very steep, indicating that the GV epizootic merely removed most of the more susceptible individuals from the population.

Briese and Mende (50) reported a significant increase in the resistance of the potato tuberworm to GV after selection for only five generations. However, the slope of the dosage-mortality regression line of the selected strain was high, suggesting that the greater resistance was due mainly to the elimination of more susceptible individuals from the population.

In contrast to the above reports, Ignoffo and Allen (51) found that resistance to a NPV did not develop through selection in laboratory populations of the cotton bollworm *Heliothis zea.* A selection pressure of LD_{50} to LD_{70} was maintained throughout 20 to 25 generations. No significant changes in LD_{50}, slope, or intercept of dose-mortality lines were detected. Laboratory populations under selection were as susceptible to the virus as nonselected or wild populations of *H. zea.* The resistance ratio (LD_{50} of selected generation/initial generation) ranged from 0.5 to 1.2.

The failure to select for resistance in certain laboratory cases does not necessarily mean that such selection is impossible for that particular species. It may be the result of the absence of a resistant gene or genes in a small laboratory population with limited genetic variability or that the appropriate agent has not been used to select the genetic trait. For example, in the silkworm, Yamafuji et al. (52, 53) attempted to select strains resistant to polyhedrosis by treating the larvae with chemicals that were effective for virus induction, but they failed to obtain a resistant strain. However, Aizawa et al. (54) and Aizawa and Furuta (55) succeeded in selecting a strain of silkworm resistant to the induction of

NPV by treating several generations of the larvae repeatedly at low temperatures. The resistance to virus induction, however, was independent of the resistance to virus infection. Uzigawa and Aruga (56) and Funada (57) attempted the selection of silkworm resistant to IFV infection by repeated exposure of the virus, and they succeeded in obtaining a resistant strain after several generations.

Watanabe (58) attempted to select a silkworm strain resistant to CPV. The larvae were fed CPV, and the offspring of the surviving individuals were exposed to the virus. The exposure to CPV was continued for eight generations. There was no increase in resistance to CPV up to the fourth generation, but resistance increased suddenly in the fifth generation. The lack of resistance in early generations may be partly caused by the application of a low dosage of virus which caused a low percent mortality. The study suggested that greater than 60% mortality was required as a selection pressure in order to induce and retain the resistance to CPV infection. Despite continued rigorous selection, there was no further increase in resistance after a certain plateau of about 10- to 16-fold of that of the unselected strain.

When two selected strains were crossed, the hybrid was more resistant than the hybrid from crosses of unselected strains. However, when the two hybrids were compared on the basis of heterosis in resistance, several different features were observed. The resistance of control hybrids appeared greater than that of the unselected parent strains, and this indicated the existence of marked heterosis. On the other hand, none of the hybrids from crosses of selected strains was more resistant than either of the parent strains. The difference in heterosis observed in the two hybrids might be explained as follows. The different genetic combinations of the two control strains were more likely to develop marked heterosis in resistance, while in the selected strains resistant genes of similar genetic constitution became integrated during the course of selection, and the genetic combination of the two selected strains tended to show poor heterosis.

Kulincevic and Rothenbuhler (16) carried out a two-way selection experiment for resistance and susceptibility to hairless-black syndrome, a probable viral disease in the honeybee, for four generations. From the parental generation of 15 colonies (treated from three basic stock colonies) tested for disease resistance, one-fifth was selected to start a resistant line and one-fifth to start a susceptible one. Thereafter, these two groups were kept genetically separated and, from each tested generation, an average of about 22% of the best colonies were used as parents of the next generation. Bacteria-free inoculum, which was fed to challenge these bees, was prepared by macerating and centrifuging specific numbers of bees showing symptoms of hairless-black syndrome and collected from a number of colonies in apiaries. Resistant and susceptible lines did not differ significantly in the first selected generation. In the second, third, and fourth generations the two lines diverged increasingly, and in each generation they differed statistically at the 1% level of probability.

A number of attempts have been made to select insect strains resistant to bacteria. Harvey and Howell (59) induced resistance to *Bacillus thuringiensis* in the house fly *Musca domestica* by selecting survivors during 50 generations of rearing. The resistance in *M. domestica* was against the exotoxin. The degree of resistance, based on the ratios of LD_{50} values, varied from 8- to 14-fold between the 27th and 50th generation, inclusive. When the selection pressure was removed over 20 generations from the selected resistant strain, there was a significant decline in resistance. These results suggest that the resistance of *M. domestica* to *B. thuringiensis* exotoxin may be controlled genetically by a polygenic system and that the increased resistance may have been accompanied by a reduction in fitness under ordinary rearing conditions.

Feigin (60) also tried to select a resistant strain of housefly to *B. thuringiensis* but failed because he maintained the strain at a low selection pressure, though for a large number of generations. On the other hand, the two selection studies of lepidopteran insects *Anagasta kuehniella* and *Plutella maculipennis* against *B. thuringiensis* produced no change in susceptibility because of high selection pressures for only a few generations (61, 62).

The first case of insect resistance to the application of a microbial insecticide, in this case *Bacillus thuringiensis,* developed in populations of Indian meal moth *Plodia interpunctella* (258). McGaughey (258) reported that the resistance was against the spore-crystal protein complex of the bacillus and was inherited as a recessive trait. Moths isolated from treated grain bins were more resistant than strains from untreated bins, indicating that resistance to the bacillus can develop quickly in the field. In laboratory tests, resistance developed in a few generations; nearly 30-fold in two generations and reached a plateau level 100 times higher than the control level after 15 generations.

Bacillus popilliae, the causative agent of the milky disease of the Japanese beetle *Popillia japonica* and the Oriental beetle *Anomala orientalis,* has been successfully used to control these insects in the USA since 1939. In Connecticut, control was still effective in 1962, but by 1974 the insect populations increased despite an extensive buildup of disease inoculum in the soil. Dunbar and Beard (63) found that the infectivity of the *B. popilliae* spores had diminished significantly and also suggested that resistance to infection might have developed in the larvae.

Hoage and Peters (64) demonstrated that resistance of larval honeybee to American foulbrood could be rapidly selected using a hybrid of resistant and susceptible parent lines. The hybrid larvae increased the level of resistance by 45% within three generations.

There has been only one report so far of the selection for resistance to fungi or nematodes. Vansulin (65) obtained a reduction in mortality of 94% to 48% over eight generations in a strain of the mosquito *Culex pipiens molestus* treated with a constant dose of "Boverin" (a commercial preparation of *Beauveria bassiana*). In the case of nematodes, Petersen (66) demonstrated that the susceptibility of the southern house mosquito to *Romanomermis culicivorax* was reduced by 50% infection during the course of selection.

2.2.3. MECHANISMS OF GENETICAL RESISTANCE.   Natural infection of viruses in insects occurs perorally. Steps in the infectious process with viruses include the entering of virions into the gut lumen, the adsorption and fusion of virus particles to the cell plasma membrane of the midgut epithelium, the penetration of virions into the midgut cell where they replicate, and the passing of virions through the midgut epithelium into the body cavity where they eventually attack the target tissues. Accordingly, the resistance of the host to the virus might be caused by blocking any, if not all, of these steps in infection. For example, when silkworm strains that showed distinct differences in susceptibility to the peroral infection of a CPV were tested for their susceptibility to subdermal infection, the strain differences in susceptibility were quite small. The order of susceptibility of the strains to peroral infection was not correlated with that of strains exposed to subdermal infection (67). The results indicated that the host resistance to peroral infection depended on inhibitory mechanisms in the gut lumen against the invasion of the virus into midgut cells, and less on the suppression of virus multiplication within the cell.

Watanabe (68) studied differences in susceptibilities of several silkworm strains to NPV and CPV. He found that the order of susceptibility of the strains to one type of virus was significantly related to the order of susceptibilities to the second type of virus (Fig. 4.5). The results indicated that the mechanisms of resistance of the silkworm larva to the viruses were similar and that the defense mechanisms might occur in the midgut lumen.

As mentioned previously, the resistance of the silkworm to infection by NPV, CPV, and IFV is polygenic, and this is consistent with the existence of several defenses against viral infections. Antiviral substances in the silkworm gut juice, such as the red fluorescent protein, which agglutinates and inactivates virus particles (69, 70), have been found to inactivate both NPV (71) and CPV (72, 73). However, no parallel relationship could be detected between the resistance of the strains to oral infection with viruses and the antiviral activity of

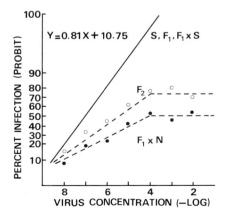

Figure 4.5. Relationships among various susceptible silkworm strains to CPV, NPV, and IFV, showing a correlation in susceptibility between CPV and NPV but not between CPV and IFV.

the gut juice. The antiviral activity of the gut juice, therefore, seems to play a minor role in the mechanism of resistance of the silkworm.

There are some reports that the peritrophic membrane may be one of the major factors preventing virus adsorption to the midgut epithelium. The peritrophic membrane is supposed to play a role similar to that of the mucus membrane in the vertebrate gut and protect the midgut cells from mechanical damage caused by abrasive food particles (74). In Lepidoptera, this sheath is formed along the length of the midgut and is presumed to be a secretory product of the epithelial cells. It is renewed at each molt. Biochemically, the peritrophic membrane is composed of chitin, protein, and different mucopolysaccharide and hyaluronic acidlike compounds. The virus receptor of a susceptible animal cell is lipoprotein or glycoprotein, which is also plentiful in the peritrophic membrane. The peritrophic membrane of the silkworm larva is able to absorb CPV in vitro (5).

Brandt et al. (75) showed by electron microscopy that there were no pores or other discontinuities in the peritrophic membrane of *Orgyia pseudotsugata* and that this membrane could function as a mechanical barrier. In an attempt to explain the low infectivity of mosquito iridescent virus (MIV) given orally to *Aedes* larvae, Stoltz and Summers (76) showed that the virus particles were destroyed rapidly (process unknown) in the anterior portion of the midgut. They also reported that the peritrophic membrane acted as an effective barrier, since it seemed to lack the gaps present in lepidopterous larvae. Presumably, infection took place when the particles passed through rare ruptures in the peritrophic membrane. On the basis of experiments with silkworm larvae fed *Serratia marcescens,* Weiser and Lysenko (77) postulated that the peritrophic membrane of insects was resistant to bacterial penetration.

When mechanical injury to the peritrophic membrane occurs, microbes may penetrate into the cell and hemocoel to produce disease. Paschke and Summers (78) have suggested that some microbial enzymes may damage the peritrophic membrane and may assist in the penetration of pathogens. Glass and abrasive particles have been fed to larvae to enhance the invasion of bacteria through the larval midgut (77, 79). Accordingly, the chemical and physical properties of the peritrophic membrane appear responsible for its action as a barrier in filtering or preventing most of the ingested pathogens from reaching the midgut epithelium.

In the case of CPV, old midgut columnar cells infected with the virus are discharged into the gut lumen and new cells develop to take their place (80). The degree of this epithelial regeneration varies with the silkworm strain, and this may be one of the basis for interstrain differences in tolerance to disease development. In the fall webworm *Hyphantria cunea* infected with the *Bombyx* CPV, the regenerated columnar cells are subsequently immune to the virus, and the infected larvae recover from the disease (81). Similar mechanisms seem to be involved with silkworm resistance to an IFV. This virus multiplies in the midgut epithelium to the same extent in both susceptible and resistant larvae. But in the resistant strain infected goblet cells are discharged into the gut lumen at each molt, and regenerative cells rapidly develop into new goblet cells, prolonging the larval period of lethal infection more than that of the susceptible strain (82). Although this mechanism for IFV is similar in action to that for CPV, it seems to have a different genetical basis. Watanabe et al. (83) have shown that there is no significant genetical correlation between interstrain differences in the susceptibility to IFV and CPV (Fig. 4.5).

Phagocytes in insect blood may have important defensive roles against viruses that have invaded the body cavity. Stairs (84) blocked the activity of phagocytes of *Galleria mellonella* larvae with India ink and increased the susceptibility to NPV by 13 times. Inoue (85) observed with the electron microscope the processes of phagocytosis of nuclear polyhedra by granular cells of the silkworm and the degradation of the virus.

The nonsusceptibility of the silkworm to a DNV is controlled by a single recessive gene (34). The gene may cause a deficiency of an enzyme involved in virus multiplication or in the receptor synthesis.

The genetical mechanisms of honeybee resistance to American foulbrood have been demonstrated as follows: (1) rate at which young larvae become innately resistant to infection with age (86), (2) efficient hygienic behavior of adults in removing diseased larvae (87), and (3) effectiveness of adults in filtering the spores of *B. larvae* from food by means of their proventriculus and/or the potency of a bactericidal factor in gland secretions of nurse bees (88).

Splittstoesser et al. (89) observed that the defense mechanisms in the European chafer *Amphimallon majalis* against *B. popilliae* involved the encapsulation of the bacterium, the extrusion of infected columnar cells from the midgut epithelium, and subsequent midgut regeneration. The midgut pH has been considered to influence susceptibility differences among insect species (90), but Kinsinger and McGaughey (91) found no correlation between midgut pH and the response of *E. cautella* and *P. interpunctella* populations to *B. thuringiensis*. Hall and Arakawa (92) and Galichet (93) reported that house fly strains resistant to chemical insecticides were not resistant to β-exotoxin of *B. thuringiensis*, indicating that the two mechanisms of resistance were not related.

Ignoffo et al. (94) demonstrated that larvae of *A. gemmatalis* were 100 and 200 times more resistant to injected conidia and blastospores of *N. rileyi*, respectively, than were larve of *Trichoplusia ni*. The results suggest that the resistance of *A. gemmatalis* to *N. rileyi* may not be solely at the integumental barrier, as is often believed, but may be a function of an internal physiological response.

Aratake (95) found that the growth of germ tubes of *B. bassiana* was greater in the hemolymph of a susceptible silkworm larvae than in that of a resistant larva, indicating that the physiological resistance mechanisms might include properties of the hemolymph. In the case of silkworm resistance to *Aspergillus* spp., the increased resistance was neither related to the percentage of spore germination nor to the length of germ tubes (96). However, Koidsumi (97) found that the removal of cuticular lipids from the silkworm larvae greatly increased their susceptibility to *Aspergillus* spp. The antifungal activity was associated with the amount of free medium-chain fatty acids. An extract of the living integument and exuviae of resistant Chinese race silkworms had a greater antifungal action than that from the susceptible Japanese race (98). This was attributed to the greater concentration of the free fatty acids in the resistant silkworm. Such a defense mechanism against fungal penetration of the cuticle varies among populations of an insect species.

There are, therefore, at least three possible mechanisms of defense against fungal attack that are subject to genetic variation: cuticular defense against fungal penetration, hemolymph properties against fungal growth, and resistance to toxic fungus metabolites.

Tissue regeneration and pathogen replication affect insect resistance. Weiser (27) has reported that individuals of susceptible strains of insect species are larger, eat more, and regenerate their tissues more slowly than those of resistant strains. The rapid regeneration of infected tissues appears to confer an increased resistance to protozoa, similar to the defense mechanisms observed in insects resistant to some viruses and bacteria. Sidorov and Mikhailenko (99) have found a significant interstrain difference in the number of spores in the mid- and hindguts of experimentally infected honeybees. They observed a nine-fold difference in the multiplication rate of spores after the infection in different populations.

Insect behavior is a factor in insect resistance. Woodard and Fukuda (100) found that larvae of a selected strain were much more effective than those of an unselected strain in defending themselves against attacking nematodes. Petersen (66), however, hypothesized that differences in the nature and thickness of the cuticle might retard the penetration of the nematode and thereby contribute to reduced susceptibility.

2.3. Internal and External Factors Affecting Susceptibility

Endogenous and exogenous factors are involved in resistance, but their separate effects may not be clear since both types of factors may act together in the association between insect and pathogen. For instance, genetical and physiological functions are endogenous factors associated with insect resistance, but these functions, especially when they are controlled by a polygenic system, may be readily modified by exogenous factors, such as temperature, chemicals, and food, that are applied naturally or artificially to the insect.

2.3.1. ROUTE OF INFECTION.   In terrestrial insects, there is so far no evidence of natural infection with viruses through the integument or spiracle; most virus infections occur perorally and some through the egg. However, insects in general are highly susceptible to viruses injected into their hemocoels. Stairs (101) reported that in the greater wax moth *G. mellonella* the larva was more susceptible to subdermal than to peroral infection of a NPV. Watanabe (67) found 100- to 10,000-fold lower LD_{50} values when silkworm larvae were given subcutaneous injections of CPV than when given perorally. This result suggested that much of the ingested virus did not enter the susceptible epithelial cells of the midgut, and the resistance might depend more on inhibitory mechanisms in the gut than on the suppression and virus multiplication within cells.

Miyajima and Kawase (102) found that the susceptibility of the silkworm larva to intrahemocoelic injection of CPV varied depending on the portion of the body that received the virus infection. Greater susceptibility occurred when the virus was injected into the posterior as compared to the anterior part of the larva. The larval response to virus injection at the posterior part increased with the age of the larva (103). The basis for the differences in response at the anterior and posterior larval parts has not been resolved.

The honeybee is very sensitive to the chronic bee paralysis virus (CBPV) and the acute bee paralysis virus (ABPV) when these viruses are injected into the hemocoel, but more than 10^{10} particles of either virus are needed to kill a bee when the virus is sprayed on it or is fed to it (15, 104, 105). Although the general belief is that adult honeybees are not susceptible to the pathogenic bacterium *B. larvae,* the causative agent of American foulbrood disease, both abdominal and thoracic injections of *B. larvae* spores produce a general septicemia in adult bees (106).

2.3.2. AGE.   There are many reports that insect larvae decrease in susceptibility to infection by microbial pathogens as they age, whether such susceptibility is measured as a decrease in mortality or as an increase in time for initial mortality. For instance, some of the newly hatched larvae of *Malacosoma disstria* can become infected when they ingest a single viral polyhedron, but most larvae in the last instar are resistant to infection by dosages as high as 10^6 polyhedra (107). Since second instar larvae of *C. fumiferana* are about 100 times more susceptible to NPV than fourth instar larvae, it is essential in microbial control that the virus be disseminated as soon as the larvae emerge from their overwintering sites to ensure a maximum number of infected larvae (108).

Ignoffo (109) demonstrated that in *H. zea* and *Heliothis virescens* there was a decrease in larval mortality with an increase in age when the larvae were exposed to a standard dose of NPV. In *Heliothis armigera* larvae, virus-induced mortality decreased with increasing larval weight and age (110). Boucias and Nordin (111) also found a correlation of an increase in larval age and weight with resistance of the fall webworm larva *Hyphantria cunea* to NPV and GV infections. On the other hand, Smith et al. (112) showed that third instar larvae of *Harrisinia brillians* were more susceptible to viral infection than first instar larvae.

The apparent increase in resistance to virus as the larvae mature and gain weight may be explained as a "dilution" effect of a constant viral dose. If the value of LD_{50}/ larval body weight increases greatly with age, there may be a specific defense mechanism.

Aruga and Watanabe (72) demonstrated in the silkworm that susceptibility to peroral infection by CPV decreased with larval age from the first to the fourth instar, but not between the fourth and fifth instars. Kobara et al. (113) investigated daily changes in the susceptibility of fourth instar silkworm larvae to peroral infection of the virus. They found that susceptibility was high early in that stadium, declined at the middle stadium, and then increased up to and prior to molt.

Tipula iridescent virus (TIV) infects all four larval instars, pupae, and adults of both sexes of *Tipula oleracea.* Third and fourth instars are more resistant to ingested TIV than first and second instars (114). When TIV is injected into the hemocoel, resistance decreases from the third instar to the pupa. Incubation periods (times from injection of TIV to the appearance of iridescence) are significantly shorter in older fourth instars than in younger fourth or third instars.

Very young larvae of honeybee are highly susceptible to *B. larvae,* the causative agent of American foulbrood, whereas larvae about 2 days of age are highly resistant. Bamrick and Rothenbuhler (86) have found that larvae of a susceptible strain are immune at 48 hours of age, whereas resistant larvae become immune at 36 hours. The maximum difference in susceptibility occurs at age 21 hours when there is greater spore germination and multiplication of the bacteria in the susceptible strain (115). Sutter et al. (116) have found that resistant bee larvae develop more rapidly than susceptible larvae of the same age. This suggests that the resistance is a genetically determined increase in the rate of early larval development that leads to more rapid maturation immunity. The mechanism of this developing immunity remains to be clarified, but it appears to be related to the ability of the bacteria to build up a population sufficient to penetrate the peritrophic membrane and midgut epithelium, and perhaps to the thickness of the peritrophic membrane (117, 118).

Rizzo (119) demonstrated that adult ages of the three flies, house fly *(M. domestica),* black blowfly *(Phormia regina),* and onion fly *(Hylemya antiqua),* had no significant effect on the average death time after the inoculation of either *B. bassiana* or *M. anisopliae.*

In microsporidia, experimental data on the peroral inoculation of spores to insect larvae indicate a sharp increase in the spore dose needed to infect later instars. In many instances, it was difficult or impossible to infect older instars. Milner (120) reported that the LD_{50} of *Nosema whitei* in *Tribolium castaneum* was 1.8×10^6 spores for the first instar and 1.0×10^{10} spores for the fifth instar in experiments in which the spores were administered in the food. Similar results were obtained upon feeding different instars of the wax moth larva *G. mellonella* with spores of *N. heterosporum* or *N. plodiae* (121). In contrast, the injection of high and low spore doses into different instars of the wax moth resulted in infection. Apparently, infection did not depend as much on the absolute number of spores introduced into the host as it did on the gut barrier.

2.3.3. MOLT, METAMORPHOSIS, AND DIAPAUSE.   As an insect larva undergoes molting, metamorphosis, and diapause, its susceptibility to most viruses seems to change. The effect of developmental changes and metamorphosis was first studied in the European spruce sawfly *D. hercyniae,* where cells in an embryonic state or during diapause were not susceptible to NPV (122). When diapause ceased, cellular development proceeded and the cells became susceptible. Similar observations were made by Stairs (123) who found that in the prepupal stages of *G. mellonella, B. mori,* and *C. fumiferana,* when the tissues regenerate extensively, a progressive infection of the respective NPV was temporarily suppressed, resulting in a delay of the lethal infection time.

During the molt of a silkworm larva infected with a NPV or a CPV, virus multiplication and development of inclusion bodies are nearly inhibited until after molting (124, 125). Watanabe and Aruga (126) reported that an experimentally produced diapausing silkworm pupa (dauerpupa) was less susceptible to infection with NPV than a nondiapausing pupa. Moreover, the virus developed more slowly in pupal tissues during diapause, and viral development was accelerated when diapause was cancelled by the injection of an ecdysone analogue.

Silkworm pupae are susceptible to intrahemocoelic infection with NPV, but are quite immune to IFV (127). The CPV multiplies in the pupal midgut, but the infected pupa is immune to a lethal infection and develops into an adult with normal fecundity (128).

In certain lepidopterous larvae, an arrested state of development can be induced experimentally by placing a tight ligature behind the head a few days before they pupate. Stairs (129) reports that the NPV develops more slowly in ligatured mature larvae (in the diapausing state) of the greater wax moth than in unligatured larvae (in the nondiapausing state). Similar results are obtained with the infection of a NPV in ligatured silkworm larvae (130). When the ligatured larvae are given an injection of ecdysone analogue to promote pupation, the rate of NPV development is greatly increased.

Alatectomy of tobacco armyworm *Spodoptera littoralis* increases its susceptibility to NPV infection, while larvae injected with a juvenile hormone are less susceptible to NPV than untreated larvae (131). Similarly, when NPV-infected larvae receive an injection of ether extracts of corpora allata, disease development is suppressed (132). Thus, the development of viral diseases may depend on the physiological state of the host, which is endocrinologically controlled by hormones.

All stages of insect development, eggs, larvae, pupae, and imago, are generally susceptible to mycosis. However, the lower susceptibility of the egg than of the other stages has been reported frequently. For example, eggs of *Eurigaster integriceps* infected by *B. bassiana* or *Aspergillus flavus* showed only low mortality (21 and 33%, respectively, against 8% in the control), whereas 100% of first instar larvae are infected with *B. bassiana* and 54% with *A. flavus* (133).

2.3.4. FOOD. There is some evidence that food quality is important in the susceptibility of insects to viral infections. For polyphagous insect species, a change of host plant is particularly effective in inducing viral diseases, and obviously such a change may be forced upon insects when they migrate or when favorable food plants become unavailable. According to Grison and de Sacy (134), *P. brassicae* larvae became infected with a granulosis virus when fed pale leaves from the interior of cabbage or old leaves, but not when fed the green external leaves of the same cabbage. Sippell (135) observed that more forest tent caterpillars *M. disstria* died from viral disease when fed maple leaves than other plant leaves.

Certain varieties of mulberry increase the incidence of silkworm mortality from viral diseases (136). The prevalence of most viral diseases in the silkworm is low in the spring but increases in autumn. The high incidence of viral diseases in autumn seems to be associated with the quality of the mulberry leaves. For example, when silkworm larvae are reared on an artificial diet containing powdered mulberry leaves obtained from different seasons of the year, those fed on artificial diet containing autumn-harvested leaves are more susceptible to viral infections than those reared on artificial diet with spring-harvested leaves (137). However, mulberry leaf alone affects susceptibility since silkworm larvae reared on an artificial diet without mulberry leaf are more susceptibile to infection by polyhedrosis viruses than those reared on a diet with mulberry leaves (138). Silkworm larvae fed on artificial diets low in protein, low in sucrose, or high in cellulose contents tend to increase in susceptibility to viral infections (139).

The susceptibility to GV of *P. brassicae* fed on synthetic diet significantly increased when either the sucrose or the casein content was reduced (140). Shvetsova (141) reported that greater wax moth larvae were most susceptible to NPV infection when fed wax enriched with nitrogen and carbohydrate. This was confirmed by Pimentel and Shapiro (142) with a high nitrogen diet but not with a standard or a carbohydrate diet.

Rose and Briggs (143) found that the brood food of a honeybee line resistant to *B. larvae* contained more factors that inhibited the germination of spores and the growth of vegetative cells than did the brood food of a susceptible line. Differences in pollen content of the brood food might cause the significantly different susceptibilities to *B. larvae* among larvae of queen (susceptible), drone (resistant), and worker honeybee (intermediate) (144). Rinderer et al. (145) studied the effect of pollen on the susceptibility of worker honeybee larvae at the time of inoculation of *B. larvae* spores. The results indicated that spore feeding preceded by water feeding results in 94.84% mortality, and spore feeding preceded by pollen feeding resulted in 71.17%, a highly significant reduction. Vitamins may affect insect susceptibility to pathogens. Pristavko and Dovzhenok (146) reported that the larvae of the codling moth *Laspeyresia pomonella* reared on artificial diets containing different concentrations of ascorbic acid varied in susceptibility to *B. bassiana.* When larvae were fed on diets with more or less than 0.6–0.8% ascorbic acid, the total hemocyte count of the larvae was reduced appreciably and the susceptibility to *B. bassiana* was greatly increased. Armstrong (147) investigated the effect of vitamin B-deficient diets on *T. castaneum* infected with *N. whitei.* The mortality of larvae fed diets deficient in vitamins, e.g., biotin, thiamine, pantothenic acid, vitamin B_{12}, pyridoxine, niacin, and riboflavin, was higher than that of larvae fed on a vitamin B-complete diet. Less than 80% of the larvae pupated on diets deficient in biotin, thiamine, and pantothenic acid, whereas less than 50% pupated on diets deficient in vitamin B_{12}, pyridoxine, niacin, and riboflavin. When infected *T. castaneum* were fed vitamin-deficient diets, less than 26% of the larvae pupated.

2.3.5. TEMPERATURE AND HUMIDITY. Temperature is the most important external physical factor for both insect susceptibility and multiplication of microbial pathogens in the host. For example, the development of milky disease in *Melolontha melolontha* is a function of temperature, 15°C being the minimum and 25° the optimum for development. The injected dosage, as long as it is above the minimum required for infection, is not important in affecting the rate of development (148). Canerday and Arant (149) confirmed that the time required for lethal infection by the cabbage-looper *(T. ni)* NPV is inversely related to dosage level and environmental temperature. *Tipula* iridescent virus readily replicated in *T. oleracea* larvae and pupae at 20°C, which was the optimum growth temperature for the insect, and was able to replicate at 3° and 27°C, which were near the growth temperature limits of the host (150).

Most silkworm varieties have been adapted to rearing at 25°C, which is most suitable for their development. Accordingly, temperatures much higher or lower than 25°C tend to act as a stressor and increase the larval susceptibility to viral infections (151). Exposure of silkworm larvae to low temperatures (5°C for several hours) before peroral infection with CPV (152), NPV (153), or an IFV (154), enhanced susceptibility to each virus.

On the other hand, high temperatures are known to increase resistance or cause the disappearance of viral infections in plants and higher animals. This is also true in insects. Tanada (155) observed that although the susceptibility of the imported cabbageworm *P. rapae* to the GV increased with an increase in temperature, the larvae became resistant to viral infection when reared at a high temperature of 36°C. The larvae of *D. hercyniae* (156), *T. ni,* and *H. zea* (157), when reared continuously at 29.4°C, resisted infection by their respective NPVs. The alfalfa caterpillar *Colias eurytheme* also resisted infection by CPV at a high temperature (158). The virus-infected silkworm larvae, when reared at an elevated temperature (36–37°C), were able to survive virus infection (159, 160).

The failure of *Sericesthis* iridescent virus (SIV) to cause lethal infection occurred when the host *Sericesthis pruinosa* was reared at 28°C (161, 162). Tanada and Tanabe (163) studied the effect of temperature on the replication of TIV in *G. mellonella* larvae. Inoculated larvae incubated at 23 to 25°C died from TIV infections, whereas those incubated at temperatures above 30°C survived and produced adults.

The nonoccluded virus of citrus red mite retained full infectivity when exposed to 40.5°C for 24 hours within intact mite bodies but was inactivated at

46°C for 6 hours and 60°C for 1 hour. Exposures to 38°C for 28 days failed to destroy infectivity (164).

Ignoffo (165) demonstrated that a high temperature (40°C) inhibited NPV multiplication in bollworm larvae *H. zea* but the normal field temperature (13 to 35°C) would not inhibit the development of viral infection. The high or low field temperatures affected larval feeding activity and, therefore, the ingestion of virus-treated leaf surfaces. Bullock (166) found that treatment of larvae of the tobacco budworm *H. virescens* with a high temperature (36°C) reduced patent infection with a CPV if it was begun within 24 hours after exposure to contaminated diet. However, if the treatment was delayed, the therapeutic effect diminished until there was none when the delay was 4 days. Inoue (154) demonstrated thermal therapy in the silkworm infected with IFV. Larvae that had been given a lethal dose of IFV just after hatching did not succumb to the disease but made cocoons when they were repeatedly exposed at each larval ecdysis to a high temperature (37°C) for 24 hours.

Watanabe and Tanada (167) provided an explanation for the resistance of the armyworm *Pseudaletia unipuncta* to NPV when reared at 37°C. Histopathological study indicated that a small number of cells of the fat body, hypodermis, and trachea of the virus-exposed larvae held at 37°C contained virus particles but no polyhedra. Autoradiographic results revealed that viral DNA synthesis occurred in a few cells, but polyhedron-protein synthesis did not occur at 37°C. These results indicated that high temperature (37°C) may prevent the lethal infection of virus-exposed larvae not only by interfering with the mechanism of adsorption and/or penetration of the virus into the cell, but also by suppressed viral replication. In the case of DNV, viral multiplication was reduced when infected silkworm larvae reared at 25 to 28°C were subsequently reared at 37°C (168). Autoradiographic results with ^{3}H-thymidine and ^{3}H-tyrosine revealed that the synthesis of both viral DNA and protein was greatly reduced in infected larvae maintained at 37°C. Fluorescent antibody studies also confirmed that the synthesis of DNV-antigen in the larvae was inhibited at 37°C. These results indicated that high temperatures (37°C) apparently reduced the activity of enzymes concerned with viral DNA and protein syntheses.

The activity of fungi in an insect population is greatly influenced by temperature and humidity. It is generally recognized that high atmospheric humidities and moderately high temperatures are essential for the germination of spores of most fungi and subsequent development of epizootics. Nonetheless, studies of temperature and humidity in relation to insect susceptibility to fungi are still limited. Mohamed et al. (169) conducted laboratory studies to determine the susceptibility of various larval instars of *H. zea* to different spore doses of *N. rileyi* at constant and variable temperatures. The fungus was most effective at 20° and 25°C, with a mortality of 80% and 71%, respectively. At 15°C, the disease progressed very slowly with larval mortality occurring in 12 to 28 days posttreatment. Conversely, at temperatures above 15°C, larval mortality occurred in 6 to 12 days.

The infectivity as well as the development of infection by microsporidia in host insects is dependent on the host-rearing temperature. Although microsporidian spores can tolerate a wide range of temperatures (170–172), this may not be the case with the infectious process in the host. The optimum temperatures of most microsporidian infections lie between 20–30°C, and no development occurs below 10°C (172). Experiments with simultaneous infections of grubs with *Nosema melolonthae* or *Adelina melolonthae* brought similar results, with increased mortality in the 20 to 25°C temperature range, compared with chronic long-lasting infections at lower temperatures (173).

Kramer (174) exposed the black blowfly to spores of *Octospora muscaedomesticae* at temperatures from 12 to 32°C. Apparently no infection took place at 12°C because the spores failed to germinate. When infection occurred, the harmful effects were directly proportional to the rise in temperature. At each temperature, the longevity of infected flies was about one-half that of uninfected flies. On the other hand, the pathogenicity of *N. whitei* for *T. castaneum* decreased as the temperature was increased from 25° ($LD_{50} = 4.2 \times 10^6$ spore/g) through 30° ($LD_{50} = 1.3 \times 10^7$ spore/g) to 35°C ($LD_{50} = 3.2 \times 10^8$ spore/g) (175). Maddox (176) demonstrated that the reproduction of *Nosema necatrix* in host tissues is optimal at a temperature of 30°C. Ishihara (177) demonstrated that *Nosema bombycis* was able to develop in primary cell cultures of mammalian and chicken embryo when the cultures were kept at the normal temperatures of insect tissues, e.g., 28°C, but there was no microsporidian growth in cultures maintained at a high temperature (37°C).

The stress of very high or very low temperatures is harmful to the infected host. Kramer (178) found that low winter and high summer temperatures killed more European corn borers infected by microsporidia than uninfected insects.

2.3.6. RADIATION AND LIGHT.   Exposure of insects to X-rays and gamma rays affects their susceptibility to microbial pathogens. Smirnoff (179) demonstrated that mortality caused by NPV was accelerated in virus-infected larvae of eastern tent caterpillar *Malacosoma americanum* irradiated with 10,000 and 50,000 rad of ^{60}Co source; death occurred after 11 to 12 days whereas nonirradiated infected larvae died after 14 to 16 days. On the other hand, Jafri (180) found a decrease in susceptibility of silkworm larvae to CPV following exposure to sublethal and lethal doses of X-rays. Jafri and Khan (181) inoculated *Heliothis* NPV into the greater wax moth larvae *G. mellonella* after exposure to 10 and 15 kr of ^{60}Co gamma-ray. In virus-infected larvae, polyhedra were few in number and appeared only in some cells of the midgut epithelium between 72 and 96 hours postinoculation, and in adipose tissues and hypodermis at the end of the first week.

Jafri and Chaudhry (182) inoculated TIV into *G. mellonella* larvae that had received several doses of gamma radiation. Infection and multiplication of TIV in several tissues were suppressed differently depending on the radiation dose. The sublethal and lethal doses of X-rays in *G. mellonella* larvae inhibited the development of a densonucleosis virus in vivo (183).

X-rays also affect the infection of *B. thuringiensis*. Adults of *T. castaneum* and *Tribolium confusus* were more susceptible to infection by *B. thuringiensis* after exposure to sublethal and lethal doses of X-rays (184). At 10 days postradiation of ^{137}Cs to *Hemerocampa leucostigma* larvae, there was a depression of larval resistance to intoxication and/or infection by *B. thuringiensis* var *thuringiensis* (185).

Smirnoff (186) exposed *Neodiprion swainei* to daylight and complete darkness to determine the role of light on the development of a viral disease. Larval mortality from NPV occurred sooner in daylight. Although this inferred that the virus was more effective in daylight, it is more likely that light had no direct influence on the viral development but that the absence of light modified the physiological condition and metabolism of the host, thus affecting the development of the virus disease. Watanabe and Takamiya (187) found, on the other hand, that silkworm larvae reared in constant light were more resistant to CPV than those reared in constant darkness. This was associated with a decrease in the ratio of the susceptible columnar to the resistant goblet cells in the midgut of larvae reared in constant light.

2.3.7. CHEMICALS.   When microbial control is attempted, microbial pathogens are sometimes combined with chemical insecticides in order to enhance pest control. Some chemical insecticides and adjuvants increase insect susceptibility to pathogens, resulting in a synergistic enhancement of infection (188). Hunter et al. (189) demonstrated that a mixture of GV and malathion was more effective against the Indian meal moth *P. interpunctella* than either material alone. The decline with time in mortalities of the merchant grain beetle *Oryzaephilus mercator* fed diet treated with GV + malathion probably occurred because of a partial breakdown of malathion. However, Ignoffo and Montoya (190) found that methylparathion adversely affected the infectivity of *Heliothis* NPV to *H. zea* larvae. A 30% reduction in larval mortality also was detected in carbonated-water adjuvant-virus mixtures, but this reduction was not statistically significant at the 5% level.

Sublethal dosages of certain insecticides enhanced the susceptibility of the silkworm to viral infection (191). When sublethal doses of DDT and Smithion, an organophosphorous insecticide, were applied topically to larvae, no signs of intoxication such as paralysis, vomiting, reduction of feeding, or growth inhibition appeared. However, larvae treated with Smithion were more susceptible to peroral infection with NPV or CPV than larvae not treated with the insecticide, and larvae treated with DDT showed an increased susceptibility to NPV.

A synergism developed when *M. melolontha* larvae, attacked by muscardine caused by *Beauveria brongniartii,* were exposed to low doses of organochlorides or organophosphates (192, 193). Under these conditions, there was an increase in the incidence of muscardine, an acceleration of the infectious process, and susceptibility of the insect to lower spore concentrations than without the insecticide.

2.3.8. OTHER PATHOGENS.   When a mixture of pathogens is inoculated into an insect, the pathogens may coexist, react synergistically, or interfere with one another during infection. In some cases, one of the ingested pathogens or its infection acts as a biological stressor and increases the susceptibility of an insect to another pathogen. For example, Ishikawa and Miyajima (194) have reported that silkworm larvae that have been exposed to bacteria show an increase in susceptibility to viral infections. When toxic bacteria, such as *Pseudomonas* spp., *Serratia* spp., or *Proteus* spp., are present together with microsporidian spores, doses of 5 to 10,000 spores per caterpillar cause primary septicemias, and the caterpillars die within the first 5 days, generally without the development of the microsporidia (195). In experiments with insects reared on semisterile artificial media, the doses of spores necessary to cause septicemias are much higher. Tamashiro (196) reports that larvae of the rice moth, *Corcyra cephalonica,* parasitized by a species of *Bracon* are more susceptible to *B. thuringiensis* than are unparasitized larvae.

In some cases, there is interference or antagonism between pathogens. Nonparasitized second-instar larvae of *T. ni* are twice as susceptible (at the LD_{50} level) to the *T. ni* NPV as those parasitized by *Hyposoter exiguae* (197). The LD_{50} values for nonparasitized and parasitized larvae are 1.58×10^3 and 3.16×10^3 polyhedra/ml of diet, respectively. The LT_{50} values for parasitized larvae also are significantly longer than those for nonparasitized larvae. Bailey (104) also reports that honeybees parasitized with the mite *Acarapis woodi* are less suceptible to infections with a bacterium *(Pseudomonas apiseptica),* acute bee paralysis virus, and chronic bee paralysis virus than unparasitized bees.

Stelzer (198) conducted field tests to evaluate the control possibilities of a NPV, *B. thuringiensis,* and a combination of the two pathogens against larvae of the Great Basin tent caterpillar *Malacosoma fragile.* The results indicated that all treatments were capable of causing lethal infections, and a combination of the two pathogens provided more efficient control than either alone.

In the armyworm *P. unipuncta,* Tanada (199) found that a GV enhanced the infection of a NPV and appeared to play an important role in epizootics in field populations. The enhancing or synergistic factor was present in the protein matrix of the capsule (200). It was found in one strain of the GV (Hawaiian) but not in another (Oregonian). The synergistic factor was isolated, purified, and found to be a lipoprotein (MW = 126,000), and the phospholipid appeared to be essential for the synergism (201). Similarly, Watanabe and Shimizu (202) reported synergism between an infectious flacherie virus and a DNV in the silkworm, but the mechanism was unknown. The interaction took place in a silkworm population susceptible to DNV and probably resulted in a severe epizootic of both viruses.

Interference phenomena occurred between two strains of CPV in the silkworm (203) and two strains of NPV in the armyworm (204). However, Lowe and Paschke (205) reported that only an additive effect resulted when a GV and a NPV were administered simultaneously to the cabbage looper *Trichoplusia ni.* Simultaneous infections of a NPV and *Nosema* spp. in larvae of the fall webworm *H. cunea* reduced larval mortality by the virus (206).

Synergism resulted when larvae of *M. melolontha* were infected first with *B. popilliae* or an entomopoxvirus, then with *Beauveria tenella* (207, 208). Sharpe and Detroy (209) reported that milky disease somehow caused Japanese beetle larvae to become susceptible to *B. thuringiensis.* They believed that the milky-diseased larvae became susceptible because the midgut digestive juice developed sufficient alkalinity to solubilize and activate the crystal.

There seems to be synergism between *Malpighamoeba mellificae* and *Nosema apis* in the honeybee. Many colonies infected with *M. mellificae* are also infected with *N. apis,* and in such colonies, there are significantly more queens infected with *N. apis* than in colonies infected with *N. apis* alone (210).

3. POPULATION PARAMETERS IN RELATION TO PATHOGEN DISSEMINATION

The dissemination of pathogens within a host population or among populations plays an important role in the development of epizootics. The pathogen may be spread not only by mechanical factors, such as wind, rain, rivers, etc., but also by biotic factors. Biotic factors that can affect disease spread include population density, behavior of infected or contaminated individuals, the primary and secondary host populations, and other nonsusceptible organisms that may harbor or carry the pathogens.

3.1. Density

Epizootics of infectious diseases generally develop or are evident at high host population densities. Such high densities increase the contact between uninfected and infected hosts, between hosts and the pathogen, and also increase the susceptibility to infection due to food competition. As more hosts become infected, more pathogens are produced and epizootics take place. Steinhaus (211) suggested that crowding acts as a stressor on individuals in a population, thereby influencing the incidence of disease. Thus, in general, pathogens act as density-dependent mortality factors, that is, they destroy more hosts as the host density increases. Because the threshold density at which the pathogens are active is generally high, epizootics in insect populations usually occur at high host densities, but epizootics may also occur at low host densities, especially if the pathogen is widely distributed in the host habitat (43, 212, 213).

Virus acts as a density-dependent factor in the regulation of insect population, but it may also act as a density-independent factor, especially when the virus is widely distributed in the insect ecosystem after an epizootic or after its application to an insect population as a microbial insecticide (214). Viral disease of sawflies is apparently not disseminated effectively in populations of low host density but is usually present in epizootic proportions in populations whose host density is high (122, 215). Density-dependent development of epizootics of nuclear polyhedroses have been reported in larval populations of the cabbage looper *T. ni* (216) and the gypsy moth *Lymantria dispar* (217).

The Indian rhinoceros beetle *Oryctes rhinoceros* is infected by a baculovirus.

When the larval and adult populations of *O. rhinoceros* in Western Samoa were monitored for several years, evidence was obtained that the prevalence of baculovirus-infected individuals was dependent on population density (218). In the armyworm, as the larval population increased in density there was an associated but slightly delayed increase in the prevalence of GV and especially NPV infections. This suggested that the virus epizootics in the armyworm were host-density dependent (213). In some cases, however, even with a very high host density, as in the case of the alfalfa caterpillar, no epizootic developed in the insect population, even though the virus was present in some individuals (214). Carter et al. (219) found in a field population of *Tipula paludosa* larvae that the proportion of NPV-infected larvae was positively correlated with population density, whereas the proportion of infected larvae with a spore-forming bacterium was negatively correlated with population density.

There are numerous reports that NPVs, CPVs, and GVs persist under low host densities. The entomopoxvirus of the spruce budworm *C. fumiferana* also persists from one year to the next at very low host density (220).

A NPV and introduced parasites are important factors controlling the European spruce sawfly *D. hercyniae* in Central New Brunswick. Bird and Elgee (43) have found that the virus is very effective in controlling the spruce sawfly at high population levels, whereas certain parasites are more effective at low population levels. The epizootic of the viral disease appears to be density dependent and to be independent of stress factors, such as weather, overcrowding, and lack of food.

The manifestation of a nonoccluded viral disease of the citrus red mite *Panonychus citri* is also influenced by population density (221). When the virus is applied to citrus red mites by spraying, the incidence of infection increases progressively with density in populations of 1, 5, 20, and 80 mites caged on lemon surface areas of 45 to 75 cm². When single, infected mites are introduced at the same population levels, there is a tenfold increase in transmission at the lowest density in contrast to spraying the virus. Density remains a factor, however, since the transmission at the 80-mite level is higher than at the one- and five-mite levels. In citrus groves, the high incidence of the nonoccluded viral disease in populations of the citrus red mite is correlated with a high density of mites. Invariably, epizootics have developed in large mite populations (222).

Doane (223) found that the larvae of *Scolytus multistriatus* carried sufficient bacteria on their mouthparts to cause infection by wounding and that this occurred more frequently when their density was high.

The microsporidia are a density-dependent mortality factor and play an important role in population regulation (170). Weiser (224) reported that microsporidia initially occurred in small foci and then spread out in the field.

3.2. Behavior

The behavior and dispersal of the host insect may affect dissemination of the pathogen. Movements of infected hosts disperse the pathogens, which are de-

posited in the environment in regurgitations, fecal deposits, and in the disintegrating bodies after death.

3.2.1. ABNORMAL BEHAVIOR DUE TO INFECTION.   Insects infected with viruses and fungi often exhibit abnormal movements just before their deaths. The
infected insects climb to elevated locations, such as the tops of trees and plants.
Death in such high positions benefits the dispersal of pathogens. Infection may
also affect the behavior of gregarious insect species and cause their abnormal
dispersal (225).

In adults of *O. rhinoceros* chronically infected with the baculovirus, the
epithelium of the midgut proliferates excessively so that the lumen becomes
filled with a mass of cells in which the virus multiplies actively. Such adults
become flying reservoirs of virus, which is excreted into the host habitat (226).
Similar effective dispersal of pathogens is known in *Lygus communis* var *novascotiensis* infected with *Entomophthora erupta* and *Cicada septendecim* infected with *Massospora cicadina* (227, 228). These fungi do not invade the
thorax of the insect but develop only in the abdomen. Eventually, the abdominal wall is ruptured and the segments slough off, leaving a mass of conidiophores attached to a still-active thorax. The flight of the diseased insect distrib-
utes spores over a considerable area.

3.2.2. OVIPOSITIONAL BEHAVIOR.   In some cases, the ovipositional behavior
of the host insect is important in determining whether or not a pathogen is
successfully introduced and can cause an epizootic. In both the European
spruce sawfly *D. hercyniae* and the European pine sawfly *Neodiprion sertifer*
the virus is transmitted from one generation to the next through eggs laid by
infected females (215, 229). Females of the former species lay their eggs singly
and fly actively during the egg-laying period. Such a virus-infected female will
distribute virus-contaminated eggs over a considerable area, establishing numerous foci from which the disease may spread. Relatively few infected females
can initiate an epizootic. Largely because of such transmission, Bird (215, 229)
has succeeded in establishing the virus in European spruce sawfly populations
through a single introduction of the virus. On the other hand, each *N. sertifer*
female lays her eggs in one cluster, and the larvae feed as a group in the early
instars and move only short distances to obtain food. For this reason, the spread
of virus between *N. sertifer* colonies is slow even though 100% mortality can
occur within an infected colony. Dissemination of this virus is largely due to
mechanical agents such as rain, parasites, and frass droppings.

3.2.3.. GROOMING.   Grooming behavior, common in termites, can spread
pathogens between interacting individuals. In termites, grooming consists of
licking the body surfaces of nestmates. The subterranean termite *Reticulitermes* sp. when exposed to cultures of *M. anisopliae* for several hours, is
capable of spreading the fungus to a much larger healthy termite population

(230). The disease spreads rapidly throughout the population since healthy termites concentrate grooming behavior on diseased individuals.

3.2.4. CANNIBALISM. Carter (231) suggested that TIV was transmitted principally within the population of *T. oleracea* larvae by cannibalism, and a new generation might become infected by first instars feeding upon injured infected fourth instars which survived from the previous generation.

In *P. interpunctella* larval populations contaminated with spores of *B. thuringiensis,* there are many spores on the surfaces of all stages of the insect and in the diseased larval cadavers (232). Cannibalism did not occur among living larvae while food was present, but they sometimes ate cadavers. This is the most important means of natural spread of the disease.

Cannibalism in termites regulates the proportion of individuals in each caste. However, in the subterranean termite *Reticulitermes* sp. the individuals that have been killed by the fungus *M. anisopliae* are avoided and are not cannibalized by healthy termites; consequently the disease is not spread effectively (230).

Cannibalism probably occurs in nearly all major groups of insects that serve as hosts for microsporidia and it constitutes another example of transmission by direct contact. For instance, spores of *Nosema plodiae* may be transferred directly from a dead or dying larva of the Indian meal moth *P. interpunctella* to a healthy one as a result of the latter devouring the former (233).

3.2.5. HYGIENIC BEHAVIOR. Woodrow and Holst (234) noticed that many infected larvae of the honeybee in colonies resistant to American foulbrood were removed about 6 days after they were sealed in their cells, whereas very few sister larvae placed in control colonies had been removed 11 days after they were sealed, by which time they were full of infective spores. This was confirmed by Rothenbuhler (235) and his colleagues, who reported that the hygienic behavior of adults in removing diseased larvae was controlled genetically. They made selections for resistance and susceptibility, increasing both about tenfold as compared to the initial strain. The efficiency of hygienic behavior was separated into two components of factors: prompt uncapping of cells and removal of the larvae (236). Both genetic factors were recessive. (Also see Section 2.1.)

3.3. Relationship among Different Populations

Insect pathogens are dispersed within the host population or among populations not only by infected or contaminated individuals of the primary host, but also by secondary hosts, parasites, predators, scavengers, and other carriers. Thus, the effective dispersal and persistence of the pathogens depend to a large extent on the ecosystem consisting of the host insect and other insects and animals.

Smirnoff (237) has observed that, when relatively small amounts of a highly virulent virus are introduced into a population of the sawfly *N. swainei* the virus eventually spreads and controls the host throughout the infested area. He has found that this spread may be brought about through various agents such as insect parasites and predators or by diseased but surviving females laying contaminated eggs. Birds feeding on diseased insects also may spread the virus in their droppings, since the virus appears to be unharmed by passage through the bird's intestine.

3.3.1. ALTERNATE HOST.   Since baculoviruses, in general, are highly specific (238), most of them do not have alternate or secondary hosts in the ecosystem. The presence of such alternate hosts, however, may play a significant role with viruses having a wide host-range, such as CPVs and iridescent viruses. There are increasing numbers of laboratory studies reporting the cross-transmission of these viruses among different insect species, some of which may occur in the same ecosystem.

Cunningham (239) speculated that several nymphalid species that appeared to be susceptible to the same NPV or GV, might acquire the virus by feeding on a common food plant contaminated with the virus. Tanada and Omi (240) found that the CPV and GV collected from the alfalfa looper, the alfalfa caterpillar, the armyworm, and the beet armyworm were transmissible to nearly all of these four species in the alfalfa ecosystem. The NPVs isolated from these insects, however, appeared to be host specific.

The principal host of *Thelohania hyphantriae* is the fall webworm *H. cunea* but this microsporidium also infects alternate hosts, *Malacosoma neustria, Euproctis chrysorrhoea,* and *Hyponomeuta malinellus,* which may aid its spread (241).

Alternate hosts in the ecosystem can vary in their susceptibility to some pathogens. Larvae of the fiery skipper *Hylephila phylaeus* are found in the same habitat as the lawn moth *Crambus bonifatellus* but the skippers are more susceptible to infection by *Nosema infesta* and may be the primary host (242). Maddox (172) has observed that closely related hosts may vary in susceptibility to microsporidia even though present in the same location. One host may have a high incidence of microsporidiosis, while another has a very low incidence, even though both species are equally susceptible in the laboratory. He speculates that the difference in the incidence of infection may be due to host behavioral differences. This situation becomes highly complex when susceptible hymenopterous parasites and hyperparasites are involved together with their habitual and alternate hosts (243, 244).

3.3.2. PARASITES.   In *M. disstria* populations, the dipterous parasite *Sarcophaga aldrichi* is consistently present, and the adult flies are efficient transmitters of a NPV (245). Epizootics develop more rapidly when a relatively dense population of flies and virus occur together. The flies consistently feed on recently dead, virus-diseased larvae, become contaminated, and are capable of depositing the virus on plant foliage. Moreover, the flies are present from the time the *M. disstria* larvae hatch until the adults emerge, and thereby disseminate virus throughout the habitat. The presence of a large number of *S. aldrichi* in the system appears to be essential in the rapid development of epizootics (246).

Insect parasites play an important role in the transmission of microsporidia in insects. Paillot (247) was the first to suggest that *Apanteles glomeratus* transmitted *Perezia legeri* to *P. brassicae* because the incidence of infection by the microsporidium in the cabbageworm was closely associated with the parasitization by the wasp. Some hymenopterous parasites transmit microsporidia mechanically to the insect hosts through their contaminated ovipositors (248).

Most insect parasites are susceptible to microsporidian pathogens of their hosts (243). Generally, parasites that develop within infected hosts fail to mature, or if they do, develop into weak adults that are short lived and exhibit reduced fecundity. The detrimental effects generally result from the direct infection of the parasite by the pathogen but may also occur when large numbers of ingested spores accumulate in the parasite's midgut. The deleterious effects can be enhanced if the parasite prefers the infected host for oviposition, as in *Sarcophaga stützei* which attacks only larvae of the nun moth or gypsy moth infected with polyhedrosis (249).

The hymenopterous parasite *Apanteles marginiventris* oviposits with equal frequency in *Nosema* spp.-infected and uninfected lawn-armyworm *Spodoptera mauritia*. However, *Nosema*-infected hosts have deleterious effects on their internal parasites. There are high parasite larval and pupal mortalities, reduced adult emergence, and the few adults that emerge are smaller and live significantly shorter than the parasites from uninfected hosts. These detrimental effects are not due to direct infection of the parasite by *Nosema* spp., but due rather to the infected host apparently not being nutritionally adequate for parasite development. There are no observable pathological changes in tissues of parasites reared in infected hosts. The parasites successfully transmit *Nosema* spp. from severely infected to noninfected hosts during oviposition (250).

A braconid parasite *Macrocentrus grandii*, which develops within *Nosema pyrausta*-infected larvae of the European corn borer *Ostrinia nubilalis*, develops systemic infections from the ingestion of spores at the time of larval emergence from the host. Infections adversely affect pupal development and longevity of the parasite adult. Infected parasite females are unable to transmit the microsporidium to other corn borer hosts. The prevalence of infection in natural parasite populations is 53.8%, which parallels the 56.7% prevalence in corn-borer populations. The results suggest that *N. pyrausta* may play a significant role in limiting *M. grandii* populations when levels of *N. pyrausta* in corn borers are high (251).

3.3.3. MITES.   Phoretic mites have been credited with adventitiously transferring various entomopathogenic fungi to a variety of insects (252). Mites were considered instrumental in transmitting *B. bassiana* to the pales weevil *Hylo-*

bius pales, a serious forest pest. The mites spread the fungal conidia through the soil.

Schabel (252) ascertained in a laboratory experiment that the two phoretic mites *Macrocheles* spp. and *Histiogaster anops* were susceptible to the green muscardine fungus *M. anisopliae,* and that they also transferred the fungal spores to host larvae of pales weevil, thus spreading the disease.

3.3.4. PREDATORS AND SCAVENGERS.   English sparrows have been observed methodically searching cabbage plants for insects. Larvae of the cabbage looper, *T. ni,* in advanced stages of nuclear polyhedrosis, lose their natural defense mechanisms and become easy prey for birds. Hostetter and Biever (253) collected and examined bird feces from a cabbage field and found high numbers of polyhedra in nine of the 12 samples. The per os inoculation of third instar *T. ni* larvae proved that all 12 suspensions contained virulent polyhedra. Bird (156) reported that the stomach contents of two of four birds collected from a Scots pine plantation that had virus-infected European pine sawfly larvae *N. sertifer* contained virus that was highly infectious for sawfly larvae. He concluded that the virus was probably transmitted through bird droppings and was distributed over long distances.

Nonsusceptible insects and vertebrate animals may serve as carriers and disseminators of microsporidia. *T. hypantriae* is dispersed vertically on a tree by ants and beetles, and laterally among trees by adults of *H. cunea* (224). The spores of *T. hyphantriae* retain their virulence after being excreted by predators and scavengers, such as *Calosoma sycophanta, Xylodrepa quadripunctata, Cantharis fusca, Formica rufa,* and the mite *Tyrophagus noxius* (254).

Scavengers also play a role in the dispersal of *N. plodiae,* a pathogen of the stored-product insect, *P. interpunctella* (255). The spores of *Pleistophora schubergi* and *Nosema polyvora* retain their virulence for their insect hosts after passage through the alimentary tract of a bird or that of an earwig (256). According to Weiser (224, 257), the microsporidium-infected insects are favorable hosts for predators and scavengers, which may aid in the dispersal of the microsporidia.

4. CONCLUSION

Progress on the properties and characteristics of host populations and their role in epizootics has been slow compared to that of the other principal factors, namely, the pathogen population, transmission, and the environment. This is due not only to the immense numbers of insect species, but also to the diversity of the insect ecosystem, biology, behavior, and type of population (solitary, gregarious, etc.). Most of our present knowledge of the properties and characteristics of host populations is obtained from studies conducted with two cultured, beneficial insects, the silkworm and honeybee. The ecosystems and biologies of these two insects differ, and they have provided valuable information about the effects of differences in strains, food, colonies, and castes. However, these studies have not provided much information about the qualitative composition of the population, in other words, the presence of various types of individuals such as the susceptible host, resistant host, carrier of latent infection, or the nonsusceptible carrier. Such information is needed to study the onset, development, and cessation of an epizootic.

Information about host populations in the field is limited, and there is a need for quantitative data involving different parameters of the population before, during, and after an epizootic. With the rapid development of computers, the collection and analysis of massive data may not be such an overwhelming task as in previous years. Such data will provide the basis for the modeling of host populations and their reactions to epizootics. Such models will expedite and enhance the development of epizootiology, especially in the ability to predict epizootics in insect populations.

REFERENCES

1. E. A. Steinhaus, *Principles of Insect Pathology,* 1st ed., McGraw-Hill, New York, 1949.

2. Y. Tanada, "Epizootiology of Infectious Diseases," in E. A. Steinhaus, Ed., *Insect Pathology: An Advanced Treatise,* Vol. 2, Academic Press, New York, 1963, p. 423.

3. D. T. Briese, "Resistance of Insect Species to Microbial Pathogens," in E. W. Davidson, Ed., *Pathogenesis of Invertebrate Microbial Diseases,* Allanheld, Osmun Publishers, Totowa, New Jersey, 1981, p. 511.

4. Y. Tanada, Factors affecting the susceptibility of insects to viruses, *Entomophaga,* **10,** 139 (1965).

5. H. Watanabe, "Resistance of the Silkworm to Cytoplasmic-Polyhedrosis Virus," in H. Aruga and Y. Tanada, Eds., *The Cytoplasmic-Polyhedrosis Virus of the Silkworm,* Univ. Tokyo Press, Tokyo, 1971, p. 169.

6. L. L. J. Ossowski, Variation in virulence of a wattle bagworm virus, *J. Insect Pathol.,* **2,** 35 (1960).

7. W. A. L. David and B. O. C. Gardiner, Resistance of *Pieris brassicae* (Linnaeus) to granulosis virus and the virulence of the virus from different host races, *J. Invetebr. Pathol.,* **7,** 285 (1965).

8. D. K. Hunter and D. F. Hoffmann, Susceptibility of two strains of Indian meal moth to a granulosis virus, *J. Invertebr. Pathol.,* **21,** 114 (1973).

9. D. T. Briese, The incidence of parasitism and disease in field populations of the potato moth *Phthorimaea operculella* (Zeller) in Australia, *J. Aust. Entomol. Soc.,* **20,** 319 (1981).

10. Y. Aratake, Difference in the resistance to infectious flacherie virus between the strains of the silkworm, *Bombyx mori* L., *Sansi-Kenkyu,* **86,** 48 (1973).

11. Y. Aratake, Strain differences of the silkworm, *Bombyx mori* L., in the resistance to a nuclear polyhedrosis virus, *J. Seric. Sci. Japan,* **42,** 230 (1973).

12. H. Watanabe, Genetic resistance to peroral infection with the cytoplasmic-polyhedrosis virus in the silkworm, *Bombyx mori* L., *J. Seric. Sci. Japan,* **35,** 27 (1966).

13. T. E. Rinderer, W. C. Rothenbuhler, and J. M. Kulincevic, Responses of three genetically different stocks of the honey bee to a virus from bees with hairless-black syndrome, *J. Invertebr. Pathol.,* **25,** 297 (1975).

14. L. Bailey, The incidence of virus diseases in the honey bee, *Ann. Appl. Biol.,* **60,** 43 (1967).

15. L. Bailey, Paralysis of the honey bee, *Apis mellifera* Linnaeus, *J. Invertebr. Pathol.,* **7,** 132 (1965).

16. J. M. Kulincevic and W. C. Rothenbuhler, Selection for resistance and susceptibility to hairless-black syndrome in the honey bee, *J. Invertebr. Pathol.,* **25,** 289 (1975).

17. W. C. Rothenbuhler and V. C. Thompson, Resistance to American foulbrood in honey bee. I. Differential survival of larvae of different genetic lines, *J. Econ. Entomol.,* **49,** 470 (1956).

18. T. R. Hoage and W. C. Rotherbuhler, Larval honey bee response to various doses of *Bacillus larvae* spores, *J. Econ. Entomol.,* **59,** 42 (1966).

19. V. T. Cherepov, Resistance to European foulbrood disease among the main honey bee races, *Veterinariya (Moscow),* **6,** 54 (1970).

20. B. Papierok and N. Widing, Mise en évidence d'une différence de sensibilité entre 2 clones du Puceron du Pois, *Acyrthosiphon pisum* Harr. (Homoptères *Aphididae*), exposés à 2 souches du champignon Phycomycète: *Entomophthora obscura* Hall & Dunn, *C.R. Acad. Sci. Paris, Sér. D.* **288,** 93 (1979).

21. C. P. Milne, Jr., Laboratory measurement of brood disease resistance in the honeybee. 1. Uncapping and removal of freeze-killed brood by newly emerged workers in laboratory test cages, *J. Apic.,* **21,** 111 (1982).

22. C. P. Milne, Jr., Honey bee (Hymenoptera: Apidae) hygienic behavior and resistance to chalkbrood, *Ann. Entomol. Soc. Am.,* **76,** 384 (1983).

23. J. V. Bell and R. J. Hamalle, Comparative mortalities between field-collected and laboratory-reared wireworm larvae, *J. Invertebr. Pathol.,* **18,** 150 (1971).

24. F. E. Moeller and P. H. Williams, Chalkbrood research at Madison, Wisconsin, *Am. Bee J.,* **116,** 484 (1976).

25. D. M. Menapace, Chalkbrood infection and detection in colonies of honey bee *Apis mellifera, Am. Bee J.,* **118,** 158 (1978).

26. D. De Jong, Experimental enhancement of chalkbrood infections, *Bee World,* **57,** 114 (1976).

27. J. Weiser, "Immunity of Insects to Protozoa," in G. J. Jackson, R. Herman, and I. Singer, Eds., *Immunity to Parasitic Animals,* Appleton-Century-Crofts, New York, 1963, p. 129.

28. J. C. M. L'Arrivee, The effect of sampling sites on *Nosema* determination, *J. Insect Pathol.,* **5,** 349 (1963).

29. C. E. Burnside and I. L. Revell, Observations on nosema disease of honey bees, *J. Econ. Entomol.,* **41,** 603 (1948).

30. B. Furgala, The effect of the intensity of *Nosema* inoculum on queen supersedure in the honey bee, *Apis mellifera* Linnaeus, *J. Insect Pathol.,* **4,** 429 (1962).

31. L. Kartman, Factors influencing infection of the mosquito with *Dirofilaria immitis* (Leidy, 1956), *Exp. Parasitol.,* **2,** 27 (1953).

32. M. Tsukamoto, The log dosage-probit mortality curve in genetic researches of insect resistance to insecticides, *Botyu-Kagaku,* **28,** 91 (1963).

33. H. Watanabe, Resistance to peroral infection by the cytoplasmic-polyhedrosis virus in the silkworm, *Bombyx mori* (Linnaeus), *J. Invertebr. Pathol.,* **7,** 257 (1965).

34. H. Watanabe and S. Maeda, Genetically determined nonsusceptibility of the silkworm, *Bombyx mori,* to infection with a densonucleosis virus (*Densovirus*), *J. Invertebr. Pathol.,* **38,** 370 (1981).

35. C. F. Reichelderfer and C. V. Benton, Some genetic aspects of the resistance of *Spodoptera frugiperda* to a nuclear polyhedrosis virus, *J. Invertebr. Pathol.,* **23,** 378 (1974).

36. D. T. Briese, Genetic basis for resistance to a granulosis virus in the potato moth, *Phthorimaea operculella, J. Invertebr. Pathol.,* **39,** 218 (1982).

37. G. C. Llewellyn and J. P. Chinnici, Variation in sensitivity to Aflatoxin B_1 among several strains of *Drosophila melanogaster* (Diptera), *J. Invertebr. Pathol.,* **31,** 37 (1978).

38. J. P. Chinnici and C. Llewellyn, Reduced Aflatoxin toxicity in hybird crosses of Aflatoxin B_1 sensitive and resistant strains of *Drosophila melanogaster* (Diptera), *J. Invertebr. Pathol.*, **33**, 81 (1979).

39. W. L. Kilama and G. B. Craig, Jr., Monofactorial inheritance of susceptibility to *Plasmodium gallinaceum* in *Aedes aegypti*, *Ann. Trop. Med. Parasitol.*, **63**, 419 (1969).

40. N. G. Sidorov, V. S. Koptev, and I. A. Mugalimov, Resistance of honey bees to *Nosema* disease and a genetic method of control, *Veterinariya (Moscow)*, **7**, 63 (1975).

41. W. W. MacDonald, "The Influence of Genetic and Other Factors on Vector Susceptibility to Parasites," in J. W. Wright and R. Pal, Eds., *Genetics of Insect Vectors of Disease*, Elsevier, Amsterdam, 1967, p. 567.

42. G. H. Bergold, The polyhedral disease of the spruce budworm, *Choristoneura fumiferana* (Clem.) (Lepidoptera: Tortricidae), *Can. J. Zool*, **29**, 17 (1951).

43. F. T. Bird and D. E. Elgee, Virus disease and introduced parasites as factors controlling the European spruce sawfly *Diprion hercyniae* (Htg.), in central New Brunswick, *Can. Entomol.*, **89**, 371 (1957).

44. F. T. Bird and J. M. Burk, Artificially disseminated virus as factor in controlling the European spruce sawfly, *Diprion hercyniae* (Htg.), in the absence of introduced parasites, *Can. Entomol.*, **93**, 228 (1961).

45. C. Sidor, Susceptibility of larvae of the large white butterfly *(Pieris brassicae* L.) to two virus diseases, *Ann. Appl. Biol.*, **47**, 109 (1959).

46. C. F. Rivers, Virus resistance in larvae of *Pieris brassicae* (L.), *Trans. Int. Conf. Ins. Pathol. Biol. Control (Praha, 1958)*, 205 (1959).

47. W. A. L. David and B. O. C. Gardiner, A *Pieris brassicae* (Linnaeus) culture resistant to granulosis, *J. Insect Pathol.*,**2**, 106 (1960).

48. M. E. Martignoni and P. Schmidt, Studies on the resistance to virus infections in natural populations of Lepidoptera, *J. Insect Pathol.*, **3**, 62 (1961).

49. M. E. Martignoni, Contibuto alla conoscenza di una granulosi di *Eucosma griseana* (Hübner) (Tortricidae, Lepidoptera) quale Fattore linitante il pullulanemts dell'insetts nella Engadina alta, *Mitt. Schweiz. Anst. Forstl. Versuchw.*, **32**, 371 (1957).

50. D. T. Briese and H. A. Mende, Differences in susceptibility to a granulosis virus between field populations of the potato moth, *Phthorimaea operculella* (Zelber) (Lepidoptera: Gelechiidae), *Bull. Entomol. Res.*, **71**, 11 (1981).

51. C. M. Ignoffo and G. E. Allen, Selection for resistance to a nucleopolyhedrosis virus in laboratory populations of the cotton bollworm, *Heliothis zea*, *J. Invertebr. Pathol.*, **20**, 187 (1972).

52. K. Yamafuji, M. Sato, and J. Nagata, Chemical virogenesis and virogenic treatment in silkworm, *Enzymologia*, **19**, 48 (1958).

53. K. Yamafuji, M. Sato, and J. Kishikawa, Chemical virogenesis and remote infection in silkworm, *Enzymologia*, **19**, 151 (1958).

54. K. Aizawa, Y. Furuta, and K. Nakamura, Selection of a resistant strain to virus induction in the silkworm, *Bombyx mori*, *J. Seric. Sci. Japan*, **30**, 405 (1961).

55. K. Aizawa and Y. Furuta, Resistance to virus induction in F_1 hybrids between resistant and common strains in the silkworm, *Bombyx mori*, *J. Seric. Sci. Japan*, **31**, 245 (1962).

56. K. Uzigawa and H. Aruga, On the selection of resistant strains to the infectious flacherie virus in the silkworm, *Bombyx mori* L., *J. Seric. Sci. Japan*, **35**, 23 (1965).

57. T. Funada, Genetic resistance of the silkworm, *Bombyx mori* L., to an infection of a flacherie virus, *J. Seric. Sci. Japan*, **37**, 281 (1968).

58. H. Watanabe, Development of resistance in the silkworm, *Bombyx mori*, to peroral infection of a cytoplasmic-polyhedrosis virus, *J. Invertebr. Pathol.*, **9**, 474 (1967).

59. T. L. Harvey and D. E. Howell, Resistance of the house fly to *Bacillus thuringiensis* Berliner, *J. Invertebr. Pathol.*, **7**, 92 (1965).

60. J. M. Feigin, Exposure of the housefly to selection by *Bacillus thuringiensis, Ann. Entomol. Soc. Am.*, **56**, 878 (1963).

61. C. Yamvrias, Contribution a l'étude du mode d'action de *Bacillus thuringiensis* Berliner vis-à-vis de la Teigne de la Farine *Anagasta (Ephestia) kuehniella* Zeller (Lepidoptera), *Entomophaga*, **7**, 101 (1962).

62. M. Devriendt and D. Martouret, Absence de résistance a *Bacillus thuringiensis* chez la teigne des cruciferes, *Plutella maculipennis* (Lep.: Hyponomeutidae), *Entomophaga*, **21**, 189 (1976).

63. D. M. Dunbar and R. L. Beard, Present status of milky disease of Japanese and Oriental beetles in Connecticut, *J. Econ. Entomol.*, **68**, 453 (1975).

64. T. R. Hoage and D. C. Peters, Selection for American foulbrood resistance in larval honey bees, *J. Econ. Entomol.*, **62**, 896 (1969).

65. S. A. Vansulin, On the resistance of *Culex pipiens molestus* Forsk. to Boverin, *Parzitologiya*, **8**, 274 (1974).

66. J. J. Petersen, Development of resistance by the southern house mosquito to the parasitic nematode *Romanomermis culicivorax, Environ. Entomol.*, **7**, 518 (1978).

67. H. Watanabe, Some aspects of the mechanism of resistance to peroral infection by cytoplasmic-polyhedrosis virus in the silkworm, *Bombyx mori* L., *J. Seric. Sci. Japan*, **35**, 411 (1966).

68. H. Watanabe, Relative virulence of polyhedrosis viruses and host-resistance in the silkworm, *Bombyx mori* L. (Lepidoptera: Bombycidae), *Appl. Entomol. Zool.*, **1**, 139 (1966).

69. K. Hayashiya, Y. Uchida, and M. Himeno, Mechanism of antiviral action of red fluorescent protein (RFP) on nuclear-polyhedrosis virus (NPV) in silkworm larvae, *Japan J. Appl. Entomol. Zool.*, **22** 238 (1978).

70. Y. Uchida, F. Kawamoto, M. Himeno, and K. Hayashiya, A virus-inactivating protein isolated from the digestive juice of the silkworm, *Bombyx mori, J. Invertebr. Pathol.*, **43**, 182 (1984).

71. Y. Aratake and H. Ueno, Inactivation of a nuclear-polyhedrosis virus by the gut-juice of the silkworm, *Bombyx mori* L., *J. Seric. Sci. Japan*, **42**, 279 (1973).

72. H. Aruga and H. Watanabe, Resistance to *per os* infection with cytoplasmic-polyhedrosis virus in the silkworm, *Bombyx mori* (Linnaeus), *J. Insect Pathol.*, **6**, 387 (1964).

73. Y. Aratake, T. Kayamura, and H. Watanabe, Inactivation of a cytoplasmic-polyhedrosis virus by gut-juice of the silkworm, *Bombyx mori* L., *J. Seric. Sci. Japan*, **43**, 41 (1974).

74. A. G. Richards and P. A. Richards, The peritrophic membrane of insects, *Annu. Rev. Entomol.*, **22**, 219 (1977).

75. C. R. Brandt, M. J. Adang, and K. D. Spence, The peritrophic membrane: ultrastructure analysis and functions as a mechanical barrier to microbial infection to *Orygia pseudotsugata, J. Invertebr. Pathol.*, **32**, 12 (1978).

76. D. B. Stoltz and M. D. Summers, Pathway of infection of mosquito iridescent virus. I. Preliminary observations on the fate of ingested virus, *J. Virol.*, **8**, 900 (1971).

77. J. Weiser and O. Lysenko, Septisemie bsource morusoveho, *Ceskoslov. Mikrobiol.*, **1**, 216 (1956).

78. J. D. Paschke and M. D. Summers, "Early Events in the Infection of the Arthropod Gut by Pathogenic Insect Viruses," in K. I. Maramorosch and R. E. Shope, Eds., *Invertebrate Immunity,* Academic Press, New York, 1975, p. 75.

79. E. A. Steinhaus, Stress as a factor in insect disease. *Proc. Int. Congr. Entomol. 10th Congr. Montreal 1956*, **4**, 725 (1958).

80. H. Inoue and M. Miyagawa, Regeneration of midgut epithelial cells in the silkworm, *Bombyx mori,* infected with viruses, *J. Invertebr. Pathol.*, **32**, 373 (1978).

81. K. Yamaguchi, Natural recovery of the fall webworm, *Hyphantria cunea* to infection by a cytoplasmic-polyhedrosis virus of the silkworm, *Bombyx mori, J. Invertebr. Pathol.*, **33**, 126 (1979).

82. H. Inoue, Multiplication of an infectious-flacherie virus in the resistant and susceptible strains of the silkworm, *Bombyx mori, J. Seric. Sci. Japan*, **43**, 318 (1974).

83. H. Watanabe, S. Tanaka, and T. Shimizu, Interstrain difference in the resistance of the silkworm, *Bombyx mori*, to a flacherie and a cytoplasmic-polyhedrosis virus, *J. Seric. Sci. Japan*, **43**, 98 (1974).

84. G. R. Stairs, Changes in the susceptibility of *Galleria mellonella* (Linnaeus) larvae to nuclear-polyhedrosis virus following blockage of the phagocytes with India ink, *J. Insect Pathol.*,**6**, 373 (1964).

85. K. Inoue, Phagocytosis of nuclear polyhedra and the localization of the acid phosphatase in the granular cell of the silkworm, *Bombyx mori* L., *J. Seric. Sci. Japan*, **43**, 394 (1974).

86. J. F. Bamrick and W. C. Rothenbuhler, Resistance to American foulbrood in honey bees. IV. The relationship between larval age at inoculation and mortality in a resistant and in a susceptible line, *J. Insect Pathol.*, **3**, 381 (1961).

87. W. C. Rothenbuhler, Behavior genetics of nest-cleaning in honey bees. I. Responses of four inbred line to disease-killed larvae, *Anim. Behav.*, **12**, 578 (1964).

88. V. C. Thompson and W. C. Rothenbuhler, Resistance to American foulbrood in honey bees. II. Differential protection of larvae by adults of different genetic lines, *J. Econ. Entomol.*, **50**, 731 (1957).

89. G. M. Splittstoesser, C. Y. Kawanishi, and H. Tashiro, Infection of the European chafer, *Amphimallon majalis*, by *Bacillus popilliae*: light and electron microscope observations, *J. Invertebr, Pathol.*, **31**, 84 (1978).

90. A. Burgerjon and D. Martouret, "Determination and Significance of the Host Spectrum of *Bacillus thuringiensis,*" in H. D. Burges and N. W. Hussey, Eds., *Microbial Control of Insects and Mites,* Academic Press, London and New York, 1971, p. 305.

91. R. A. Kinsinger and W. H. McGaughey, Susceptibility of populations of Indian meal moth and almond moth to *Bacillus thuringiensis, J. Econ. Entomol.*, **72**, 346 (1979).

92. I. M. Hall and K. Y. Arakawa, The susceptibility of the housefly, *Musca domestica* Linnaeus, to *Bacillus thuringiensis* var. *thuringiensis* Berliner, *J. Insect Pathol.*, **1**, 351 (1959).

93. P. F. Galichet, Sensitivity to the soluble heat-stable toxin of *Bacillus thuringiensis* of strains of *Musca domestica* tolerant to chemical insecticides, *J. Invertebr. Pathol.*, **9**, 261 (1967).

94. C. M. Ignoffo, C. Garcia, and M. J. Kroha, Susceptibility of larvae of *Trichoplusia ni* and *Anticarsia gemmatalis* to intrahemocoelic injections of conidia and blastospores of *Nomuraea rileyi, J. Invertebr. Pathol.*, **39**, 198 (1982).

95. Y. Aratake, Genetical analyses of the infection with muscardines of *Bombyx mori* L. I. Infection with the white muscardine, *Bull. Seric. Exp. Stn. (Tokyo)*, **17**, 136 (1961).

96. K. Kawakami, Susceptibiltiy of several varieties of the silkworm, *Bombyx mori* L., to *Aspergillus* disease and germination of fungus spores in larval hemolymph, *J. Seric. Sci. Japan*, **44**, 39 (1975).

97. K. Koidsumi, Antifungal action of cuticular lipids in insects, *J. Insect Physiol.*, **1**, 40 (1957).

98. K. Koidsumi and Y. Wada, Studies on the antimicrobial function of insect lipids. IV. Racial differences in the antifungal activity in the silkworm integument, *Japan, J. Appl. Zool.*, **20**, 184 (1955).

99. N. G. Sidorov and G. P. Mikhailenko, Resistance of different races of bees to experimental infection with *Nosema* disease, *Pchelovodstvo*, **89**, 14 (1969).

100. D. B. Woodard and T. Fukuda, Laboratory resistance of the mosquito, *Anopheles quadrimaculatus* to the mermithid nematode, *Diximermis peterseni, Mosq. News*, **37**, 192 (1977).

101. G. R. Stairs, Dosage-mortality response of *Galleria mellonella* (Linnaeus) to a nuclear-poly-hedrosis virus, *J. Invertebr. Pathol.,* **7,** 5 (1965).

102. S. Miyajima and S. Kawase, Different infection response of silkworm larvae to the locality of cytoplasmic-polyhedrosis virus injection, *J. Invertebr. Pathol.,* **9,** 441 (1967).

103. S. Miyajima and S. Kawase, Different infection response of silkworm larvae to the cytoplas-mic-polyhedrosis virus with the special reference to the locality of virus injection. I. Larval age, *J. Seric. Sci. Japan,* **38,** 237 (1969).

104. L. Bailey, Susceptibility of the honey bee, *Apis mellifera* Linnaeus, infected with *Acarapis woodi* (Rennie) to infection by airborne pathogens, *J. Invertebr. Pathol.,* **7,** 141 (1965).

105. L. Bailey and A. J. Gibbs, Infection of bees with acute paralysis virus, *J. Insect Pathol.,* **6,** 395 (1964).

106. W. T. Wilson and W. C. Rothenbuhler, Resistance to American foulbrood in honey bees. VIII. Effects of infecting *Bacillus larvae* spores into adults, *J. Invertebr. Pathol.,* **12,** 418 (1968).

107. G. R. Stairs, Quantitative differences in susceptibility to nuclear-polyhedrosis virus among larval instars of the forest tent caterpillar, *Malacosoma disstria* (Hübner), *J. Invertebr. Pathol.,* **7,** 427 (1965).

108. F. T. Bird, Infection and mortality of spuce budworm, *Choristoneura fumiferana,* and forest tent caterpillar, *Malacosoma disstria,* caused by nuclear and cytoplasmic-polyhedrosis vi-ruses, *Can. Entomol.,* **101,** 1269 (1969).

109. C. M. Ignoffo, Effects of age on mortality of *Heliothis zea* and *Heliothis virescens* larvae exposed to a nuclear-polyhedrosis virus, *J. Invertebr. Pathol.,* **8,** 279 (1966).

110. R. A. Daoust, Weight-related susceptibility of larvae of *Heliothis armigera* to a crude nuclear-polyhedrosis virus preparation, *J. Invertebr. Pathol.,* **23,** 400 (1974).

111. D. G. Boucias and G. L. Nordin, Interinstar susceptibility of the fall webworm, *Hyphantria cunea,* to its nucleopolyhedrosis and granulosis viruses, *J. Invertebr. Pathol.,* **30,** 68 (1977).

112. O. J. Smith, K. M. Hughes, P. H. Dunn, and I. M. Hall, A granulosis virus disease of western grape leaf skeletonizer and its transmission, *Can. Entomol.,* **88,** 507 (1956).

113. R. Kobara, H. Aruga, and H. Watanabe, Effect of larval growth on the susceptibility of silkworm, *Bombyx mori* L., to a cytoplasmic-polyhedrosis virus, *J. Seric. Sci. Japan,* **36,** 165 (1967).

114. J. B. Carter, *Tipula* iridescent virus infection in the developmental stages of *Tipula oleracea, J. Invertebr. Pathol.,* **24,** 271 (1974).

115. J. F. Bamrick, Resistance to American foulbrood in honey bees. VI. Spore germination in larvae of different ages, *J. Invertebr. Pathol.,* **9,** 30 (1967).

116. G. R. Sutter, W. C. Rothenbuhler, and E. S. Raun, Resistance to American foulbrood in honey bees. VII. Growth of resistant and susceptible larvae, *J. Invertebr. Pathol.,* **12,** 25 (1968).

117. J. F. Bamrick, Resistance to American foulbrood in honey bees. V. Comparative pathogen-esis in resistant and susceptible larvae, *J. Insect Pathol.,* **6,** 284 (1964).

118. E. W. Davidson, Ultrastructure of American foulbrood disease pathogenesis in larvae of the worker honey bee, *Apis mellifera, J. Invertebr. Pathol.,* **21,** 53 (1973).

119. D. C. Rizzo, Age of three dipteran hosts as a factor governing the pathogenicity of *Beauveria bassiana* and *Metarhizium anisopliae, J. Invertebr. Pathol.,* **30,** 127 (1977).

120. R. J. Milner, *Nosema whitei,* a microsporidian pathogen of some species of *Tribolium.* III. Effect of *T. castaneum, J. Invertebr. Pathol.,* **19,** 248 (1972).

121. J. Weiser and O. Lysenko, Protein changes in the hemolymph of *Galleria mellonella* infected with virus and protozoan pathogen, *Acta Entomol. Bohemoslov,* **69,** 97 (1972).

122. F. T. Bird, The use of a virus disease in the biological control of the European pine sawfly, *Neodiprion sertifer* (Geoffr.), *Can. Entomol.,* **85,** 437 (1953).

123. G. R. Stairs, The effect of metamorphosis on nuclear-polyhedrosis virus infection in certain Lepidoptera, *Can. J. Microbiol.,* **11,** 509 (1965).

124. H. Watanabe and H. Aruga, The effect of molt on the development of nuclear-polyhedrosis in the silkworm, *Bombyx mori, J. Seric. Sci. Japan,* **40,** 37 (1971).

125. H. Watanabe, H. Aruga, and H. Namura, An effect of molt on the development of cytoplasmic polyhedra in the silkworm, *Bombyx mori* L., *J. Seric. Sci. Japan,* **40,** 275 (1971).

126. H. Watanabe and H. Aruga, Susceptibility of dauerpupa of the silkworm, *Bombyx mori* L. (Lepidoptera: Bombycidae), to a nuclear-polyhedrosis virus, *Appl. Entomol. Zool.,* **5,** 118 (1970).

127. H. Inoue, Effect of the metamorphosis of the silkworm, *Bombyx mori* on the multiplication of a flacherie virus, *J. Seric. Sci. Japan,* **46,** 20 (1977).

128. H. Watanabe and H. Namura, Fecundity of adult of the silkworm, *Bombyx mori,* infected with a cytoplasmic-polyhedrosis virus in the pupal stage, *J. Seric. Sci. Japan,* **40,** 145 (1971).

129. G. R. Stairs, The development of nuclear-polyhedrosis virus in ligatured larvae of the greater wax moth, *Galleria mellonella, J. Invertebr. Pathol.,* **15,** 60 (1970).

130. M. Kobayashi and S. Yamaguchi, Effects of inokosterone and ecdysterone on the susceptibility of the silkworm, *Bombyx mori* L., to a nuclear-polyhedrosis virus, *J. Seric. Sci. Japan,* **39,** 33 (1970).

131. M.I. El-Ibrashy and M. Sadek, Hormonal control of larval sensitivity to a nuclear-polyhedrosis virus in noctuids, *Appl. Entomol. Zool.,* **8,** 44 (1973).

132. Y. Ben-Sshaked and I. Harpaz, Protection of susceptible insect host against a nuclear-polyhedrosis virus by ether extracts from insect larvae, *J. Invertebr. Pathol.,* **8,** 283 (1966).

133. J. Fargues, Sensibilité des larves de *Leptinotarsa decemlineata* Say à *Beauveria bassiana* (Bals.) Vuill. (Fungi imperfecti) en présence de doses réduites d'insecticides, *Ann. Zool. Ecol. Anim.,* **6,** 231 (1973).

134. P. Grison and R. S. de Sacy, L'élevage de *Pieris brassicae* L. pour les essais de traitements microbiologiques, *Ann. Inst. Natl. Rech. Agron., Sér. C. (Ann. Epiphyt.),* **7,** 661 (1956).

135. W. L. Sippell, Winter rearing of the forest tent caterpillar, *Malacosoma disstria* Hbn., *Can. Dept. Agric. For. Biol. Div. Bi-Monthly Prog. Rep.,* **8,** 1 (1952).

136. C. Vago, Facteurs alimentaires at activation des viroses latentes chez les insectes, *Sixth Int. Microbiol.,* **5,** 556 (1953).

137. T. Ebihara, Effect of mulberry leaf quality on the resistance of silkworm to a cytoplasmic-polyhedrosis virus, *Rep. Ibaraki Seric. Exp. Stn.,* **1,** 61 (1966).

138. F. Matsubara and K. Hayashiya, The susceptibility to the infection with nuclear-polyhedrosis virus in the silkworm reared on artificial diet, *J. Seric. Sci. Japan,* **38,** 43 (1969).

139. H. Watanabe and S. Imanishi, The effect of the content of certain ingredients in an artificial diet on the susceptibility to virus infection in the silkworm, *Bombyx mori, J. Seric. Sci. Japan,* **49,** 404 (1980).

140. W. A. L. David, S. Ellaby, and G. Taylor, The effect of reducing the content of certain ingredients in a semisynthetic diet on the incidence of granulosis virus disease in *Pieris brassicae, J. Invertebr. Pathol.,* **20,** 332 (1972).

141. O. I. Shvetsova, The polyhedrosis disease of the greater wax moth *(Galleria mellonella* L.) and the role of the nutritional factor in virus diseases of insects, *Mikrobiologiia,* **19,** 532 (1950).

142. D. Pimentel and M. Shapiro, The influence of environment on a virus-host relationship, *J. Insect Pathol.,* **4,** 77 (1962).

143. R. Rose and J. D. Briggs, Resistance to American foulbrood in honey bees. IX. Effects of honey bee larval food on the growth and viability of *Bacillus larvae, J. Invertebr. Pathol.,* **13,** 74 (1969).

144. T. E. Rinderer and W. C. Rothenbuhler, Resistance to American foulbrood in honey bees. X. Comparative mortality of queen, worker and drone larvae, *J. Invertebr. Pathol.,* **13,** 81 (1969).

145. T. E. Rinderer, W. C. Rothenbuhler, and T. A. Gochnauer, The influence of pollen on the susceptibility of honey bee larvae to *Bacillus larvae, J. Invertebr. Pathol.,* **23,** 347 (1974).

146. V. P. Pristavko and N. V. Dovzhenok, Ascorbic acid influence on larval blood cell number and susceptibility to bacterial and fungal infection in the codling moth, *Laspeyresia pomonella, J. Invertebr. Pathol.,* **24,** 165 (1974).

147. E. Armstrong, The effects of vitamin deficiencies on the growth and mortality of *Tribolium castaneum* infected with *Nosema whitei, J. Invertebr. Pathol.,* **31,** 303 (1978).

148. B. Hurpin, Etude de diverses souches de maladie laiteuse sur les larves de *Melolontha melolontha* L. et sur celles de quelques éspèces voisines, *Entomophaga,* **4,** 233 (1959).

149. T. D. Canerday and F. S. Arant, Pathogen-host-environment relationships in a cabbage-looper nuclear polyhedrosis, *J. Invertebr. Pathol.,* **12,** 344 (1968).

150. J. B. Carter, The effect of temperature upon *Tipula* iridescent virus in *Tipula oleracea, J. Invertebr. Pathol.,* **25,** 115 (1975).

151. H. Watanabe, Temperature effects on the manifestation of susceptibility to peroral infection with cytoplasmic polyhedrosis in the silkworm, *Bombyx mori* L., *J. Seric. Sci. Japan,* **33,** 286 (1964).

152. H. Aruga, N. Yoshitake, and M. Owada, Factors affecting the infection *per os* in the cytoplasmic polyhedrosis of the silkworm, *Bombyx mori* L., *J. Seric. Sci. Japan,* **32,** 41 (1963).

153. Y. Abe and C. Ayuzawa, Susceptibility of larval tissues in the silkworm, *Bombyx mori* L., to the nuclear polyhedrosis virus. I. Changes of the susceptibility to virus in the midgut epithelial cells of the silkworm larvae treated with low temperature, *J. Seric. Sci. Japan,* **43,** 200 (1974).

154. H. Inoue, Effect of low temperature on the multiplication of a flacherie virus in the silkworm, *Bombyx mori, J. Seric. Sci. Japan,* **46,** 453 (1977).

155. Y. Tanada, Description and characteristics of a granulosis virus of the imported cabbage-worm, *Proc. Hawaii. Entomol. Soc.,* **15,** 235 (1953).

156. F. T. Bird, Virus diseases of sawflies, *Can. Entomol.,* **87,** 124 (1955).

157. C. G. Thompson, Thermal inhibition of certain polyhedrosis virus disease, *J. Invertebr. Pathol.,* **1,** 189 (1959).

158. Y. Tanada and G. Y. Chang, Resistance of the alfalfa caterpillar, *Colias eurytheme,* at high temperatures to a cytoplasmic-polyhedrosis virus and thermal inactivation point of the virus, *J. Invertebr. Pathol.,* **10,** 79 (1968).

159. Y. Tanada, Effect of high temperatures on the resistance of insects to infectious diseases, *J. Seric. Sci. Japan,* **36,** 333 (1967).

160. S. Miyajima and S. Kawase, Effect of a high temperature on the incidence of cytoplasmic polyhedrosis of the silkworm, *Bombyx mori* L., *J. Seric. Sci. Japan,* **37,** 390 (1968).

161. M. F. Day and E. H. Mercer, Properties of an iridescent virus from the beetle, *Sericesthis pruinosa, Aust. J. Biol. Sci.,* **17,** 892 (1964).

162. M. F. Day and M. L. Dudzinski, The effect of temperature on the development of *Sericesthis* iridescent virus, *Aust. J. Biol. Sci.,* **19,** 481 (1966).

163. Y. Tanada and A. M. Tanabe, Resistance of *Galleria mellonella* (Linnaeus) to the *Tipula* iridescent virus at high temperatures, *J. Invertebr. Pathol.,* **7,** 184 (1965).

164. D. K. Reed, Effects of temperature on virus-host relationships and on activity of the noninclusion virus of citrus red mites, *Panonychus citri, J. Invertebr. Pathol.,* **24,** 218 (1974).

165. C. M. Ignoffo, Effects of temperature on mortality of *Heliothis zea* larvae exposed to sublethal doses of a nuclear-polyhedrosis virus, *J. Invertebr. Pathol.,* **8,** 290 (1966).

166. H. R. Bullock, Therapeutic effect of high temperature on tobacco budworms to a cytoplasmic-polyhedrosis virus, *J. Invertebr. Pathol.,* **19,** 148 (1972).

167. H. Watanabe and Y. Tanada, Infection of a nuclear-polyhedrosis virus in armyworm, *Pseudaletia unipuncta* Haworth (Lepidoptera: Noctuidae), reared at a high temperature, *Appl. Entomol. Zool.,* **7,** 43 (1972).

168. H. Watanabe and S. Maeda, Multiplication of densonucleosis virus in the silkworm, *Bombyx mori* L., reared at high temperature, *Japan. J. Appl. Ent. Zool.,* **23,** 151 (1979).

169. A. K. A. Mohamed, P. P. Sikorowski, and J. V. Bell, Susceptibility of *Heliothis zea* larvae to *Nomuraea rileyi* at various temperatures, *J. Invertebr. Pathol.,* **30,** 414 (1977).

170. J. Weiser, Die Microsporidien als Parasiten der Insekten, *Monographien Z. Angew. Entomol.,* **17,** 149 (1961).

171. J. Weiser, Influence of environmental factors on protozoan diseases of insects, *Proc. 12th Int. Congr. Entomol.,* 726 (1963).

172. J. V. Maddox, The persistence of the microsporida in the environment, *Entomol. Soc. Am. Misc. Publ.,* **9,** 99 (1973).

173. B. Hurpin, La léthargie, nouvelle virose des larves de *Melolontha melolontha* L. (Col., Scarab.), *Entomophaga,* **12,** 311 (1967).

174. J. P. Kramer, An octosporeosis of the black blowfly, *Phormia regina:* effect of temperature on the longevity of diseased adults, *Texas Rep. Biol. Med.,* **26,** 199 (1968).

175. R. J. Milner, *Nosema whitei,* a microsporidian pathogen of some species of *Tribolium.* IV. The effect of temperature, humidity and larval age on pathogenicity for *T. castaneum, Entomophaga,* **18,** 305 (1973).

176. J. V. Maddox, Generation time of the microsporidian *Nosema necatrix* in larvae of the armyworm, *Pseudaletia unipuncta, J. Invertebr. Pathol.,* **11,** 90 (1968).

177. R. Ishihara, Growth of *Nosema bombycis* in primary cell cultures of mammalian and chicken embryos, *J. Invertebr. Pathol.,* **11,** 328 (1968).

178. J. P. Kramer, Observations on the seasonal incidence of microsporidiosis in European corn borer populations in Illinois, *Entomophaga,* **4,** 37 (1959).

179. W. A. Smirnoff, The effect of gamma radiation on the larvae and the nuclear-polyhedrosis virus of the eastern tent caterpillar, *Malacosoma americanum, J. Invertebr. Pathol.,* **9,** 264 (1967).

180. R. H. Jafri, A preliminary report on the susceptibility of irradiated *Bombyx mori* larvae to *Borrelinavirus bombycis* infection, *J. Sci. Res., Panjab Univ. Lahore Pakistan,* **1,** 83 (1966).

181. R. H. Jafri and A. A. Khan, A study of the development of a nuclear-polyhedrosis virus of *Heliothis* in *Galleria mellonella* larvae exposed to gamma rays, *J. Invertebr. Pathol.,* **14,** 104 (1969).

182. R. H. Jafri and M. B. Chaudhry, Development of *Tipula* iridescent virus (TIV) in *Galleria mellonella* larvae exposed to gamma radiation, *J. Invertebr. Pathol.,* **18,** 46 (1971).

183. R. H. Jafri, The susceptibility of irradiated larvae of *Galleria mellonella* to VND virus, *J. Invertebr. Pathol.,* **10,** 355 (1968).

184. R. H. Jafri, Influence of pathogens on the life span of irradiated insects, *J. Invertebr. Pathol.,* **7,** 66 (1965).

185. H. W. Rossmoore and E. A. Hoffman, The effect of gamma radiation on larval resistance to *Bacillus thuringiensis* infection, *J. Invertebr. Pathol.,* **17,** 282 (1971).

186. W. A. Smirnoff, Influence of light on the development of a nuclear-polyhedrosis of *Neodiprion swainei, J. Invertebr. Pathol.,* **9,** 269 (1967).

187. H. Watanabe and K. Takamiya, Susceptibility of the silkworm larvae, *Bombyx mori,* reared under different light conditions to polyhedrosis viruses, *J. Seric. Sci. Japan,* **45,** 403 (1976).

188. G. Benz, "Synergism of Micro-Organisms and Chemical Insecticides," in D. H. Burges and N. W. Hussey, Eds., *Microbial Control of Insects and Mites,* Academic Press, New York, 1971, p. 327.

189. D. K. Hunter, S. J. Collier, and D. F. Hoffman, Compatibility of malathion and the granulosis virus of the Indian meal moth, *J. Invertebr. Pathol.,* **25,** 389 (1975).

190. C. M. Ignoffo and E. L. Montoya, The effects of chemical insecticidal adjuvants of a *Heliothis* nuclear-polyhedrosis virus, *J. Invertebr. Pathol.,* **8,** 409 (1966).

191. H. Watanabe, Susceptibility to infection in the silkworm, *Bombyx mori*, applied topically with sublethal dosages of insecticides, *J. Seric. Sci. Japan*, **40**, 350 (1971).

192. P. Ferron, Augmentation de la sensibilité des larves de *Melolontha melolontha* L. à *Beauveria tenella* (Delacr.) Siemszko au moyen de quantités réduites de HCH, *Proc. 5th Int. Colloq. Insect Pathol., College Park, Maryland*, 1970, p. 66.

193. P. Ferron, Modification of the development of *Beauveria tenella* mycosis in *Melolontha melolontha* larvae, by means of reduced doses of organophosphorous insecticides, *Entomol. Exp. Appl.*, **14**, 457 (1971).

194. Y. Ishikawa and S. Miyajima, Interaction between infectious flacherie virus and some bacteria in the silkworm, *Bombyx mori* L., *J. Seric. Sci. Japan*, **37**, 471 (1968).

195. Z. Hostounsky and J. Weiser, Production of spores of *Nosema plodiae* Kellen *et* Lindegren in *Mamestra brassicae* L. after different infective dosage, I, *Vestn. Cesk. Spol. Zool.*, **36**, 97 (1972).

196. M. Tamashiro, The susceptibility of *Bracon*-paralyzed *Corcyra cephalonica* (Stainton) to *Bacillus thuringiensis* var. *thuringiensis* Berliner, *J. Invertebr. Pathol.*, **2**, 209 (1960).

197. C. C. Beagle and E. R. Oatman, Differential susceptibility of parasitized and nonparasitized larvae of *Trichoplusia ni* to a nuclear polyhedrosis virus, *J. Invertebr. Pathol.*, **24**, 188 (1974).

198. M. J. Stelzer, Susceptibility of the great basin tent caterpillar, *Malacosoma fragile* (Stretch), to a nuclear-polyhedrosis virus and *Bacillus thuringiensis* Berliner, *J. Invertebr. Pathol.*, **7**, 122 (1965).

199. Y. Tanada, Synergism between two viruses of the armyworm, *Pseudaletia unipuncta* (Haworth) (Lepidoptera, Noctuidae), *J. Insect Pathol.*, **1**, 215 (1959).

200. Y. Tanada and T. Hukuhara, A non-synergistic strain of a granulosis virus of the armyworm, *Pseudaletia unipuncta*, *J. Invertebr. Pathol.*, **12**, 263 (1968).

201. T. Yamamoto and Y. Tanada, Phospholipid, an enhancing component in the synergistic factor of a granulosis virus of the armyworm, *Pseudaletia unipuncta*, *J. Invertebr. Pathol.*, **31**, 48 (1978).

202. H. Watanabe and T. Shimizu, Epizootiological studies on the occurrence of densonucleosis in the silkworm, *Bombyx mori*, reared at sericultural farms, *J. Seric. Sci. Japan*, **49**, 485 (1980).

203. H. Aruga, T. Hukuhara, N. Yoshitake, and A. Israngkul, Interference and latent infection in the cytoplasmic polyhedrosis of the silkworm, *Bombyx mori* (Linnaeus), *J. Insect Pathol.*, **3**, 81 (1961).

204. K. S. Ritter and Y. Tanada, Interference between two nuclear polyhedrosis viruses of the armyworm, *Pseudaletia unipuncta* (Lep.: Noctuidae), *Entomophaga*, **23**, 349 (1978).

205. R. E. Lowe and J. D. Paschke, Simultaneous infection with the nucleopolyhedrosis and granulosis viruses of *Trichoplusia ni*, *J. Invertebr. Pathol.*, **12**, 86 (1968).

206. G. L. Nordin and J. V. Maddox, Effects of simultaneous virus and microsporidian infections on larvae of *Hyphantria cunea*, *J. Invertebr. Pathol.*, **20**, 66 (1972).

207. P. Ferron and B. Hurpin, Effects de la contamination simultanée ou successive par *Beauveria tenella* et par *Entomopoxvirus melolontha*, *Ann. Soc. Entomol. Fr.*, **10**, 771 (1974).

208. P. Ferron, B. Hurpin, and P. H. Robert, Sensibillisation des larves de *Melolontha melolontha* L. à la mycosa à *Beauveria tenella* par une infection préalable à *Bacillus popilliae*, *Entomophaga*, **14**, 429 (1969).

209. E. S. Sharpe and R. W. Detroy, Susceptibility of Japanese beetle larvae to *Bacillus thuringiensis:* associated effects of diapause, midgut pH, and milky disease, *J. Invertebr. Pathol.*, **34**, 90 (1979).

210. W. Fyg, Anomalies and diseases of the queen honey bee, *Annu. Rev. Entomol.*, **9**, 207 (1964).

211. E. A. Steinhaus, Stress as a factor in insect disease, *Proc. Int. Congr. Entomol. 10th Congr. Montreal 1956*, 4 (1958).

212. E. C. Clark and C. G. Thompson, The possible use of microorganisms in the control of the great basin tent caterpillar, *J. Econ. Entomol.*, **47**, 268 (1954).

213. Y. Tanada, The epizootiology of virus diseases in field populations of the armyworm, *Pseudaletia unipuncta* (Haworth), *J. Insect Pathol.,* **3,** 310 (1961).

214. Y. Tanada, Epizootics of virus disease of insects, *1st Intersection. Congr. Int. Assoc. Microbial. Soc., Tokyo, Japan, 1974,* Vol. 2, 1975, p. 621.

215. F. T. Bird, Transmission of some insect viruses with particular reference to ovarial transmission and its importance in the development of epizootics, *J. Insect Pathol.,* **3,** 352 (1961).

216. R. P. Jaques, Stress and nuclear polyhedrosis in crowded populations of *Trichoplusia ni* (Hübner), *J. Insect Pathol.,* **4,** 1 (1962).

217. C. C. Doane, Primary pathogens and their role in the development of an epizootic in the gypsy moth, *J. Invertebr. Pathol.,* **15,** 21 (1970).

218. B. Zelazny, *Oryctes rhinoceros* populations and behavior influenced by a baculovirus, *J. Invertebr. Pathol.,* **29,** 210 (1977).

219. J. B. Carter, E. I. Green, and A. J. Kirkham, A. *Tipula paludosa* population with a high incidence of two pathogens, *J. Invertebr. Pathol.,* **42,** 312 (1983).

220. F. T. Bird, J. C. Cunningham, and C. M. Howse, The possible use of viruses in the control of spruce budworm, *Proc. Entomol. Soc. Ontario,* **103,** 69 (1972).

221. J. E. Gilmore and F. Munger, Influence of population density on the incidence of a noninclusion virus disease of the citrus red mite in the laboratory, *J. Invertebr. Pathol.,* **7,** 156 (1965).

222. J. G. Shaw, H. Tashiro, and E. J. Dietrick, Infection of the citrus red mite with virus in central and southern California, *J. Econ. Entomol.,* **61,** 1492 (1968).

223. C. C. Doane, Bacterial pathogens of *Scolytus multistriatus* Marsham as related to crowding, *J. Insect Pathol.,* **2,** 24 (1960).

224. J. Weiser, "Sporozoan Infections," E. A. Steinhaus, Ed., *Insect Pathology: An Advanced Treatise,* Vol. 2, Academic Press, New York, 1963, p. 291.

225. W. A. Smirnoff, Observations on the effect of virus infection on insect behavior, *J. Invertebr. Pathol.,* **7,** 387 (1965).

226. A. M. Hüger, Grundlagen zur biologischen Bekämpfung des Indischen Nashornkafers *Oryctes rhinoceros* (L.) mit *Rhabdinovirus oryctes:* Histopathologie der Virose bei Käfern, *Z. Angew. Entomol.,* **72,** 309 (1973).

227. A. T. Speare, *Massospora cicadina* Peck, a fungus parasite of the periodical cicada, *Mycologia,* **13,** 72 (1921).

228. A. G. Dustan, The natural control of the green apple bug *(Lygus communis* var. *novascotiensis* Knight) by a new species of *Empusa, Proc. Quebec Soc. Prot. Plants 1922–1923,* 1924, p. 3.

229. F. T. Bird, The use of virus diseases against sawflies, *Rept. Commonwealth Entomol. Conf. London, July, 1954, 6th,* 1954, p. 122.

230. K. R. Kramm, D. F. West, and P. G. Rockenbach, Termite pathogens: transfer of the entomopathogen *Metarhizium anisopliae* between *Reticulitermes* sp. termites, *J. Invertebr. Pathol.,* **40,** 1 (1982).

231. J. B. Carter, The mode of transmission of *Tipula* iridescent virus. II. Route of infection, *J. Invertebr. Pathol.,* **21,** 136 (1973).

232. H. D. Burges and J. A. Hurst, Ecology of *Bacillus thuringiensis* in storage moths, *J. Invertebr. Pathol.,* **30,** 131 (1977).

233. W. R. Kellen and J. E. Lindegren, Biology of *Nosema plodiae* sp. n., a microsporidian pathogen of the Indian meal moth, *Plodia interpunctella* (Hübner), (Lepidoptera: Phycitidae), *J. Invertebr. Pathol.,* **11,** 104 (1968).

234. A. W. Woodrow and E. C. Holst, The mechanism of colony resistance to American foulbrood, *J. Econ. Entomol.,* **35,** 327 (1942).

235. W. C. Rothenbuhler, Genetics and breeding of the honey bee, *Annu. Rev. Entomol.,* **3,** 161 (1958).

236. W. C. Rothenbuhler, Behavior genetics of nest cleaning in honey bees. IV. Responses of F_1 and backcross generations to disease-killed brood, *Am. Zool.*, **1**, 111 (1964).

237. W. A. Smirnoff, A virus disease of *Neodiprion swainei* Middleton, *J. Insect Pathol.*, **3**, 29 (1961).

238. C. M. Ignoffo, Specificity of insect viruses, *Bull. Entomol. Soc. Am.*, **14**, 265 (1968).

239. J. C. Cunningham, Serological and morphological identification of some nuclear-polyhedrosis and granulosis viruses, *J. Invertebr. Pathol.*, **11**, 132 (1968).

240. Y. Tanada and E. M. Omi, Persistence of insect viruses in field populations of alfalfa insects, *J. Invertebr. Pathol.*, **23**, 360 (1974).

241. J. Weiser and J. Veber, Moznosti biolgického boje s prastevnickem americkym (*Hyphantria cunea* Drury)-II, *Cesk. Parasitol.*, **2**, 191 (1955).

242. I. M. Hall, Studies of microorganism pathogenic to sod webworm, *Hilgardia*, **22**, 535 (1954).

243. W. M. Brooks, Protozoa: host-parasite-pathogen interrelationships, *Entomol. Soc. Am. Misc. Publ.*, **9**, 105 (1973).

244. J. N. McNeil and W. M. Brooks, Interactions of the hyperparasitoids *Catolaccus aenooviridis* (Hym: Pteromalidae) and *Spilochalis side* (Hym.: Chalcididae) with the microsporidians, *Nosema heliothidis* and *N. campoletidis, Entomophaga*, **19**, 195 (1974).

245. G. R. Stairs, Transmission of virus in tent caterpillar populations, *Can. Entomol.*, **98**, 1100 (1966).

246. G. R. Stairs, "Use of Viruses for Microbial Control of Insects," in H. D. Burges and N. W. Hussey, Eds., *Microbial Control of Insects and Mites,* London, Academic Press, 1971, p. 97.

247. A. Paillot, "L'infection chez les insectes," G. Patissier, Trévoux (1933).

248. J. J. Lipa, Studia inwazjologiczne i epizootiologiczne nad kilkoma gatunkami pier wotniakow z rzedu Microsporidia pasozytujacymi w owadach, *Pr. Nauk. Inst. Ochr. Rosl.*, **5**, 103 (1963).

249. K. Gosswald, Zur Biologie und Okologie von *Parasitigena segragata* Rond. und *Sarcophaga schutzei* Kram. (Dipt.) nebst Bemerkungen über die forstliche Bedeutung der beiden Arten, *Z. Angew. Entomol.*, **21**, 1 (1934).

250. F. M. Laigo and M. Tamashiro, Interaction between a microsporidian pathogen of the lawn-armyworm and the hymenopterous parasite *Apanteles marginiventris, J. Invertebr. Pathol.*, **9**, 546 (1967).

251. T. G. Andreadis, *Nosema pyrausta* infection in *Macrocentrus grandi*, a branconid parasite of the European corn borer, *Ostrinia nubilalis, J. Invertebr. Pathol.*, **35**, 229 (1980).

252. H. G. Schabel, Phoretic mites as carriers of entomopathogenic fungi, *J. Invertebr. Pathol.*, **39**, 410 (1982).

253. D. L. Hostetter and K. D. Biever, The recovery of virulent nuclear-polyhedrosis virus of the cabbage looper, *Trichoplusia ni*, from the feces of birds, *J. Invertebr. Pathol.*, **15**, 173 (1970).

254. J. Weiser, Moznosti biologického boje s prastevnickem americkym (*Hyphantria cunea* Drury)-III, *Cesk. Parasitol.*, **4**, 359 (1957).

255. W. R. Kellen and J. E. Lindegren, Modes of transmission of *Nosema plodiae* Kellen and Lindegren, a pathogen of *Plodia interpunctella* (Hübner), *J. Stored Prod. Res.*, **7**, 31 (1971).

256. S. Günther, Über die Auswirkung auf die infektiosität bei der Passage insekten Microspoidien durch den Darm von Vögeln und Insekten, *Nachrichtenbl. Dtsch. Pflanzenschutzdienst (Berlin), (N. F.)*, **13**, 19 (1959).

257. J. Weiser, Protozoäre Infektionen im Kampfe gegen Insekten, *Z. Pflanzenkrankr, (Pflanzenpathol.) Pflanzenschutz.*, **63**, 625 (1956).

258. W. H. McGaughey, Insect resistance to the biological insecticide *Bacillus thuringiensis, Science*, **229**, 193 (1985).

5

THE PATHOGEN POPULATION

YOSHINORI TANADA

Department of Entomological Sciences
University of California
Berkeley, California

JAMES R. FUXA

Department of Entomology
Louisiana State University Agricultural Center
Louisiana State University
Baton Rouge, Louisiana

1. INTRODUCTION

An epizootic develops when a pathogen population produces disease in an insect host population, the members of which may or may not die from the infection. The intensity and extent of the interaction between the two populations determine the state of the epizootic, that is, an epizootic or an enzootic. The pathogen-host interaction is influenced by the properties of the pathogen and host populations, the methods of transmission (see Ch. 6), and environmental factors (see Ch. 7). In general, the development of an epizootic by the pathogen population is host-density dependent; the epizootic is more apt to develop as the host population increases (see Section 3.1). There are situations, however, when the epizootic develops at a low host density depending on the spatial distribution of the pathogen (see Section 5). Accordingly, pathogens have been considered to be imperfectly density-dependent mortality factors (1).

    In this chapter, we shall be concerned mainly with the role of the pathogen population in epizootics.

2. PATHOGEN PROPERTIES OF SIGNIFICANCE IN EPIZOOTICS

To understand and evaluate the role of pathogens in epizootics, a knowledge of the individual pathogen properties is required, and the role that such properties play in the pathogen population may be the basis for the development of an epizootic. The properties of the individual pathogen (e.g., virulence and infectivity) in association with the properties of the host (e.g., physical properties, growth stage, and resistance) and methods of transmission are responsible for the initiation, manifestation, and development of the disease (2). Pathogen properties vary with different types of pathogens—viruses, bacteria, fungi, protozoa, and nematodes—and we shall refer to some specific examples of various pathogens to illustrate the principles of pathogen involvement in epizootics. Readers should refer to chapters dealing with the various pathogens for other examples.

2.1. Latency

There is ample evidence for inapparent infections occurring in insect populations, but evidence for latent infections is still incomplete. Latency is closely associated with environmental stress factors discussed in Chapter 7. We wish to point out, however, that the development of a frank infection after an insect is exposed to a stressor is not proof that the disease is latent or that the pathogen is in an occult state; the pathogen may be causing an inapparent chronic infection. The proof depends on whether the pathogen in the host (cell or tissue) is in a noninfective and nonreplicative state and whether it is transformed to an infective and replicative state when the insect is stressed.

Latency may occur in most, if not all, groups of entomopathogens. Viruses have received most attention but other pathogens may have occult forms that remain inactive until the host is exposed to an incitant or stressor. Latency in bacterial infections has been reported but only rarely (see 1, 2). Certain microsporidia cease to develop when their insect hosts undergo hibernation, and Weiser (3) has suggested that the microsporidioses are in a latent state during this period. Andreadis (4) has detected sporadic horizontal transmission of the microporidium *Amblyospora* during early larval stages in which occult infection develops in epithelial cells of the gastric caeca.

Before 1970, there were numerous reports, based primarily on disease outbreaks resulting from stressors or incitants, that occult viruses were common in insects (see 5, 1, 6, 2). However, the actual demonstration of the presence of an occult virus has been lacking in nearly all cases. Among the most often cited case is the Sigma virus of *Drosophila* spp. which is transmitted maternally and causes a lethal sensitivity in flies exposed to carbon dioxide (see 7, 8). The occult virus is detectable by superinfection. Grace (9) reported, in apparently uninfected cell lines, the appearance of a nuclear polyhedrosis virus (NPV) and concluded that the "physiological shock," brought about by an abrupt change from one medium to another, activated the occult virus.

Workers had suggested that an occult virus was activated when an insect was fed a foreign virus or during virus cross-transmission. For example, Longworth and Cunningham (10) failed to cross-transmit NPVs of *Aglais urticae* and *Lymantria dispar (Porthetria dispar)*, but activated occult viruses in both species. The question of latency has been studied also with virus cross-transmission in insect tissue culture and with the analysis of viral genomes (11, 12). By identifying the viral genomes, McKinley et al. (12) demonstrated that an insect species fed a NPV from a different host species became infected instead with its own virus.

Many workers have attempted to demonstrate latency with temperature stress. For example, Himeno et al. (13) inoculated low doses of NPV (10^3 particles) into fifth instar silkworm larvae which remained uninfected at 25°C, but which became infected when the larvae were placed at 5°C for 24 hours. Cold treatment before virus inoculation did not induce polyhedrosis. They concluded that the virus combined with biochemical substances to become inactive and only the low temperature treatment freed the virus from the complex and changed it to an active state.

Kislev et al. (14) isolated and characterized the host and viral DNAs from healthy and virus-infected larvae of *Spodoptera littoralis*. Since the viral DNA was detected also in noninoculated larvae, they concluded that a latent virus infection was present.

2.2. Biological Properties

Major biological properties of a pathogen involved in causing diseases are method of infection, virulence, pathogenicity (infectivity), and replication. Infection is closely associated with transmission, the mode or pathway of invasion (see Ch. 6). Infection may occur through the mouth and digestive tract, integument, trachea, reproductive system, and through the action of vectors.

2.2.1. INFECTIVITY, VIRULENCE, PATHOGENICITY, AND REPLICATION. Infectivity or the ability to cause infection varies with different insect pathogens. A number of factors are involved in infectivity, and the mode of invasion is one of them. A pathogen with more than one method of invasion is more apt to cause epizootics than one with a single method of invasion. Examples of pathogens with multiple invasion routes include *Metarhizium anisopliae,* through the integument, mouth, and anal spiracle of the mosquito (see 357) or through the larval integument and egg chorion (358); and microsporidian species through the egg, larval midgut, and by vectors (see 3, 15).

Virulence is the disease-producing power of the pathogen, that is, the ability to invade and injure the host's tissues. It is often measured by the response of the host to a known pathogen inoculum; for example, death from a median lethal concentration (LC_{50}), median lethal dose (LD_{50}), or after a median specified time (LT_{50}).

Virulence has been used interchangeably with pathogenicity. The latter refers usually to groups or species of pathogens, whereas virulence is used in the sense of degree of pathogenicity within a group or species. Both virulence and pathogenicity are often linked with the replication of the pathogen within the insect; a rapidly replicating pathogen is generally more virulent than one that replicates slowly in the host, though other factors, such as the vital importance of the infected tissues, toxins, and enhancing factors, also may play an important role. The effects of these biological properties vary with the number of pathogens invading a host. A single pathogen rarely causes infection, and a number below the lethal infectious dose may cause a chronic nonlethal infection. High virulence generally is selected for pathogens to be used in microbial control, but this may not be advantageous for long-term introductions.

The cultural method may affect the virulence of entomopathogens, especially facultative forms. This has been well documented with bacteria, e.g., *Bacillus thuringiensis* (see 16) and with fungi (see 17), e.g., *Verticillium lecanii* (18). Less is known with obligate entomopathogens, such as microsporidia and viruses, but the loss in virulence through passage in alternate hosts (cited below) may be associated with nutritional factors.

Pathogens have been manipulated in the laboratory to alter their virulence by (1) passage through more or less susceptible hosts, (2) dissociation in culture into strains of different virulence, (3) introduction of pathogens together with substances (mucin, boric acid, etc.) that may aid in increasing their invasive powers, and (4) mutualistic or antagonistic association with other microorganisms (see 19). These methods have been applied especially with bacteria and fungi (see 17). Less is known regarding the protozoa (see 20), viruses (see 21, 22), and nematodes. Examples of the enhancement of microsporidia through host passage are *Nosema* sp. in *Anopheles quadrimaculatus* (23) and *Thelohania pristiphora* in sawflies (24) and in tent caterpillars (25). Loss in virulence, however, has resulted when microsporidia are propagated in alternate hosts (26, see 27).

The virulence of NPVs has been enhanced through host passages in sawflies (28) and in *Trichoplusia ni* (29). Loss in the virulence of NPV has been reported after prolonged passage through insect cell cultures (30–32). Whether alterations in pathogen virulence occur during an epizootic has received little, if any, attention (see Section 2.3.2).

2.2.2. TOXIN PRODUCTION.   Insect pathogens produce toxins that enhance their pathogenicity. These toxins have been identified in fungi and bacteria (see 33–36), but little is known about whether protozoa and viruses produce toxins or factors that enhance pathogenicity.

There are several methods of classifying microbial toxins (see 33), but we shall classify them as exotoxins (ectotoxins) and endotoxins, a common, but not necessarily the best method. Many exotoxins are enzymes, and the most thoroughly studied are proteinases and lecithinases of bacteria (see 34, 35). Toxins that are produced after the invasion and multiplication by bacteria in the hemocoel result in septicemia. Both types of toxins may be produced by a single entomopathogen; the most notable example is *Bacillus thuringiensis*, which produces four or more toxins, one of which is the delta endotoxin derived from the parasporal body (see 37, 38). The milky disease bacterium *Bacillus popilliae* causes septicemia, but the prolonged chronic infection and no detectable toxin suggests a bacteremia. The parasporal body of this bacillus is generally considered to be nontoxic, however, Weiner (39) has shown it to be toxic when solubilized and inoculated into the hemocoel. Some bacteria, which do not invade the hemocoel, produce toxins in the lumen of the digestive tract and cause toxemia; for example, *Clostridium* spp. (brachytosis) in tent caterpillars (40), *Streptococcus faecalis* in gypsy moth (41), *Melissococcus pluton* (= *Streptococcus pluton*) (European foulbrood) in honeybee (see 42), and *Bacillus sphaericus* in mosquito (see 43). Other examples are in Bucher (44). The nature of the exotoxins formed in the gut lumen has not been established in the above examples except partially for *B. sphaericus* (see 43).

Exotoxins produced by fungi are involved in invasion through the integument and in the destruction of insect internal tissues. These toxins are mainly enzymes such as chitinase, lipase, and proteinase for integumental invasion, and depsidases, proteinases, and lipases for the digestion of intrahemocoelic tissues (see 36, 45). *Aspergillus flavus* and related species produce aflatoxins that are toxic to insects and also are mutagenic for mammals (see 46).

Few endotoxins have been reported from entomopathogens. The crystalline parasporal body of *Bacillus thuringiensis* is a protoxin which when dissolved in the insect's midgut forms the delta endotoxin, the most thoroughly studied of the endotoxins of entomogenous bacteria (see 47, 48). *Bacillus thuringiensis* has a number of H-serotypes or subspecies (49) which produce in the sporangium one or more crystalline parasporal bodies that vary in activity with different insects. The delta endotoxin alone can kill susceptible insects, but in certain cases death results from the replication of the bacteria in the hemocoel and the associated septicemia (see 50). Thus in *B. thuringiensis,* larval death results from toxemia and/or septicemia.

Bacillus sphaericus produces endotoxins, associated with vegetative cell walls, spores, and proteinaceous parasporal crystals that are toxic to mosquito larvae (see 51, 43, 52).

Although both *B. thuringiensis* and *B. sphaericus* replicate in the host, *B. sphaericus* sporulates readily whereas *B. thuringiensis* often does not sporulate in infected and dead hosts, depending on the bacillus subspecies and host susceptibility. Accordingly, *B. sphaericus* is known to persist in the host environment and causes epizootics, whereas *B. thuringiensis* rarely causes naturally occurring epizootics (see Section 3.2).

2.2.3. ENHANCING FACTORS.   Infectivity can be enhanced by factors produced by pathogens and by factors present in the environment. Not much is known of enhancing factors produced by microorganisms. The toxins described in Section 2.2.2 play a role in enhancing infectivity and pathogenicity. Nontoxic enhancing substances are produced in certain infections. In the granuloses of the armyworm *Pseudaletia unipuncta* (53) and of the clover cutworm *Scotogramma trifolii* (54), the viruses have synergistic factors (SF) that are present in the matrix of the virus occlusion body (capsule). The SF enhances baculovirus infections, in vivo and in vitro, through the adhesion of the viral envelope to the cell plasma membrane (see 55).

According to Smirnoff (56), chitinase enhances the infection of *Bacillus thuringiensis* in larvae of the spruce budworm *Choristoneura fumiferana* but Fast (57) has not been able to confirm such enhancement. In gypsy moth larvae, the chitinase is active provided the midgut pH is near the optimum pH 5 of the chitinolytic bacteria (58).

Physical factors, such as temperature, humidity, and pH, affect the biological properties of the pathogens and may thereby influence infectivity. Certain chemical substances have direct enhancing effect, for example, mucin (59), boric acid (60–62), adjuvants, and chemical insecticides combined with pathogens to enhance infectivity (see Section 7).

An important biological factor is host plant nutrition which is closely associated with host resistance (see Ch. 4). Insects reared on certain plants may vary in their susceptibility to pathogens. Some plants produce substances that adversely affect the pathogens (63) (see Section 3.2.2).

2.3. Genetics and Strains of Pathogens

Pathogen genetics and strains have rarely, if ever, been included in epizootiological studies, so it will be necessary to speculate about their effects on epizootics based on laboratory data. All entomopathogens, including viruses, have members (strains or varieties) which differ in infectivity and pathogenicity. We shall not discuss the basis and principles involved in separating the classification of entomopathogens, for example, species, subspecies, varieties, strains, etc.

2.3.1. GENETICS.   Classical genetics of entomopathogens is little known and up to this point has been limited to basic descriptions of genotypes and phenotypes. This is a necessary, basic beginning and will undoubtedly lead to studies of crosses, interactions with hosts, population genetics, and epizootiology. Descriptions of phenotypes have included *Bacillus thuringiensis* mutants (see 64) based on sporulation and crystals, different NPVs based on shapes of polyhedra or single versus multiple virions in envelopes (65), and morphologically distinct groups of fungi (66). Other authors have carried the concept of phenotype of the same or similar pathogens down to more minor differences (see 67–70).

Descriptions of entomopathogen genotypes have been largely in the realm of molecular genetics. One exception has been in chromosome analyses of certain microporidia (see 71). Molecular genetic studies have been directed mostly toward viruses and *B. thuringiensis* and already have had important epizootiological implications (72). Molecular genetics of *B. thuringiensis* has revealed that the delta endotoxin is a product of a single plasmid or chromosomal gene, which is unusual for pathogenic bacteria (see 73). Molecular genetics of viruses have revealed that heterogeneity of viral isolates is common whether the virus source is one insect or insects from different geographical areas. Geographical isolates distinguishable at the genome level are NPVs of *Spodoptera frugiperda* (74), *Heliothis* spp. (75), and *Lymantria dispar* (76). Variation has been demonstrated within NPV isolates from *S. frugiperda, L. dispar,* and *Autographa californica* (65, 76, 77), and one of these, the *A. californica* NPV, is known to have both heterogenous and homogenous genotypes within different isolates (78). The genomes of several entomogenous viruses have been partially mapped (e.g., 79, 80, 81, 82, 77), which may lead to a better understanding of several factors important to epizootiology, such as virulence (see 93) and susceptibility to environmental factors. Finally, viruses from different host species have been compared for DNA homology (84–86), an area of study of great potential importance to epizootiology due to implications in cross-infectivity and coevolution.

Several types of genetic changes are known, again mostly in *B. thuringiensis* and NPVs plus certain fungi. Sporeless mutants have been observed in *B. thuringiensis* (87), mutants resistant to antibiotics in *B. thuringiensis* and *Bacillus popilliae* (88), and mutations of various cell surface components in *Pseudomonas aeruginosa* (89). In fungi, mutants with increased virulence are known in *Beauveria bassiana* (90), *B. brongniartii* (91), and *Metarhizium anisopliae* (92), and increased sporulation and toxin production as well as faster germination were observed in the *M. anisopliae* mutant (92). Mutants with increased virulence or temperature sensitivity occur in the *A. californica* NPV (93, 359). There is a good possibility that RNA entomopathogenic viruses will have particularly high rates of mutation (83). Reichelderfer (94) discussed entomopathogenic viral mutants and inferences about them based on observations of host range and mixed infections. Genetic engineering (e.g., transduction, transformation, translocation, and recombination) has resulted in new forms of bacteria, in particular *B. thuringiensis* (64, see 95, 361), NPVs (96, 97), and the fungus *Paecilomyces fumoso-roseus* (98, 199).

Unfortunately, little is known about genetic changes in entomopathogens outside of the laboratory except that which can be inferred from observed differences in isolates (see Section 2.3.2). The laboratory studies, however, leave little doubt that such changes occur, and the nature of those changes, particularly in virulence, indicates that they can greatly influence epizootiology and microbial control. For example, the homologous DNA sequences among subgroups A, B, and C baculoviruses have led Smith and Summers (100) to speculate that recombination may be important to genetic exchange between closely related viruses but not distantly related ones.

2.3.2. STRAINS OF ENTOMOPATHOGENS.   We shall restrict ourselves to the use of "strain" since other terms, such as "genotypic variants," "biotypes," "pathotypes," and "varieties" have evolutionary or other implications that could confuse their usage for our purposes (see 101, 102). Strains that differ in some property or other have commonly been isolated within an entomopathogen species. There are even a few examples of different strains being tested in an insect's habitat for microbial control; such examples permit conclusions relating to epizootiology. We shall briefly discuss differences in entomopathogen strains and the implications for epizootiology.

At this point we mention some morphological and biochemical ways of recognizing or differentiating strains before considering other biological differences perhaps more important to epizootiology. Differentiating strains depends on the type of microorganism. The nematodes lend themselves well to morphological differences and measurement of body parts (see 103). Protein analysis and serology of spores have the most promise for the microsporidia (see 71). Strains of entomopathogenic fungi have been identified by enzyme analysis (104) and by pyrolysis-gas chromatography (105). There is a variety of methods for entomopathogenic bacteria, including biochemical analysis (see 44) and bacteriophage typing (360); however, serotyping has been most generally useful (see 49, 106), with occasional exception due to culture difficulty (see 107). Restriction endonuclease analysis of the genomes has been the best method to identify entomopathogenic virus strains (see 108), though again, several other methods are possible (see 108 – 110).

One of the most important differences among strains is in their pathogenicity. Though there is little evidence, it is intuitively obvious that pathogenicity is important to epizootics. For example, a strain pathogenic to three or four host species in a particular habitat is likely to produce greater numbers of infectious units and more frequent or severe epizootics than a strain pathogenic to only one of those hosts. Strain differences in pathogenicicty are known. Certain varieties of *Bacillus popilliae* and *B. thuringiensis* have host ranges different from other varieties (see 107, 111). One strain of *Lymantria dispar* NPV infects *Orgyia pseudotsugata* whereas another does not (see 112), though there is at least some doubt that both strains originated in *L. dispar* (see 76). There are no clear examples in fungi. In one case, the virulence of some strains of *Nomuraea rileyi* was so low to certain hosts compared to that of other strains that they were virtually nonpathogenic (113), though, in the strict sense, this was a difference in virulence rather than pathogenicity. In another, different pathogenicities as well as other characteristics led researchers to conclude that two strains (pathotypes) of *Entomophthora grylli* should not be in the same species (104). Similarly, two strains of the nematode *Reesimermis nielseni (Romanomermis culicivorax)* differed in pathogenicity, but the author felt that they might be different species (114).

The best studied differences among strains are in their virulence. As with pathogenicity, there is little doubt that virulence affects epizootics. It is widely believed that the use of more virulent strains enhances epizootics over the short term. The comparative virulence in a host spectrum also affects epizootics; strains of the fungus *Nomuraea rileyi* produce earlier epizootics if they are virulent to an early-season, relatively inocuous host as well as later-season, pestiferous species of caterpillars (115). The long-term effect of releasing virulent strains is more questionable. It is conceivable that such releases will upset relationships that evolved over long time periods, with the host insect responding to such strains with increased resistance and less severe epizootics likely from the low pathogen density in the host environment (see Section 3.1).

There are many examples, beyond the scope of this chapter, of strain differences in virulence, many of them relating to toxin production, particularly in *B.*

thuringiensis. Geographical differences are to be expected because, presumably, the more closely the host and pathogen are found, the greater their exposure to one another and their opportunity for coevolution. Coevolution is usually expected to cause less virulence (see 116) but sometimes more virulence (see 117). *Nomuraea rileyi* strains from Brazil and Ecuador were more virulent in the USA to *Anticarsia gemmatalis* than strains from the USA to this insect (118, 113). On the other hand, Daoust and Roberts (119) found no correlation between virulence and geography in *Metarhizium anisopliae*. Similarly, with the NPVs, strains isolated at a distance from the host origin can be more (120, 121) or less virulent (122) than viruses isolated nearer to where the host was found. It seems likely that strains of an entomopathogen isolated at one location but at different times (i.e., years) would differ in virulence, but the authors are not aware of this having been demonstrated yet. Different strains of NPV also can differ in virulence to different host species (123). Differences in virulence among isolates according to hosts also are known in fungi (see 124) and in bacteria (see 125, 43).

Other differences in entomopathogenic strains demonstrated in the laboratory are likely to affect epizootiology, though perhaps not to the same degree as differences in pathogenicity and virulence. Strain differences in morphology or life cycle have been observed in bacteria, fungi, and NPVs. Asporogenous strains are known for *B. thuringiensis* but do not occur naturally (see 126). Similarly, different culture methods can produce strains of *Beauveria* spp. with different growth rates, amount of mycelium and sporulation, and color (127), and a spontaneous mutant strain of *Metarhizium anisopliae* had more dense sporulation and faster spore germination than the wild strain (92). Better examples in fungi are those that differ in types, color, and size of spores; for example, *Verticillium lecanii* (see 128), *Conidiobolus megalotocus* (see 129), and *M. anisopliae* (see 130). Similarly, the NPV of *Autographa californica* produces a strain with larger polyhedra containing more virions than normal after several passages through hosts (29), and natural isolates of a NPV are known with differently shaped occlusion bodies (131). Different strains of three baculoviruses differed in susceptibility to sunlight or ultraviolet radiation; in all three, the resistant strain was selected after a series of radiation treatments and was not a wild isolate (132–134). Nutritional requirements and biochemistry, including susceptibility to antibiotics, commonly differ among strains of entomopathogenic bacteria and fungi (135, 88, 69, 136), and a nutritional difference also is known in a nematode (see 103). Temperature affects spore germination differently in strains of the fungus *Ascosphaera aggregata* (137).

Again, all these factors have been limited to laboratory studies, so we can only speculate about their effects, if any, on epizootics. If such strain differences do exist in the field, they could influence the pathogen population density by affecting either the rate of growth or rate of mortality of the pathogens due to environmental factors. In turn, pathogen population density affects epizootics (see Section 3.1).

Field studies of entomopathogen strains are few and usually inadequate to be of any epizootiological significance. For example, geographical isolates of entomopathogens have been distinguished biochemically, but the results were not related to any biological parameter such as virulence (138, see 101; 74). However, microbial control research has produced evidence that different strains can affect epizootics. A *Bacillus sphaericus* strain had better virulence and residual activity than another for mosquito control (139). A strain of NPV of *Lymantria dispar* initially produced higher laboratory LC_{50}s and lower field mortality than another, but after one host generation both strains produced similar LC_{50}s and field mortality (see 140, 141).

There is ample laboratory evidence that the passage of pathogens through insect hosts either increases or decreases their virulence (see above and Section 2.2.1). However, such alterations in the field, particularly during an epizootic, have rarely been reported. The survey conducted by Dunbar and Beard (142) in Japanese beetle populations in Connecticut indicated that *Bacillus popilliae* became less effective in certain locations apparently because of the appearance of less virulent strains. Moreover, the spores from diseased grubs collected at different locations varied in their rates of infectivity.

2.3.3. PATHOGEN-HOST COEVOLUTION. In insect pathology, conclusions about evolution or coevolution of entomopathogens have not been common. Humber (143) discussed possible ancestral forms of Entomophthorales and outlined patterns of the fungal taxa associated with certain insect taxa. He hypothesized that the group evolved from a rapid killing of the host to slower parasitism. Much has been written about the evolution of viruses (e.g., see 144), though usually without treating the insect viruses separately. Insects may be the natural hosts of many arthropod-borne animal viruses (see 144). May (117) has proposed that the NPVs in insects are an example of a pathogen that becomes more virulent the longer it is associated with the host rather than less virulent or harmful. Information is accumulating about relationships among these viruses (see Ch. 10). Authors have suggested that *Spodoptera* baculoviruses speciated along with their hosts (see 108); that baculoviruses of *Orgyia pseudotsugata* have separate origins (108); that certain NPVs from different host genera are closely related (100); and that there is evidence both for and against a common ancestor for the vertebrate and insect poxviruses (85).

Coevolution is an area ripe for further work in insect epizootiology. The "gene-for-gene" hypothesis proposed in epiphytology is turning out to be widely applicable. It has stimulated thought, research, and additional theories, all leading to interesting theoretical and practical approaches to research (see 145). Brooks and Mitter (146) proposed approaches to the study of coevolution of ecologically associated species. May and Anderson (147) have considered this aspect in detail, especially along mathematical analysis. They conclude " . . . that 'successful' parasites need not necessarily evolve to be harmless: both theory and some empirical evidence . . . indicate that many coevolutionary paths are possible, depending on the relation between virulence and the transmissibility of the parasite or pathogen."

3. CHARACTERISTICS OF THE PATHOGEN POPULATION

In this section we shall consider characteristics intrinsic to groups of pathogen units rather than those intrinsic to the individual pathogen as described above. There are basically only three such characteristics — density, dispersion, and composition or quality of the individuals — though others, such as the capacity to survive, are measured as group characteristics.

3.1. Pathogen Population Density

The pathogen population density is one of the most important factors that determine whether a disease becomes epizootic. Closely associated with pathogen density is the spatial distribution (see Section 5). Pathogen population density relates to several characteristics intrinsic to pathogen individuals or units, such as reproductive rate and capacity to survive, but it is a population parameter.

A high density and a widespread distribution of the pathogen population in the host habitat increase the likelihood of the pathogen coming in contact with individuals of the host population. At a low pathogen density or sparse distribution, no host individual may become infected or an enzootic develops which may intensify to an epizootic depending on various factors, such as pathogenicity, transmission, and properties of the host population. In some cases where the pathogen distribution is concentrated in limited sites, even at a relatively low density with respect to the entire host habitat, an enzootic or an epizootic may be initiated. Such is the case with *Oryctes* baculovirus which replicates prodigiously in the digestive tracts of adult female rhinoceros beetle; the virus is transmitted during copulation or it is deposited in the feeding and breeding sites of the beetle (148, see 149). Outside of the host, this virus has limited persistence in the environment.

High pathogen density occurs often after a severe epizootic has decimated a large host population and when the pathogen has replicated extensively in that population (see 150). Such high pathogen density has been reported for the NPVs of the tussock moth (151), the gypsy moth (152), and the European spruce sawfly (153). After a severe epizootic year, the populations of these insects often fail to develop a damaging outbreak in the subsequent year since virus infections occur mainly in young larvae which die without causing much damage. High pathogen density also occurs after the inundative application of pathogens in microbial control, but the pathogen must be distributed in areas occupied by the susceptible host population (see Ch. 15). Timing of the application is also important for insect control.

Very little is known of the critical inoculum threshold of the pathogen population at which an epizootic can develop in an insect population. This threshold value can be deduced by the replicative or inundative application of the pathogen, but such value (e.g., LD_{50}) is generally much higher than that occurring in nature. The curves resulting from LD_{50} values for dosage mortality or the ID_{50} values for dosage infection usually have very low slopes with insect pathogens, indicating greater variability than with most toxic chemical insecticides (see 154).

The voluminous literature of LD_{50} determined by log-dose probit (LPD) experiments provides powerful laboratory evidence of the importance of pathogen population density. The basis of LDP regression lines is that the response (e.g., percentage infection or mortality) is directly dependent on dosage (i.e., pathogen population density). Burges and Thomson (155) and Huber and Hughes (156) discussed the theory of such assays. There are numerous examples, mostly from microbial control, ranging from the use of LDP lines in host resistance (see 157), to studies of virulence in the case histories of development of control agents (see 44), to studies of pathogen interactions (158, 159), and to application in habitat components such as soil (160, 161).

The dependence of epizootics on the survival of pathogen population density is evident in microbial control. Such experiments often begin with various dosages introduced into the host insect's habitat, and there usually is a dosage-related response, though it may not be as straightforward as in the laboratory due to complicating environmental factors. Even in one-dose experiments, there is usually a greater response in treated host insects than in those from control plots. Effects of increasing the pathogen inoculum in the host insect's habitat, the majority of microbial control research, have been thoroughly reviewed in recent years for various pathogen groups (see 162, 163).

The pathogen population is usually increased in a part of the habitat where it comes into direct contact with the host insect population, such as on leaf surfaces for foliage-feeding insects. However, pathogen density-related responses also have been observed in a more indirect manner, for example, the application of various doses of baculovirus to soil and subsequent dosage-related mortality rates in foliage-feeding host insects (164, 165).

A much more adequate method for establishing threshold values is the introduction of a pathogen into a host population to initiate an epizootic (see Ch. 15). Successful and outstanding examples of introductory methods have been obtained with the *Bacillus popilliae* of Japanese beetle (see 166), the NPVs of sawflies (167, 168), and the *Oryctes* baculovirus of rhinoceros beetle (169, see 149). In the Douglas fir tussock moth *Orgyia pseudotsugata*, nuclear polyhedrosis prevalence as low as 0.1 to 1.0% in the first instar successfully initiated an epizootic that took two years to run its course (see 151). A prevalence of 50% in first instar larvae resulted in a complete collapse by the end of the fourth instar in the same year.

An enzootic may be stimulated to become an epizootic by the application of the pathogen to the host population. For example, the granulosis virus of the potato tuberworm *Phthorimaea operculella* is enzootic at a low frequency throughout the tuber moth's range in Australia, but the virus when applied artificially becomes a highly catastrophic mortality factor (170).

Occasionally there is no increased response with an increase in pathogen population density. Presumably, once a certain threshold of pathogen units is reached, further units will not significantly increase the number of hosts contacting the infectious units. For example, in certain cases there was no increase in infection or mortality with an increase in the dosage of microsporidia (171), virus (172), or nematodes (173, 174). Gottwald and Tedders (175) have observed a decline in mortality with higher doses of *Beauveria bassiana*, perhaps indicating self-inhibition. Dosage effects in microbial control can sometimes be masked due to the activity of the pathogen after its introduction (e.g., 176).

In addition to the percentage of hosts infected, the pathogen density or dosage can cause different types of disease that in turn could affect epizootiology, for example, heavy doses can cause mortality more quickly (177, 178) than light ones. Rapid mortality, on the other hand, reduces the replication of the pathogen, and the subsequent pathogen density when the insect host dies. Certain microsporidia cause acute disease with no spore production at high doses and chronic disease with spore production at low doses (e.g., see 154), and both types have been induced in field microbial control experiments (179).

Pathogen population density has been the subject of general theorists as well as researchers of specific systems. Pinnock and Brand (180) and Brand and Pinnock (181) proposed a disease forecast model based largely on the effective pathogen dose. Pathogen numbers also figured heavily in a model by Anderson and May (182) explaining the cyclic behavior of disease in populations of forest insects. Further examples can be found in Chapter 3.

In addition to microbial control research, pathogen population density has been studied in naturally occurring epizootics. These observational studies have ranged from direct and indirect measurements of pathogen density (160, 183–186) to models predicting density and subsequent disease prevalence (187, 188). Pathogen population density in epizootics in all studies has been chosen by multiple regression analyses as one of the most important or the most important factor in fungal (183, 186) and viral (185) disease prevalence.

Effect of pathogen "load" or population density within hosts on epizootics has been studied little for insect diseases compared to general parasitology, except as an indirect way to estimate pathogen units. This topic is of interest in parasitology because it is fairly well accepted that the dispersion of parasite numbers per host affects density-dependent regulation of host and parasite populations (189). Such data have been collected only for nematode infections in insects (e.g., 190).

The pathogen population density has an intimate relationship with host population density (see Ch. 4). There is no question that the two factors are intertwined in nature or that the pathogen population is generally host-density dependent, though it is also becoming clear that density dependence must be carefully defined within a space-time framework and that over a limited space or time (even years) that dependence may not be apparent (191, 185; see 192; 168, 193). Early workers assumed that high host densities spontaneously initiated epizootics in nature. This concept is apparently incorrect since careful studies indicate that at a low host density an enzootic is already occurring, and an increase in host density results in an obvious epizootic (194; see 195, 196; 152; see 151). Pathogen-host relationship is independent of host density when the pathogen threshold density is high and widely distributed, as in the case of the application of pathogens in microbial control or after a severe and extensive epizootic (see Section 5). After the initial independence of host density, the pathogen may act as a host-density dependent factor with the subsequent increase or invasion of new hosts in the habitat (152; see 151). Theoretically, heavy host population density contributes to host stress (see 197, 198) and pathogen transmission and production (see 199, 182). There are examples of pathogen population density depending on host population density, particularly with viruses (see 196; 184, 200, 201) and fungi (187, 118, 202, 186).

3.2. Capacity to Survive in the Host's Habitat

Pathogens may persist and survive in the abiotic (physical) and biotic environments of the host's habitat (see 203 for viruses). Nonresistant and resistant forms of the pathogens are involved in both environments, but the former is more common in the biotic and the latter in the physical environment.

There is probably a better understanding of the capacity of a pathogen population to survive or persist in the host's habitat in insect epizootiology than in medical epidemiology or any other field due to the extensive studies in microbial control. The literature on this subject is so extensive that only selected examples can be cited here. There are three types of such studies: laboratory dosage-mortality or dosage-infection studies (including application in habitat components, such as soil); field microbial control research; and observational field epizootiology that includes estimates of pathogen population density.

3.2.1. ABIOTIC OR PHYSICAL ENVIRONMENT.   Various pathogens have different capabilities of surviving in the abiotic environment. Factors in the physical environment above the soil surface that most significantly affect the persistence of pathogens are sunlight, moisture, and humidity (see 205). In the aquatic environments, the pH, dissolved minerals, etc., are of importance (see 206). The short survival capacity of spores of *Bacillus thuringiensis* var *israelensis* apparently is caused by chemical degradation in waters containing muddy sediment (207). In the soil, the temperature, moisture, pH, minerals, humus, and roots of plants may be involved (see below).

Although there is considerable information on the effect of sunlight, temperature, and humidity on pathogens in the environment above the soil surface (see Ch. 7), less is known about the effect of these factors in other physical environments. Sunlight, especially ultraviolet light, is the most destructive factor (see 205). Most pathogens can survive only under conditions or in areas that protect them from sunlight, for example, in cadavers, feces, etc.

Moisture and relative humidity do not generally affect the persistance of occluded viruses and spore-forming bacteria, but they greatly affect fungi, protozoa, and nematodes. Temperature encountered in the field is unlikely to cause harm to most pathogens but may affect their rate of development.

Soil is highly favorable for entomopathogen survival and is a natural reservoir. Its physicochemical properties affect pathogen survival. Fungitasis in the soil may affect the germination of certain entomopathogenic fungi (208). Root secretions are also known to affect certain microorganisms. However, the information on the role and effect of soil on pathogen persistence is limited. The spores of the milky disease organism *Bacillus popilliae* persist in the soil for about two to seven years, but the pathogen remains in colonization sites for over 25 years, suggesting that the infection of beetles invading the sites maintain the inoculum in the soil (see 209). Nuclear polyhedrosis viruses persist over winter in the soil (191, see 205; 201). Certain facultative pathogens, mainly bacteria and fungi, are capable of growing in the soil, for example, *Beauveria bassiana* (210). Viruses are known to accumulate in the litter and soil beneath trees (see 211; 212, see 151) and beneath field crops (213, 214, 191).

3.2.2. BIOTIC ENVIRONMENT.   In the biological environment of the host habitat, the various pathogens have different capabilities for surviving and persist in primary, secondary and alternate hosts, and in carriers. Pathogens with a wide host range would be expected to persist more effectively than those specific for a single host. Studies of such persistence have been reported for the NPV of the tussock moth complex (see 151), the granulosis virus of lepidopterous insects in alfalfa (215), and the microsporidian species of the insect complex of deciduous trees (216). The importance of alternate hosts is brought out in outbreaks of densovirus in silkworms being cultured in sericultural farms in Japan. Watanabe and Shimizu (204) have demonstrated that the source of the densovirus is the susceptible mulberry pyralid *Glyphodes pyloalis* which feeds on mulberry in the field. The virus is apparently brought in with contaminated mulberry leaves. It also persists in dust formed from mulberry leaves.

Pathogens may survive on plants that are fed upon by the insects provided they are protected or excluded from the detrimental effects of the physical factors, for example, through the structure, texture, and volatile substances emitted by leaves (see 217). Pathogens such as fungi and nematodes persist under barks of trees and infect hosts that enter this niche. The capsules of the granulosis virus of the potato tuberworm *Phthorimaea operculella* may penetrate through the stomata into potato leaves, and the tuberworm larvae feeding on such leaves become infected with the virus (218).

Plants are known to contain antimicrobial substances that can inhibit the growth of entomopathogens on plant foliage (219, 220). Most studies are laboratory tests, and the significance of these studies on epizootics is relatively unknown.

Surface contamination of eggs by NPVs is well documented (213, 221, 222), but the source of contamination is mostly unestablished. In the gypsy moth, the most heavily contaminated material is debris taken from sites used for larval shelter and pupation, but egg surface and especially the hairs surrounding the egg mass are also contaminated, and serve as inocula of the NPV for the infection of young larvae (223). The cocoons of pupae of Douglas fir tussock moth which had died from NPV are heavily contaminated with virus, and overwintering rains and melting snow leach the virus from the cocoons to contaminate the eggs (151). Thompson (151) used the occurrence of virus-contaminated eggs to predict an impending epizootic and population collapse.

It is obvious that the host population is the major source of survival of the pathogen population, and the pathogen may persist in infected hosts into the next host generation in the same or subsequent years. In the application of pathogens in microbial control to the host habitat, persistence results from the survival capacity of the applied pathogen and the replication of the pathogen in the initial and subsequent populations of the insect hosts. The persistence of the NPV and the entomopoxvirus of the spruce budworm *Choristoneura fumiferana* was high during the second year following field applications (224). The NPV of the velvetbean caterpillar *Anticarsia gemmatalis* existed in the environment for three years (97); the granulosis virus of *Adoxophyes orana* persisted for two generations of the tortricid (225).

Survival and replication of most pathogens occur in "primary hosts" at the initial level of a food chain. The longer the period of infection as in the case of a chronic or less virulent infection, the more advantageous it is for pathogen survival, since the pathogen has a longer period to replicate. This has been shown for *Oryctes* virus (226), *Massospora* fungus (227), the milky disease organisms (see 166, 209), and microsporidia (see 228). Upon the death of the primary hosts, the pathogens may continue to persist in cadavers, in decomposed remains, and in feces. In virus epizootics of the Douglas fir tussock moth, the virus survives in cocoons of virus-killed pupae for at least three years (151).

We shall consider "secondary hosts" of pathogens to be those hosts which are parasites, predators, or sarcophagous feeders of the primary host, but we shall not discuss the susceptible insect vectors of vertebrate diseases, for example, vectors of arboviruses and vertebrate protozoa. There are numerous examples of microsporidia infecting the parasitoids of their hosts, but such information is only few or lacking for other pathogens. These infected parasitoids are capable of transmitting the microsporidia to primary hosts. *Bacillus thuringiensis* has been reported to infect the parasitoids of certain lepidopterous hosts (229).

The term "alternate host" refers to heteroecism in which a pathogen requires another dissimilar, obligate host to complete its development. The first case of heteroecism in insect pathology was reported by Whisler et al. (230) who discovered that *Coelomomyces* spp. required a primary mosquito host and an alternate copepod host to complete its life cycle. This has been confirmed with other species of *Coelomomyces* (231). A similar situation exists with microsporidia, *Amblyospora* spp., which infect mosquito larvae but require copepod hosts to complete one type of life cycle (232, 233). The role played by the obligate alternate hosts in the development of epizootics by the above fungi and protozoa awaits thorough investigation since these pathogens are known to produce epizootics in mosquito species in nature.

Pathogens can survive in infected primary, secondary, and alternate hosts which then serve as "carriers." In addition, nonsusceptible invertebrates and vertebrates may serve also in this capacity when they occupy and invade the pathogen environment and come in contact with the pathogens. Such carriers may transport the pathogens on their body surface or may acquire them through feeding on infected hosts and cadavers and eliminate the pathogens during defecation or regurgitation (Section 4.3). Insect parasitoids and predators serve as carriers when they distribute the pathogens in the host habitat. Among the vertebrates, birds have served as effective carriers of pathogens, mainly baculoviruses, and may be closely involved in virus epizootics.

4. CAPACITY TO DISPERSE

There is a close interrelation between dispersal and spatial distribution of the insect pathogen population (see Section 5) and this aspect has also been discussed in Chapter 2.

Insect pathogens rely mainly on physical and biotic factors for their dispersal. Pathogens in general have very limited capacity to disperse through their own actions. Fungi and nematodes can disperse on their own within a limited area, but bacteria, protozoa, and virus depend on biological and physical agents for dispersal. Many fungi possess conidiophores that forcibly discharge spores (conidia), and such primary conidia when they land on a nonhost substrate are capable of producing secondary conidiophores which form secondary conidia. This process may continue until the reserve energy in the conidia is depleted. Some fungal conidia are released in slime drops or with slime coats that serve in attachment to insects as well as in dispersal (see 36); others such as those of entomophthorales are not discharged from the conidiophore but are anadhesive conidia that may stick to passing insects (234, 235). The infective nematode juveniles of *Neoaplectana carpocapsae* migrate along a temperature gradient to zones favorable for their growth and incubation (236), and they are also capable, within short distances, of finding a host through the gradient of CO_2 expelled by the host (237).

4.1. Physical Factors

The physical factors that pathogens take advantage of for dispersal are rain, streams, wind, gravity, etc. Rain is an important factor in the dispersal of viruses in trees, such as viruses of the tussock moths (see 151) and of sawflies (238). Thompson (151) reported that in one example the percentage of contaminated current year's foliage increased from 12% on the day before a light rain to 100% two days after the rain, and the percentage of virus-infected larvae increased from 23 to 99.8% in this period.

Some pathogens have resistant stages with modified structures; for example, certain microsporidia that infect aquatic insects possess spores with filaments or hairs or are surrounded by gelatinous substances that enable them to be carried some distance in the stream or river currents. Most fungus conidia are small and when discharged into the air, are carried by wind currents. Wind blown dusts also disperse pathogens which persist in the soil. This has been shown for viruses (213, 151). There is a need for quantitative studies on the significance of wind, in association with sunlight, temperature, and humidity, in pathogen dispersal.

4.2. Biological Factors (Hosts)

Biotic factors, such as infected primary, secondary and alternate hosts, and contaminated carriers have received the most attention in the dispersal of insect pathogens. The airborne migratory behavior of newly hatched larvae (e.g., Douglas fir tussock moth and gypsy moth) that are infected may aid in the distribution of NPVs. Infected hosts have been introduced into host populations and have initiated epizootics that spread into surrounding areas (see Ch. 15), for example, the NPVs of sawflies (167, 168) and the forest tent caterpillar *Malacosoma disstria* (239) and the *Oryctes* baculovirus of the rhinoceros beetle (240, 241, 169). The dispersal by nonsusceptible carriers is discussed in Section 4.3.

The behavior of hosts infected with certain pathogens contributes to the dispersal of pathogens (see 242). It would be interesting to learn whether these pathogens have evolved to induce such behavior. Lepidopterous and sawfly larvae infected with baculoviruses tend to climb to elevated places such as tops of trees before dying, and the disruption of the fragile integuments releases the fluid internal contents containing the virus occlusion bodies. Concentrations of virus-containing cadavers in the tops of trees have been observed in the nun moth *Lymantria monacha* (243), in the gypsy moth (152), and in sawflies (238). A similar negative geotropism occurs in insects infected with bacteria and entomophthorous fungi (see 242). In the case of fungi, the dead insect is often attached to the substrate (usually plants) by fungal filaments (rhizoids). The fungus *Entomophthora lampyridarum* not only attaches the dead soldier beetle *Chauliognathus pennsylvanicus* but also causes the beetle wings to open and remain in this position, exposing the upper surface of the abdomen where most conidiophores of the fungus develop and enhance spore dispersal (235). Certain gregarious insects tend to disperse when infected with NPV (224, 184) and cytoplasmic polyhedrosis virus (245). Some soil-inhabiting insects, for example several species of scarabaeid infected with rickettsia (246), and the hepialid moth *Wiseana cervinata* infected with an iridescent virus (247), tend to migrate to the soil surface and may be exposed to predators or other dispersal agents.

Movements of infected insects are important in pathogen dispersal. The above-mentioned behaviors of infected insects are examples, but more striking cases are the movements of chronically infected individuals throughout the host habitat. Such dispersal may occur commonly when adult insects are infected, but information is lacking in most cases. In the milky disease of the

Japanese beetle, the subterranean larval habitat poses a question as to how the bacillus spreads from one widely separated area to another. Dutky (166) has reported that infected adults rarely occur in nature, and other factors (see Section 4.3) may be involved in the dispersal of the bacteria. Langford et al. (248), however, have suggested that adults may be important in the dissemination of the bacteria. The adult 17-year cicada becomes infected by the spores of the fungus *Massospora cicadina* when the subterranean nymph emerges from the soil to transform to an adult (227, 249). The infection, which is confined to the adult abdomen, causes the terminal segment to break off from the pressure of the fungal growth, and the conidia are released into the environment (250, 249). The adult cicada is not killed and continues to fly even though more and more abdominal segments break away exposing and distributing additional conidia. These conidia infect other adults in which are formed resting spores that are also dispersed by the loss of abdominal segments.

In the adult rhinoceros beetle, the baculovirus causes tremendous proliferation of midgut cells (comparable to a carcinoma) infected with the virus (251, 241). These infected cells are discharged into the gut lumen, and the virus emerges with the feces. The infected adults visit breeding and feeding sites of the beetle and spread the virus (252, 148). The virus is also transmitted during copulation. Sidor et al. (253) have reported that adults of *Heliothis armigera* are susceptible to a NPV and transmit the virus to the offspring. Such adults may disperse the virus in nature.

Uninfected adult insects are known to disperse baculoviruses which are present on their body surfaces. An interesting case is the virus dispersal by adults of the European pine and spruce sawflies. The virus dispersal by the European pine sawfly, which lays its eggs in clusters, is limited as compared to that of the European spruce sawfly which lays its eggs singly and in separated locations (238). Adults of the tent caterpillar *Malacosoma disstria* disperse virus over wide areas (239, 211). Workers have succeeded in disseminating viruses by contaminating adult bodies (see 254), for example, the adults of the alfalfa caterpillar *Colias eurytheme* (255), of the cabbage looper *Trichoplusia ni* (256), of the European spruce sawfly *Diprion hercyniae* (257), and of the tomato fruitworm *Heliothis zea* (258).

In social insects, such as the honeybee (259) and the fire ants (260), the adults, which generally are either not susceptible or only lightly infected by the brood diseases, may serve as carriers in the intracolonial dispersal of the pathogens.

4.3. Biological Factors (Carriers)

Vertebrates and invertebrates that come in contact with the insect host population undergoing an epizootic, especially those caused by baculoviruses, would be expected to become contaminated and disperse the pathogens (see 203, 261). Parasites and predators may serve as vectors of insect viruses through their contaminated body parts; e.g., mouth parts, ovipositors, etc. (262, see 263–

271). Parasites may have been responsible for the introduction of the NPV of the European spruce sawfly in Canada (153). Saprophagous invertebrates [e.g., sarcophagid flies (272) and mites (273)] feeding on virus-killed insects can disperse the virus. Vertebrates, such as cows, sheep, mice, and moles, feeding on contaminated foliage are known to deposit pathogens in their feces (274, 190, 275).

More important carriers are winged parasites and predators (insects and birds) that feed on infected insects. Baculoviruses have been detected in predaceous insects (276, 264, see 263). The adult parasitoid *Sarcophaga aldrichi*, after feeding on virus-killed tent caterpillar larva, contaminates the foliage with its mouthparts and feet (276). According to Bird (277), the NPV of the red-headed pine sawfly *Neodiprion lecontei*, a native insect, spreads more rapidly than that of the European pine sawfly *N. sertifer*, an introduced species, because the native sawfly has greater numbers of well-established parasites and predators which are important agents in virus dispersal and transmission. Both Bird (238, 277) and Stairs (276) have concluded that an epizootic develops most rapidly when a high parasitoid-predator population and high virus density occur together. The pentatomid predator *Podisus maculiventris*, which has been contaminated with the NPV of the cabbage looper, has initiated an epizootic in the cabbage looper population (278). Smirnoff (279) has attracted predatory wasps using pieces of fresh meat smeared with a NPV, and the wasps dispersed the virus and initiated an epizootic. Similarly, traps contaminated with viruses have been utilized to attract flying adult insects which become carriers upon leaving the traps (see 254). Birds have been shown to disperse effectively insect viruses. Examples are the NPVs of the European spruce sawfly (280–282), of the cabbage looper (283), of *Wiseana* spp. (284), and of the gypsy moth (285, 275), and the granulosis virus of the potato tuberworm (218, 286).

Aside from baculoviruses, very little is known of the dispersal of other pathogens by nonsusceptible carriers. Birds disperse the milky disease organism *B. popilliae* since viable spores have been detected in their droppings (194). Ants, skunks, and moles feed on the Japanese beetle larvae and may distribute the bacteria.

5. SPATIAL DISTRIBUTION OR DISPERSION

We have already touched on pathogen dispersal in Section 4, and additional discussions are presented in Chapters 2 and 6. The spatial distribution of insect pathogens is potentially important but little studied in insect epizootiology. The foci of disease in a host population are considered important in population dynamics of disease (see 287, 168), and the study of such foci can be considered a study of aggregation patterns.

The distribution of the pathogen population determines to a large extent whether the pathogen acts as a host dependent or independent factor. Host density dependency results from the replication of the pathogen and its transmission (vertical and/or horizontal) to increasing numbers as the host density rises. When the pathogen population is widely distributed at or above the threshold density of infection, an epizootic develops rapidly irregardless of the host density (see 150). This is the case in microbial control when pathogens are applied thoroughly in the host habitat (see Ch. 15). A similar situation may result after a very severe and extensive epizootic when a large host population is decimated and results in a wide distribution and high density of the pathogen population.

During the years following a severe virus epizootic in populations of the European spruce sawfly, the NPV was effective in controlling the sawfly at very low host densities (288). In the tussock moth, a large outbreak of the insect was destroyed by a virus epizootic, and the virus inoculum in the forest habitat was so high and widely distributed that the subsequent generation died mainly in the early instars (151).

In general, the pathogen population is distributed at a low density or discontinuously in the host environment and must be introduced by vectors or other means into the host population. Under such conditions, an enzootic may be initiated which may remain at this level or increase rapidly to an epizootic depending on the density and properties of the host population and efficiency of transmission.

Information on the spatial distribution of pathogen units in the host's habitat is essential because environmental contamination is the route through which most entomopathogens infect their hosts. If a host population is distinctly aggregated, then it is intuitively obvious that the degree of coincident aggregation of a natural control agent in the host's habitat would partially determine the number of encounters between the two. Forgetting the host for a moment, pathogen population distribution should be described simply because it is a basic characteristic of any species (see 289).

The most rigorous description of spatial distribution or dispersion, namely, mathematical, is rare in the study of entomopathogen populations. Generally, pathogen units or infected hosts would be expected to be clumped around infected individual hosts, whether alive or dead, or their feces if transmission is by that route. Thus, such data would be expected to most often fit the negative binomial distribution. Dispersion of entomopathogens in the host's habitat has been described mathematically only twice (183, 290), both by the sampling of fungus-infected hosts. In both cases the distribution was clumped or aggregated.

There have been numerous, nonmathematical descriptions of entomopathogen distribution, including studies with other purposes. There are examples of the latter from microbial control research. For example, droplet size and hence, degree of clumping, affects control of forest lepidopterous pests by *Bacillus thuringiensis* (291), and clumping of pathogen units by use of baits has increased percentage infection (292, 179). Also, the NPV sprayed onto different parts of cotton plants demonstrated that intra-plant distribution affects disease prevalence (293).

There are studies of the distribution of entomopathogens in nature. Vertical distribution in soil has been quantified for a nematode (294) and viruses (295, 212, 296), with the pathogen occurring near the surface in all cases. Hukuhara (160) reported that the median infectious concentration (IC_{50}) of a NPV of the fall webworm *Hyphantria cunea* in the soil was 7.14×10^7 polyhedra/g of soil. He calculated that nearly 1000 virus-killed seventh-instar larvae/m² from several years of epizootics were necessary to obtain such a high soil contamination. Other studies have found differences in the horizontal distribution of NPVs in the soil (212, 160); in the vertical distribution of viruses on trees (see 297); virus concentrations in leaf, bark, litter, and soil components of the host's habitat (212, 298); and between two fungi in bark versus soil (299). Also, the distribution has been studied in feces of cattle for a nematode (190), and in feces of wild birds (281, 275) and of mammals for viruses.

Distribution of entomopathogen populations over broad areas has been studied frequently. This topic could be extended to include continental or worldwide patterns, which are discussed in Chapter 8, so we shall limit our examples to a few studies of distribution patterns over less-than-continental-size areas. Several researchers have mapped pathogen distributions in small countries or individual states of the USA. Members of Entomophthorales often have been the subjects (186, 300–304), though there also are examples of such studies for viruses (200), protozoa (305, 306), and nematodes (307, 308).

Temporal distribution of entomopathogen units in the host's habitat is as important to the development of epizootics as spatial distribution. Susceptible host and infectious pathogen must make contact in both space and time for infection to occur. Entomopathogen units have been sampled or estimated over time periods and related to environmental conditions, production in relation to various densities of host population (see above), subsequent disease prevalence, or a combination of these factors. Concentrations of viruses (200) and a nematode (294) in the soil varied seasonally in relation to the numbers of infected hosts, and the numbers of the nematode also varied seasonally with temperature. Numbers of conidia of certain fungi of Entomophthorales have characteristic daily (309, 310, 202) as well as seasonal patterns (202). In the case of daily manifestations, field and laboratory experiments indicate that diel periodicity in entomophthorous fungi depends on environmental conditions, mainly light and relative humidity (202, 311–313). The fungus produces conidia during the early morning hours, when light conditions, humidity, and temperature are optimum, and at which time transmission to susceptible hosts must occur, since the conidia succumb to sunlight and high temperature.

6. PATHOGEN POPULATION COMPOSITION

The composition of the pathogen population is the least known of the three field population characteristics. The composition entails the proportions of the population with various degrees of certain characteristics. The most obvious individuals in a pathogen population are those affected by the physical environment, that is, the susceptible and resistant individuals. The nonresistant individuals are forms which replicate in the hosts and occasionally persist external to the host. They cannot survive in the harsh physical environment, except when protected by being in dead insects, feces, etc. The dormant forms, often the resistant stages, are the ones usually persisting in the physical environment and external to the host.

In addition, individual pathogens of different genomes, as described in Section 2.3, occur in the pathogen population, but very little is known of their presence and the role they play in epizootics. The few studies on virulent strains have been concentrated on geographically separated isolates rather than isolates within a relatively localized pathogen population. Differences have been reported among geographical isolates of entomopathogenic viruses (120, 314, 134), a fungus (113), and bacteria (see 125, 315). Moreover, we do not know the role played by pathogens whose virulence is increased or decreased as they pass through different hosts. Such cases are known mainly from laboratory tests (316).

7. INTERACTIONS BETWEEN PATHOGENS, OTHER BIOLOGICAL AND CHEMICAL AGENTS

Careful studies of the progess of an insect outbreak indicate that the decline in the insect population, especially in a forest ecosystem, is caused by a multitude of natural mortality factors operating in compensatory ways against the high host outbreak (317). Such studies reveal that the interactions of pathogens (e.g., viruses) with other biological control agents (e.g., parasitoids and predators) play a major regulatory role.

As pointed out in this and other chapters, biological agents may affect the replication, dispersal, and transmission of pathogens. In addition, they may participate in other ways during an epizootic. Some parasitoids may discriminate between uninfected and infected hosts, and prefer uninfected to virus-infected larvae (318); others show no preference (270); still others may parasitize more virus-infected than uninfected hosts in the field (319). Moreover, in the cabbage looper, the unparasitized larva is twice as susceptible to the NPV as larva parasitized by the ichneumonid *Hyposoter exiguae* (320).

Pathogens may interact with other parasitoids and predators by infecting them as secondary hosts or by preventing their development. In general, internal parasitoids die when their hosts are killed before the parasitoids have completed their development (see 150; 262, 321–323). Some tachinid flies are an exception since they are capable of sustaining themselves on dead insect tissues (167, 319). The parasitoid *Cotesia marginiventris* fails to complete development when the fall armyworm *Spodoptera frugiperda* is infected with a nonoccluded *Ascovirus* (324). In this case, the parasitoid mortality is not due to the virus killing the host before the parasitoid can complete development since the virus-infected larva lives longer than parasitized larva. This may be associated with toxic factors produced during the virus infection, as in the case of the granulosis virus and NPV infections in the armyworm larva and the parasitoid *Glyptapanteles militaris (= Apanteles militaris)* (325, 326).

Biological agents may also interfere with the pathogen population. In the ant-mealybug mutualism on guava trees, the presence of the ant *Anoplolepis custodiens* limits the spread of the fungus *Cladosporium* sp. which infects the mealybug *Planococcus citri* (327).

The interaction between pathogens and chemical agents (primarily insecticides and fungicides) is receiving increased attention especially in recent approaches in integrated pest management (328). Under this approach the favorable aspects of biological, chemical and cultural controls are utilized.

Chemical agents, insecticides, fungicides, and bacteriocides vary in their effects on insect pathogens by being benign, detrimental, or enhancing (see 329–332). Most chemical insecticides and adjuvants do not adversely affect entomopathogens during the brief period of application (see 205). However, each situation should be tested to determine the effect.

Mixed infections of different pathogens occur frequently in insects and may result in independent coexistence, complementation (synergism), or interference (antagonism). These interactions are known in microsporidia (see 3, 228), viruses (see 196), bacteria, and fungi (see 333). Very little is known regarding the bases for these interactions, and they occur more commonly in chronic infections (see 333). One or more pathogens may occur in the same insect without greatly affecting each other. This is especially the case when pathogens infect different organs or tissues, but some viruses appear to coexist in the same cell and even together in the same nucleus (see 333, 196). Many pathogens are host specific, but some may infect another host when combined with a pathogen of that host. This has been reported with microsporidian species (see 334–336).

Enhancement has been observed between different bacteria, protozoa, and viruses. Some of the dual infections of microsporidian species could be considered as enhancement of one species by another. Moreover, microsporidian infections may enhance certain viral infections (337). In mixed bacterial infections, enhancement results if one bacterium attacks the midgut and the other pathogen the tissues in the hemocoel (see 333). Under aseptic culture of the armyworm *P. unipuncta* the addition of another bacterium to the bacterium *Serratia marcescens* increases the infection of the latter (338). Steinhaus (339) has reported that the infection of *S. marcescens* in *Galleria mellonella* may inhibit the development of *B. thuringiensis*, but the latter enables *S. marcescens* to develop more freely or to act more easily as a secondary invader (other examples in (2)). In the gypsy moth larva, the activity of a chitinolytic bacterium is limited by the high alkaline pH (above pH 5) of the midgut, but the addition of a nonpathogenic, acid-producing bacterium increases the mortality caused by the former (58). Facultative bacteria which occur in the digestive tract may invade into the hemocoel when the midgut wall is damaged by physical agents (340, 341), by the delta-endotoxin of *B. thuringiensis* (342), and by a microsporidium (343). *Xenorhabdus nematophilus (=Achromobacter ne-*

matophilus) enhances the infection of the nematode *Neoaplectana carpocapsae* (see 344, 345). The white muscardine fungus *Beauveria bassiana* enhances the infection of *Bacillus popilliae* (346).

Among viruses, enhancement has been reported between the granulosis viruses and NPVs (53, 54), the NPVs and cytoplasmic polyhedrosis viruses (347), the sacbrood virus and bee virus x (348), the mosquito iridescent virus and picornavirus (349), and a cytoplasmic polyhedrosis virus and the *Chilo* iridescent virus (350). The basis for the enhancement has not been established in these cases, except for the granulosis viruses of the armyworm *Pseudaletia unipuncta* and the clover cutworm *Scotogramma trifolii* (see Section 2.2.3). Interference between different pathogens has been more frequently reported than enhancement. The interference may occur at the cellular and organismal (individual) levels. The more virulent and pathogenic form tends to prevent or suppress the infection of the less virulent form. However, there are reports that the cytoplasmic polyhedrosis virus inactivated by heat or ultraviolet light inhibits the infection of a second virus (351). In some cases, antibiotics produced by the fungi or bacteria are involved. In others, the rate of microbial replication appears to be the determining factor, especially in the case of pathogens, such as viruses, that do not produce antibiotics. The less virulent pathogen is able to establish an infection if given an advantage in time of infection or in dosage (352–355). In larvae of the fall webworm *Hyphantria cunea* fed a mixture of a NPV and a *Nosema* sp., there was less mortality (70%) than when fed the virus alone (98%); the midgut infection of the microsporidium apparently affected virus invasion (356). On the other hand, a microsporidium appeared to predispose the larvae of tent caterpillar *Malacosoma americana* to the infection of a NPV (25).

8. CONCLUSION

The subject of the pathogen population in insect epizootiology is very broad and extensive because it includes all types of pathogenic microorganisms and also insect populations whose species and numbers exceed all other animals. A comprehensive treatise would require a complete text on this subject alone.

Throughout the present volume, the often repeated complaint is the lack of appropriate data, mainly quantitative, in the epizootiology of insect diseases. This is also the case with the pathogen population. The most serious deficiency occurs in the qualitative and quantitative compositions of the pathogen population in the host habitat. Others are in areas of spatial distribution and density. In addition, other properties and factors affecting the pathogen population need considerable investigation to establish the bases for the interaction of the pathogen and host populations to establish an epizootic.

Our chapter has not attacked the question, what are the ideal characteristics of a pathogen population for establishing an epizootic in a host population? We shall speculate on this question. In general, the desirable characteristics will depend on the economic importance of the insect pest. With an insect pest of low economic threshold where only a few individuals can cause an economic loss, a highly virulent pathogen population is needed to destroy the host population rapidly in order to prevent or reduce the loss. Under this situation, inundative or repeated applications of the virulent pathogens may be required. Such virulent pathogens tend to have a low level of persistance in the host habitat; for example, failure to replicate in sufficient numbers in the rapidly killed hosts and to persist in the host habitat. This may be the case with *Bacillus thuringiensis* which is highly virulent mainly because of the delta endotoxin, and which produces few or no spores in certain susceptible insect species. Moreover, the bacillus has a short persistence on the exposed plant areas fed upon by the susceptible pest.

When the economic threshold is high (the case for most insect pests), a less virulent pathogen population that is maintained in chronically infected host population is desirable. In this case, the insect pest is controlled, provided the infection reduces insect feeding and reproduction. Pest mortality is not significant, and in the chronic or prolonged infection, the pathogen replicates continuously, often to very high numbers and may establish a high density in the habitat. Much more important is the dispersal of the pathogen by infected hosts throughout the habitat. Such is the case of the *Oryctes* baculovirus which has successfully controlled the rhinocerous beetle.

Present advancement in gene manipulatin and hybridization may provide means for improving the quality of a pathogen to become more effective in controlling insect pests. Such genetic transformations, however, should also consider the role of the pathogen population in insect epizootiology.

ACKNOWLEDGMENTS

We thank Clara K. Lee for her assistance in preparing the manuscript.

REFERENCES

1. J. M. Franz, "Biologische Schädlingsbekämpfung," in H. Richter, Ed., *Handbuch der Pflanzenkrankheiten,* Vol. 6, Paul Parey, Berlin, 1961, p. 1.

2. Y. Tanada, "Epizootiology of Infectious Diseases," in E. A. Steinhaus, Eds., *Insect Pathology: An Advanced Treatise,* Vol. 2, Academic Press, New York, 1963, p. 423.

3. J. Weiser, Die Mikrosporidien als Parasiten der Insekten, *Monogr. Angew, Entomol.,* **17,** 1 (1961).

4. T. G. Andreadis, Experimental transmission of a microsporidian pathogen from mosquitoes to an alternate copepod host, *Proc. Natl. Acad. Sci. USA,* **82,** 5574 (1985).

5. G. H. Bergold, "Viruses of Insects," in C. Hallauer and K. F. Meyer, Eds., *Handbuch der Virusforschung,* Vol. 4, Springer, Vienna, 1958, p. 60.

6. H. Aruga, "Induction of Virus Infections," in E. A. Steinhaus, Ed., *Insect Pathology: An Advanced Treatise,* Vol. 1, Academic Press, New York, 1963, p. 499.

7. R. Seecoff, The sigma virus infection of *Drosophila melanogaster, Curr. Top. Microbiol. Immunol.,* **42,** 59 (1968).

8. E. S. Sylvester, "Rhabdoviruses of Insects (Sigma Virus of *Drosophila*)," in K. Maramorosch, Ed., *The Atlas of Insect and Plant Viruses Including Mycoplasmaviruses and Viroids,* Academic Press, New York, 1977, p. 131.

9. T. D. C. Grace, Induction of polyhedral bodies in ovarian tissues of the tussock moth in vitro, *Science,* **128,** 249 (1958).

10. J. F. Longworth and J. C. Cunningham, The activation of occult nuclear-polyhedrosis virus by foreign nuclear polyhedra, *J. Invertebr. Pathol.,* **10,** 361 (1968).

11. D. C. Kelly, T. Lescott, M. D. Ayres, D. Carey, A. Coutts, and K. A. Harrap, Induction of a nonoccluded baculovirus persistently infecting *Heliothis zea* cells by *Heliothis armigera* and *Trichoplusia ni* nuclear polyhedrosis virus, *Virology,* **112,** 174 (1981).

12. D. J. McKinley, D. A. Brown, C. C. Payne, and K. A. Harrap, Cross-infectivity and activation studies with four baculoviruses, *Entomophaga,* **26,** 79 (1981).

13. M. Himeno, F. Matsubara, and K. Hayashiya, The occult virus of nuclear polyhedrosis of the silkworm larvae, *J. Invertebr. Pathol.,* **22,** 292 (1973).

14. N. Kislev, M. Edelman, and I. Harpaz, Nuclear polyhedrosis viral DNA: characterization and comparison to host DNA, *J. Invertebr. Pathol.,* **17,** 199 (1971).

15. J. P. Kramer, "The Extra-corporeal Ecology of Microsporidia," in L. A. Bulla, Jr. and T. C. Cheng, Eds., *Comparative Pathology,* Vol. 1, Plenum Press, New York, 1976, p. 127.

16. H. T. Dulmage and R. A. Rhodes, "Production of Pathogens in Artificial Media," in H. D. Burges and N. W. Hussey, Eds., *Microbial Control of Insects and Mites,* Academic Press, New York, 1971, p. 507.

17. E. Müller-Kögler, *Pilzkrankheiten bei Insekten,* Paul Parey, Berlin, 1965.

18. G. Galani, Studies on the variation of pathogenicity of *Verticillium lecanii* (Zimm.) Viégas to larvae of *Trialeurodes vaporariorum, Westw. An. ICPP, Bucarest,* **15,** 244 (1979).

19. E. A. Steinhaus, *Principles of Insect Pathology,* McGraw-Hill, New York, 1949.

20. Y. Tanada, "Epizootiology and Microbial Control," in L. A. Bulla, Jr. and T. C. Cheng, Eds., *Comparative Pathobiology,* Vol. 1, Plenum Press, New York, 1976, p. 247.

21. K. Aizawa, "Strain Improvement and Preservation of Virulence of Pathogens," in H. D. Burges and N. W. Hussey, Eds., *Microbial Control of Insects and Mites,* Academic Press, New York, 1971, p. 655

22. K. Aizawa, "Selection and Strain Improvement of Insect Pathogenic Micro-organisms for Microbial Control," in K. Yasumatsu and H. Mori, Eds., *Approaches to Biological Control,* Vol. 7, JIBP Synthesis, Tokyo, 1975, p. 99.

23. E. I. Hazard and C. S. Lofgren, Tissue specificity and systematics of a *Nosema* in some species of *Aedes, Anopheles,* and *Culex, J. Invertebr. Pathol.,* **18,** 16 (1971).

24. W. A. Smirnoff, Adaptation of *Thelohania pristiphorae* on ten species of Diprionidae and Tenthredinidae, *J. Invertebr. Pathol.,* **23,** 114 (1974).

25. W. A. Smirnoff, Adaptation of the microsporidian *Thelohania pristiphorae* to the tent caterpillars *Malacosoma disstria* and *Malacosoma americanum, J. Invertebr. Pathol.,* **11,** 321 (1968).

26. I. M. Hall, Studies of microorganisms pathogenic to the sod webworm, *Hilgardia,* **22,** 535 (1954).

27. J. Weiser, "Immunity of Insects to Protozoa," in G. J. Jackson, R. Herman, and I. Singer, Eds., *Immunity to Parasitic Animals,* Vol. 1, Appleton-Century-Crofts, New York, 1969, p. 129.

28. W. A. Smirnoff, Adaptation of a nuclear-polyhedrosis virus of *Trichiocampus viminalis* (Fallén) to larvae of *Trichiocampus irregularis* (Dyar), *J. Insect Pathol.,* **5,** 104 (1963).

29. G. J. Tompkins, J. L. Vaughn, J. R. Adams, and C. F. Reichelderfer, Effects of propagating

Autographa californica nuclear polyhedrosis virus and its *Trichoplusia ni* variant in different hosts, *Environ. Entomol.,* **10,** 801 (1981).

30. E. A. MacKinnon, J. F. Henderson, D. B. Stoltz, and P. Faulkner, Morphogenesis of nuclear polyhedrosis virus under conditions of prolonged passage in vitro, *J. Ultrastruct. Res.,* **49,** 419 (1974).

31. H. Hirumi, K. Hirumi, and A. H. McIntosh, Morphogenesis of a nuclear polyhedrosis virus of the alfalfa looper in a continuous cabbage looper cell line, *Ann. N. Y. Acad. Sci.,* **266,** 302 (1975).

32. K. Yamada, K. E. Sherman, and K. Maramorosch, Serial passage of *Heliothis zea* singly embedded nuclear polyhedrosis virus in a homologous cell line, *J. Invertebr. Pathol.,* **39,** 185 (1982).

33. O. Lysenko and M. Kučera, "Micro-organisms as Sources of New Insecticidal Chemicals: Toxins," in H. D. Burges and N. W. Hussey, Eds., *Microbial Control of Insects and Mites,* Academic Press, New York, 1971, p. 205.

34. O. Lysenko, Bacterial exoenzymes toxic for insects; proteinase and lecithinase, *J. Hyg. Epidemiol. Microbiol. Immunol.,* **18,** 347 (1974).

35. O. Lysenko, "Principles of Pathogenesis of Insect Bacterial Diseases as Exemplified by the Nonsporeforming Bacteria," in E. W. Davidson, Ed., *Pathogenesis of Invertebrate Microbial Diseases,* Allanheld, Osmum Publ., Totowa, New Jersey, 1981, p. 163.

36. D. W. Roberts and R. A. Humber, "Entomopathogenic Fungi," in D. W. Roberts and J. R. Aist, Eds., *Infection Processes of Fungi,* A Bellagio Conference Report, March 21–25, Rockefeller Foundation, 1983, p. 1.

37. A. Krieg, Neues über *Bacillus thuringiensis* und seine Anwendung, *Mitt. Biol. Bundensan. Land-Forstwirt.,* **125,** 1 (1967).

38. A. M. Heimpel, A criticial review of *Bacillus thuringiensis* var. *thuringiensis* Berliner and other crystalliferous bacteria, *Annu. Rev. Entomol.,* **12,** 287 (1967).

39. B. A. Weiner, Isolation and partial characterization of the parasporal body of *Bacillus popilliae, Can. J. Microbiol.,* **24,** 1557 (1978).

40. G. E. Bucher, Artificial culture of *Clostridium brevifaciens* n. sp. and *C. malacosomae* n. sp., the causes of brachytosis of tent caterpillars, *Can. J. Microbiol.,* **7,** 641 (1961).

41. C. C. Doane and J. J. Redys, Characteristics of motile strains of *Streptococcus faecalis* pathogenic to larvae of the gypsy moth, *J. Invertebr. Pathol.,* **15,** 420 (1970).

42. L. Bailey, *Honey Bee Pathology,* Academic Press, New York, 1981.

43. E. W. Davidson, "Bacterial Diseases of Insects Caused by Toxin-Producing Bacilli Other than *Bacillus thuringiensis,"* in E. W. Davidson, Ed., *Pathogenesis of Invertebrate Microbial Diseases,* Allanheld, Osmun Publ., Totowa, New Jersey, 1981, p. 269.

44. G. E. Bucher, "Identification of Bacteria Found in Insects," in H. D. Burges, Ed., *Microbial Control of Pests and Plant Diseases 1970–1980,* Academic Press, New York, 1981, p. 7.

45. J.-M. Quiot, A. Vey, and C. Vago, "Effects of Mycotoxins on Invertebrate Cells in vitro," in K. Maramorosch, Ed., *Advances in Cell Culture,* Academic Press, New York, 1985, p. 199.

46. T. D. Wyllie and L. G. Morehouse, *Mycotoxic Fungi, Mycotoxins, and Mycotoxicoses: An Encyclopedic Handbook,* Vol. 1, Marcel Dekker, New York, 1977, p. 1.

47. P. Luthy, Insecticidal toxins of *Bacillus thuringiensis, FEMS Microbiol. Lett.,* **8,** 1 (1980).

48. P. G. Fast, "The Crystal Toxin of *Bacillus thuringiensis,"* in H. D. Burges, Ed., *Microbial Control of Pests and Plant Diseases 1970–1980,* Academic Press, New York, 1981, p. 223.

49. H. de Barjac, "Identification of H-Serotypes of *Bacillus thuringiensis,"* in H. D. Burges, Ed., *Microbial Control of Pests and Plant Diseases 1970–1980,* Academic Press, New York, 1981, p. 35.

50. A. M. Heimpel and T. A. Angus, "Diseases Caused by Certain Sporeforming Bacteria," in

E. A. Steinhaus, Ed., *Insect Pathology: An Advanced Treatise,* Vol. 2, Academic Press, New York, 1963, p. 21.

51. P. S. Myers and A. A. Yousten, Localization of a mosquito-larval toxin of *Bacillus sphaericus* 1593, *Appl. Environ. Microbiol.,* **39,** 1205 (1980).

52. E. W. Davidson, Alkaline extraction of toxin from spores of the mosquito pathogen, *Bacillus sphaericus* strain 1593, *Can. J. Microbiol.,* **29,** 271 (1983).

53. Y. Tanada, Synergism between two viruses of the armyworm. *Pseudaletia unipuncta* (Haworth) (Lepidoptera, Noctuidae), *J. Insect Pathol.,* **1,** 215 (1959).

54. P. J. Stoddard, *Persistence and transmission of baculoviruses in insect populations in alfalfa,* PhD. Dissertation, Dep. Entomol. Sci., Univ. California, Berkeley, 1980.

55. Y. Tanada, Founders lecture, a synopsis of studies on the synergistic property of an insect baculovirus: a tribute to Edward A. Steinhaus, *J. Invertebr. Pathol.,* **45,** 125 (1985).

56. W. A. Smirnoff, Effect of chitinase on the action of *Bacillus thuringiensis, Can. Entomol.,* **103,** 1829 (1971).

57. P. G. Fast, Laboratory bioassays of mixtures of *Bacillus thuringiensis* and chitinase, *Can. Entomol.,* **110,** 201 (1978).

58. R. A. Daoust and H. B. Gunner, Microbial synergists pathogenic to *Lymantria dispar:* chitinolytic and fermentative bacterial interactions, *J. Invertebr. Pathol.,* **33,** 368 (1979).

59. J. M. Stephens, Mucin as an agent promoting infection by *Pseudomonas aeruginosa* (Schroeter) Migula in grasshoppers, *Can. J. Microbiol.,* **5,** 73 (1959).

60. C. C. Doane and R. C. Wallis, Enhancement of the action of *Bacillus thuringiensis* var. *thuringiensis* Berliner on *Porthetria dispar* (Linnaeus) in laboratory tests, *J. Insect Pathol.,* **6,** 423 (1964).

61. R. Govindarajan, S. Jayaraj, and K. Narayanan, Mortality of tobacco caterpillar, *Spodoptera litura* (F.) when treated with *Bacillus thuringiensis* combinations with boric acid and insecticides, *Phytoparasitica,* **4,** 193 (1976).

62. M. Shapiro and R. A. Bell, Enhanced effectiveness of *Lymantria dispar* (Lepidoptera: Lymantriidae) nucleopolyhedrosis virus formulated with boric acid, *Ann. Entomol. Soc. Am.,* **75,** 346 (1982).

63. W. A. Ramoska and T. Todd, Variation in efficacy and viability of *Beauveria bassiana* in the chinch bug (Hemiptera: Lygaeidae) as a result of feeding activity on selected host plants, *Environ. Entomol.,* **14,** 146 (1985).

64. A. W. Martin and D. H. Dean, "Genetics and Genetic Manipulation of *Bacillus thuringiensis,"* in H. D. Burges, Ed., *Microbial Control of Pests and Plant Diseases 1970–1980,* Academic Press, New York, 1981, p. 299.

65. G. E. Smith and M. D. Summers, Analysis of baculovirus genomes with restriction endonucleases, *Virology,* **89,** 517 (1978).

66. R. A. Samson, C. W. McCoy, and K. L. O'Donnell, Taxonomy of acarine parasite, *Hirsutella thompsonii, Mycologia,* **72,** 359 (1980).

67. P. Faulkner, "Baculovirus," in E. W. Davidson, Ed., *Pathogenesis of Invertebrate Microbial Disease,* Allanheld, Osmun Publ., Totowa, New Jersey, 1981, p. 3.

68. D. G. Boucias, C. W. McCoy, and D. J. Joslyn, Isozyme differentiation among 17 geographical isolates of *Hirsutella thompsonii, J. Invertebr. Pathol.,* **39,** 329 (1982).

69. L. Baumann, K. Okamoto, B. M. Unterman, M. J. Lynch, and P. Baumann, Phenotypic characterization of *Bacillus thuringiensis* and *Bacillus cereus, J. Invertebr. Pathol.,* **44,** 329 (1984).

70. J. E. Monroe and W. J. McCarthy, Polypeptide analysis of genotypic variants of occluded *Heliothis* spp. baculoviruses, *J. Invertebr. Pathol.,* **43,** 32 (1984).

71. E. I. Hazard, E. A. Eillis, and D. J. Joslyn, "Identification of Microsporidia," in H. D. Burges, Ed., *Microbial Control of Pests and Plant Diseases 1970–1980,* Academic Press, New York, 1981, p. 163.

72. L. K. Miller, A. J. Lingg, and L. A. Bulla, Jr., Bacterial, viral, and fungal insecticides, *Science,* **219,** 715 (1983).

73. P. F. Sparling, Applications of genetics to studies of bacterial virulence, *Phil. Trans. R. Soc. London, B,* **303,** 199 (1983).

74. L. C. Loh, J. J. Hamm, C. Kawanishi, and E.-S. Huang, Analysis of the *Spodoptera frugiperda* nuclear polyhedrosis virus genome by restriction endonucleases and electron microscopy, *J. Virol.,* **44,** 747 (1982).

75. R. R. Gettig and W. J. McCarthy, Genotypic variation among wild isolates of *Heliothis* spp. nuclear polyhedrosis viruses from different geographical regions, *Virology,* **117,** 245 (1982).

76. E. M. Dougherty, "A Comparison of the Gypchek and Virin-ENSh Preparations of a Multiple Embedded Nuclear Polyhedrosis Virus of the Gypsy Moth, *Lymantria dispar* Utilizing Restriction Endonuclease Analysis, in C. M. Ignoffo, M. E. Martignoni, and J. L. Vaughn, Eds., *A. Comparison of the US (Gypchek) and USSR (Virin-ENSh) Preparations of the Nuclear Polyhedrosis Virus of the Gypsy Moth, Lymantria dispar,* Am. Soc. Microbiol., Washington, DC, 1983, p. 21.

77. J. E. Maruniak, S. E. Brown, and D. L. Knudson, Physical maps of SfMNPV baculovirus DNA and its genomic variants, *Virology,* **136,** 221 (1984).

78. L. K. Miller, S. G. Franzblau, H. W. Homan, and L. P. Kish, A new variant of *Autographa californica* nuclear polyhedrosis virus, *J. Invertebr. Pathol.,* **36,** 159 (1980).

79. G. E. Smith and M. D. Summers, Restriction map of *Rachiplusia ou* and *Rachiplusia ou-Autographa californica* baculovirus recombinants, *J. Virol.,* **33,** 311 (1980).

80. L. K. Miller, Construction of a genetic map of the baculovirus *Autographa californica* nuclear polyhedrosis virus by marker rescue of temperature-sensitive mutants, *J. Virol.,* **39,** 973 (1981).

81. B. M. Arif and W. Doerfler, Identification and localization of reiterated sequences in the *Choristoneura fumiferana* MNPV genome, *EMBO J.,* **3,** 525 (1984).

82. J. D. Knell and M. D. Summers, A physical map for the *Heliothis zea* SNPV genome, *J. Gen. Virol.,* **65,** 445 (1984).

83. B. N. Fields and K. Byers, The genetic basis of viral virulence, *Trans. R. Soc. London, B,* **303,** 209 (1983).

84. J. E. Jewell and L. K. Miller, DNA sequence homology relationships among six lepidopteran nuclear polyhedrosis viruses, *J. Gen. Virol.,* **48,** 161 (1980).

85. W. H. R. Langridge, Detection of DNA base sequence homology between entomopoxviruses isolated from Lepidoptera and Orthoptera, *J. Invertebr. Pathol.,* **43,** 41 (1984).

86. F. P. Wiegers and J. M. Vlak, Physical map of the DNA of a *Mamestra brassicae* nuclear polyhedrosis virus variant isolated from *Spodoptera exigua, J. Gen. Virol.,* **65,** 2011 (1984).

87. J. Nishiitsutsuji-Uwo, Y. Wakisaka, and M. Eda, Sporeless mutants of *Bacillus thuringiensis, J. Invertebr. Pathol.,* **25,** 355 (1975).

88. R. M. Faust and R. S. Travers, Occurrence of resistance to neomycin and kanamycin in *Bacillus popilliae* and certain serotypes of *Bacillus thuringiensis:* mutation potential in sensitive strains, *J. Invertebr. Pathol.,* **37,** 113 (1981).

89. K. F. Jarrell and A. M. Kropinski, The virulence of protease and cell surface mutants of *Pseudomonas aeruginosa* for the larvae of *Galleria mellonella, J. Invertebr. Pathol.,* **39,** 395 (1982).

90. A. Samšiňáková and S. Kálalová, The influence of a single -spore isolate and repeated subculturing on the pathogenicity of conidia of the entomophagous fungus *Beauveria bassiana, J. Invertebr. Pathol.,* **42,** 156 (1983).

91. S. Paris and P. Ferron, Study of the virulence of some mutants of *Beauveria brongniartii* (= *Beauveria tenella), J. Invertebr. Pathol.,* **34,** 71 (1979).

92. K. Al-Aidroos and D. W. Roberts, Mutants of *Metarhizium anisopliae* with increased virulence toward mosquito larvae, *Can. J. Genet. Cytol.,* **20,** 211 (1978).

93. H. A. Wood, P. R. Hughes, L. B. Johnston, and W. H. R. Langridge, Increased virulence of *Autographa californica* nuclear polyhedrosis virus by mutagenesis, *J. Invertebr. Pathol.,* **38,** 236 (1981).

94. C. F. Reichelderfer, "Mutation Potential of Insect Viruses," in M. Summers, R. Engler, L. A. Falcon, and P. V. Vail, Eds., *Baculoviruses for Insect Pest Control: Safety Considerations,* Am. Soc. Microbiol., Washington, DC, 1975, p. 73.

95. B. D. Clark, F. J. Perlak, C.-Y. Chu, and D. H. Dean, "The *Bacillus thuringiensis* Genetic Systems," in T. C. Cheng, Ed., *Comparative Pathobiology.* Vol. 7. *Pathogens of Invertebrates. Application in Biological Control and Transmission Mechanisms,* Plenum Press, New York, 1984, p. 155.

96. M. Brown and P. Faulkner, A partial genetic map of the baculovirus, *Autographa californica* nuclear polyhedrosis virus, based on recombination studies with *ts* mutants, *J. Gen. Virol.,* **48,** 247 (1980).

97. A. R. Richter and J. R. Fuxa, Timing, formulation, and persistence of a nuclear polyhedrosis virus and a microsporidium for control of the velvetbean caterpillar (Lepidoptera: Noctuidae) in soybeans, *J. Econ. Entomol.,* **77,** 1299 (1984).

98. G. Riba, Recombinaison apres heterocaryose chez le champignon entomopathogene *Paecilomyces fumosoroseus* (Deuteromycete), *Entomophaga,* **23,** 417 (1978).

99. G. Riba and A. M. Ravelojoanna, The parasexual cycle in the entomopathogenic fungus, *Paecilomyces fumoso-roseus* (Wize) Brown and Smith, *Can. J. Microbiol.,* **30,** 922 (1984).

100. G. E. Smith and M. D. Summers, DNA homology among subgroup A, B, and C baculoviruses, *Virology,* **123,** 393 (1982).

101. L. K. Miller, "The Isolation and Characterization of Genotypic Variants of *Autographa californica* Nuclear Polyhedrosis Virus," in C. M. Ignoffo, M. E. Martignoni, and J. L. Vaughn, Eds., *Characterization, Production and Utilization of Entomopathogenic Viruses,* Am. Soc. Microbiol., Washington, DC, 1980, p. 194.

102. M. F. Claridge and J. D. Hollander, The biotype concept and its application to insect pests of agriculture, *Crop Protect.,* **2,** 85 (1983).

103. G. O. Poinar, Jr., *Nematodes for Biological Control of Insects,* CRC Press, Boca Raton, Florida, 1979.

104. R. S. Soper, B. May, and B. Martinell, *Entomophaga grylli* enzyme polymorphism as a technique for pathotype identification, *Environ. Entomol.,* **12,** 720 (1983).

105. C. L. Messias, D. W. Roberts, and A. T. Grefig, Pyrolysis-gas chromatography of the fungus *Metarhizium anisopliae:* an aid to strain identification, *J. Invertebr. Pathol.,* **42,** 393 (1983).

106. H. de Barjac, I. Larget-Thiery, V. C. Dumanoir, and H. Ripouteau, Serological classification of *Bacillus sphaericus* strains in relation with toxicity to mosquito larvae, *Appl. Microbiol. Biotechnol.,* **21,** 85 (1985).

107. R. J. Milner, "Identification of the *Bacillus popilliae* Group of Insect Pathogens," in H. D. Burges, Ed., *Microbial Control of Pests and Plant Diseases 1970–1980,* Academic Press, New York, 1981, p. 35.

108. G. Rohrmann, "Biochemical and Biophysical Methods for Baculovirus Detection and Identification," in C. M. Ignoffo, M. E. Martignoni, and J. L. Vaughn, Eds., *Characterization, Production and Utilization of Entomopathogenic Viruses,* Am. Soc. Microbiol., Washington, DC, 1980, p. 176.

109. C. C. Payne and D. C. Kelly, "Identification of Insect and Mite Viruses," in H. D. Burges, Ed., *Microbial Control of Pests and Plant Diseases 1970–1980,* Academic Press, New York, 1981, p. 61.

110. I. P. Griffith, "A New Approach to the Problem of Identifying Baculoviruses," in E. Kurstak, Ed., *Microbial and Viral Pesticides,* Marcel Dekker, New York, 1982, p. 507.

111. P. Lüthy and H. R. Ebersold, *"Bacillus thuringiensis* Delta-Endotoxin: Histopathology and

Molecular Mode of Action," in E. W. Davidson, Ed., *Pathogenesis of Invertebrate Microbial Diseases,* Allanheld, Osmun Publ., Totowa, New Jersey, 1981, p. 235.

112. M. E. Martignoni, "In Vivo Host Specificity of Virin-ENSh," in C. M. Ignoffo, M. E. Martignoni, and J. L. Vaughn, Eds., *A Comparison of the US (Gypchek) and USSR (Virin-ENSh) Preparations of the Nuclear Polyhedrosis Virus of the Gypsy Moth, Lymantria dispar,* Am. Soc. Microbiol., Washington, DC, 1983, p. 43.

113. C. M. Ignoffo and C. Garcia, Host spectrum and relative virulence of an Ecuadoran and a Mississippian biotype of *Nomuraea rileyi, J. Invertebr. Pathol.,* **45**, 346 (1985).

114. J. J. Petersen, Comparative biology of the Wyoming and Louisiana populations of *Reesimermis nielseni,* parasitic nematode of mosquitoes, *J. Nematol.,* **8**, 273 (1976).

115. C. M. Ignoffo, Strategies to increase the use of entomopathogens, *J. Invertebr. Pathol.,* **31**, 1 (1978).

116. J. R. Harlan, Diseases as a factor in plant evolution, *Annu. Rev. Phytopathol.,* **14**, 31 (1976).

117. R. M. May, Parasitic infections as regulators of animal populations, *Am. Sci.,* **71**, 36 (1983).

118. C. M. Ignoffo, N. L. Marston, D. L. Hostetter, and B. Puttler, Natural and induced epizootics of *Nomuraea rileyi* in soybean caterpillars, *J. Invertebr. Pathol.,* **27**, 191 (1976).

119. R. A. Daoust and D. W. Roberts, Virulence of natural and insect-passaged strains of *Metarhizium anisopliae* to mosquito larvae, *J. Invertebr. Pathol.,* **40**, 107 (1982).

120. L. L. J. Ossowski, Variation of virulence of a wattle bagworm virus, *J. Insect Pathol.,* **2**, 35 (1960).

121. M. Shapiro, J. L. Robertson, M. G. Injac, K. Katagiri, and R. A. Bell, Comparative infectivities of gypsy moth (Lepidoptera: Lymantriidae) nucleopolyhedrosis virus isolates from North America, Europe, and Asia, *J. Econ. Entomol.,* **77**, 153 (1984).

122. J. R. Fuxa, *Spodoptera frugiperda* susceptibility to nuclear polyhedrosis virus isolates with reference to insect migration, *Environ. Entomol.,* in press.

123. P. V. Vail, J. D. Knell, M. D. Summers, and D. K. Cowan, In vivo infectivity of baculovirus isolates, variants, and natural recombinants in alternate hosts, *Environ. Entomol.,* **11**, 1187 (1982).

124. C. M. Ignoffo, "The Fungus *Nomuraea rileyi* as a Microbial Insecticide," in H. D. Burges, Ed., *Microbial Control of Pests and Plant Diseases 1970–1980,* Academic Press, New York, 1981, p. 513.

125. H. T. Dulmage and Cooperators, "Insecticidal Activity of Isolates of *Bacillus thuringiensis* and Their Potential for Pest Control," in H. D. Burges, Ed., *Microbial Control of Pests and Plant Diseases 1970–1980,* Academic Press, New York, 1981, p. 193.

126. H. E. Huber and P. Lüthy, "*Bacillus thuringiensis* Delta-Endotoxin: Composition and Activation," in E. W. Davidson, Ed., *Pathogenesis of Invertebrate Microbial Diseases,* Allanheld, Osmun, Publ., Totowa, New Jersey, 1981, p. 209.

127. D. M. MacLeod, Natural and cultural variation in entomogenous Fungi Imperfecti, *Ann. N. Y. Acad. Sci.,* **60**, 58 (1954).

128. R. A. Hall, "The Fungus *Verticillium lecanii* as a Microbial Insecticide against Aphids and Scales," in H. D. Burges, Ed., *Microbial Control of Pests and Plant Diseases 1970–1980,* Academic Press, New York, 1981, p. 483.

129. D. S. King and R. A. Humber, "Identification of the Entomophthorales," in H. D. Burges, Ed., *Microbial Control of Pests and Plant Diseases 1970–1980,* Academic Press, New York, 1981, p. 107.

130. Anonymous, Data Sheet on the Biological Control Agent *Metarhizium anisopliae* (Metschnikoff), Sorokin, 1883, *World Health Organization* VBC/80.758, VCB/BCDS/80.04, 1980.

131. J. C. Cunningham, Strains of nuclear polyhedrosis viruses displaying different inclusion body shapes, *J. Invertebr. Pathol.,* **16**, 299 (1970).

132. J. Brassel and G. Benz, Selection of a strain of the granulosis virus of the codling moth with

improved resistance against artificial ultraviolet radiation and sunlight, *J. Invertebr. Pathol.,* **33,** 358 (1979).

133. D. J. Witt and W. F. Hink, Selection of *Autographa californica* nuclear polyhedrosis virus for resistance to inactiviation by near ultraviolet, far ultraviolet, and thermal radiation, *J. Invertebr. Pathol.,* **33,** 222 (1979).

134. M. Shapiro and R. A. Bell, Selection of a UV-tolerant strain of the gypsy moth, *Lymantria dispar* (L.) (Lepidoptera: Lymantriidae), nucleopolyhedrosis virus, *Environ. Entomol.,* **13,** 1522 (1984).

135. G. C. M. Latch, *Metarrhizium anisopliae* (Metschnikoff) Sorokin strains in New Zealand and their possible use for controlling pasture-inhabiting insects, *N. Z. J. Agric. Res.,* **8,** 384 (1965).

136. D. G. Boucias and J. C. Pendland, Nutritional requirements for conidial germination of several host range pathotypes of the entomopathogenic fungus *Nomuraea rileyi, J. Invertebr. Pathol.,* **43,** 288 (1984).

137. L. P. Kish, The effect of high temperatures on spore germination of *Ascosphaera aggregata, J. Invertebr. Pathol.,* **42,** 244 (1983).

138. D. S. King, "Strain Variation in *Conidiobolus,"* in *Proc. First Int. Colloq. Invertebr. Pathol. IXth Annu. Meet., Soc. Invertebr. Pathol.,* Printing Dept., Queen's Univ., Kingston, Canada, 1976, p. 277.

139. F. S. Mulligan III, C. H. Schaefer, and T. Miura, Laboratory and field evaluation of *Bacillus sphaericus* as a mosquito control agent, *J. Econ. Entomol.,* **71,** 774 (1978).

140. C. F. B. Lewis, "Comparison of Spray Tower Applications of Virin-ENSh and Gypchek," in C. M. Ignoffo, M. E. Martignoni, and J. L. Vaughn, Eds., *A Comparison of the US (Gypchek) and USSR (Virin-ENSh) Preparations of the Nuclear Polyhedrosis Virus of the Gypsy Moth, Lymantria dispar,* Am. Soc. Microbiol., Washington, DC, 1983, p. 50.

141. M. Shapiro, "Comparative Infectivity of Gypchek L-79 and Virin-ENSh to *Lymantria dispar,"* in C. M. Ignoffo, M. E. Martignoni, and J. L. Vaughn, Eds., *A Comparison of the US (Gypchek) and USSR (Virin-ENSh) Preparations of the Nuclear Polyhedrosis Virus of the Gypsy Moth, Lymantria dispar,* Am. Soc. Microbiol., Washington, DC, 1983, p. 38.

142. D. M. Dunbar and R. L. Beard, Present status of milky disease of Japanese and Oriental beetles in Connecticut, *J. Econ. Entomol.,* **68,** 453 (1975).

143. R. A. Humber, "Foundations for an Evolutionary Classification of the Entomophthorales (Zygomycetes)," in Q. Wheeler and M. Blackwell, Eds., *Fungus-Insect Relationships. Perspectives in Ecology and Evolution,* Columbia Univ. Press, New York, 1984, p. 166.

144. A. J. Nahmias and D. C. Reanney, The evolution of viruses, *Annu. Rev. Ecol. Syst.,* **8,** 29 (1977).

145. A. H. Ellingboe, Changing concepts in host-pathogen genetics, *Annu. Rev. Phytopathol.,* **19,** 125 (1981).

146. D. R. Brooks and C. Mitter, "Analytical Approaches to Studying Coevolution," in Q. Wheeler and M. Blackwell, Eds., *Fungus-Insect Relationships, Perspectives in Ecology and Evolution,* Columbia Univ. Press, New York, 1984, p. 42.

147. R. M. May and R. M. Anderson, Epidemiology and genetics in the coevolution of parasites and hosts, *Proc. R. Soc. London,* **B219,** 281 (1983).

148. B. Zelazny, Transmission of a baculovirus in populations of *Oryctes rhinoceros, J. Invertebr. Pathol.,* **27,** 221 (1976).

149. G. O. Bedford, "Control of the Rhinoceros Beetle by Baculovirus," in H. D. Burges, Ed., *Microbial Control of Pests and Plant Diseases 1970–1980,* Academic Press, New York, 1981, p. 409.

150. E. A. Steinhaus, The effects of disease on insect populations, *Hilgarida,* **23,** 197 (1954).

151. C. G. Thompson, "Nuclear Polyhedrosis Epizootiology," in M. H. Boorks, R. W. Stark, and R. W. Campbell, Eds., *The Douglas-Fir Tussock Moth: A Synthesis,* U. S. Dep. Agric., For. Ser. Sci. Educ. Agency Tech. Bull. 1585, 1978, p. 136.

152. C. C. Doane, Primary pathogens and their role in the development of an epizootic in the gypsy moth, *J. Invertebr. Pathol.,* **15,** 21 (1970).

153. R. E. Balch and F. T. Bird, A disease of the European spruce sawfly, *Gilpinia hercyniae* (Htg.), and its place in natural control, *Sci. Agric.,* **25,** 65 (1944).

154. J. V. Maddox, "Use of Insect Pathogens in Pest Management," in R. L. Metcalf and W. H. Luckmann, Eds., *Introduction to Insect Pest Management,* John Wiley & Sons, New York, 1982, p. 175.

155. H. D. Burges and E. M. Thomson, "Standardization and Assay of Microbial Insecticides," in H. D. Burges and N. W. Hussey, Eds., *Microbial Control of Insects and Mites,* Academic Press, New York, 1971, p. 591.

156. J. Huber and P. R. Hughes, Quantitative bioassay in insect pathology, *Bull. Entomol. Soc. Am.,* **30,** 31 (1984).

157. D. T. Briese, "Resistance of Insect Species to Microbial Pathogens," in E. W. Davidson, Ed., *Pathogenesis of Invertebrate Microbial Diseases,* Allanheld, Osmun Publ., Totowa, New Jersey, 1981, p. 511.

158. J. R. Fuxa, Interactions of the microsporidium *Vairimorpha necatrix* with a bacterium, virus, and fungus in *Heliothis zea, J. Invertebr. Pathol.,* **33,** 316 (1979).

159. A. R. Richter and J. R. Fuxa, Pathogen-pathogen and pathogen-insecticide interactions in velvetbean caterpillar (Lepidoptera: Noctuidae), *J. Econ. Entomol.,* **77,** 1559 (1984).

160. T. Hukuhara, Further studies on the distribution of a nuclear-polyhedrosis virus of the fall webworm, *Hyphantria cunea,* in soil, *J. Invertebr. Pathol.,* **22,** 345 (1973).

161. J. R. Fuxa, G. W. Warren, and C. Y. Kawanishi, Comparison of bioassay and enzyme-linked immunosorbent assay for quantification of *Spodoptera frugiperda* nuclear polyhedrosis virus in soil, *J. Invertebr. Pathol.,* **46,** 133 (1985).

162. H. D. Burges, Ed., *Microbial Control of Pests and Plant Diseases 1970–1980,* Academic Press, New York, 1981.

163. E. Kurstak, Ed., *Microbial and Viral Pesticides,* Marcel Dekker, New York, 1982.

164. R. P. Jaques, Occurrence and accumulation of viruses of *Trichoplusia ni* in treated field plots, *J. Invertebr. Pathol.,* **23,** 140 (1974).

165. S. Y. Young and W. C. Yearian, Soil application of *Pseudoplusia* NPV: persistence and incidence of infection in soybean looper caged on soybean, *Environ. Entomol.,* **8,** 860 (1979).

166. S. R. Dutky, "The Milky Diseases," in E. A. Steinhaus, Ed., *Insect Pathology: An Advanced Treatise,* Vol. 2, Academic Press, New York, 1963, p. 75.

167. F. T. Bird and J. M. Burk, Artificially disseminated virus as a factor controlling the European spruce sawfly, *Diprion hercyniae* (Htg.) in the absence of introduced parasites, *Can. Entomol.,* **93,** 228 (1961).

168. P. F. Entwistle, P. H. W. Adams, H. F. Evans, and C. F. Rivers, Epizootiology of a nuclear polyhedrosis virus (Baculoviridae) in European spruce sawfly *(Gilpinia hercyniae):* spread of disease from small epicentres in comparison with spread of baculovirus diseases in other hosts, *J. Appl. Ecol.,* **20,** 473 (1983).

169. E. C. Young, The epizootiology of two pathogens of the coconut palm rhinoceros beetle, *J. Invertebr. Pathol.,* **24,** 82 (1974).

170. D. T. Briese, The incidence of parasitism and disease in field populations of the potato moth *Phthorimaea operculella* (Zeller) in Australia, *J. Aust. Entomol. Soc.,* **20,** 319 (1981).

171. G. G. Wilson and W. J. Kaupp, *Application of Nosema fumiferanae* and *Pleistophora schubergi (Microsporida) against the Spruce Budworm in Ontario, 1976,* Can. For. Serv. Inf. Rep. No. IP-X-15, Sault Ste. Marie (1976).

172. J. D. Podgwaite, P. Rush, D. Hall, and G. S. Walton, Efficacy of the *Neodiprion sertifer* (Hymenoptera: Diprionidae) nucleopolyhedrosis virus *(Baculovirus)* product, Neochek-S, *J. Econ. Entomol.,* **77,** 525 (1984).

173. R. K. Washino and B. B. Westerdahl, Influence of soil type and inoculation rate on population dynamics of *Romanomermis culicivorax, Univ. Calif. Mosq. Control Res. Annu. Rep.* (1981).

174. O. Sosa, Jr. and J. B. Beavers, Entomogenous nematodes as biological control organisms for *Ligyrus subtropicus* (Coleoptera: Scarabaeidae) in sugarcane, *Environ. Entomol.,* **14,** 80 (1985).

175. T. R. Gottwald and W. L. Tedders, Suppression of pecan weevil (Coleoptera: Curculionidae) populations with entomopathogenic fungi, *Environ. Entomol.,* **12,** 471 (1983).

176. R. A. Hall, Control of aphids by the fungus, *Verticillium lecanii:* effect of spore concentration, *Entomol. Exp. Appl.,* **27,** 1 (1980).

177. W. R. Kellen and D. F. Hoffmann, Dose-mortality and stunted growth responses of larvae of the navel orangeworm, *Amyelois transitella,* infected by chronic stunt virus, *Environ. Entomol.,* **11,** 214 (1982).

178. R. F. Sheppard and G. R. Stairs, Dosage-mortality and time-mortality studies of a granulosis virus in a laboratory strain of codling moth, *Laspeyresia pomonella, J. Invertebr. Pathol.,* **29,** 216 (1977).

179. J. R. Fuxa and W. M. Brooks, Effects of *Vairimorpha necatrix* in sprays and corn meal on *Heliothis* species in tobacco, soybeans, and sorghum, *J. Econ. Entomol.,* **72,** 462 (1979).

180. D. E. Pinnock and R. J. Brand, "A Quantitative Approach to the Ecology of the Use of Pathogens for Insect Control," in H. D. Burges, Ed., *Microbial Control of Pests and Plant Diseases 1970–1980,* Academic Press, New York, 1981, p. 655.

181. R. J. Brand and D. E. Pinnock, "Application of Biostatistical Modelling to Forecasting the Results of Microbial Control Trials," in H. D. Burges, Ed., *Microbial Control of Pests and Plant Diseases 1970–1980,* Academic Press, New York, 1981, p. 667.

182. R. M. Anderson and R. M. May, Infectious diseases and population cycles of forest insects, *Science,* **210,** 658 (1980).

183. R. S. Soper and D. M. MacLeod, *Descriptive Epizootiology of an Aphid Mycosis,* U. S. Dep. Agric. Tech. Bull. No. 1632, 1981.

184. H. F. Evans and G. P. Allaway, Dynamics of baculovirus growth and dispersal in *Mamestra brassicae* L. (Lepidoptera: *Noctuidae)* larval populations introduced into small cabbage plots, *Appl. Environ. Microbiol.,* **45,** 493 (1983).

185. J. R. Fuxa and J. P. Geaghan, Multiple-regression analysis of factors affecting prevalence of nuclear polyhedrosis virus in *Spodoptera frugiperda* (Lepidoptera: Noctuidae) populations, *Environ. Entomol.,* **12,** 311 (1983).

186. R. I. Carruthers, D. L. Haynes, and. D. M. MacLeod, *Entomophthora muscae* (Entomophthorales: Entomophthoraceae) mycosis in the onion fly, *Delia antiqua* (Diptera: Anthomyiidae), *J. Invertebr. Pathol.,* **45,** 81 (1985).

187. L. P. Kish and G. E. Allen, *The Biology and Ecology of* Nomuraea rileyi *and a Program for Predicting its Incidence on* Anticarsia gemmatalis *in Soybean,* Univ. Fla. Agric. Exp. Stn. Tech. Bull. 795, 1978.

188. C. R. Flückiger, Untersuchungen über drei Baculovirus-Isolate des Schalenwicklers, *Adoxophyes orana* F. v. R. (Lep., Tortricidae), dessen Phänologie und erste Feldversuche, als Grundlagen zur mikrobiologischen Bekämpfung dieses Obstschädlings, *Mitt. Schweiz. Entomol. Ges. Bull. Soc. Entomol. Suisse,* **55,** 241 (1982).

189. R. M. Anderson and D. M. Gorden, Processes influencing the distribution of parasite numbers within host populations with special emphasis on parasite-induced host mortalities, *Parasitology,* **85,** 373 (1982).

190. H. K. Kaya and R. D. Moon, The nematode *Heterotylenchus autumnalis* and face fly *Musca autumnalis:* a field study in northern California, *J. Nematol.,* **10,** 333 (1978).

191. Y. Tanada and E. M. Omi, Persistence of insect viruses in field populations of alfalfa insects, *J. Invertebr. Pathol.,* **23,** 360 (1974).

192. B. A. Federici, "Disease Prevalence and Epizootics in Insect Populations," in C. M. Ignoffo, M. E. Martignoni, and J. L. Vaughn, Eds., *Characterization, Production and Utilization of Entomopathogenic Viruses,* Proc. 2nd Conf. Proj. V, Microbial Control Insect Pests, US/ USSR Joint Working Group on the Production of Substances by Microbiological Means, Am. Soc. Microbiol., Washington, DC, 1980, p. 17.

193. T. G. Andreadis, Epizootiology of *Nosema pyrausta* in field populations of the European corn borer (Lepidoptera: Pyralidae), *Environ. Entomol., 13,* 882 (1984).

194. R. T. White and S. R. Dutky, Effect of the introduction of milky diseases on populations of Japanese beetle larvae, *J. Econ. Entomol., 33,* 306 (1940).

195. Y. Tanada, The role of viruses in the regulation of the population of the armyworm, *Pseudaletia unipuncta* (Haworth), *Proc. Joint U.S.-Japan Semin. Microbial Control Insect Pests, U.S.-Japan Committee Sci. Cooperation, Panel 8,* p. 25, Fukuoka, Japan (1968).

196. Y. Tanada, "Ecology of Insect Viruses," in J. F. Anderson and H. K. Kaya, Eds., *Perspectives in Forest Entomology,* Academic Press, New York, 1976, p. 265.

197. E. A. Steinhaus, Symposium: selected topics in microbial ecology. II. The importance of environmental factors in the insect-microbe ecosystem, *Bacteriol. Rev., 24,* 365 (1960).

198. R. M. May and R. M. Anderson, Population biology of infectious diseases: part II, *Nature, 280,* 455 (1979).

199. R. M. Anderson and R. M. May, Regulation and stability of host-parasite population interactions. I. Regulatory processes, *J. Anim. Ecol., 47,* 219 (1978).

200. R. P. Jaques and D. G. Harcourt, Viruses of *Trichoplusia ni* (Lepidoptera: Noctuidae) and *Pieris rapae* (Lepidoptera: Pieridae) in soil in fields of crucifers in southern Ontario, *Can. Entomol., 103,* 1285 (1971).

201. W. J. Kaupp, Estimation of nuclear polyhedrosis virus produced in field populations of the European pine sawfly, *Neodiprion sertifer* (Geoff.) (Hymenoptera: Diprionidae), *Can. J. Zool., 61,* 1857 (1983).

202. J. D. Harper, D. A. Herbert, and R. E. Moore, Trapping patterns of *Entomophthora gammae* (Weiser) (Entomophthorales: Entomophthoraceae) conidia in a soybean field infested with the sobyean looper, *Pseudoplusia includens* (Walker) (Lepidoptera: Noctuidae), *Environ. Entomol., 13,* 1186 (1984).

203. Y. Tanada, "Persistence of Entomogenous Viruses in the Insect Ecosystem," in S. Asahina, J. L. Gressitt, Z. Hidaka, T. Nishida, and K. Nomura, Eds., *Entomological Essays to Commemorate the Retirement of Professor K. Yasumatsu,* Hokuryukan, Tokyo, 1971, p. 367.

204. H. Watanabe and T. Shimizu, Epizootiological studies on the occurrence of densonucleosis in the silkworm, *Bombyx mori,* reared at sericultural farms, *J. Sericult. Sci. Japan, 49,* 485 (1980).

205. C. M. Ignoffo and D. L. Hostetter, Summary of the environmental stability of microbial insecticides, *Misc. Publ. Entomol. Soc. Am., 10,* 117 (1977).

206. Y. Tanada, "Persistence of Pathogens in the Aquatic Environment," in A. W. Bourquin, D. G. Ahearn, and S. P. Meyers, Eds., *EPA 660-3-75-001, U.S. Environ. Prot. Agency,* 1975, p. 83.

207. J. Margalit, A. Markus, and Z. Pelah, Effect of encapsulation on the persistence of *Bacillus thuringiensis* var. *israelensis,* serotype H-14, *Appl. Microbiol. Biotechnol., 19,* 382 (1984).

208. V. M. Sharapov and T. K. Kalvish, Effect of soil fungistasis on zoopathogenic fungi, *Mycopathologia, 85,* 121 (1984).

209. M. G. Klein, "Advances in the Use of *Bacillus popilliae* for Pest Control," in H. D. Burges, Ed., *Microbial Control of Pests and Plant Diseases 1970–1980,* Academic Press, New York, 1981, p. 183.

210. T. R. Gottwald and W. L. Tedders, Colonization, transmission, and longevity of *Beauveria bassiana* and *Metarhizium anisopliae* (Deuteromycotina: Hyphomycetes) on pecan weevil larvae (Coleoptera: Curculionidae) in the soil, *Environ. Entomol., 13,* 557 (1984).

211. G. R. Stairs, Quantitative aspects of virus dispersion and the development of epizootics in insect populations, *Proc. Joint U.S.-Japan Semin. Microbial Control Insect Pests, U.S.-Japan Committee Sci. Cooperation. Panel 8,* p. 19, Fukuoka, Japan (1968).

212. T. Hukuhara and H. Namura, Distribution of a nuclear-polyhedrosis virus of the fall webworm, *Hyphantria cunea,* in soil, *J. Invertebr. Pathol.,* **19,** 308 (1972).

213. C. G. Thompson and E. A. Steinhaus, Further tests using a polyhedrosis virus to control the alfalfa caterpillar, *Hilgardia,* **19,** 411 (1950).

214. R. P. Jaques, Natural occurrence of viruses of the cabbage looper in field plots, *Can. Entomol.,* **102,** 36 (1970).

215. Y. Tanada and E. M. Omi, Epizootiology of virus diseases in three lepidopterous insect species of alfalfa, *Res. Popul. Ecol.,* **16,** 59 (1974).

216. J. Weiser and J. Veber, Die Mikrosporidie *Thelohania hyphantriae* Weiser des weissen Bärenspinners und anderer Mitglieder seiner Biocönose, *Z. Angew. Entomol.,* **40,** 55 (1957).

217. J. M. Franz, "Influence of Environment and Modern Trends in Crop Management on Microbial Control," in H. D. Burges and N. W. Hussey, Eds., *Microbial Control of Insects and Mites,* Academic Press, New York, 1971, p. 407.

218. E. M. Reed, Factors affecting the status of a virus as a control agent for the potato moth *(Phthorimaea operculella)* (Zell.) (Lep., Gelechiidae), *Bull. Entomol. Res.,* **61,** 207 (1971).

219. D. J. Kushner and G. T. Harvey, Antibacterial substances in leaves: their possible role in insect resistance to disease, *J. Insect Pathol.,* **4,** 155 (1962).

220. B. Maksymiuk, Occurrence and nature of antibacterial substances in plants affecting *Bacillus thuringiensis* and other entomogenous bacteria, *J. Invertebr. Pathol.,* **15,** 356 (1970).

221. D. E. Leonard and C. C. Doane, An artificial diet for the gypsy moth, *Porthetria dispar* (Lepidoptera: Lymantriidae), *Ann. Entomol. Soc. Am.,* **59,** 462 (1966).

222. C. C. Doane, Trans-ovum transmission of a nuclear-polyhedrosis virus in the gypsy moth and the inducement of virus susceptibility, *J. Invertebr. Pathol.,* **14,** 199 (1969).

223. C. C. Doane, Infectious sources of nuclear polyhedrosis virus persisting in natural habitats of the gypsy moth, *Environ. Entomol.,* **4,** 392 (1975).

224. F. T. Bird, J. C. Cunningham, and G. M. Howse, The possible use of viruses in the control of spruce budworm, *Proc. Entomol. Soc. Ontario,* **103,** 69 (1972).

225. M. Shiga, H. Yamada, N. Oho, H. Nakaguwa, and Y. Ito, A granulosis virus, possible biological agent for control of *Adoxophyes orana* (Lepidoptera: Tortricidae) in apple orchards, II. Semipersistent effect of artificial dissemination into an apple orchard, *J. Invertebr. Pathol.,* **21,** 149 (1973).

226. B. Zelazny, Studies on *Rhabdionvirus oryctes,* III. Incidence in the *Oryctes rhinoceros* population of Western Samoa, *J. Invertebr. Pathol.,* **22,** 359 (1973).

227. R. S. Soper, L. F. R. Smith, and A. J. Delyzer, Epizootiology of *Massospora levispora* in an isolated population of *Okanagana rimosa, Ann. Entomol. Soc. Am.,* **69,** 275 (1976).

228. J. Weiser, "Sporozoan Infections," in E. A. Steinhaus, Ed., *Insect Pathology: An Advanced Treatise,* Vol. 2, Academic Press, New York, 1963, p. 291.

229. A. R. Hamed, Zur Wirkung von *Bacillus thuringiensis* auf Parasiten ünd Pradatoren von *Yponomeuta evonymellus* (Lep., Yponomeutidae), *Z. Angew, Entomol.,* **87,** 294 (1979).

230. H. C. Whisler, S. L. Zebold, and J. A. Shemanchuk, Life history of *Coelomomyces psorophorae, Proc. Nat. Acad. Sci. USA,* **72,** 693 (1975).

231. B. A. Federici and D. W. Roberts, Experimental laboratory infection of mosquito larvae with fungi of the genus *Coelomomyces.* II. Experiments with *Coelomomyces punctatus* in *Anopheles quadrimaculatus, J. Invertebr. Pathol.,* **27,** 333 (1976).

232. A. W. Sweeney, E. I. Hazard, and M. F. Graham, Intermediate host for an *Amblyospora* sp. (Microspora) infecting the mosquito, *Culex annulirostris, J. Invertebr. Pathol.,* **46,** 98 (1985).

233. T. G. Andreadis, Life cycle, epizootiology, and horizontal transmission of *Amblyospora*

(Microspora: Amblyosporidae) in a univoltine mosquito, *Aedes stimulans, J. Invertebr. Pathol.,* **46,** 31 (1985).

234. S. Bitton, R. G. Kenneth, and I. Ben-Ze'ev, Zygospore overwintering and sporulative germination in *Triplosporium fresenii* (Entomophthoraceae) attacking *Aphis spiraecola* on citrus in Israel, *J. Invertebr. Pathol.,* **34,** 295 (1979).

235. G. R. Carner, *Entomophthora lampyridarum,* a fungal pathogen of the soldier beetle, *Chauliognathus pennsylvanicus, J. Invertebr. Pathol.,* **36,** 394 (1980).

236. M. Burman and A. E. Pye, *Neoaplectana carpocapsae:* movements of nematode populations on a thermal gradient, *Exp. Parasitol.,* **49,** 258 (1980).

237. R. Gaugler, L. LeBeck, B. Nakagaki, and G. M. Boush, Orientation of the entomogenous nematode, *Neoaplectana carpocapsae* to carbon dioxide, *Environ. Entomol.,* **9,** 649 (1980).

238. F. T. Bird, Transmission of some insect viruses with particular reference to ovarial transmission and its importance in the development of epizootics, *J. Insect Pathol.,* **3,** 352 (1961).

239. G. R. Stairs, Artificial initiation of virus epizootics in forest tent caterpillar populations, *Can. Entomol.,* **97,** 1059 (1965).

240. K. J. Marschall, Introduction of a new virus disease of the coconut rhinoceros beetle in Western Samoa, *Nature,* **225,** 288 (1970).

241. A. M. Huger, Grundlagen zur biologischen Bekämpfung des Indischen Nashornkäfers, *Oryctes rhinoceros* (L.), mit *Rhabdionvirus oryctes:* Histopathologie der Virose bei Käfern, *Z. Angew. Entomol.,* **72,** 309 (1973).

242. G. Benz, "Physiopathology and Histopathology," in E. A. Steinhaus, Ed., *Insect Pathology: An Advanced Treatise,* Vol. 1, Academic Press, New York, 1963, p. 299.

243. J. Komarek and V. Breindl, Die Wipfelkrankheit der Nonne und der Erreger derselben, *Z. Angew. Entomol.,* **10,** 99 (1924).

244. W. A. Smirnoff, Observations on the migration of larvae of *Neodiprion swainei* Midd. (Hymenoptera: Tenthredinidae), *Can. Entomol.,* **92,** 957 (1960).

245. M. Tamashiro and S.-S. Huang, A cytoplasmic polyhedrosis of *Cactoblastis cactorum* (Berg), *J. Invertebr. Pathol.,* **5,** 397 (1963).

246. O. F. Niklas, Vertikalbewegungen Rickettsiose-kranker Larven von *Amphimallon solstitiale* (Linnaeus), *Anomala dubia aenea* (DeGeer) und *Maladera brunnea* (Linnaeus) (Col., Lamellicornia) *Anz. Schaedlingsk.,* **37,** 22 (1964).

247. M. Fowler and J. S. Robertson, Iridescent virus infection in field populations of *Wiseana cervinata* (Lepidoptera: Hepialidae) and *Witlesia* sp. (Lepidoptera: Pyralidae) in New Zealand, *J. Invertebr. Pathol.,* **19,** 154 (1972).

248. G. S. Langford, R. H. Vincent, and E. N. Cory, The adult Japanese beetle as host and disseminator of type A milky disease, *J. Econ. Entomol.,* **35,** 165 (1942).

249. R. S. Soper, A. J. Delyzer, and L. F. R. Smith, The genus *Massospora* entomopathogenic for cicadas, Part II, Biology of *Massospora levispora* and its host *Okanagana rimosa,* with notes on *Massospora cicadina* on the periodical cicadas, *Ann. Entomol. Soc. Am.,* **69,** 89 (1976).

250. A. T. Speare, *Massospora cicadina* Peck: a fungous parasite of the periodical cicada, *Mycologia,* **13,** 72 (1921).

251. A. M. Huger, Report on the activities of the "Institut fur Biologische Schadlingsbekampfung," Darmstadt, Germany. Further diagnostic and histopathological studies on *Oryctes rhinoceros* and two of its predators, UNDP/SPC project for research on the control of the coconut palm rhinoceros beetle, *Semi-Annu. Rep. Proj. Manager, Nov. 1969 to May 1970,* p. 16 (1970).

252. B. Zelazny, Studies on *Rhabdionvirus oryctes,* II. Effect on adults of *Orcytes rhinoceros, J. Invertebr. Pathol.,* **22,** 122 (1973).

253. C. Sidor, L. J. Dušánic, B. Zamola, B. Todoroviski, L. J. Vasilev, and F. Kajefez, A study of

viral disease in *Heliothis armigera* Hbn. caused by polyhedral viruses from *Macedonia, Yugoslavia, Redia,* **60,** 317 (1977).

254. I. E. Gard and L. A. Falcon, "Autodissemination of Entomopathogens: Virus," in G. E. Allen, C. M. Ignoffo, and R. P. Jaques, Eds., *Microbial Control of Insect Pests: Future Strategies in Pest Management Systems,* NSF—USDA—Univ. Fla. Gainesville, 1978, p. 45.

255. M. E. Martignoni and J. E. Milstead, Trans-ovum transmission of the nuclear polyhedrosis virus of *Colias eurytheme* Boisduval through contamination of the female genitalia, *J. Insect Pathol.,* **4,** 113 (1962).

256. J. C. Elmore and A. F. Howland, Natural versus artificial dissemination of nuclear-polyhedrosis virus by contaminated adult cabbage loopers, *J. Insect Pathol.,* **6,** 430 (1964).

257. M. M. Neilson and D. E. Elgee, The method and role of vertical transmission of a nucleopolyhedrosis virus in the European spruce sawfly, *Diprion hercyniae, J. Invertebr. Pathol.,* **12,** 132 (1968).

258. I. E. Gard, *Utilization of light traps to disseminate insect viruses for pest control,* Ph.D. Dissertation, Dept. Entomol. Sci., Univ. California, Berkeley, 1975.

259. L. Bailey, The multiplication and spread of sacbrood virus of bees, *Ann. Appl. Biol.,* **63,** 483 (1969).

260. D. P. Jouvenaz, E. A. Ellis, and C. S. Lofgren, Histopathology of the tropical fire ant, *Solenopsis geminata,* infected with *Burenella dimorpha* (Microspora: Microsporida), *J. Invertebr. Pathol.,* **43,** 324 (1984).

261. P. F. Entwistle, Passive carriage of baculoviruses in forests, *Proc. IIIrd Int. Coll. Invertebr. Pathol. XVth Annu. Meet. Soc. Invertebr. Pathol. Sept. 6–10, 1982,* Univ. Sussex, Brighton, United Kingdom, 344 (1982).

262. F. M. Laigo and M. Tamashiro, Virus and insect parasite interaction in the lawn armyworm, *Spodoptera mauritia acronyctoides* (Guenée), *Proc. Hawaii. Entomol. Soc.,* **19,** 233 (1966).

263. W. A. Smirnoff, "Parasites and Predators as Vectors of Insect Viruses," in M. Summers, R. Engler, L. A. Falcon, and P. V. Vail, Eds., *Baculovirus for Insect Pest Control: Safety Considerations,* Am. Chem. Soc. Microbiol., Washington, D. C., 1975, p. 131.

264. J. L. Capinera and P. Barbosa, Transmission of nuclear-polyhedrosis virus to gypsy moth larvae by *Calosoma sycophanta, Ann. Entomol. Soc. Am.,* **68,** 593 (1975).

265. R. C. Reardon and J. D. Podgwaite, Disease-parasitoid relationships in natural populations of *Lymantria dispar* [Lep.: Lymantriidae] in the northeastern United States, *Entomophaga* **21,** 333 (1976).

266. B. Raimo, R. C. Reardon, and J. D. Podgwaite, Vectoring gypsy moth nuclear polyhedrosis virus by *Apanteles melanoscelus* [Hym.: Braconidae], *Entomophaga,* **22,** 207 (1977).

267. D. B. Levin, J. E. Laing, and R. P. Jaques, Transmission of granulosis virus by *Apanteles glomeratus* to its host *Pieris rapae, J. Invertebr. Pathol.,* **34,** 317 (1979).

268. A. G. B. Beekman, The infectivity of polyhedra of nuclear polyhedrosis virus (N.P.V.) after passage through gut of an insect predator, *Experientia,* **36,** 858 (1980).

269. D. J. Cooper, The role of predatory Hemiptera in disseminating a nuclear polyhedrosis virus of *Heliothis punctiger, J. Aust. Entomol. Soc.,* **20,** 145 (1981).

270. D. B. Levin, J. E. Laing, R. P. Jaques, and J. E. Corrigan, Transmission of the granulosis virus of *Pieris rapae* (Lepidoptera: Pieridae) by the parasitoid *Apanteles glomeratus* (Hymenoptera: Braconidae), *Environ. Entomol.,* **12,** 166 (1983).

271. M. S. T. Abbas and D. G. Boucias, Interaction between nuclear polyhedrosis virus-infected *Anticarsia gemmatalis* (Lepidoptera: Noctuidae) larvae and predator *Podisus maculiventris* (Say) (Hemiptera: Pentatomidae), *Environ. Entomol.* **13,** 599 (1984).

272. D. L. Hostetter, A virulent nuclear polyhedrosis virus of the cabbage looper, *Trichoplusia ni,* recovered from the abdomens of sarcophagid flies, *J. Invertebr. Pathol.,* **17,** 130 (1971).

273. L. Szalay-Marzsó and C. Vago, Transmission of baculovirus by mites. Study of granulosis

virus of codling moth *(Laspeyresia pomonella* L.), *Acta Phytopathol. Acad. Sci. Hung.,* **10,** 113 (1975).

274. A. M. Crawford and J. Kalmakoff, A host-virus interaction in a pasture habitat *Wiseana* spp. (Lepidoptera: Hepialidae) and its baculoviruses, *J. Invertebr. Pathol.,* **29,** 81 (1977).

275. R. A. Lautenschlager, J. D. Podgwaite, and D. E. Watson, Natural occurrence of the nucleo-polyhedrosis virus of the gypsy moth, *Lymantria dispar* (Lep.: Lymantriidae) in wild birds and mammals, *Entomophaga,* **25,** 261 (1980).

276. G. R. Stairs, Transmission of virus in tent caterpillar populations, *Can. Entomol.,* **98,** 1100 (1966).

277. F. T. Bird, The use of viruses in biological control, *Entomophaga Mém. Hors Sér. No.* **2,** 465 (1964).

278. K. D. Biever, P. L. Andrews, and P. A. Andrews, Use of a predator, *Podisus maculiventris,* to distribute virus and initiate epizootics, *J. Econ. Entomol.,* **75,** 150 (1982).

279. W. A. Smirnoff, Predators of *Neodiprion swainei* Midd. (Hymenoptera: Tenthredinidae) larval vectors of virus diseases, *Can. Entomol.,* **91,** 246 (1959).

280. P. F. Entwistle, P. H. W. Adams, and H. F. Evans, Epizootiology of a nuclear-polyhedrosis virus in European spruce sawfly *(Gilpinia hercyniae):* the status of birds as dispersal agents of the virus during the larval season, *J. Invertebr. Pathol.,* **29,** 354 (1977).

281. P. F. Entwistle, P. H. W. Adams, and H. F. Evans, Epizootiology of a nuclear-polyhedrosis virus in European spruce sawfly, *Gilpinia hercyniae:* birds as dispersal agents of the virus during winter, *J. Invertebr. Pathol.,* **30,** 15 (1977).

282. P. F. Entwistle, P. H. W. Adams, and H. F. Evans, Epizootiology of a nuclear polyhedrosis virus in European spruce sawfly *(Gilpinia hercyniae):* the rate of passage of infective virus through the gut of birds during cage tests, *J. Invertebr. Pathol.,* **31,** 307 (1978).

283. D. L. Hostetter and K. D. Biever, The recovery of virulent nuclear-polyhedrosis virus of the cabbage looper, *Trichoplusia ni,* from the feces of birds, *J. Invertebr. Pathol.,* **15,** 173 (1970).

284. A. M. Crawford and J. Kalmakoff, Transmission of *Wiseana* spp. nuclear polyhedrosis virus in the pasture habitat, *N. Z. J. Agric. Res.,* **21,** 521 (1978).

285. R. A. Lautenschlager and J. D. Podgwaite, Passage of nucleopolyhedrosis virus by avian and mammalian predators of the gypsy moth, *Lymantria dispar, Environ. Entomol.,* **8,** 210 (1979).

286. J. N. Matthiessen and B. P. Springett, The food of the silvereye, *Zosterops gouldi* (Aves: Zosteropidae), in relation to its role as a vector of a granulosis virus of the potato moth, *Phthorimaea operculella* (Lepidoptera: Gelechiidae), *Aust. J. Zool.,* **21,** 533 (1973).

287. J. E. Van Der Plank, *Principles of Plant Infection,* Academic Press, New York, 1975.

288. F. T. Bird and D. E. Elgee, A virus disease and introduced parasites as factors controlling the European spruce sawfly, *Diprion hercyniae* (Htg.), in central New Brunswick, *Can. Entomol.,* **89,** 371 (1957).

289. L. R. Taylor, Assessing and interpreting the spatial distributions of insect populations, *Annu. Rev. Entomol.,* **29,** 321 (1984).

290. J. R. Fuxa, Dispersion and spread of the entomopathogenic fungus *Nomuraea rileyi* (Moni-liales: Moniliaceae) in a soybean field, *Environ. Entomol.,* **13,** 252 (1984).

291. C. G. Thompson, J. Neisess, and H. O. Batzer, Field tests of *Bacillus thuringiensis* and aerial application strategies on western mountainous terrain, *U. S. Dep. Agric. For. Serv. Res. Pap. PNW-230,* 1977.

292. J. E. Henry, E. A. Oma, and J. A. Onsager, Relative effectiveness of ULV spray applications of spores of *Nosema locustae* against grasshoppers, *J. Econ. Entomol.,* **71,** 629 (1978).

293. A. L. Stacey, S. Y. Young, and W. C. Yearian, *Baculovirus heliothis:* effect of selective placement on *Heliothis* mortality and efficacy in directed sprays on cotton, *J. Georgia Ento-mol. Soc.,* **12,** 167 (1977).

294. Z. Mracek, Horizontal distribution in soil, and seasonal dynamics of the nematode *Steinernema kraussei*, a parasite of *Cephalcia abietis*, *Z. Angew. Entomol.*, **94**, 110 (1982).

295. R. P. Jaques, Leaching of the nuclear-polyhedrosis virus of *Trichoplusia ni* from soil, *J. Invertebr. Pathol.*, **13**, 256 (1969).

296. C. G. Thompson and D. W. Scott, Production and persistence of the nuclear polyhedrosis virus of the Douglas fir tussock moth, *Orgyia pseudotsugata* (Lepidoptera: Lymantriidae), in the forest ecosystem, *J. Invertebr. Pathol.*, **33**, 57 (1979).

297. H. F. Evans and P. F. Entwistle, "Epizootiology of the Nuclear Polyhedrosis Virus of European Spruce Sawfly with Emphasis on Persistence of Virus Outside the Host," in E. Kurstak, Ed., *Microbial and Viral Pesticides*, Marcel Dekker, New York, 1982, p. 449.

298. J. D. Podgwaite, K. S. Shields, R. T. Zerillo, and R. B. Bruen, Environmental persistence of the nucleopolyhedrosis virus of the gypsy moth, *Lymantria dispar, Environ. Entomol.*, **8**, 528 (1979).

299. J. W. Doberski and H. T. Tribe, Isolation of entomogenous fungi from elm bark and soil with reference to ecology of *Beauveria bassiana* and *Metarhizium anisopliae, Trans. Br. Mycol. Soc.*, **74**, 95 (1980).

300. J. A. Hutchison, The genus *Entomophthora* in the western hemisphere, *Trans. Kan. Acad. Sci.*, **66**, 237 (1963).

301. J. D. Vandenberg and R. S. Soper, Prevalence of Entomophthorales mycoses in populations of spruce budworm, *Choristoneura fumiferana, Environ. Entomol.*, **7**, 847 (1978).

302. B. Puttler, D. L. Hostetter, S. H. Long, R. E. Munson, and J. L. Huggans, Distribution of the fungus *Entomophthora phytonomi* in larvae of the alfalfa weevil in Missouri, *J. Econ. Entomol.*, **72**, 220 (1979).

303. R. J. Barney, P. L. Watson, K. Black, J. V. Maddox, and E. J. Armbrust, Illinois distribution of the fungus *Entomophthora phytonomi* (Zygomycetes: Entomophthoraceae) in larvae of the alfalfa weevil (Coleoptera: Curculionidae), *Great Lakes Entomol.*, **13**, 149 (1980).

304. B. McDaniel and R. A. Bohls, The distribution and host range of *Entomophaga grylli* (Fresenius), the fungal parasite of grasshoppers in South Dakota, *Proc. Entomol. Soc. Wash.*, **86**, 864 (1984).

305. T. Soldan, Host and tissue specificity of *Spiriopsis adipophila* (Arvy et Delage) (Protozoa, Coccidia) and its distribution in the Elbe Basin of Czechoslovakia, *Fol. Parasitol. (Praha)*, **27**, 77 (1980).

306. A. B. Ewen, Extension of the geographic range of *Nosema locustae* (Microsporidia) in grasshoppers (Orthoptera: Acrididae), *Can. Entomol.*, **115**, 1049 (1983).

307. Z. Mracek, The use of "*Galleria* traps" for obtaining nematode parasites of insects of Czechoslovakia (Lepidoptera: Nematoda, Steinernematidae), *Acta Entomol. Bohemoslov.*, **77**, 378 (1980).

308. R. J. Akhurst and W. M. Brooks, The distribution of entomophilic nematodes (Heterorhabditidae and Steinernematidae) in North Carolina, *J. Invertebr. Pathol.*, **44**, 140 (1984).

309. N. Wilding, *Entomophthora* conidia in the air-spora, *J. Gen. Microbiol.*, **62**, 149 (1970).

310. G. G. Newman and G. R. Carner, Diel periodicity of *Entomophthora gammae* in the soybean looper, *Environ. Entomol.*, **3**, 888 (1974).

311. G. G. Newman and G. R. Carner, Environmental factors affecting conidial sporulation and germination of *Entomophthora gammae, Environ. Entomol.*, **4**, 615 (1975).

312. J. Aoki, Pattern of conidial discharge of an *Entomophthora* species ("*grylli*" type) (Entomophthorales: Entomophthoraceae) from infected cadavers of *Mamestra brassicae* L. (Lepidoptera: Noctuidae), *Appl. Entomol. Zool.*, **16**, 216 (1981).

313. M. Yamamoto and J. Aoki, Periodicity of conidial discharge of *Erynia radicans, Trans. Mycol. Soc. Japan*, **24**, 487 (1983).

314. P. R. Hughes, R. R. Gettig, and W. J. McCarthy, Comparison of the time-mortality response

of *Heliothis zea* to 14 isolates of *Heliothis* nuclear polyhedrosis virus, *J. Invertebr. Pathol.,* **41,** 256 (1983).

315. S. Singer, "Potential of *Bacillus sphaericus* and Related Spore-Forming Bacteria for Pest Control," in H. D. Burges, Ed., *Microbial Control of Pests and Plant Diseases 1970–1980,* Academic Press, New York, 1981, p. 283.

316. J. F. Fargues and P. H. Robert, Effects of passaging through scarabeid hosts on virulence and host specificity of two strains of entomopathogenic hyphomycete *Metarhizium anisopliae, Can. J. Microbiol.,* **29,** 576 (1983).

317. R. R. Mason, Life tables for a declining population of the Douglas-fir tussock moth in northeastern Oregon, *Ann. Entomol. Soc. Am.,* **69,** 948 (1976).

318. P. L. Versoi and W. G. Yendol, Discrimination by the parasite, *Apanteles melanoscelus,* between healthy and virus-infected gypsy moth larvae, *Environ. Entomol.,* **11,** 42 (1982).

319. P. V. Vail, Cabbage looper nuclear polyhedrosis virus-parasitoid interactions, *Environ. Entomol.,* **10,** 517 (1981).

320. C. C. Beegle and E. R. Oatman, Differential susceptibility of parasitized and nonparasitized larvae of *Trichoplusia ni* to a nuclear polyhedrosis virus, *J. Invertebr. Pathol.,* **24,** 188 (1974).

321. T. A. Ibaragon and W. M. Brooks, Interaction of *Campoletis sonorensis* and a nuclear polyhedrosis virus in larvae of *Heliothis virescens, J. Econ. Entomol.,* **67,** 229 (1974).

322. C. C. Beegle and E. R. Oatman, Effect of a nuclear polyhedrosis virus on the relationship between *Trichoplusia ni* (Lepidoptera: Noctuidae) and the parasite, *Hyposoter exiguae* (Hymenoptera: Ichneumonidae), *J. Invertebr. Pathol.,* **25,** 59 (1975).

323. D. B. Levin, J. E. Laing, and R. P. Jaques, Interactions between *Apanteles glomeratus* (L.) (Hymenoptera: Braconidae) and granulosis virus in *Pieris rapae* (L.) (Lepidoptera: Pieridae), *Environ. Entomol.,* **10,** 65 (1981).

324. J. J. Hamm, D. A. Nordlung, and O. G. Marti, Effects of a nonoccluded virus of *Spodoptera frugiperda* (Lepidoptera: Noctuidae) on the development of a parasitoid, *Cotesia marginiventris* (Hymenoptera: Braconidae), *Environ. Entomol.,* **14,** 258 (1985).

325. H. K. Kaya and Y. Tanada, Response of *Apanteles militaris* to a toxin produced in a granulosis-virus-infected host, *J. Invertebr. Pathol.* **19,** 1 (1972).

326. H. K. Kaya and Y. Tanada, Hemolymph factor in armyworm larvae infected with a nuclear-polyhedrosis virus toxic to *Apanteles militaris, J. Invertebr. Pathol.,* **21,** 211 (1973).

327. M. J. Samways, Interrelationship between an entomogenous fungus and two ant-homopteran (Hymenoptera: Formicidae-Hemiptera: Pseudococcidae + Aphidae) mutualisms on guava trees, *Bull. Entomol. Res.,* **73,** 321 (1983).

328. R. L. Metcalf and W. H. Luckmann, Eds., *Introduction to Insect Pest Management,* 2nd ed., John Wiley & Sons, New York, 1982.

329. V. P. Pristavko, On the use of entomopathogenic bacteria together with insecticides in the control of insect pests, *Entomol. Rev.,* **46,** 443 (1967).

330. G. Benz, "Synergism of Micro-organisms and Chemical Insecticides," in H. D. Burges and N. W. Hussey, Eds., *Microbial Control of Insects and Mites,* Academic Press, New York, 1971, p. 327.

331. R. P. Jaques and O. M. Morris, "Compatibility of Pathogens with Other Methods of Pest Control and with Different Crops," in H. D. Burges, Ed., *Microbial Control of Pests and Plant Diseases, 1970–1980,* Academic Press, New York, 1981, p. 695.

332. Y. Tanada, "*Bacillus thuringiensis:* Integrated Control—Past, Present, and Future," in T. C. Cheng, Ed., *Comparative Pathobiology,* Vol. 7, Plenum Press, New York, 1984, 59.

333. C. Vago, "Predisposition and Interrelations in Insect Diseases," in E. A. Steinhaus, Ed., *Insect Pathology: An Advanced Treatise,* Vol. 1, Academic Press, New York, 1963, p. 339.

334. J. Weiser, Mikrosporidien des Schwammspinners und Goldafters, *Z. Angew. Entomol.,* **40,** 509 (1957).

335. W. R. Kellen, T. B. Clark, and J. E. Lindegren, Two previously undescribed *Nosema* from mosquitoes of California (Nosematidae: Microsporidia), *J. Invertebr. Pathol.,* **9,** 19 (1967).

336. W. R. Kellen and J. E. Lindegren, Host-pathogen relationships of two previously undescribed microsporidia from the Indian-meal moth, *Plodia interpunctella* (Hübner), (Lepidoptera: Phycitidae), *J. Invertebr. Pathol.,* **14,** 328 (1969).

337. A. M. Huger, Untersuchungen zur Pathologie einer Microsporidiose von *Agrotis segetum* (Schiff.) *(Lepidopt., Noctuidae),* verursacht durch *Nosema perezioides* nov, spec., *Z. Pflanzenkr. (Pflanzenpathol.) Pflanzenschutz,* **67,** 65 (1960).

338. R. H. Goodwin, Nonsporeforming bacteria in the armyworm, *Pseudaletia unipuncta,* under gnotobiotic conditions, *J. Invertebr. Pathol.,* **11,** 358 (1968).

339. E. A. Steinhaus, *Serratia marcescens* Bizio as an insect pathogen, *Hilgardia,* **28,** 351 (1959).

340. J. Weiser and O. Lysenko, Septikemie bource morušového, *Česk. Mikrobiol.,* **1,** 216 (1956).

341. E. A. Steinhaus, Stress as a factor in insect disease, *Proc. Tenth Int. Congr. Entomol.,* **4,** 725 (1958).

342. H. J. Somerville, Y. Tanada, and E. M. Omi, Lethal effect of purified spore and crystalline endotoxin preparations of *Bacillus thuringiensis* on several lepidopterous insects, *J. Invertebr. Pathol.,* **16,** 241 (1970).

343. J. R. Fuxa, Susceptibility of lepidopterous pests to two types of mortality caused by the microsporidium *Vairimorpha necatrix, J. Econ. Entomol.,* **74,** 99 (1981).

344. S. R. Dutky, Insect microbiology, *Adv. Appl. Microbiol.,* **1,** 175 (1959).

345. G. O. Poinar, Jr. and G. M. Thomas, Significance of *Achromobacter nematophilus* Poinar and Thomas (Achromobacteraceae: Eubacteriales) in the development of the nematode, DD-136 *(Neoaplectana* sp. Steinernematidae), *Parasitology,* **56,** 385 (1966).

346. P. Ferron, B. Hurpin, and P. H. Robert, Sensibilisation des larves de *Melolontha melolontha* L. à la mycose à *Beauveria tenella* par une infection préalable à *Bacillus popilliae, Entomophaga,* **14,** 429 (1969).

347. T. B. Clark and T. Fukuda, Field and laboratory observations of two viral diseases in *Aedes sollicitans* (Walker) in southwestern Louisiana, *Mosq. News,* **31,** 193 (1971).

348. L. Bailey and R. D. Woods, Three previously undescribed viruses from the honey bee, *J. Gen. Virol.,* **25,** 175 (1974).

349. G. W. Wagner, S. R. Webb, J. D. Paschke, and W. R. Campbell, A picornavirus isolated from *Aedes taeniorhynchus* and its interaction with mosquito iridescent virus, *J. Invertebr. Pathol.,* **24,** 380 (1974).

350. M. Arella, G. Devauchelle, and S. Belloncik, Dual infection of a lepidopteran cell line with the cytoplasmic polyhedrosis virus (CPV) and the *Chilo* iridescent virus (CIV), *Ann. Virol. (Inst. Pasteur),* **134E,** 455 (1983).

351. H. Aruga and Y. Hashimoto, Interference between the UV-inactivated and active cytoplasmic-polyhedrosis viruses in the silkworm, *Bombyx mori* (Linnaeus), *J. Sericult. Sci. Japan,* **34,** 351 (1965).

352. F. T. Bird, Polyhedrosis and granulosis viruses causing single and double infections in the spruce budworm, *Choristoneura fumiferana* Clemens, *J. Insect Pathol.,* **1,** 406 (1959).

353. F. T. Bird, Infection and mortality of spruce budworm, *Choristoneura fumiferana,* and forest tent caterpillar, *Malacosoma disstria,* caused by nuclear and cytoplasmic polyhedrosis viruses, *Can. Entomol.,* **101,** 1269 (1969).

354. K. S. Ritter and Y. Tanada, Interference between two nuclear polyhedrosis viruses of the armyworm, *Pseudaletia unipuncta [Lep.: Noctuidae], Entomophaga,* **23,** 349 (1978).

355. D. C. Kelly, Suppression of baculovirus and iridescent virus replication in dually infected cells, *Microbiologica,* **3,** 177 (1980).

356. G. L. Nordin and J. V. Maddox, Effects of simultaneous virus and microsporidian infections on larvae of *Hyphantria cunea, J. Invertebr. Pathol.,* **20,** 66 (1972).

357. D. W. Roberts, *Coelomomyces, Entomophthora, Beauveria,* and *Metarrhizium* as parasites of mosquitoes, *Misc. Publ. Entomol. Soc. Am.,* **7,** 140 (1970).

358. S. Marcandier, *Sensibilite de la pyrale du mais (Ostrinia nubilalis Hübner) aux hyphomycetes entomopathogenes: analyse des phenomenes de mortalite et de morbidite,* Ph. D. Thesis, Univ. Pierre et Marie Curie, Paris, 1982.

359. L. K. Miller, R. E. Trimarchi, D. Browne, and G. D. Pennock, A temperature-sensitive mutant of the baculovirus *Autographa californica* nuclear polyhedrosis virus defective in an early function required for further gene expression, *Virology,* **126,** 376 (1983).

360. A. A. Yousten, Bacteriophage typing of mosquito pathogenic strains of *Bacillus sphaericus, J. Invertebr. Pathol.,* **43,** 124 (1984).

361. A. I. Aronson, W. Beckman, and P. Dunn, *Bacillus thuringiensis* and related insect pathogens, *Microbiol. Rev.,* **50,** 1 (1986).

6

TRANSMISSION

THEODORE G. ANDREADIS

Department of Entomology
Connecticut Agricultural Experiment Station
New Haven, Connecticut

1. INTRODUCTION

Transmission may be defined as the process by which a pathogen or parasite is passed from a source of infection to a new host (1). Transmission is direct when the pathogen is transferred from an infected to a susceptible host without the intervention of any other living agent, or indirect when one or more species of intermediate hosts or vectors is involved. The most dominant form of transmission among insect pathogens is direct and typically follows a host-host pathway of infection (via physical contact or congenital transfer) or a host-environment-host pathway.

The key to the occurrence of any disease in a host population is the ability of the pathogen to be transmitted, and without an efficient means of dispersal to susceptible populations, a pathogen cannot persist (2). The mechanism(s) by which transmission is achieved is also an important determinant of changes in host populations (3) and is primary factor affecting the spread of a disease (4).

Insect pathogens, like those in humans, have evolved a wide variety of mechanisms and adaptations which ensure their survival. A knowledge of these methods and transmission pathways is fundamental to understanding disease dynamics and insect epizootiology. It is the purpose of this chapter to describe the great diversity of transmission mechanisms displayed by insect pathogens and to characterize the relationships between transmission and the initiation and maintenance of enzootic and epizootic infections in host insect populations.

2. METHODS OF PATHOGEN TRANSMISSION AND ROUTES OF ENTRY

Transmission may be separated into two broad categories according to the manner in which the pathogen is transferred within a host population. Transmission is horizontal when the pathogen is transferred from individual to individual but not directly from parent to offspring (5). This can occur within or between host generations and is irrespective of the mode of entry. Horizontal transmission is the most common method of transmission among certain groups of insect pathogens.

Transmission is vertical when there is direct transfer of the pathogen from a parent organism to his or her progeny (6). This serves to transfer the pathogen from one host generation to the next and has also been described as congenital, parental, or hereditary. Once thought to be rare, vertical transmission is now recognized as a major pathway for the dissemination of many pathogens in nature.

2.1. Routes of Horizontal Transmission

There are two major ways in which insect pathogens gain entrance into their hosts and are transmitted horizontally: (1) through natural body openings (mouth, anus, spiracles), and (2) through the integument (7, 8). Infection through the mouth *(per os)* is by far the most common mode of entry by insect pathogens under natural conditions. It occurs to some degree in all pathogen groups, including fungi (9, 10) and nematodes (11), but is most prevalent among the bacteria (12), viruses (13 – 16), and protozoa (17). Oral transmission normally occurs by ingestion of food or fecal material contaminated with infectious stages of the pathogen but may also take place when susceptible individuals feed directly on infected cadavers or cannibalize infected hosts (7, 8).

The ingested pathogen penetrates the gut wall and establishes itself within epithelial cells of the intestinal tract, body cavity (hemolymph), or other host tissues (i.e., Malpighian tubules, fat body, muscle). Successful transmission through oral ingestion often is dependent on the acquisition of sufficient quantities of the pathogen (2) as well as on appropriate conditions within the host's alimentary tract (pH, enzymes) (18) which facilitate entry into susceptible host tissues.

Pathogen entry through other body openings is less common but does occur. Infective-stage juveniles of certain neoaplectanid nematodes, for example, may enter their hosts and reach the hemocoel via the spiracles and anus (11). Studies also have demonstrated that conidial spores of at least two fungal pathogens, *Beauveria bassiana* (19, 20) and *Metarhizium anisopliae* (21, 22), can enter spiracular openings, germinate, and penetrate the walls of the tracheae to initiate infection in some terrestrial insects.

Infection through the integument via direct contact of the pathogen with the host occurs commonly with many nematodes and is the principal mode of transmission of most fungi. In nematodes, entry into the host is normally achieved by direct penetration of the host cuticle by infective stages equipped with styles or piercing teeth (23). In fungi, infection is initiated when conidia or zoospores, as in the case of many aquatic *Phycomycetes,* germinate on the host integument, producing germ tubes that either penetrate directly or form specialized structures called appressoria that subsequently penetrate the epicuticle of the host (9, 10).

Direct cuticular invasion is very rare among the Protozoa. It is known to occur only among aquatic ciliates of the genus *Lambornella.* These gain entry into the hemocoel of larval mosquitoes by forming "invasion" cysts on the host's cuticle (24). Bacteria, on the other hand, may invade the host cuticle only through wounds or sites of mechanical injury (passive penetration) and according to Lysenko (25), active penetration of the unbroken cuticle has never been conclusively demonstrated for any bacterial pathogen.

Some bacteria (26 – 28), microsporidia (18, 29), and baculoviruses, particularly nucleopolyhedrosis (NPV) and granulosis (GV) viruses (30 – 33), also may be transmitted indirectly through the integument by oviposition of hymenopterous parasites. In most cases, this method of transmission is mechanical, that is, the pathogen neither changes form nor multiplies on or within the parasite. Entry into the host usually occurs by direct intrahemocoelic inoculation of the pathogen by means of a contaminated ovipositor. Parasites also may mechanically transfer pathogens onto the bodies of new hosts during oviposition wherein infections then arise through oral ingestion (32, 34).

Initial contamination of adult parasites results most frequently from prior oviposition in diseased hosts but also may take place when parasites successfully develop within infected host larvae. In the latter case, the parasite may or may not develop systemic infections.

Viral pathogens also have been reported to gain entry into their hosts by the bites of predatory insects with contaminated mouthparts (35). However, the extent to which this phenomenon occurs in nature and in other pathogen groups is not known.

2.2. Routes of Vertical Transmission

Direct transfer of infection from parent insects to their progeny is an important mode of transmission of many viruses and protozoa. In certain microsporidian pathogens of mosquitoes, for example, it is the principal means by which the pathogen is maintained from one host generation to the next (36–38). In other host insects, it augments horizontal routes of infection and facilitates the persistence of the pathogen when host densities are low or when no susceptible stages of the insect host are present (39).

In the majority of host insects, vertical transmission occurs entirely through the female line and is termed matroclinal or maternal-mediated (6). Such infections may arise in two distinct ways depending upon whether passage of the pathogen occurs within the ovary (transovarial) or on the surface of the egg (transovum) (40). In transovarial transmission, the pathogen gains entry to the egg while within the female host via infection of the ovaries and associated reproductive structures. Infection is achieved by direct invasion of the embryo (38, 41) or by oral ingestion of a pathogen which occurs in the yolk by the embryo at or near the time of eclosion (5, 42). According to Canning (5), the latter represents an important adaptation by the pathogen which ensures that hosts do not succumb to infection while still within the egg and thus defeat the purpose for which transovarial transmission has evolved. In transovum transmission, infective stages contaminate the external surface of the egg and are consumed by host larvae at eclosion (5). Although transovum transmission has previously been used as a general term to describe any transmission via the egg, regardless of whether it occurs within or outside the egg, it is now more commonly used when referring to external contamination only.

Transovum transmission appears to be the principal method of vertical transfer of most baculoviruses and cytoplasmic polyhedrosis viruses (CPV) of Lepidoptera and Hymenoptera (2, 13, 16). Transovarial transmission, on the other hand, predominates among nonoccluded iridescent viruses of mosquitoes (43) and most microsporidia (5, 17), many of which have distinct developmental cycles in female hosts that lead to the formation of specialized spores whose sole function is ovarian infection (38, 41).

Maternal-mediated vertical transmission has also been reported for some

TABLE 6.1

Major Transmission Pathways of Insect Pathogens and Relative Frequency of Occurrence Among Members of Each Group

| Pathogen Group | Horizontal Routes | | | | | Vertical Routes | | | |
| | Through Body Openings | | | Through the Integument | | Maternal-mediated | | Paternal-mediated | |
	Mouth	Anus	Spiracle	Direct	"Vector"-mediated	Transovarial	Transovum	Direct	Venereal
Viruses	+++	−	−	−	+	+	++	+	+
Bacteria	+++	−	−	−	+	−	+	−	−
Fungi	++	−	+	+++	−	−	−	−	−
Protozoa	+++	−	−	+	+	++	+	−	+
Nematodes	++	+	+	++	−	−	−	−	−

+++ = high, ++ = moderate, + = low, − = unknown

bacterial pathogens of insects (44–46). In these cases, however, infection appears to occur by oral ingestion of contaminated fecal or meconial discharges of adults or the mucilaginous matter that covers the surface of the egg (44, 46).

Paternal-mediated vertical transmission has been observed but is not common. It can result from direct parental infection of ova via infected sperm, as in the case of sigma virus of *Drosophila* (47, 48), or more frequently, by venereal transfer of infection to the female parent during mating and subsequent transfer to the egg, as in some microsporidia (49, 50) and NPVs (51). Although rare, paternal-mediated transmission is important epizootiologically because it provides a mechanism by which a pathogen can be maintained within a host population by vertical transmission alone, provided there is sufficient assistance from the female parent (i.e., some degree of maternal-mediated transmission occurs) and infections are not too pathogenic (52).

2.3. Transstadial Transmission

The term transstadial transmission describes the transfer of a pathogen from one host stage to the next throughout all or part of the host's life cycle (40). It occurs within all pathogen groups and is characteristic of pathogens with low pathogenicity that produce chronic infections in their hosts and are vertically transmitted.

Transstadial transmission typically takes place when larvae or nymphs acquire benign or sublethal infections, either through horizontal or vertical routes, and survive to adulthood wherein they then succumb to infection or live to transmit the pathogen to their progeny. This mode of transmission has important epizootiological implications because it enables those pathogens with low transmission efficiency to persist in relatively low host densities, since the lifespan of infected hosts is comparatively long (39). It also allows a pathogen to multiply to higher numbers in older, larger larvae.

3. METHODS OF PATHOGEN DISSEMINATION

Dispersal or dissemination is defined as the capacity of a pathogen to spread and distribute itself within a host population and throughout the environment (7). It is an important aspect of transmission and is fundamental to understanding the natural dynamics of infection in host insect populations. The successful long-term persistence of any pathogen within a host population is directly related to its ability to disperse, and pathogens with low dispersal capabilities typically have a very low potential for developing epizootics, even though they may be highly virulent or survive efficiently in the environment (7).

There are four principal ways in which insect pathogens are disseminated to susceptible individuals in nature (8): (1) by their own motility and actions, (2) by the behavior and movements of infected primary hosts, (3) by the behavior and movements of secondary hosts and nonhost carriers, and (4) by climatic and physical agents.

3.1. Pathogen Motility

Automobility is rare among insect pathogens and apparently is limited to certain parasitic nematodes and a few fungi. Many nematodes actively search for hosts over relatively short distances. Infective larval stages may move randomly or orient themselves in response to chemical stimuli emitted by the host (e.g., CO_2 and fecal components), as in the case of *Neoaplectana carpocapsae* (53, 54). These methods of dispersal appear to be effective for host-to-host transmission within a population but probably play a minor role in disseminating these nematodes from one host population to another.

Automobility also is characteristic of the aquatic fungal genera *Lagenidium* (Oomycetes) and *Coelomomyces* (Chytridiomycetes) (10, 55). Species of both genera produce motile zoospores which actively seek and penetrate the cuticle of susceptible mosquito larvae and other aquatic Diptera. The involvement of chemical stimuli in the host-seeking behavior of *Coelomomyces* spp. is unknown, but in *Lagenidium giganteum* zoospores appear to be attracted by some chemotactic agent emitted by the larval host (55).

The forcible discharge of conidial spores from the surface of infected hosts is an important method of dissemination among the entomophthoraceous fungi. This may contribute to explosive epizootics when host densities are high and environmental conditions are favorable (56). The formation of resistant resting spores with the onset of unfavorable conditions ensures the survival of these fungi over prolonged periods when no susceptible hosts are present and provides an effective means of dispersal to new host populations in subsequent years.

3.2. Behavior and Movements of Primary Hosts

One of the most important ways in which insect pathogens are disseminated within a host population and are dispersed to new habitats is by the activities and movements of their primary hosts (7, 8). Dispersal within a population may occur by: (1) direct congenital transfer of infection to progeny via transovarial or transovum transmission (6, 17, 36, 38, 41, 51); (2) cannibalism, including killing and ingestion of infected hosts (57) as well as ingestion of infected cadavers (57, 58); and (3) transfer of infection during mating, either by venereal transmission of the pathogen to the uninfected mate (49–51) or by oral ingestion of the pathogen excreted by one infected partner (59). Among social insects, intracolonial transmission may also occur by (4) grooming diseased individuals, as in subterranean termites infected with *M. anisopliae* (60), or by (5) feeding infective spores to the developing brood, which has been observed with the microsporidium *Burenella dimorpha* in the tropical fire ant *Solenopsis geminata* (61).

Other methods of pathogen dispersal by primary hosts involve contamination of the insect's habitat. These methods of dissemination follow a host-environment-host pathway and include: (1) excretion of infective stages of fecal and meconial discharges, which is characteristic of many bacterial, viral, and protozoan pathogens that infect the alimentary tract (7, 8, 17); (2) disintegration of infected tissues and release of the pathogen following death of the host, as in most pathogens that produce systemic infections (7, 8); (3) regurgitation of infective stages, which has been observed in some bacterial (62) and viral (63) pathogens; and (4) elimination of infectious stages in cast exuviae during molting (64).

Pathogen dispersal within a host population also is affected by the distinctive movements or "migrations" of diseased hosts. The most well-known example of this is the tree-top or "Wipfelkrankheit" disease of many lepidopterous and hymenopterous hosts infected with NPV (4). Larvae with these diseases often migrate to the upper portions of the plants or trees on which they are feeding. Here, they typically hang by their prolegs and die. Following death, their bodies disintegrate and the remains, along with infectious stages of the virus, are dispersed onto adjacent foliage. This increases the chance of contact with healthy larvae and thus facilitates natural dissemination of the pathogen to the healthy portion of the population. Similar migratory behavior to tops of plants is known in insect hosts infected with certain conidial fungi. In these cases, the elevated positions favor the dissemination of spores via the wind (8, 56) (see Section 3.4).

Pathogen dispersal to new habitats and host populations normally depends upon movements and long-distance migrations of infected hosts, especially winged adults. The frequency and efficiency with which this occurs depends upon the pathogenicity of the organism, the longevity of infected adults, and the routes of transmission. Where horizontal routes of infection are involved, long-range dispersal and survival from year to year normally are achieved via resistant spores or other stages of the pathogen which are capable of persisting within the environment or in the body of the host until contact is made with a suitable new host. With vertical routes, pathogen dispersal to new habitats and survival from one host generation to the next normally are achieved by passage of the pathogen within the living host. This usually takes place in the egg or adult stage but may also occur in larvae, pupae, or nymphs when these are the overwintering stage of the host.

3.3. Behavior and Movements of Secondary Hosts and Nonhost Carriers

Viral and protozoan pathogens may be disseminated within a host population by the ovipositional and feeding activities of insect parasites and predators.

Hymenopterous parasites may transfer pathogens during oviposition by direct insertion of a contaminated ovipositor into the body of a healthy host or by external contamination of the host or host's environment (28, 34, 65). Parasites normally acquire the pathogen by prior oviposition or development within infected hosts. According to Entwistle (66), the method by which the parasite acquires the pathogen is important because, in general, hymenopterous parasites that develop within infected hosts do not transmit infections to new hosts as effectively as do parasites that acquire infections by first ovipositing in infected larvae. This phenomenon may be due to the elimination of the pathogen by the parasite in its cast exuvium or meconial discharge.

Dipterous parasites also may acquire pathogens by developing within infected host larvae (30) but do so more frequently when adults feed on diseased and dying hosts (67, 68). They are known to disperse these pathogens, which are still virulent, to new hosts by ovipositional transmission as in some Hymenoptera (30) or by spreading the pathogen via contaminated mouthparts and legs onto the foliage on which the host is feeding (34, 67).

Natural dissemination of insect viruses by predatory insects has been observed in several insect orders including the Coleoptera (69), Hemiptera (35, 70), Hymenoptera (35), and Orthoptera (71). Predators normally acquire the virus by feeding on diseased hosts and then disseminate the unaffected pathogen in the environment by defecation.

The extent to which natural enemies contribute to the dissemination of disease and development of epizootics varies with each individual host-parasite-pathogen system. In those instances where there is a correlation between outbreaks of disease and the presence of the parasite or predator (65, 69, 72), or where there is a high prevalence of pathogen-induced mortality in hosts after oviposition by contaminated parasites (33), or where predators are significantly attracted to diseased hosts (67), the parasites or predators may be important in establishing foci of infection from which the disease can subsequently spread by other means (72).

On the other hand, when infections are detrimental to the parasite, as are some viruses (30) and microsporidia (29, 73, 74), or the parasite shows no preference for infected hosts or prefers to oviposit in healthy individuals (75), dispersal by these secondary hosts and nonhost carriers is probably not significant.

It has become increasingly evident (76–81) that predatory birds and small mammals that feed on virus-infected insects are capable of "passive" dissemination of these viruses to new habitats and host populations by depositing infective excreta in the environment. Other studies have demonstrated that grazing sheep and cattle can mechanically transfer baculoviruses from soil surfaces to vegetation and thus help spread the disease throughout a field or pasture (82, 83).

In most instances the extent to which these methods of dissemination contribute to epizootics is unknown. However, Entwistle et al. (78) are of the opinion that "birds may be the single most important carriers" of NPV of the European spruce sawfly, *Gilpinia hercyniae,* "over long distances as well as being substantial contributors to more local carriage." Lautenschlager and Podgwaite (80) also feel that birds transport NPV among gypsy moths, *Lymantria dispar,* because birds (1) are highly mobile and pass large amounts of infective inocula through their alimentary tracts, (2) have high metabolic rates and feed frequently, (3) prefer to feed on diseased insects, and (4) are likely to deposit inoculum on foliage where host larvae feed. They also believe that small mammals such as mice, shrews, and opossums may play a more significant role than birds in "long distance" transport of this virus because they pass active NPV more slowly through their alimentary tracts.

3.4. Physical Agents

Wind, rain, and running water also disperse insect pathogens. This occurs mainly among fungi and viruses and may involve the physical transport of infected hosts to new locales or the circulation of infectious stages of the pathogen within a host population. These actions may result in an increased prevalence of disease within a host population by bringing the pathogen into contact with new hosts, or a decrease by removing inocula from the environment or foliage on which the insect is feeding (e.g., rain or running water) (84, 85).

The frequency and degree with which physical agents contribute to the natural dynamics of infection in host insect populations is not known. However, in a few instances (86, 87) correlations have been made between the numbers of fungal conidia in the air and the prevalence of disease in field populations. Wind and rain have similarly been reported to play major roles in distributing viruses (88) and fungi (89) to susceptible individuals in local populations of some terrestrial insects, thereby initiating epizootics.

4. FACTORS GOVERNING TRANSMISSION RATES AND THEIR RELATION TO THE INITIATION AND MAINTENANCE OF ENZOOTIC AND EPIZOOTIC INFECTIONS

The ability of any pathogen to persist and flourish within a host population is inherently dependent upon the efficiency with which it is transmitted from one host to another (39). There are many factors that affect the transmissibility of insect pathogens in wild host populations. These include innate qualities of the pathogen and host that facilitate or retard transmission rates as well as many environmental factors that directly influence host susceptibility or pathogen persistence and infectivity. Interdependent with these, and possibly the most important factor underlying the dynamics of disease in any host population, is the route or method of transmission itself.

4.1. Methods of Transmission

The evolution of vertical transmission, whether maternal- or paternal-mediated, provides a unique mechanism whereby a pathogen can be maintained in low host densities and persist when host populations oscillate from year to year, as most do. This method of transmission also facilitates pathogen survival in those insect hosts that possess distinct, nonoverlapping generations or where development and multiplication of the pathogen is limited to certain life stages of the host (39).

Pathogens that rely on horizontal routes of transmission, on the other hand, are dependent on host density for their survival and disperal. These pathogens are normally transmitted by direct or indirect physical contact. Therefore, the rates at which infections are acquired and disseminated within a host population are directly proportional to the number of encounters made between susceptible and infected hosts or susceptible hosts and the pathogen itself (39). Reliance on horizontal routes also means that the pathogen must either maintain itself within the living host or cadaver throughout the year or produce a resistant, free-living stage that is capable of surviving in the environment when susceptible stages of the host are absent or inactive (5).

As a result, in those host-pathogen relationships where horizontal transmission predominates, disease prevalence is usually density dependent and typically increases steadily or dramatically during a host generation or season. This often culminates in large-scale epizootics. In those relationships where vertical transmission prevails, prevalence rates operate in a more density-independent manner and the pathogen is usually maintained in a steady, enzootic state. Vertical transmission also may function in establishing initial foci of infection from which the pathogen subsequently can be disseminated by horizontal means and in a density-dependent manner.

Although some investigators (36, 37) have suggested that certain pathogens, most notably polymorphic microsporidia of the genus *Amblyospora* from mosquitoes, can be maintained within a host population by vertical transmission alone, most available evidence now indicates that this is not the case (5, 6, 39, 41, 52, 90, 94). Recent studies (95, 96) have now shown that many and possibly all of these microsporidia, require an alternate copepod host to complete their development and that horizontal transmission via this intermediate host is a necessary prerequisite for pathogen survival. Pathogen maintenance by vertical routes alone can occur only if sufficient assistance is obtained from the male parent, as in sigma virus of *Drosophila,* or if infections impart some selective advantage to the host (6, 52). In as much as most insect pathogens cause some pathology in their hosts and are not effectively transmitted from males to progeny, some degree of horizontal transmission is essential for the maintenance of most insect pathogens in nature (5, 6, 39, 52).

4.2. Pathogen Factors

There are a number of strategies employed by insect pathogens that directly facilitate their transmissibility and survival within a host population. An excellent in-depth analysis of those factors that enhance pathogen persistence is presented by Anderson (91) and Anderson and May (39, 92). Such factors include: (1) high infectivity (i.e., the ability to initiate and maintain infection in the host); (2) low pathogenicity (i.e., the ability to injure the host once an infection is established); (3) short latency (i.e., incubation period within a host before beginning to produce transmissible stages); (4) high reproductive rate with the production of large numbers of infective stages; and (5) long-lived infective stages.

These factors generally operate in an interdependent fashion and are frequently counterbalanced with one another (39, 92). For example, in many horizontally transmitted diseases that produce systemic infections within the host (e.g., microsporidia, NPV, bacteria), the rate at which the pathogen is transmitted is often dependent upon the rate at which the pathogen kills its hosts, i.e., how pathogenic it is. However, if the organism is too pathogenic and the host succumbs before a sufficient number of infected stages can be produced or effective transmission can take place, then the pathogen is unable to persist and gradually disappears from the host population. Therefore, the reproductive success and transmission efficiency of the pathogen is usually maximal at some intermediate level of pathogenicity which allows for a maximum production of transmissible stages (39, 92). Anderson and May (92) also note, however, that "the ability to multiply rapidly within the host, or produce large numbers of transmission stages (both attributes frequently being correlated with parasite pathogenicity) will often be beneficial to reproductive success even if the host is eventually killed by such action."

4.3. Host Factors

Several factors associated with the host also have a direct influence on transmission rates and help facilitate pathogen dispersal within a population. One of the most important of these is host density. High host density is of primary importance for the dissemination of those pathogens that rely on direct, horizontal transmission. It increases the likelihood of contact between susceptible hosts and the pathogen. As more hosts become infected, more inoculum is produced and epizootics will often ensue (7, 8).

Pathogen persistence also is affected by the rate at which susceptible hosts enter a population (39). As a result, the reproductive rate of the host is an important factor in the spread of infection, and high birth rates usually enhance transmission efficiency and lead to increased prevalences (39).

Of equal importance in the spread and development of epizootics is the spatial and temporal distribution of susceptible hosts. These factors may provide for (1) greater host exposure to infective inoculum or (2) more favorable physical conditions (e.g., temperature, moisture) that enhance the infectious process (7, 8). The latter is an important factor with many fungi since they are usually dependent upon specific climatic factors for germination and host penetration.

The rapidity with which infections spread within an insect population often depends on the dispersal capabilities of infected hosts (7) (see Section 3.2). Consequently, host mobility and behavior play significant roles in initiating epizootics, especially in those host-pathogen associations where infections are located within the alimentary tract and infective stages of the pathogen are released throughout the lifetime of the host. An additional factor of importance in these relationships is host longevity which is interrelated with pathogenicity of the organism.

The presence of innate host resistance within a portion of the population may also influence the rate at which a pathogen is transmitted. However, it is not known to what degree host resistance influences transmission processes in wild populations.

4.4. Environmental Factors

There are several physical factors in the environment or microhabitat surrounding the host and pathogen that directly affect the transmissibility of many insect pathogens. These include: (1) temperature, (2) moisture, (3) solar radiation, and (4) physiochemical conditions (e.g., pH, organic content and texture of soil in terrestrial environments, or salinity and organic pollution in aquatic environments) (7, 8). In most instances, these factors either enhance or inhibit a pathogen's development in the host or persistence in the environment. For example, among fungal pathogens of terrestrial insects, high humidity or moisture is almost always a necessary prerequisite for the production and germination of infective spores. Because of this, effective transmission cannot be achieved unless appropriate climatic conditions exist. Epizootics of fungi therefore usually are associated with periods of high moisture, often in the form of rain or dew.

Solar radiation and temperature extremes are physical factors that inactivate pathogens in all groups. Their inhibitory influence on pathogen survival may alter or stop the dissemination of a particular disease within a host population and thereby reduce the rate of development of an epizootic.

Physical factors may also influence pathogen transmission by affecting host susceptibility or behavior which in turn affects pathogen acquisition. For example, feeding rates of susceptible hosts may increase with rising temperature and result in the ingestion of greater quantities of a particular pathogen per unit time (93). High temperatures may also place the host insect under stress and thereby make it more susceptible to infection (7, 8). These subjects are dealt with in more detail in Chapters 4 and 7.

5. CONCLUSION

In reviewing the literature, it is apparent that insect pathogens display a great variety of methods by which they are transferred among individual hosts and are disseminated to new host populations. It is also apparent, however, that in most host-pathogen relationships the extent to which these various modes of transmission contribute to the natural dynamics of infection and occurrence of epizootics in wild host populations is incompletely known or poorly understood.

There is clearly a need for more quantitative field studies that carefully assess the contribution of major and minor routes of transmission to the initiation and development of both enzootic and epizootic infections. There is an even greater urgency to determine the way in which many pathogens with poorly understood modes of transmission (e.g., polymorphic microsporidia of aquatic Diptera) are maintained and perpetuated in wild host popuations. We must also have a further elucidation of the many factors surrounding the host, pathogen, and environment that directly influence the transmissibility of each particular pathogen.

Studies of this nature will provide a better understanding of disease dynamics and the role of pathogens in natural population reduction. This will hopefully culminate in an increased utilization of naturally occurring pathogens in pest management programs.

REFERENCES

1. R. M. Anderson and R. M. May, Eds., *Population Biology of Infectious Diseases,* Springer-Verlag, Berlin, Heidelberg, New York, 1982.
2. C. C. Payne, Insect viruses as control agents, *Parasitology,* **84,** 35 (1982).
3. R. M. Anderson, "Transmission Dynamics and Control of Infectious Disease Agents," in R. M. Anderson and R. M. May, Eds., *Population Biology of Infectious Diseases,* Springer-Verlag, Berlin, Heidelberg, New York, 1982, p. 67.
4. E. A. Steinhaus, *Principles of Insect Pathology,* McGraw-Hill, New York, 1949.
5. E. U. Canning, An evaluation of protozoal characteristics in relation to biological control of pests, *Parasitology,* **84,** 119 (1982).
6. P. E. M. Fine, Vectors and vertical transmission: an epidemiologic perspective, *Ann. N.Y. Acad. Sci.,* **266,** 173 (1975).
7. Y. Tanada, "Epizootiology of Infectious Diseases," in E. A. Steinhaus, Ed., *Insect Pathology: An Advanced Treatise,* Vol. 2, Academic Press, New York, 1963, p. 423.
8. Y. Tanada, "Epizootiology of Insect Diseases," in P. DeBach, Ed., *Biological Control of Insect Pests and Weeds,* Reinhold, New York, 1964, p. 548.
9. R. Y. Zacharuk, "Fungal Diseases of Terrestrial Insects," in E. W. Davidson, *Pathogenesis of Invertebrate Microbial Diseases,* Allanheld, Osmun Publishers, Totowa, New Jersey, 1981, p. 367.
10. A. W. Sweeney, "Fungal Pathogens of Mosquito Larvae," in E. W. Davidson, Ed., *Pathogenesis of Invertebrate Microbial Diseases,* Allanheld, Osmun Publishers, Totowa, New Jersey, 1981, p. 403.
11. G. O. Poinar, Jr., *Entomogenous Nematodes. A Manual and Host List of Insect-Nematode Assocations,* E. J. Brill, Leiden, Netherlands, 1975.
12. H. D. Burges, Control of insects by bacteria, *Parasitology,* **84,** 79 (1982).
13. J. F. Longworth, "Viruses and Lepidoptera," in A. J. Gibbs, Eds., *Viruses and Invertebrates,* North-Holland, Amsterdam, London, 1973.
14. P. Faulkner, "Baculovirus," in E. W. Davidson, Ed., *Pathogenesis of Invertebrate Microbial Diseases,* Allanheld, Osmun Publishers, Totowa, New Jersey, 1981, p. 3.
15. D. C. Kelly, "Non-Occluded Viruses," in E. W. Davidson, Ed., *Pathogenesis of Invertebrate Microbial Diseases,* Allanheld, Osmun Publishers, Totowa, New Jersey, 1981, p. 39.

16. C. C. Payne, "Cytoplasmic Polyhedrosis Viruses," in E. W. Davidson, Ed., *Pathogenesis of Invertebrate Microbial Diseases,* Allanheld, Osmun Publishers, Totowa, New Jersey, 1981, p. 61.

17. E. U. Canning, Transmission of Microsporidia, *Proc. Int., Colloq. Insect Pathol.,* 4th, 415 (1971).

18. Y. Tanada, "Epizootiology and Microbial Control," in L. A. Bulla, Jr., and T. C. Cheng, Eds., *Comparative Pathobiology. Biology of the Microsporidia,* Vol. 1, Plenum Press, New York, 1976, p. 247.

19. R. C. Hedlund and B. C. Pass, Infection of the alfalfa weevil, *Hypera postica,* by the fungus, *Beauveria bassiana, J. Invertebr. Pathol.,* **11,** 25 (1968).

20. S. Pekrul and E. A. Grula, Mode of infection of the corn earworm *(Heliothis zea)* by *Beauveria bassiana* as revealed by scanning electron microscopy, *J. Invertebr. Pathol.,* **34,** 238 (1979).

21. M. F. Madelin, "Diseases Caused by Hyphomycetous Fungi," in E. A. Steinhaus, Ed., *Insect Pathology: An Advanced Treatise,* Vol. 2, Academic Press, New York, 1963, p. 233.

22. V. J. E. McCauley, R. Y. Zacharuk, and R. D. Tinline, Histopathology of green muscardine in larvae of four species of Elateridae (Coleoptera). *J. Invertebr. Pathol.,* **12,** 444 (1968).

23. H. E. Welch, "Nematode Infections," in E. A. Steinhaus, Ed., *Insect Pathology: An Advanced Treatise,* Vol. 2, Academic Press, New York, 1963, p. 363.

24. T. B. Clark and D. G. Brandl, Observations on the infection of *Aedes sierrensis* by a tetrahymenine ciliate, *J. Invertebr. Pathol.,* **28,** 341 (1976).

25. O. Lysenko, "Principles of Pathogenesis of Insect Bacterial Diseases as Exemplified by the Nonsporeforming Bacteria," in E. W. Davidson, Ed., *Pathogenesis of Invertebrate Microbial Diseases,* Allanheld, Osmun Publishers, Totowa, New Jersey, 1981, p. 163.

26. G. E. Bucher, Transmission of bacterial pathogens by the ovipositor of a hymenopterous parasite, *J. Insect Pathol.,* **5,** 277 (1963).

27. E. S. Kurstak, Etude des relations entre l'infection a *Bacillus thuringiensis* Berliner et le parasitisms par *Nemeritis canescens* (Gravenhorst) (Ichneumonidae) Chez *Ephestia kuhniella* Zeller (Pyralidae), *Ann. Epiphyties,* **17,** 451 (1966).

28. J. V. Bell, E. G. King, and R. J. Hamalle, Interactions between bollworms, a braconid parasite, and the bacterium *Serratia marcescens, Ann. Entomol., Soc. Am.,* **67,** 712 (1974).

29. W. M. Brooks, Protozoa: host-parasite-pathogen interrelationships, *Entomol. Soc. Am. Misc. Publ.,* **9,** 105 (1973).

30. T. A. Irabagon and W. M. Brooks, Interaction of *Campoletis sonorensis* and a nuclear polyhedrosis virus in larvae of *Heliothis virescens, J. Econ. Entomol.,* **67,** 229 (1974).

31. C. C. Beegle, and E. R. Oatman, Effect of nuclear polyhedrosis virus on the relationship between *Trichoplusia ni* (Lepidoptera: Noctuidae) and the parasite, *Hypersoter exiguae* (Hymenoptera: Ichneumonidae), *J. Invertebr. Pathol.,* **25,** 59 (1975).

32. B. Raimo, R. C. Reardon, and J. D. Podgwaite, Vectoring gypsy moth nuclear polyhedrosis virus by *Apanteles melanoscelus* (Hym: Braconidae), *Entomophaga,* **22,** 207 (1977).

33. D. B. Levin, J. E. Laing, R. P. Jaques, and J. E. Corrigan, Transmission of the granulosis virus of *Pieris rapae* (Lepidoptera: Pieridae) by the parasitoid *Apanteles glomeratus* (Hymenoptera: Braconidae), *Environ. Entomol.,* **12,** 166 (1983).

34. P. V. Vail, Cabbage looper nuclear polyhedrosis virus-parasitoid interactions, *Environ. Entomol.,* **517** (1981).

35. W. A. Smirnoff, Predators of *Neodiprion swainei* Midd. (Hymenoptera: Tenthredinidae) larval vectors of virus diseases, *Can. Entomol.,* **91,** 246 (1959).

36. W. R. Kellen, H. C. Chapman, T. B. Clark, and J. E. Lindegren, Host-parasite relationships of some *Thelohania* from mosquitoes (Nosematidae: Microsporidia), *J. Invertebr. Pathol.,* **7,** 161 (1965).

37. H. C. Chapman, D. B. Woodard, W. R. Kellen, and T. B. Clark, Host-parasite relationship of *Thelophania* associated with mosquitoes in Louisiana (Nosematidae: Microsporidia), *J. Invertebr. Pathol.,* **8**, 452 (1966).

38. T. G. Andreadis and D. W. Hall, Development, ultrastructure, and mode of transmission of *Amblyospora* sp. (Microspora) in the mosquito, *J. Protozool.,* **26**, 444 (1979).

39. R. M. Anderson and R. M. May, The population dynamics of microparasites and their invertebrate hosts, *Phil. Trans. R. Soc. London B,* **291**, 451 (1981).

40. E. A. Steinhaus and M. E. Martignoni, *An Abridged Glossary of Terms Used in Invertebrate Pathology,* 2nd ed., Pac. N.W. For. Range Exp. Stn., USDA For. Ser., 1970.

41. T. G. Andreadis, Life cycle and epizootiology of *Amblyospora* sp. (Microspora: Amblyosporidae) in the mosquito, *Aedes cantator, J. Protozool.,* **30**, 509 (1983).

42. H. M. Thomson, The effect of a microsporidian parasite on the development, reproduction and mortality of the spruce budworm, *Choristoneura fumiferana* (Clem.), *Can. J. Zool.,* **36**, 499 (1958).

43. B. A. Federici, "Virus Pathogens of Mosquitoes and Their Potential Use in Mosquito Control," in A. Aubin et al., Eds., *Le Controle des Moustiques/Mosquito Control,* Quebec Univ. Press, Canada, 1974, p. 93.

44. F. d'Herelle, Le coccobacille des sauterelles, *Ann. Inst. Pasteur (Paris),* **28**, 280 (1914).

45. G. E. Bucher and J. M. Stephens, A disease of grasshoppers caused by the bacterium *Pseudomonas aeruginosa* (Schroeter) Migula, *Can. J. Microbiol.,* **3**, 611 (1957).

46. G. E. Bucher, Survival of populations of *Streptococcus faecalis* Andrewes and Horder in the gut of *Galleria mellonella* (Linnaeus) during metamorphosis, and transmission of the bacteria to the filial generation of the host, *J. Insect Pathol.,* **5**, 336 (1963).

47. R. Seecof, "The Sigma Virus Infection of *Drosphila melanogaster,"* In. K. Maramorosh, Ed., *Current Topics in Microbiology and Immunology, Vol. 42,* Springer-Verlag, New York, 1968, p. 59.

48. Ph. L'Heritier, *Drosophila* viruses and their role as evolutionary factors, *Evolution. Biol.,* **4**, 185 (1970).

49. H. M. Thomson, Some aspects of the epidemiology of a microsporidian parasite of the spruce budworm, *Choristoneura fumiferana* (Clem.), *Can. J. Zool.,* **36**, 309 (1958).

50. W. R. Kellen and J. E. Lindegren, Modes of transmission of *Nosema plodiae* Kellen and Lindegren, a pathogen of *Plodia interpunctella* (Hubner). *J. Stored Prod. Res.,* **7**, 31 (1971).

51. J. J. Hamm and J. R. Young, Mode of transmission of nuclear-polyhedrosis virus to progeny of adult *Heliothis zea, J. Invertebr. Pathol.,* **24**, 70 (1974).

52. P. E. M. Fine, "Vertical Transmission of Pathogens of Invertebrates," in T. C. Cheng, Ed., *Comparative Pathobiology,* Vol. 7, Plenum Press, New York, 1984, p. 205.

53. J. Schmidt and J. N. All, Attraction of *Neoaplectana carpocapsae* (Nematoda: Steinernematidae) to common excretory products of insects, *Environ. Entomol.,* **8**, 55 (1979).

54. R. Gaugler, L. LeBeck, B. Nakagaki, and G. M. Boush, Orientation of the entomogenous nematode *Neoaplectana carpocapsae* to carbon dioxide, *Environ. Entomol.,* **9**, 649 (1980).

55. A. J. Domnas, "Biochemistry of *Lagenidium giganteum* Infection of Mosquito Larvae," in E. W. Davidson, Ed., *Pathogenesis of Invertebrate Microbial Diseases,* Allanheld, Osmun Publishers, Totowa, New Jersey, 1981, p. 425.

56. D. M. MacLeod, "Entomophthorales Infections," in E. A. Steinhaus, Ed., *Insect Pathology: An Advanced Treatise,* Vol. 2, Academic Press, New York, 1963, p. 189.

57. J. B. Carter, The mode of transmission of *Tipula* iridescent virus II. Route of infection, *J. Invertebr. Pathol.,* **21**, 136 (1973).

58. H. D. Burges and J. A. Hurst, Ecology of *Bacillus thuringiensis* in storage moths, *J. Invertebr. Pathol.,* **30**, 131 (1977).

59. B. Zelazny, Transmission of a baculovirus in populations of *Oryctes rhinoceros, J. Invertebr. Pathol.,* **27**, 221 (1976).

60. K. R. Kramm, D. F. West, and P. G. Rockenbach, Termite pathogens: transfer of the entomopathogen *Metarhizium anisopliae* between *Reticulitermes* sp. termites, *J. Invertebr. Pathol.,* **40**, 1 (1982).

61. D. P. Jouvenaz, C. S. Lofgren, and G. E. Allen, Transmission and infectivity of spores of *Burenella dimorpha* (Microsporida: Burenellidae), *J. Invertebr. Pathol.,* **37**, 265 (1981).

62. A. M. Heimpel and T. A. Angus, "Diseases Caused by Certain Sporeforming Bacteria" in E. A. Steinhaus, Ed., *Insect Pathology: An Advanced Treatise,* Vol. 1, Academic Press, New York, 1963, p. 21.

63. R. R. Granados, "Entomopox Infections in Insects," in E. W. Davidson, Ed., *Pathogenesis of Invertebrate Microbial Diseases,* Allanheld, Osmun Publishers, Totowa, New Jersey, 1981, p. 101.

64. J. N. Talukdar, The dispersion of spores of a *Nosema* attacking *Philosamia ricini* Hutton, *J. Insect Pathol.,* **4**, 128 (1962).

65. R. C. Reardon and J. D. Podgwaite, Disease-parasitoid relationships in natural populations of *Lymantria dispar* (Lep.: Lymantriidae) in the Northeastern United States, *Entomophaga,* **21**, 333 (1976).

66. P. F. Entwistle, Passive carriage of baculoviruses in forests, *Proc. 3rd Int. Coll. Invertebr. Pathol., Brighton, U.K.,* 344 (1982).

67. G. R. Stairs, Artificial initiation of virus epizootics in forest tent caterpillar populations, *Can. Entomol.,* **97**, 1059 (1965).

68. D. L. Hostetter, A virulent nuclear-polyhedrosis virus of cabbage looper, *Trichoplusia ni,* recovered from the abdomens of sarcophagid flies, *J. Invertebr. Pathol.,* **17**, (1971).

69. J. L. Capinera and P. Barbosa, Transmission of nuclear-polyhedrosis virus to gypsy moth larvae by *Calosoma sycophanta, Ann. Entomol. Soc. Am.,* **68**, 593 (1975).

70. M. S. T. Abbas and D. G. Boucias, Interaction between nuclear polyhedrosis virus-infected *Anticarsia gemmatalis* (Lepidoptera: Noctuidae) larvae and predator *Podisus maculiventris* (Say) (Hemiptera: Pentatomidae), *Environ. Entomol.,* **13**, 599 (1984).

71. C. Vago, J. Fosset, and M. Bergoin, Dissemination des virus de polyedries par les ephippigeres predateur d'insects. *Entomophaga,* **11**, 177 (1966).

72. O. J. Smith, K. M. Hughes, P. H. Dunn, and I. M. Hall, A granulosis virus disease of the western grape leaf skeletonizer and its transmission, *Can. Entomol.,* **88**, 507 (1956).

73. T. G. Andreadis, *Nosema pyrausta* infection in *Macrocentrus grandii,* a braconid parasite of the European corn borer, *Ostrinia nubilalis, J. Invertebr. Pathol.,* **35**, 229 (1980).

74. T. G. Andreadis, Impact of *Nosema pyrausta* on field populations of *Macrocentrus grandii,* an introduced parasite of the European corn borer, *Ostrinia nubilalis, J. Invertebr. Pathol.,* **39**, 298 (1982).

75. P. L. Versoi and W. G. Yendol, Discrimination by the parasite, *Apanteles melanoscelus* between healthy and virus-infected gypsy moth larvae, *Environ. Entomol.,* **11**, 42 (1982).

76. D. L. Hostetter and K. D. Biever, The recovery of virulent nuclear-polyhedrosis virus of the cabbage looper, *Trichoplusia ni,* from the feces of birds, *J. Invertebr., Pathol.,* **15**, 173 (1970).

77. P. F. Entwistle, P. H. W. Adams, and H. F. Evans, Epizootiology of a nuclear-polyhedrosis virus in European spruce sawfly *(Gilpinia hercynia):* the status of birds as dispersal agents of the virus during the larval season, *J. Invertebr. Pathol.,* **29**, 354 (1977).

78. P. F. Entwistle, P. H. W. Adams, and H. F. Evans, Epizootiology of a nuclear-polyhedrosis virus in European spruce sawfly, *Gilpinia hercyniae:* birds as dispersal agents of the virus during winter, *J. Invertebr. Pathol.,* **30**, 15 (1977).

79. P. F. Entwistle, P. H. W. Adams, and H. F. Evans, Epizootiology of a nuclear-polyhedrosis

virus in European spruce sawfly *(Gilpinia hercyniae):* the rate of passage of infective virus through the gut of birds during cage tests, *J. Invertebr. Pathol.,* **31,** 307 (1978).

80. R. A. Lautenschlager and J. D. Podgwaite, Passage of nucleopolyhedrosis virus by avian and mammalian predators of the gypsy moth, *Lymantria dispar, Environ. Entomol.,* **8,** 210 (1979).

81. R. A. Lautenschlager, J. D. Podgwaite, and D. E. Watson, Natural occurrence of the nucleopolyhedrosis virus of the gypsy moth, *Lymantria dispar* (Lep.: Lymantriidae) in wild birds and mammals, *Entomophaga,* **25,** 261 (1980).

82. A. M. Crawford and J. Kalmakoff, A host-virus interaction in a pasture habitat: *Wiseana* spp. (Lepidoptera: Hepialidae) and its baculovirus, *J. Invertebr. Pathol.,* **29,** 81 (1977).

83. J. R. Fuxa and J. P. Geaghan, Multiple-regression analysis of factors affecting prevalence of nuclear polyhedrosis virus in *Spodoptera frugiperda* (Lepidoptera: Noctuidae) populations, *Environ. Entomol.,* **12,** 311 (1983).

84. J. R. Harkrider and I. M. Hall, The dynamics of an entomopoxvirus in a field population of larval midges of *Chironomus decorus* complex, *Environ. Entomol.* **7,** 858 (1978).

85. J. D. Vandenberg and R. S. Soper, Prevalence of entomophthorales mycoses in populations of spruce budworm, *Choristoneura fumiferana, Environ. Entomol.,* **7,** 847 (1978).

86. N. Wilding, *Entomophthora* species infecting pea aphid, *Trans. R. Entomol. Soc.,* **127,** 171 (1975).

87. L. P. Kish and G. E. Allen, The biology and ecology of *Nomuraea rileyi* and a program for predicting its incidence on *Anticarsia gemmatalis* in soybean, *Fla. Agr. Exp. Stn. Bull.* **795,** 1978.

88. F. T. Bird, Transmission of some insect viruses with particular reference to ovarial transmission and its importance in the development of epizootics, *J. Insect Pathol.,* **3,** 352 (1961).

89. C. M. Ignoffo, C. Garcia, D. L. Hostetter, and R. E. Pinnell, Laboratory studies of the entomophathogenic fungus *Nomuraea rileyi:* soil-borne contamination of soybean seedlings and dispersal of diseased larvae of *Trichoplusia ni, J. Invertebr. Pathol.,* **29,** 147 (1977).

90. T. G. Andreadis and D. W. Hall, Significance of transovarial infections of *Amblyospora* sp. (Microspora: Thelohaniidae) in relation to parasite maintenance in the mosquito *Culex salinarius, J. Invertebr. Pathol.,* **34,** 152 (1979).

91. R. M. Anderson, Theoretical basis for the use of pathogens as biological control agents of pest species, *Parasitology,* **84,** 3 (1982).

92. R. M. Anderson and R. M. May, Coevolution of hosts and parasites, *Parasitology,* **85,** 411 (1982).

93. S. P. Wraight, D. Molloy, H. Jamnback, and P. McCoy, Effects of temperature and instar on the efficacy of *Bacillus thuringiensis* var. *israelensis* and *Bacillus sphaericus* strain 1593 against *Aedes stimulans* larvae, *J. Invertebr. Pathol.,* **38,** 78 (1981).

94. T. G. Andreadis, Life cycle, epizootiology, and horizontal transmission of *Amblyospora* (Microspora: Amblyosporidae) in a univoltine mosquito, *Aedes stimulans, J. Invertebr. Pathol.,* **46,** 31 (1985).

95. A. W. Sweeney, E. I. Hazard, and M. F. Graham, Intermediate host for an *Amblyospora* sp. (Microspora) infecting the mosquito, *Culex annulirostris, J. Invertebr. Pathol.,* **46,** 98 (1985).

96. T. G. Andreadis, Experimental transmission of a microsporidian pathogen from mosquitoes to an alternate copepod host, *Proc. Natl. Acad. Sci. USA,* **82,** 5574 (1985).

7

ENVIRONMENT

GEORG BENZ

Department of Entomology
Swiss Federal Institute of Technology
Zurich, Switzerland

1. INTRODUCTION

Long before diseases of insects were known to be caused by microorganisms, Vida of Cremona (1) and Merian (2) described the diseases of the silkworm and stated that they were caused by environmental factors such as crowding, excessive heat and humidity, thunderstorms, and abnormal nutrition. Later, when most of the diseases of the silkworm were found to be caused primarily by microorganisms, the various environmental factors were believed to aggravate or predispose the insects to diseases. This is still accepted today.

One of the first to experiment with the interactions between pathogens and environmental factors in the silkworm and then in other insects was Vago in France during the 1950s (3). Temperature, humidity, lack of oxygen, abnormal or qualitatively insufficient nutrition, and sublethal doses of toxic substances like lead arsenate were found to induce epizootics in the silkworm and in many other insects in different orders. Furthermore he found that subacute infections with some bacteria or viruses could induce epizootics of other infectious diseases. Vago used the term "concatenation" of insect disease (l'enchaînement des maladies) to describe the phenomenon that was later verified by many authors. Thus *Bacillus thuringiensis* may activate granulosis in *Hyphantria cunea* (4) and *Zeiraphera diniana* (5).

It is evident that disease dynamics in free living insects that live in a more variable environment than the silkworm depend even to a higher degree on environmental conditions. According to Steinhaus (6), Franz (7), Tanada (8), and Grison (9) there are three primary factors contributing to the epizootics of infectious diseases: the host population, the pathogen population, and an efficient means of transmission. Environmental factors may directly or indirectly influence all of these primary factors. Hurpin (10) acknowledged this fact by stating that "epizootics depend essentially on . . . the host, the germ, and the environmental factors." Since the primary factors of epizootics cannot be separated from the environment, both the biotic and physical environmental factors are expected to influence all stages of disease outbreaks. The situation is always so complex that it is often difficult to separate the influence of one environmental factor from others or from factors of the host and pathogen populations. Climatic factors may have a profound influence on the host population dynamics as well as the pathogen population. Other physical factors such as solar irradiation, wind, rain, and characteristics of the soil may influence the pathogens more than the host, whereas the biotic factors may influence the host population more than the pathogen.

2. ENVIRONMENT AND THE HOST POPULATION

Host population resistance or susceptibility to a pathogen is crucial to epizootiology. There is little information about climatic and physical factors that may increase the resistance of insect populations, although one would expect environmental conditions that favor the host population to increase host resistance until the density-dependent antagonistic factors reduce resistance and enhance the development of disease. The frequent observations of epizootics in dense populations support this assumption (11 – 14). Such density dependence may lag as with *Entomophthora muscae* in the onion fly, *Delia antiqua,* where host and pathogen density were the only significant variables (15). However, epizootics are more likely to be overlooked in populations of low density than in dense populations. According to Bird and Elgee (16) and Bird (14) virus epizootics may be initiated at low and at high sawfly densities. Indeed, if epizootics did not occur at low densities of sawflies, long-lasting control by these viruses would not be possible. Granulosis prevalence of over 90%, regardless of larval density, was also reported for the clover cutworm, *Scotogramma trifolii,* in Southern California (17). Benz (18), too, found that granulosis prevalence in the larch bud moth, *Z. diniana,* varied on individual trees from 0 to 25% but was not correlated with larval density on the trees. A 4-year study in alfalfa fields revealed no correlation between virosis prevalence and the densities of *Autographa californica, Colias eurytheme,* and *Spodoptera exigua* populations (19). Viral epizootics also have been observed at low population density in the Great Basin tent caterpillar (20, 21), the armyworm (22), and the larch bud moth, *Z. diniana* (Benz, unpublished). The multiple regression analysis of factors affecting prevalence of nuclear polyhedrosis virus (NPV) in *Spodoptera frugiperda* revealed that overwintering virus in the soil, but not *S. frugiperda* density, was an important variable relative to prevalence of the disease (23). Epizootics of *Bacillus popilliae* occurred in light populations of *Popillia japonica* (24) and *Melolontha melolontha* (25). Similar observations were made with certain fungi, for example, *Beauveria bassiana (= B. globulifera)* on the chinch bug (26), and *Entomophthora* spp. on the spotted alfalfa aphid (27) and the diamondback moth, *Plutella maculipennis* (28). These fungi were distributed throughout the host habitat and depended on favorable climatic conditions for their epizootic outbreaks. Ullyett and Schonken (28) concluded from this that fungal diseases should be classified as density-independent mortality factors. Ullyett (29) later modified this viewpoint and stated that these factors were neither wholly density-dependent nor -independent but pass through both phases. Since all infectious diseases may spread more easily the closer the host individuals are to one another, a certain degree of density dependence will always occur. Thus Carruthers et al. (15) found host and pathogen densities to be the predominantly stimulating variables in the disease system of the fungus *E. muscae* in the onion fly, *D. antiqua.* The effect of abiotic variables on *E. muscae* infection and development was probably important but was not the limiting or controlling factor because conditions conducive to infection were usually present in the microenvironment. Pathogens may therefore be regarded as imperfectly density-dependent mortality factors in the sense of Milne (30 – 32). This point of view has already been taken by Steinhaus (6), Franz (7), and Grison (9). Indeed, since epizootics are not only dependent on the host populations but also on the density and the virulence of the pathogens, this conclusion is evident. How imperfectly density-dependent infectious diseases may be is perhaps best illustrated by crowding effects that may or may not activate latent diseases (33, 34).

The complex effect of the environment that generally is termed "stress" must be briefly discussed. A number of factors, so-called stressors, can alter the general condition of organisms and predispose them to infection or activate already-present inapparent diseases. According to Selye (35) stress is the unspecific response of the body to all types of environmental demands or stressors. Stress is thus neither avoidable nor negative. Whether it predisposes the organism to disease or enhances its defense depends on the constitution of the organism on the one hand and on the intensity of the stressor(s) and the duration of its (their) action on the other. This makes the experimental approach to the role of stress in epizootiology difficult, and reproducible results rare. No universal stressor is therefore known. Those investigated by Vago (3) have already been mentioned. Of these, temperature and humidity seem to act most commonly in nature (36). Crowding may also be regarded as a universal stressor (33), and, because of its action on the neurosecretory system, may be the stressor that best conforms to the primary concept of stress by Selye.

Steinhaus (37) concluded from his experiments that stressors, or activators (38), may activate microorganisms present in a latent state in an insect host; that stress may condition an insect to make it susceptible to microorganisms ordinarily not pathogenic to it; and that stress, or depressors (38), may cause an insect to be more susceptible to a known pathogen than it is ordinarily. At the same time he concluded that the provocation of disease by stress may not be as regular and uniform as had been indicated by some authors.

3. ENVIRONMENT AND THE PATHOGEN POPULATION

Pathogen infectivity and virulence on the one hand and survival outside the hosts and capacity to disperse on the other are crucial in epizootiology. It has been already mentioned that stress may activate latent disease in an insect host (37, 38), that is, stress or the so-called activators may change the virulence of a pathogen. Environmental factors, especially the climatic factors, act foremost on the capacity of the pathogens to survive in the biotope and, as in the case of mycoses, the germination of the pathogens on their hosts (39, 40) as well as conidial discharge (41, 42). The detrimental influence of high temperature, desiccation, and sunlight on free pathogens is well known. Pathogens with a resting stage in their life cycles (such as the spores of bacteria, protozoa, and fungi, the inclusion bodies of viruses, and the ensheathed stages of nematodes) generally are better able to survive and persist in the environments of their hosts. However, the case of most strains of the sporogenic *B. thuringiensis* with their low capacity to survive in host populations shows that this rule has its exceptions. This is probably due to the fact that *B. thuringiensis* in most Lepidoptera populations acts as a poison rather than an infecting pathogen.

Wind and water may influence the dispersal and transmission of certain pathogens (14, 43). The same is true for biotic factors, such as parasites and predators that disperse diseases and even act as vectors. Several authors maintain that the preservation of the floral and faunal diversity of an ecosystem may contribute to the dispersion of the pathogens, because the diversity of highly mobile vectors is also increased (see 44). The topics of dispersion and spatial distribution are treated in detail elsewhere (see Ch. 5).

4. ENVIRONMENTAL FACTORS

Environmental factors act not discretely but as a complex whole. Therefore, the analysis of single environmental factors on the epizootics of infectious diseases is not always successful. This is probably the reason why, in spite of numerous observations on the importance of environmental factors in epizootics, quantitative data on the specific factors are rather scarce. The complexity of the problem is best illustrated by the factors that have received the most attention — temperature and humidity. The German entomologist Zwölfer, investigating the ecological parameters of the pine beauty moth *Panolis flammea* and the nun moth *Lymantria monacha,* developed the concept of "ecological value" and "vital range" of species which he defined by temperature and relative humidity (45, 46). He showed that the influence of the two factors on an organism always leads to a combined effect. This must of course be true also for diseased organisms. Therefore, studies on heat resistance of the conidia of *Metarhizium anisopliae* showed a clear correlation to moisture conditions (47). Tanada (8) pointed out that temperature within the normal growth ranges of the insect hosts appears to have only a limited effect on epizootics but is of more importance in combination with the other environmental factors. This is probably true for most other factors, though in certain diseases a single factor may play a dominant role, for example, high humidity, in most fungus and nematode diseases.

Many available quantitative data originated in the laboratory and might not fully correspond with natural conditions. Thus Milner et al. (48) investigating the development of milky disease under laboratory and field temperature regimes in larvae of *Rhopaea verreauxi* found only small differences among the probit lines of percentage infection against dose under constant 18°, 23°, and 28°C as well as alternating 13 to 23°C and 18 to 28°C. However, the more natural alternating temperatures reduced the time to mortality when compared with constant temperatures; and a higher prevalence of disease was obtained under field conditions than under any of the laboratory conditions tested. Moreover, true quantitative values are difficult to obtain in nature, since the microenvironment of an organism and/or a pathogen may differ considerably from the average conditions of the environment that are usually measured, in particular when the microenvironment is more important for infection and disease development. However, stepwise multiple regression analysis may sometimes provide better insight into the importance of the variables relative to prevalence of disease (23). This analysis showed for example that a seasonal influence on polyhedrosis prevalence in a *S. frugiperda* population was to a large degree a temperature variable.

The reader should keep these complications in mind in the following sections in which environmental factors are considered separately. It is not the purpose of this chapter to review the large body of literature about multiple interactions between insects and climatic and nutritional factors. However, since the enivronment profoundly influences the development and physiological condition of the host populations as well as their density and distribution in space and time, we have to consider environmental factors in relation to disease, that is, to host/pathogen interrelations.

4.1. Temperature

4.1.1. THE INFLUENCE ON THE HOST POPULATION.   Most laboratory insects are more susceptible to infections under the stress of high temperatures, though high temperature may inhibit the development of viroses (49–54), whereas certain insects are more susceptible to fungi, bacteria, and viruses when reared at low temperatures (7, 55). In the waxmoth *Galleria mellonella* on the other hand, the more or less normal temperatures of its natural environment (the beehive) promote nuclear polyhedrosis, whereas both low and high temperatures suppress it (56). However, the same insect is more susceptible to infection with *Tipula* iridescent virus at suboptimal temperatures (57). Hurpin (58) investigated the effect of temperature on cockchafer larvae *(M. melolontha)* in soil treated with the fungus *Beauveria bassiana* (= *B. tenella),* the bacterium *Bacillus popilliae* var. *melolonthae,* the rickettsia *Rickettsiella melolonthae,* the protozoa *Nosema melolonthae* and *Adelina melolonthae,* and an entomopoxvirus. The mycosis, bacteriosis, and nosematosis had optima at 20°C, the rickettsiosis at 17°C, and the coccidiosis at the highest temperature, whereas the virus acted independently of temperature.

According to Hurpin (59), third instar larvae of *M. melolontha* can only be controlled by the milky disease organism *B. popilliae* var. *melolonthae* between 15 and 27°C. This temperature range allows sufficient feeding and thus ingestion of enough spores for infection as well as normal larval metabolism for bacterial multiplication. The example shows that temperature acts neither on the host nor the pathogen alone but on the pathological interrelation as a whole.

Field temperatures affect larval feeding activity and therefore may influence the pathogen dose ingested (60, 61). However, low temperature (and rain), by stopping the feeding activity of an insect, might not only reduce the doses of pathogens ingested, but might, because of the inactivity of the intestinal tract (allowing an undisturbed contact of fungal germ tubes with the gut membrane), increase the incidence of fungal infections via the gut. Rozsypal (62) claimed that beet weevils that had stopped feeding on cold and rainy days easily succumbed to infection by *Beauveria bassiana*. Temperature also affects the growth of the plant on which a herbivorous insect feeds. The growth of the leaf surface in spring may decrease the pathogen density per unit leaf area (61, 63). Such threefold insect-plant-pathogen interrelationships may be crucial for the success or failure of the application of a pathogen in the field for pest suppression. Because of the lower thermal threshold of most plants, the leaves may continue to grow at low temperatures and dilute the pathogen while the insect host ceases to feed (61).

Few data have been collected concerning temperature and *Bacterioses* of insects. Higher temperatures favored bacterioses of *B. thuringiensis* var. *kenyae* and *israelensis* in blackflies (64, 65) and of *B. t. israelensis* and *B. sphaericus* in *Aedes stimulans* (66). Temperatures of 28 to 30°C after a light rain were optimal for epizootics of "*Coccobacillus acridiorum* d'Herelle" (= *Cloaca cloacae*) in grasshoppers (67). However, high as well as low temperatures favored bacterial septicemia in *Protoparce sexta* (68). *Bombyx mori* exposed to small doses of *B. thuringiensis* var. *alesti* had chronic infections that were activated to acute and rapidly fatal bacteriosis after a short thermal shock of 30°C (3). Heat and high humidity cause "flacherie" in the silkworm (see 69). Vago (70) has divided "flacherie" into different etiological types. One of these is associated with respiratory troubles due to heat and humidity shocks, followed by reduced function of the alimentary canal, which in turn leads to enterococcus multiplication.

Contrary to these observations Beard (71) and Dutky (72) found that *B. popilliae* infections killed *P. japonica* only in the temperature range of 15.5 to 36°C. The same was true for *B. popilliae* in *M. melolontha* (59), whereas Milner (73) found an upper limit of 30°C in *R. verreauxi*. An even narrower range of 25 to 33°C has been found for the septicemias of the armyworm *Pseudaletia unipuncta* caused by *Aerobacter aerogenes*, *Pseudomonas aeruginosa*, and *Serratia marcescens* (74).

Regarding *Rickettsioses,* the *Rickettsiella* species pathogenic for beetle larvae have temperature minima of 14 to 20°C, an optimum between 25 to 28°C, and a maximum at 32°C (75). In autumn, the grubs of *M. melolontha* infected with *R. melolonthae* crawl to the soil surface instead of moving deeper into the soil (76). Experiments by Niklas (76) could not demonstrate a definite reaction of diseased grubs to falling temperatures as against constant or rising temperatures. Disease prevalence showed two yearly maxima, one in summer and one in late autumn. If the upward movement of the diseased grubs was caused by factors other than temperature, one would expect them also to appear on the surface in summer. Since this has never been observed, the reverse orientation of the grubs in autumn may by directed by a temperature gradient in the soil.

Many authors have cited the influence of temperature on the development of *Viroses,* though field data are scarce. Fuxa and Geaghan (23), investigating the NPV of *S. frugiperda* populations in Louisiana and analyzing the factors affecting disease by multiple regression analysis, found that time of the year was a consistently important independent variable that was partially compensated by a temperature variable when time was dropped from the model. According to Aizawa (77, 78), the higher the temperature the shorter is the latent period of the NPV of *B. mori*. The same was found for *Neodiprion sertifer* infected by its baculovirus (79). Similarly, the granulosis virus (GV) of the red-banded leaf roller, *Argyrotaenia velutinana,* is more efficient in the southern USA than in the north (80), and Escherich and Miyajima (81) maintained that nun moth larvae contracted nuclear polyhedrosis more often during hot weather than in normal or cool weather. Yokokawa and Yamaguchi (82) reported a higher prevalence of cytoplasmic polyhedrosis in *B. mori* when the larvae were reared at 30° than at 25°C. The temperature optimum for a NPV of the velvetbean caterpillar *Anticarsia gemmatalis* was 30°C; inhibition occurred at 10 and 40°C (83). Ignoffo (60), too, found in *Heliothis zea* infected with NPV an increased and accelerated mortality at higher temperatures but interpreted it as a consequence of the increased feeding activity of the larvae rather than an effect on virus development. McLeod et al. (84) on the other hand found that a rise in temperature from 15 to 30°C had little effect on the activity of the NPV of *H. zea,* though a significant drop in mortality due to NPV was detected as temperature was increased to 45°C.

Because of the intimate physiological relationship between viruses and their host cells it will always be difficult to distinguish the effect of temperature (or another environmental factor) on the insect from its effect on the virus. This is especially true for positive temperature effects. However, there can be little doubt that the host insects are directly affected in those cases where latent viral infections are activated by cold stress (81, 85–91). But except for the work of Escherich and Miyajima (81), who worked with *L. monacha,* all positive results have been obtained with *B. mori.* Activation of latent viral infection with low temperature shocks was not possible in *Pieris rapae* (92), *C. eurytheme, H. zea, Junonia coenia, Laphygma exigua, Nymphalis antiopa, Peridroma margaritosa, Prodenia praefica* (38, 93), *Mamestra brassicae,* or *Hyphantria cunea* (94). It seems therefore highly probable that the immediate result of thermal shock is a cellular metabolic deviation that induces the multiplication of viruses already present in the cell, possibly in an occult state (95). Activation of nuclear and cytoplasmic polyhedroses by high temperature treatment also was effective only in Japanese strains of *B. mori* (88, 96, 97), but not in a Hungarian strain of *B. mori,* nor in *C. eurytheme, H. zea, J. coenia, L. exigua, N. antiopa,* or *P. praefica* (93).

Concerning high temperatures it should be added that several cases of induced immunity to viroses are known, for example, the granulosis of *P. rapae* at 36°C (98), the nuclear polyhedroses of *Diprion hercyniae* at 29°C (49), *Trichoplusia ni* and *H. zea* at 39°C (99), the *Tipula* iridescent virus in *G. mellonella* at 30°C (100), and the cytoplasmic polyhedrosis of *C. eurytheme* at 35°C (51). Investigations of Kobayashi et al. (53) on the inhibiting effect of 35°C on the development of nuclear polyhedrosis in *B. mori* pupae and isolated abdomens indicate that the inhibition may be attributed to the restricted accumulation of infectious virions, suggesting a temperature sensitivity of the virus replication mechanism itself. Exposure of natural insect populations to high summer temperatures may thus hinder the outbreak of a viral epizootic.

Humidity is more important than temperature to the *Mycoses* of insects of aereal habitats. This is not true, however, for insects living in water or soil, where humidity is never or rarely a limiting factor and thus temperature is most important (101). Because the entomophagous fungi usually invade the hosts through their integument rather than the mouth, the feeding activity of the insect is not important and high infection rates are possible even in relatively low temperatures. According to Schaerffenberg (102) the optimum temperature for the infection of the Colorado potato beetle by *B. bassiana* is between 20 and 30°C, but the minimum and maximum limits are near 0 and 40°C. Thus infection is possible when the beetle is completely inactive. The same was found for larvae of the elm bark beetle *Scolytus scolytus* infected with one strain each of *B. bassiana* or *Paecilomyces farinosus* at 2°C (103). On the other hand, an *Aspergillus* sp. was unable to infect mealybugs below 17°C (104). Similarly, a strain of *M. anisopliae* caused no infection in larvae of *S. scolytus* below 10°C (103), and the susceptibility of *H. zea* larvae to *Nomuraea rileyi* was lower at 15°C than at 20 to 30°C (105). *Culicinomyces clavosporus* infected and killed more than 95% of larvae of different mosquitoes, black flies, and other aquatic Diptera at 15 to 27°C; infection of the foregut was possible at 30°C, but hyphal growth was limited and the hyphae were discarded when the larvae molted (106).

Since most insect parasitic protozoa must be ingested by the host, *Protozoonoses* usually are possible only at the temperature range at which the insect feeds. On the other hand, nosematosis is not transmitted to young honeybees in summer because there is less infective fecal matter when the adults leave the hive, not because of direct disease suppression by high temperature (107). Similarly, the epizootic cycles of the disease caused by *Malpighamoeba mellificae* in the honeybee may depend on the more rapid development of bees in summer in contrast to the slow multiplication and spread of the amoeba (108). According to Kramer (109) more microsporidia-infected *Pyrausta nubilalis* are killed by low winter and high summer temperatures in Illinois than uninfected individuals.

4.1.2. THE INFLUENCE ON THE PATHOGEN POPULATION.   Many data on the influence of temperature on diseases were collected under laboratory conditions. In addition, high temperatures that would occur in nature only rarely were used in these tests. However, high temperatures may occur on the soil surface, depending on the type, color, plant cover, humidity, and geographical latitude. Soil temperatures in summer may reach 30 to 40°C in moderate climates and 30 to 70°C in tropical and subtropical climates (110). According to Johnpulle (111) temperatures of 70°C are common in compost heaps. Unfortunately only few data for temperature effects on insect disease have been collected in nature. Previous research reveals differences in the effect of temperature on pathogens in aqueous suspension and in the dry state (47).

Since the multiplication of *Bacteria* is temperature dependent, a rise in temperature is expected to accelerate the development of bacterial diseases in insects. However, only few results on this topic are available. The LT_{50} values of *B. thuringiensis* septicemia in *Pectinophora gossypiella* become, within the limits of 8 to 40°C, smaller as temperature rises (112). McLaughlin (74) maintained that bacterial septicemia in *P. unipuncta* as a result of high temperature is caused by the activation of the bacteria rather than a depressing action on the host larvae. There are also only few data on the heat resistance of bacterial suspensions: *S. marcescens* can survive 50°C for 30 minutes (113), sporulated *B. popilliae* 70°C for 1 hour (71, 114), and *B. thuringiensis* 80°C for 2 hours (115). Practically nothing is known about the resistance of desiccated bacterial pathogens to temperature extremes. Only Sidorenko (113) mentioned that *S. marcescens* keeps its virulence after a threefold freezing at − 10°C during 24 hours and thawing at 16°C in a solution of 4.5% sodium chloride. The few data available may suffice to conclude that natural temperatures are rarely detrimental to the survival of bacterial pathogens, though low temperatures will of course slow down or stop their development, whereas the temperature for optimum growth may perhaps but rarely be reached in nature. *Streptococcus pluton* with optimum growth at 35°C may be an exception, since this temperature prevails in the beehive (116).

The *Rickettsiae* vary in their response to temperature. *Rickettsiella popilliae* is killed in water suspension after 10 minutes at 60°C (75), whereas *R. melolonthae* survives 85°C for over 1 hour and several minutes at 100°C (117). The longevity of different species in aqueous suspensions differs widely in the temperature range 4 to 20°C; in some cases *R. melolonthae* is infective for a month at the higher temperature and more than a year at the lower. Frozen *R. popilliae* retain their activity for more than 3 years at − 80°C, and dried *R. melolonthae* survive several months at room temperature.

Most investigations of *Viruses* concerned baculoviruses and cytoplasmic polyhedrosis viruses (CPV). Baculoviruses in their occlusion bodies withstand freezing and thawing (36, 98, 118, 119) and can retain activity during prolonged exposure to normal field temperatures (120, 121). There is a difference between suspensions of viruses in the hemolymph or pure water and between films with or without hemolymph. Thus the NPV of *B. mori* in diluted hemolymph at pH 7.5 is thermoresistant at 60° but not 70° C during 30 minutes, whereas an aqueous suspension at the same pH is inactivated at 60°C in 10 minutes, 55°C in 30 minutes, 45°C in a day, and 37°C in 5 days (122). In general, insect viruses in occlusion bodies are resistant to temperatures up to 70°C for 20 to 30 minutes in suspension and to 100°C for 1 hour in dry hemolymph smears (36, 78, 123). They may remain infective at room temperature for several years and for more than 20 years in a pure suspension kept at 4°C in a sealed ampule (38). Suspensions of the GV of *Z. diniana* have not lost their infectivity after 7 years at 4°C (13). The same is true for the GV of the codling moth *Cydia pomonella* (Benz, unpublished). It should be mentioned here that the same viruses lose their infectivity quickly if they are not occluded in their inclusion bodies. Thus a fresh suspension of the GV of *Z. diniana* kept at 4°C lost 90% of its activity within 1 month and another 50 to 90% during the next year, but almost none during the next 3 years (Benz, unpublished). This indicated a quick inactivation of the naked virions in the primary suspension, followed by a slow inactivation of the virions with defective capsules, and hardly any inactivation of the perfectly encapsulated virions.

Since David and Gardiner (118) showed that crude GV of *Pieris brassicae* was still highly active after being kept for 1 year at variable temperatures between 15 to 28°C in darkness and that the purified virus was not entirely inactivated after 5 days at 50°C or 20 days at 40°C, temperature in most regions is not a major factor in baculovirus inactivation. These viruses would "survive" with little loss of activity from one season to the next under naturally occurring temperatures provided there was no other inactivating factor.

In general the environment acts more on the *Fungi* than on their hosts. Although Martin (124) could not find a correlation between water temperature and the infection of eggs of *Chironomus attenuatus* by the phycomycetous fungi *Catenaria circumplexa* and *C. spinosa*, these fungi appeared and disappeared in response to changing water temperatures, the first being present at weekly mean temperatures ranging from 14.2 to 16.6°C and the second at 14.2 to 20.6°C. Temperature acts not only on the germination of fungal spores and the development of the mycelia but also on the speed and the quantity of spore production. The latter sometimes has temperature optima other than germination (respectively 30 to 24°C in *B. bassiana*) (125). *Entomophthora gammae* sporulates and germinates at 10 to 26.7°C (126), whereas the spores of *B. bassiana* germinate between 10 and 38°C (40). Similarly the hyphomycete *Tolypocladium cylindrosporum*, a pathogen of mosquitoes, germinates from 8 to 33°C; optimal germination and growth are between 24 and 27°C (127). However, there are not only differences in the temperature requirements of different species (128), but also among different strains of the same species (103). In the typhlocladiosis of mosquitoes and in other mycoses (15) the LT_{50} values and/or the incubation period are inversely temperature dependent. Nevertheless natural disease prevalence in *Choristoneura fumiferana* caused by *Entomophthora sphaerosperma, E. egressa,* and a *Conidiobolus* species was enhanced by cool, wet weather (129).

Concerning high and low temperatures, the spores of *B. bassiana* survive 45°C for 48 hours but are killed if exposed longer to this temperature or for the same time to higher temperatures, for example, 48°C (130). Glaser (131) reported that conidia of *M. anisopliae* tolerated a temperature of 60°C for 1 hour in a suspension, whereas Vouk and Klas (132) found that their thermal death point in a moist atmosphere was reached after 5 minutes at 55 to 60°C, and Walstad et al. (133) found a temperature of 49°C for 10 minutes to be lethal. The different results can be explained by the fact that the sensitivity of conidia to temperature is influenced by humidity (47). Some spores survive heating up to 80°C for 2 hours, and freezing at −21°C for 2 weeks conserves the conidial vitality. However, when Newman and Carner (126) froze cadavers of *Pseudo-*

plusia includens that had been killed by *E. gammae*, the pathogen was inactivated in less than 1 hour. Spores of *B. bassiana, M. anisopliae*, and *P. farinosus* had respectively 99%, 85%, and 14% survival after a year at 8°C, and 29% and 0% at 25°C under the best conditions of humidity and darkness (134). Of the *B. bassiana* spores, 90% survived 2 years at 8°C. Spores of *M. anisopliae* survived longest at low temperatures and low humidity (4°C—0% RH) or at moderate temperatures and high humidity (19 to 26°C—97% RH) (135).

Regarding *Protozoa,* Schulz-Langner (136) found that temperatures of 37°C and higher suppressed *Nosema apis* in the honeybee, whose body temperature may reach 44°C; the number of hours the bees spent at 37°C in the hive proportionately retarded the development of the nosema infection.

4.2. Humidity

Humidity is a second important environmental factor affecting disease in insect populations and may be the single physical factor of importance in many epizootics of fungi and nematodes. Thus, Nordin et al. (137) proposed that accumulated moisture above a base humidity threshold could serve as a key factor for explaining epizootics of the fungus *Erynia (Entomophthora) phytonomi* in the alfalfa weevil *Hypera postica.* However, the interpretation of some results concerning humidity is not easy because of the interrelations with temperature, as mentioned before.

Little is known about the influence of humidity on *Bacterioses.* Bucher (138) stated that high humidity increased infection of grasshoppers by *P. aeruginosa,* whereas White (139) reported that neither excessive wetness nor dryness reduced the infection of the Japanese beetle grubs with *B. popilliae.* The same was reported by Steinhaus (140) for *S. marcescens* infections in larvae of *Galleria* and *Peridroma.*

Many investigators found high humidity to be an important factor in the development of *Virus* epizootics, for example, in *Phryganidia californica* (141), *Plusia gamma* (142), *H. cunea* (143), *Lymantria (Porthetria) dispar* (144), and *P. unipuncta* (145). This seems not to be true in other cases such as the nuclear polyhedroses of *Colias* (146), the European spruce and pine sawflies (49), and *G. mellonella* (147) in which humidity appears to play a minor role, if any. It is possible that humidity acts as a stressor in the positive cases.

Most authors agree that humidity is of paramount importance in *Mycoses,* especially with hyphomycetous fungi. Therefore, most reports specify high relative humidity (RH) as the limiting factor in fungal epizootics. Brown and Nordin (148) wrote an empirical model parameterized from the data of Watson et al. (149) to forecast field epizootics based on relative humidities. Sites having a relatively high air humidity have more fungal epizootics, in general, than dry sites. Thus Michelbacher et al. (150) found that more bark-inhabiting larvae succumbed to mycoses on the north side of trees than on the south, and Doane (151) observed that 92% of the larvae of *Scolytus multistriatus* overwintering beneath the bark of elms in a damp habitat were infected with *B. bassiana,* whereas in a dry habitat only 4% were. Furthermore control of citrus scale insects by fungi is effective in humid Florida but not in California with its dry summer months (152). On the other hand, the occurrence of fungus epizootics in Californian insects that live on irrigated low vegetation, such as the spotted alfalfa aphid (27), emphasizes the importance of the microclimate. The many reports on spore germination at different relative humidities may therefore reflect measurements made at the microclimate and the macroclimate levels. There are numerous examples of infections by different fungus species being influenced by the duration of high humidity, by changes of humidity due to microclimatic conditions on the insect or on plants, and by the type of crop (130, 153). Also, *E. gammae* in *P. includens* sporulated only at relative humidities of 80 to 100% and was inactivated by humidities of 50% and below, and the conidia germinated only at 98 and 100% RH and could not survive exposure to humidities below 75% (126). Modifying the microclimate with shade trees in coffee plantations may facilitate fungal epizootics in *Coccus viridis* and *Saissetia coffeae* by *Cephalosporium* sp. (44).

Notwithstanding what has been said so far, humidity seems not always to prevail in epizootics caused by some fungi. Thus, Schaefer (154) reported that the red locust *Nomadacris septemfasciata* can be infected with *Entomophthora grylli* at relative humidities below 60%. Pospelov (155) reported that excessive drought provoked an epizootic of the red muscardine, *Sorosporella uvella,* in the beet weevil *Cleonus punctiventris.* Teng (125) found that 25 to 50% RH favored the formation of *B. bassiana* spores. However, although infection of bean weevil adults *Anthoscelides obtectus* (156) and of the chinch bug *Blissus leucopterus* (157) with *B. bassiana* was possible at any RH between 30 and 100%, normally replication and conidiogenesis occurred only on host incubated at 75% or higher RH. *Erynia (Entomophthora) phytonomi* in the alfalfa weevil, too, would only sporulate if RH was high (42, 149). According to Millstein et al. (42, 158), conidia discharge (showering) occurred when RH in the alfalfa canopy exceeded 91%. This result is in agreement with the earlier observation of Wilding (41) that conidial discharge in the two entomophthoran species *E. aphidis* and *E. thaxteriana* increased with increasing humidity, and that more conidia were produced from cadavers in contact with liquid water than from those in a saturated atmosphere. Gottwald and Tedders (159), on the other hand, found that conidia release by *B. bassiana* and *M. anisopliae* from diseased adult pecan weevils was stimulated by lowering RH from near saturation to below 50%. Sustained periods of constant RH below 40% also favored some conidia release but always less than that accompanying RH changes.

As mentioned before, conidia of *M. anisopliae* survive longest when kept at moderate temperature and high RH or low temperature and low RH, while RH values between 33 and 75% are most lethal (135).

Humidity and moisture are the most critical environmental factors for insect pathogenic *Nematodes,* especially during the dispersal phase outside of their host (160–163). Schvester (164) found parasitization of *Ruguloscolytus rugulosus* by *Parasitylenchus dispar* to exceed 30% in humid years and to be as low as 12% in dry years.

4.3. Moisture, Wetness, and Rain

Not only high humidity but also moisture and wetness may act as stressors. According to Wahl (165), Glaser (166), and Wellenstein (167) damp weather leads to increased prevalence of polyhedrosis in *L. monacha* and *L. dispar*. These observations are in agreement with the fact that in the recent outbreaks of *L. monacha* in Poland during the abnormally dry years 1981 to 1984 the famous polyhedrosis ("Wipfelkrankheit") did not appear at all or only in isolated spots. This led to the defoliation of more than 2 million hectares of Polish pine forest (personal communication, K. Borusiewicz).

Some authors report that water on plant or insect surfaces is a requirement for optimal spore production and germination (e.g., 41). Free water must be present for the zoospores of certain entomogenous fungi to emerge from the infected host and transfer to new hosts. It is therefore interesting that the mosquito fungus *Lagenidium giganteum* persisted through two winters and a summer in rice fields without the availability of irrigation water (168). In this case the fungus probably survived as resting spores. Water is also necessary for mosquito nematodes such as the mermithid *Hydromermis churchillensis* which is irregularly distributed in forest pools and seemingly not correlated with the physical features of the pools (169).

Dew may extract substances from plants and may thus inactivate viruses. McLeod et al. (84) found cotton dew to have pH of 7.4 to 9.3, whereas soybean dew had pH 7.2. Exposure of the NPV of *H. zea* to pH 9.3 cotton dew resulted in loss in activity during the drying of the dew. Exposure to cotton dews at pH 7.4 to 8.8 or soybean dew at pH 7.2 produced no significant viral inactivation.

Rain may wash pathogens from plant surfaces (170, 171). However, when crude or purified preparations of the GV of *P. brassicae* were applied to cabbage leaves, the virus was not significantly removed by exposure for 5 hours to simulated rain (172). Under more natural conditions of weathering the virus has been shown to persist up to 4 months. The unimportance of rain is also stated by Ossowski (173), Burgerjon and Grison (174), Bullock (175), and Jaques (170). The dilution effect of rain is known in water habitats; Martin (124) found significant negative correlations of rainfall over a 3-year period with prevalence of the fungi *C. spinosa* and *C. circumplexa* in *C. attenuatus*.

4.4. Desiccation

Desiccation is one of the decisive factors for the persistence of pathogens in nature. Numerous pathogens cannot survive desiccation, for example, *P. aeruginosa* (176) and *Achromobacter eurydice* (138), as well as the ensheathed larval resting stages of some nematodes (163). However, such pathogens may survive for long periods under certain conditions, such as moist soil, dried cadavers, and feces. Thorough investigations, especially concerning desiccation in nature, are still missing. There are also interactions with temperature and other factors. Thus *Rickettsiella melolonthae* in dry hemolymph smears is still infective after 1 year at 4°C but less than 6 months at 20°C (117). Some nonsporulating bacteria also may survive relatively long periods of desiccation: 1 year or more for *S. pluton* in larval gut contents smeared over brood cells at pupation (138), and 3 months for *S. marcescens* when dried on sand at 35°C, (113). The latter bacterium dies rapidly on drying when sprayed as an aerosol of a pure suspension (177). As in the case of *R. melolonthae, S. pluton,* and *S. marcescens,* a support (preferably proteinaceous) seems to increase resistance to desiccation. According to David and Gardiner (118) this holds also for baculoviruses which remain active longer in a dry cadaver or a dried blood smear than in a dry film of purified material.

Many pathogens have special stages that can resist desiccation, such as the occlusion bodies of baculoviruses, the spores of *Bacillus* and *Clostridium* species, the cysts of amoebae, the spores and cysts of some sporozoa, the resting spores and, to a certain degree, the conidia of fungi, and the eggs or larval resting stages of some nematodes. The thick-walled resting sporangia of *Coelomomyces indiana* from *Anopheles* spp. germinate only after being incubated dry at 28°C for at least 2 weeks before being wetted again (178). Dry spores of *B. thuringiensis* remain viable up to 10 years (94), and the NPV of *B. mori* and *L. dispar* as well as the capsules of the *P. brassicae* GV are still active after more than 5 years storage as dry pills (179). It is therefore interesting that the spores of some microsporidia survive but 1 month in a dry smear, 3 to 4 months in insect cadavers, and 13 months in aqueous suspensions at laboratory temperatures (180). Also, freshly extracted spores of *Thelohania californica* are not infectious to *Culex tarsalis,* but the spores become infectious when alternately dried and hydrated (181).

Some fungal conidiospores may also survive long periods of desiccation, especially at low temperature. *Beauveria bassiana* survived two and a half years at 4°C but not more than 12 weeks at 23°C (94). Conidia of *M. anisopliae* have survived for 1 to 3 years (131, 182) and dry spores of *Spicaria farinosa* and *Aspergillus ochraceus* about 1 year (183, 184). The latter fungus will sporulate within the exoskeleton of dead honeybees if they are kept in too dry an atmosphere (184).

Desiccation increases the resistance of baculoviruses to light (170, 185). It also increases the resistance of microorganisms to high temperatures. Thus, the thermal death point for 30 minutes exposure of *M. anisopliae* conidia is about 50°C in wet spores, 50 to 55°C in dry spores at 100% RH, 70 to 75°C at 76% RH, and 80 to 85°C at 33% RH (47).

4.5. Light

In nature, light may affect organisms by its daily rhythm, its frequency (infrared, visible, or ultraviolet), and its energy. It may influence the distribution of microorganisms and their persistence in nature.

The duration of photoperiod controls diapause in many insects and may thus influence the hormonal milieu (186, 187) and disease development in these organisms (188–190). It may even assure the transmission of the GV by the overwintering larval population of the summer fruit tortrix, *Adoxophyes orana* (63). No other direct influence of light on the susceptibility of insects to disease has been reported. Steinhaus and Dineen (191) and Aruga and Yoshitake (192) could not induce respectively granulosis in *Peridroma margaritosa* or nuclear polyhedrosis in *B. mori* with ultraviolet (uv) light. Moreover, Pimentel and Shapiro (147) found that nuclear polyhedrosis prevalence in the larvae of *G. mellonella* was the same independent of whether the insects were kept in continuous light, continuous darkness, or irradiated with uv light for 65 minutes.

Positive stimulation of pathogens by light is reported, though rarely. Thus, Teng (125) found light to stimulate three developmental phases of the fungus *B. bassiana:* the growth of mycelium, the intensity of sporulation (positively correlated to light intensity and depending on the spectrum, with an optimum at 500 nm), and the germination of spores. Muspratt (178) suggested that sunlight stimulated the formation of *Coelomomyces* sporangia in *Anopheles gambiae.* Müller-Kögler (130) cited other examples of positive effects of light on fungi, especially Entomophthorales. Ege (193) confirmed the stimulating action of the blue part of the light spectrum on certain Entomophthorales of aphids. Conidia release was stimulated in *B. bassiana* by red-infrared (159). However, these few positive reports are in contrast to a number of reports of negative effects of visible light on the fructification of fungi and the release of conidia or a stimulation of these processes by darkness (15, 132, 134, 159).

Since unprotected protoplasm is easily destroyed by the impact of energy, the action of light may be destructive. Therefore many organisms have evolved light-absorbing protective pigments. This is especially striking in the spores of many fungi. On the other hand, baculovirus deposits on foliage retain their activity only for relatively short periods (49), usually less than 2 to 3 weeks (120, 170, 175, 194–197), depending largely on the exposure to sunlight (34, 84, 119, 170, 185, 195, 196, 198–201). Jaques (170) reported 40 and 0% activity of the NPV of *T. ni* after respectively 4 and 7 days of exposure on foliage. The GV of *P. brassicae* was totally inactivated between 12 and 19 hours exposure to direct sunlight (196), and the NPV of *T. ni* and the GV of *Estigmene acraea* in only 3 hours (194). However, persistence of up to 16 weeks was reported for the GV of *P. brassicae* when the virus was applied to the lower surfaces of cabbage leaves in the winter (172). Yearian and Young (202) showed also that the NPV of *H. zea* was more persistent on lower than on upper surfaces of leaves. Clark (21, 203) presented evidence that the polyhedrosis virus of *Malacosoma fragile* persisted over winter on trees in California, and Benz (198) showed that branches of larch trees in the Engadin (1800 m above sea level) treated with the GV of *Z. diniana* were still contaminated after 1 year, despite the fact that larch drops its needles in autumn and thus part of the virus deposit. David and Gardiner (118) found that dry films of crude GV stored up to 4 years were less stable when kept in light than in darkness. The visible as well as the uv components of sunlight inactivated the viruses (34, 170, 185, 200), though the uv component was especially detrimental (119, 122, 194, 199, 201, 204). The inactivating effect of high uv irradiation was probably responsible for the unsatisfactory control by artificially applied GV of *Z. diniana* in larch forests at altitudes of 1800 to 1900 m (18, 34, 185), though the virus was virulent in the laboratory (205). Ignoffo et al. (206) found the following relative ranking in stability of different viruses irradiated with uv light: entomopoxvirus $>$NPV $=$ CPV $>$ GV. Also, spores of the bacterium *B. thuringiensis* were more stable than those of the fungus *N. rileyi* (both ranking higher than the viruses) and the microsporidium *Nosema necatrix* (the latter ranking between the CPV and the GV).

Hurpin (117) was the only investigator who studied the light resistance of a rickettsia. It is high in *R. melolonthae;* exposure of a dry smear to uv light (253.7 nm, 8 ergs/mm^2/sec) for 1 week hardly changed its infectivity.

The effects of light on bacteria were discussed by Beard (71), Vago and Busnel (207), Burgerjon and Yamvrias (208), Cantwell and Franklin (209), Pinnock et al. (210), and Krieg et al. (211). It is not surprising that most papers deal with *B. thuringiensis.* Its infective half-life varies from half a day on soybean foliage (212) to 4 days on forest trees (213). Hostetter et al. (214) and Beegle et al. (215) found intermediate values of 2 days or less on foliage of cotton or *Juniperus virginiana,* respectively.

Leach (216) and Roberts and Campbell (217) reviewed the effects of visible and uv light on fungi. Toumanoff (218) and Teng (125) reported that *B. bassiana* conidia were killed by sunlight within 2 to 3 hours and more than 4 hours respectively, whereas the reaction to uv light varied in different strains. Clerk and Madelin (134) found survival rates of 13 and 72% when they exposed dry *B. bassiana* conidia for 6 months at 18°C to light or darkness, respectively. Ignoffo et al. (206), who studied the effects of sunlight on *N. rileyi,* found an interpolated half-life of 2.4 hours for uv irradiation from a source emitting 0.14 W/cm^2 at 215–260 nm and 1.8 mW/cm^2 at 290–400 nm. This value is comparable to that of Zimmermann (47) who exposed the spores of *M. anisopliae* to "artificial sunlight" and calculated the maximum half-life to correspond to 165 minutes of natural sunlight. The value of about 4 hours, reported by Roberts and Campbell (217), for the same spores exposed to direct sunlight lies also in this range.

There are not many data available on the influence of light on protozoa, and none on nematodes. Weiser (219) stated that dry spores of microsporidia were killed within 3 or 4 days in full sunshine. *N. apis* spores remained viable after 24 hours of sunlight, whereas *N. necatrix* spores did not survive 5 hours (220). Kaya (221), reporting on the persistence of *Plistophora schubergi* spores on oak leaves under field conditions (solar energy input of 960 to 2700 Joules/cm^2/ day), concluded that treatment to the point of runoff with 2×10^8 spores/ml left enough surviving spores after 4 days to infect 10% of second and third instar larvae of *Anisota senatoria* and none after 6 days. The addition of the uv protectant "Shade" increased the infection rates to 88 and 31%, respectively. The quick inactivation of *N. necatrix* spores by uv irradiation reported by

Ignoffo et al. (206) has been mentioned. Dry spores of the same species were inactivated under a germicidal lamp with uv of 254 nm within 4 to 10 minutes, whereas the same treatment on cotton leaves did not fully inactivate the spores within 3 hours (222). The same was true for dry spores on glass exposed to uv of 366 nm. Sunlight inactivated most spores within 24 hours and all within 78 hours. The spores of *Nosema heliothidis* on cotton leaves were almost completely killed after 7 hours of sunshine (223), whereas purified spores of *Octosporea muscaedomesticae* did not survive even 3 hours in sunlight (224). However, the spores within their dead host, the blow fly *Phormia regina,* were not affected by 12 hours of sunlight.

4.6. Soil

Soil is the most complex of the abiotic environmental factors. Moreover, since natural soils always contain large numbers of microorganisms, they have an intermediate status between the abiotic and the biotic factors.

Because of its water content and its light protecting nature, the soil protects microorganisms from desiccation and/or inactivation, for example, the bacterium *S. marcescens* (68), several baculoviruses (23, 146, 225–237), CPVs (235), and entomopoxviruses (236). The most thorough of these investigations, especially under field conditions, have been conducted by Jaques on the NPV of the cabbage looper in Canada. Tanada (238) reviewed virus persistence in soil up to 1970. There is general agreement that the persistence of occluded viruses is good in soils. Jaques (227) found 25% of the original activity of the NPV of *T. ni* more than 5 years after application in agricultural plots, and Thompson et al. (237) recorded active NPV of the Douglas-fir tussock moth *Orgyia pseudotsugata* in forest mineral soil (but not in the duff overlayer) more than 40 years after the last tussock moth outbreak. The soil is therefore a major reservoir for occluded insect viruses and thus an important epizootiological factor.

Concerning the influence of soil on sporeforming bacteria, almost all information relates to the milky diseases of soil grubs and to *B. thuringiensis.* The spores of *B. popilliae* can survive for several years in the soil (10, 25, 59, 71), but a low pH may damage them (71). Survival of the spores and crystals in soil is a key factor in the ecology of *B. thuringiensis.* Saleh et al. (239, 240), Kiselek (241), and Sekijima et al. (242) studied the fate of the spores in the soil, but not that of the crystals. However, the fates of both were investigated by Pruett et al. (243) and West et al. (244). After an application of *B. thuringiensis* onto trees, viable spores were found in the soil for up to 6 months, but not after 12, and spores survived in the bark for 2 years (241), whereas Saleh et al. (239) and Sekijima et al. (242) observed spore survival in soil for 3 to 16 months. Survival of the spores depended on the pH (240). Numbers of both vegetative cells and spores of *B. thuringiensis* increased in soil supplemented with dried powdered alfalfa at a neutral pH but not at acidic pHs. Since the other soil bacteria appeared to increase as well this would suggest that, under favorable nutrient conditions, *B. thuringiensis* was not inhibited by other bacteria of the natural environment. However, of 50 species of soil bacteria tested by Kiselek (241), 24% antagonized *B. thuringiensis.* Moreover, Pruett et al. (243) and West et al. (244), working with respectively a clay loam soil at pH 6.8 and acidic natural or autoclaved soil of pH 5.0, found that loss of spore viability was greater in natural than autoclaved soil, that the pathogenicity due to the δ-endotoxin was more quickly lost than viability, and that in autoclaved soil no significant loss of the δ-endotoxin activity was detected over a period of more than 2 years. These results suggested that soil microorganisms were responsible for the loss of the δ-endotoxin activity and partly responsible for the inactivation of the spores.

Concerning mycoses, the physical structure of the soil as well as its pH seems important (130). Larvae of the beet weevil, *Cleonis punctiventris,* were killed chiefly by *M. anisopliae* in acidic soil and by *S. uvella* in alkaline soil (245, 246). However, Bünzli and Büttiker (247) found that soils with high organic content (which bind more water) were more favorable for fungal diseases of soil-inhabiting insects than sandy soils with low organic matter. Spores of entomopathogenic fungi may survive in the soil longer than one year (93, 130, 134, 217). Most of these studies were done with *B. bassiana* and *M. anisopliae.* In a more recent paper (248) the half-lives of *B. bassiana* conidia in soil ranged from 14 days at 25°C and 75% water saturation to 276 days at 10°C and 25% water saturation. Conidia held at $-$ 15°C exhibited little or no loss in viability, regardless of water content and pH. Conidia were not recoverable after 10 days at 55°C. Conidia survival in nonsterile soil that was amended with carbon and nitrogen sources was low, and loss was often complete in less than 3 weeks, whereas the number of conidia increased in sterile soil treated in the same manner because the fungus grew under these conditions and produced new conidia. The fungistatic effect in amended nonsterile soils was possibly related to *Penicillium urticae* which was routinely isolated from the soils and produced a water-soluble inhibitor of *B. bassiana.* The naked blastospores of *B. bassiana* were inactivated after 3-week incubation at 20°C (249); coating blastospores with clay made them much more resistant since they were still active after 2 months in soil. This demonstrated the protective role of clay against biodegradation by soil bacteria (250, 251) and protozoa (252, 253). The conidia of *N. rileyi* layered directly on soil were still infectious for third instar larvae of the tobacco budworm after 138 days (254). Ignoffo et al. (255) found half-life values of 40 and 65 days in two experiments with the same pathogen; 99.9% of the conidia were inactivated after 250 days. The results indicated that the conidia of *N. rileyi* overwintered in the soil and that enough viable conidia produced during the season survived to initiate an epizootic in the next season. Soil or sediment from the area where there were parasitized larvae seemed necessary for the infection of mosquitoes with *Coelomomyces* zoospores (178, 256, 257).

There is little information concerning the so-called nonoccluded viruses, microsporidia, and nematodes. The activity of the iridescent virus of *Aedes taeniorhynchus* was lost within 1 day when placed on soil surface similar to that found in the natural breeding grounds of the mosquito (258). Microsporidian spores survived in the soil for at least 12 months (259). Some nematodes persisted as eggs, cysts, or ensheathed stages for long periods in the soil (163, 260, 261).

4.7. Pesticides

Pesticides are anthropogenic factors that often are in the environment of insects and their pathogens. Since pesticides may synergize or antagonize disease in insects (262) they may be regarded as epizootiologically relevant factors. Benz (262) and Jaques and Morris (263) have reviewed the compatibility of insect pathogens with different pesticides. Mixtures of *B. thuringiensis* with different pesticides have been synergistic as well as antagonistic, especially with some acaricides and insecticides, but less so with fungicides and herbicides (264–271). Telenga (272), Ramaraje et al. (273), Olmert and Kenneth (274), Johnson et al. (275), Livingston et al. (276), and Merriam and Axtell (277) investigated the effect of different pesticides on some entomopathogenic fungi and mycosis prevalence, whereas Fritz (278), Tedders (279), Loria et al. (280), and Perry and Latge (281) concentrated their research on the fungicides alone. A comprehensive review of the compatibility of pesticides with entomopathogenic fungi up to 1976 was published by Roberts and Campbell (217). Contrary to the observations with *B. thuringiensis* and baculoviruses no cases of synergism have been reported, except by Telenga (272) who observed more cases of muscardine caused by *B. bassiana* and *M. anisopliae* in beet-root weevils slightly poisoned by HCH. Most insecticides, acaricides, and fungicides had either no effect (e.g., diflubenzuron, permethrin, propoxur, sulfur, dinocap, chlorothalonyl, and metalaxal), a mild antagonism (e.g., malathion, carbaryl, metiram, and atrazin), or a clearly antagonistic effect on fungi (e.g., DDT, lindane, toxaphene, chlordimeform, chlorpyrifos, fenthion, benomyl, captan, mancozeb, triphenyltin, and zineb). The germination of resting spores was less suppressed than that of conidia. Many pesticides completely inhibited zoospore production in the mosquito pathogen *L. giganteum* at concentrations below 5 ppm in water, that is, at concentrations that inhibited the mycelial growth rate only by 10% (277). The pesticides that showed medium or less effect in these tests are probably compatible with most entomopathogenic fungi in the field. This is not the case, however, with the strongly antagonistic chemicals. The fungicides, in particular, are significant since they cannot be omitted from the production of many agricultural crops. It may be noteworthy in this context that the development of microsporidia may be suppressed by the systemic fungicide benomyl (282, 283).

4.8. Host Plants and Other Nutritional Factors

Host plants and other nutritional factors are environmental components whose roles in the epizootics of insect disease are not well understood. It has been suspected for a long time that starvation, such as can occur after nutritional resources have been used up in a mass outbreak of insects, may stimulate latent infectious bacterial diseases or activate viroses, for example the granulosis of the larch bud moth *Z. diniana* (11). However, nutritional deficiency did not stimulate granulosis in uninfected larvae and had a negative effect on granulosis prevalence in larvae infected with low doses of the virus (Benz, unpublished). The same was true for the variegated cutworm *P. margaritosa* (191). Similarly, reduced quantities of food had little effect on the mortality of *T. ni* from polyhedrosis (284). Further studies with the looper *Sterrha seriata* confirmed these results and showed additionally that starvation suppressed the development of the NPV (285). When David et al. (286) found that a lack of sucrose in the diet of *P. brassicae* enhanced granulosis prevalence, they too suggested that it might be due to starvation. However, further experiments showed that disease stimulation was more likely due to increased permeability of the gut epithelium for the virions (287). Thus the conclusion drawn by Aruga (288) ". . . that starvation may not play an important role as a stressor in the induction of insect virus disease, but acts rather as a modifier in the induction by other primary stressors . . ." remains acceptable.

The results of David et al. (286, 287) suggested that nutritional effects on disease prevalence in insects were of a qualitative rather than quantitative nature. Ripper (289) and later Vago (95, 290) noticed an increase of nuclear polyhedrosis in silkworms reared on *Scorzonera hispanica* and *Maclura aurantiaca* respectively instead of mulberry *(Morus alba)*. On the other hand, stress from feeding the larvae of the noctuid *Calophasia lunula* on snapdragon *Antirrhinum majus* instead of toadflax *Linaria vulgaris* decreased the survival rate of larvae already infected with a benign CPV but had no influence on pathogenesis of a pathogenic NPV (291). Silkworms fed mulberry leaves which had grown in the shade and contained less chlorophyll than sungrown leaves succumbed to a multibacterial dysentery (292), whereas polyhedrosis of *B. mori* was favored by feeding young larvae on hard leaves and last instar larvae on young juicy leaves of mulberry. Vago (290) found differences in the virosis curves of silkworms according to whether they had been fed on young, medium aged, or old mulberry leaves. Shvetzova (293) noted similar results with *Antheraea pernyi* and felt that there might be a relationship between polyhedrosis prevalence and the nitrogen/carbohydrate balance of the leaves. Prevalence of cytoplasmic polyhedrosis in the silkworms increased in autumn, apparently associated with the quality of the mulberry leaves used as food by the insects (294). A similar increase every autumn for several years had been observed in the author's laboratory in rearings of the looper *S. seriata* which were otherwise disease free throughout the year. These insects were fed withered leaves of dandelion which are rarely attacked by Lepidoptera in the field, and never by *S. seriata*. Contamination of the leaves with polyhedra was therefore extremely improbable. Hare and Andreadis (295) working with *Leptinotarsa decemlineata* on different host plants found that the beetles were least susceptible to infection with *B. bassiana* on the plant species most suitable for the beetle's survival and that host plant suitability declined with host plant age.

These phenomena seem to be caused less by a specific action or a specific substance in the particular plants than by a more general disturbance of the plant metabolism. This is perhaps best demonstrated by the example of the granulosis of the larch bud moth *Z. diniana* which shows cyclic gradations in the Alpine larch forests (11). Again, the virosis cannot be induced by starvation, although its prevalence is coupled with the gradation cycle and the defoliation of the larch trees. However, the quality of the larch needles, and with it the food of the bud moth larvae, changes as a result of their feeding activity (296), which induces the granulosis (34). Several such feedback mechanisms have been found during the last 10 years, and their significance for the population dynamics of insects is increasingly recognized (297, 298). As in the example of *Z. diniana,* it may also be of epizootiological significance.

Another epizootiological aspect of host plants is their possible content of antimicrobial substances. Thus the activity of the NPV of *Heliothis armigera* in Botswana was lost rapidly on cotton but remained high for up to 30 days on sorghum; the activity was still detectable on the harvested sorghum more than 80 days after spraying (299). This effect of cotton leaf surfaces has also been observed by Andrews and Sikorowski (300). Moreover, in Section 4.3 it has already been mentioned that McLeod et al. (84) found that exposure of *Baculovirus heliothidis* to pH 9.3 cotton dew resulted in loss of activity during the drying of the dew. This apparently resulted from exposure to high pH and high basic ion concentration. Exposure to soybean dew at pH 7.2 produced no significant viral inactivation. Stubblebine and Langenheim (301), who investigated the effect of leaf resin of the leguminose *Hymenaea courbaril* in the diet of *Spodoptera exigua,* demonstrated that mortality due to an unidentified virus was dependent on leaf resin concentration. Furthermore, L-canavanine (a plant allelochemical synthesized by more than 500 leguminous plants including a number of forage crops) interfered with *T. ni* nuclear polyhedrosis induction of infected cell-specific polypeptides in a *S. frugiperda* cell culture (302). Felton and Dahlman (303) demonstrated increased susceptibility of *Manduca sexta* to commercial formulations of *B. thuringiensis* when the larvae was reared on an artificial diet supplemented with a sublethal concentration of L-canavanine. Several plants' juices have an antimicrosporidian effect in *Archips cerasivorus* (304). The majority of papers deal with antibacterial substances found in plant juices (305–310) or as volatile components of some trees and shrubs (311, 312). Maksymiuk (310) isolated five antibacterial substances from pitch pine foliage which he thought were possibly hydroxy carboxylic acids. Smirnoff (312), on the other hand, found the different terpenes of the balsam fir to be responsible for the *B. thuringiensis* inactivation observed in balsam fir stands. Of the principal substances released by the *Abies balsamea* foliage Smirnoff found the following sequence of decreasing antibacterial potency: thujone > phellandrene > fenchone > limonene > β-pinene > α-pinene.

5. Conclusion

There can be no doubt that epizootics of infectious insect diseases are influenced by environmental factors and that these, occasionally, may be the most relevant elements in the epizootiological process. Our understanding of the epizootiology of disease depends largely therefore on our understanding of their influence and the processes they affect. However, although a wealth of pragmatic facts are available, very little analytical work showing the underlaying mechanisms has been done. This is especially obvious (and might be expected) in all cases where the environmental factors act as stressors; effects of stress, too often, cannot be predicted, as many contradictory results reported in this chapter show. But even where the action of single environmental factors lends itself to causal analysis, the interdependence of that factor with others may render such analysis useless. The separation of effects on the host and on the pathogens, too, as it has been tried by other authors and in this chapter, is of no great help. When dealing with infectious disease, we always deal with a host organism and host population in its biocenosis, a pathogen population, and their natural or managed environment(s). Each of these "elements" itself is a more or less complex system. Obviously the ecosystem to which the host belongs as well as the host and the pathogen themselves, are or should be self-regulating systems in which all partial functions, elements, and subsystems are properly coordinated and function in adequate harmony. That is, they are well adjusted to each other, as the result of selection processes. Disease in an ecosystem is an element of this harmony. All these systems are influenced by all the environmental factors treated in this chapter. A deeper understanding of their epizootiological significance will only be possible if two seemingly opposing approaches are increasingly chosen by insect pathologists: exact causal analyses on the one hand, and synthesis on the basis of informatics, cybernetics, and system theory on the other. Only multiple-regression analyses of as many factors as possible with subsequent systems modeling supported experimentally will help to better understand and adequately describe the sometimes very complex situations and perhaps allow reliable predictions. Such knowledge would be very useful when new strategies for insect suppression with pathogens are developed within the frame of integrated pest management systems (313).

REFERENCES

1. M. J. Vida of Cremona, *De bombice,* Rome, 1527 (French translation: M. Bonafous, *Le ver a soie: Poème de Marc–Jerome Vida,* Impr. Bonchard-Huzard, Paris, 1840, 2nd book).

2. M. S. Merian, Der Raupen wunderbare Verwandlung, und sonderbare Blumennahrung, Vol. 1. J. A. Graff, Nurnberg, 1679, p. 115.

3. C. Vago, L'enchaînement des maladies chez les insectes. *Recherches experimentales en pathologie comparée,* Comiss. Seric. Int. Edit. spec. Alès (Gard), 1956, 184 pp.

4. A. Krieg and L. Schmidt, Ueber die Möglichkeit einer mikrobiologischen Bekämpfung von

Hyphantria cunea (Drury), *Nachrichtenbl. Dtsch. Pflanzenschutzdienstes (Braunschweig),* **14,** 177 (1962).

5. A. Schmid, Interferenz zwischen dem spezifischen Granulosis-virus und zwei Bakterienprä-paraten bei Raupen des Grauen Lärchenwicklers, *Zeiraphera diniana* (Gu.), *Mitt. Schweiz. Entomol. Ges.,* **48,** 173 (1975).

6. E. A. Steinhaus, The Effects of Disease on Insect Populations, *Hilgardia,* **23,** 197 (1954).

7. J. M. Franz, "Biologische Schädlingsbekämpfung," in H. Richter, Ed., *Handbuch der Pflan-zenkrankheiten,* Vol. 6, Paul Parey, Berlin, 1961, p. 1.

8. Y. Tanada, "Epizootiology of Infectious Diseases," in E. A. Steinhaus, Ed., *Insect Pathology: An Advanced Treatise,* Vol. 2, Academic Press, New York, 1963, p. 423.

9. P. Grison, Réflexions sur l'épizootiologie. Coll. Int. Pathol. Insectes, Paris 1962, *Entomo-phaga, Mem. Hors Ser.* **2,** 483 (1964).

10. B. Hurpin, Influence des facteurs du milieu, in P. A. Van der Laan, Ed., *Insect Pathology and Microbial Control,* North-Holland Publishing Co., Amsterdam, 1967, p. 135.

11. M. E. Martignoni, Contributo alla conoscenza di una granulosi di *Eucosma griseana* (Hübner) (Tortricidae, Lepidoptera) quale fattore limitante il pullulamento dell'insetto nella Engadina alta. *Mitt. Schweiz. Zentralanst. Forstl. Versuchswesen,* **32,** 371 (1957).

12. J. P. Perron and R. Crete, Premiers observations sur le champignon, *Empusa muscae* Cohn (Phycomycetes: Entomophthoraceae), parasitant la mouche de l'oignon, *Hylemya antiqua* (Meig.) (Diptera: Anthomyiidae) dans le Québec, *Ann. Entomol. Soc. Qu.,* **5,** 25 (1960).

13. G. Benz, Über eine Polyedrose als Begrenzungsfaktor einer Population von *Malacosoma alpicola,* Stdg. *Mitt. Schweiz. Entomol. Ges.,* **34,** 382 (1961).

14. T. F. Bird, Transmission of some insect viruses with particular reference to ovarial transmis-sion and its importance in the development of epizootics, *J. Insect Pathol.,* **3,** 352 (1961).

15. R. I. Carruthers, D. L. Haynes, and D. M. McLeod, *Entomophthora muscae* (Entomophthor-ales: Entomophthoraceae) mycosis in the onion fly, *Delia antiqua* (Diptera: Anthomyiidae), *J. Invertebr. Pathol.,* **45,** 81 (1985).

16. T. F. Bird and D. E. Elgee, A virus disease and introduced parasites as factors controlling the European spruce sawfly, *Diprion hercyniae* (Htg.), in central New Brunswick, *Can. Entomol.,* **89,** 371 (1957).

17. B. A. Federici, Baculovirus epizootic in a larval population of the clover cutworm, *Scoto-gramma trifolii,* in Southern California, *Environ. Entomol.,* **7,** 423 (1978).

18. G. Benz, Bekämpfungsversuche gegen den Grauen Lärchenwickler mittels Granulosisvirus und *Bacillus thuringiensis.* Unpubl. Internal Rep., Entomol. Inst. ETH Zurich (1963).

19. Y. Tanada and E. M. Omi, Epizootiology of virus diseases in three lepidopterous insect species of alfalfa, *Res. Popul. Ecol.* **16,** 59 (1974).

20. E. C. Clark and C. G. Thompson, The possible use of microorganisms in the control of the Great Basin tent caterpillar, *J. Econ. Entomol.,* **47,** 268 (1954).

21. E. C. Clark, Observations on the ecology of a polyhedrosis of the Great Basin tent caterpillar, *Malacosoma fragilis, Ecology,* **36,** 373 (1955).

22. Y. Tanada, The epizootiology of virus diseases in field populations of the armyworm, *Pseu-daletia unipuncta* (Haworth), *J. Insect Pathol.,* **3,** 310 (1961).

23. J. R. Fuxa and J. P. Geaghan, Multiple-regression analysis of factors affecting prevalence of nuclear polyhedrosis virus in *Spodoptera frugiperda* (Lepidoptera: Noctuidae) populations, *Environ. Entomol.,* **12,** 311 (1983).

24. S. R. Dutky, "The Milky Diseases," in E. A. Steinhaus, Ed., *Insect Pathology: An Advanced Treatise,* Vol. 2, Academic Press, New York, 1963, p. 75.

25. B. Hurpin, Recherches épizootiologiques sur la maladie laiteuse à *Bacillus popilliae* "melo-lontha", *Ann. Epiphyt.,* **18,** 127 (1967).

26. F. H. Billings and P. A. Glenn, Results of the artificial use of the white-fungus disease in

Kansas: with notes on approved methods on fighting chinch bugs, *U. S. Dep. Agric. Bur. Entomol. Bull.,* **107,** 1911, 58 pp.

27. I. M. Hall and P. H. Dunn, Fungi on spotted alfalfa aphid, *Calif. Agric.,* **11,** 5 (1957).

28. G. C. Ullyett and D. B. Schonken, A fungus disease of *Plutella maculipennis* Curt. in South Africa, with notes on the use of entomogenous fungi in insect control, *Union S. Afr. Sci. Bull. Dep. Agric. Forestry,* No. 218, 1940, 24 pp.

29. G. C. Ullyett, Biomathematics and insect population problems. A critical review, *Mem. Entomol. Soc. S. Afr.,* No. 2, 1953, 89 pp.

30. A. Milne, Theories of natural control of insect populations, *Cold Spring Harbor Symp. Quant. Biol.,* **22,** 253 (1957).

31. A. Milne, The natural control of insect populations, *Can. Entomol.,* **89,** 193 (1957).

32. A. Milne, On a theory of natural control of insect populations, *J. Theor. Biol.,* **3,** 19 (1962).

33. E. A. Steinhaus, Crowding as a possible stress factor in insect disease, *Ecology,* **39,** 503 (1958).

34. A. Schmid, Unterschungen zur Trans-ovum-Uebertragung des Granulosisvirus des Grauen Lärchenwicklers, *Zeiraphera diniana* (Lep.: Tortricidae) und Auslösung der akuten Virose durch Stressfaktoren, *Entomophaga,* **19,** 279 (1974).

35. H. Selye, A syndrome produced by diverse noxious agents, *Nature,* **2,** 32 (1936).

36. G. H. Bergold, "Viruses of Insects," in K. F. Meyer and C. Hallauer, Eds., *Handbuch der Virusforschung,* Vol. 4, 1958, p. 60.

37. E. A. Steinhaus, Stress as a factor in insect disease, *Proc. 10th Int. Congr. Entomol. Montreal 1956,* **4,** 725 (1958).

38. E. A. Steinhaus, The importance of environmental factors in the insect-microbe ecosystem, *Bacteriol. Rev.,* 24, 365 (1960).

39. R. Schneider, Unterschungen über Feuchtigkeitsansprüche parasitischer Pilze, *Phytopathol. Z.,* **21,** 63 (1953).

40. M. P. Hart and D. M. McLeod, An apparatus for determining the effects of temperature and humidity on germination of fungous spores, *Can. J. Bot.,* **33,** 289 (1955).

41. N. Wilding, Effect of humidity on the sporulation of *Entomophthora aphidis* and *E. thaxteriana, Trans. Br. Mycol. Soc.,* **53,** 126 (1969).

42. J. A. Millstein, G. C. Brown, and G. L. Nordin, Microclimatic moisture and conidial production in *Erynia* sp. (Entomophthorales: Entomophthoraceae). In vivo production rate and duration under constant and fluctuating moisture regimes, *Environ. Entomol.,* **12,** 1344 (1983).

43. R. N. Hofmaster, Seasonal abundance of the cabbage looper as related to light trap collections, precipitation, temperature and incidence of a nuclear polyhedrosis virus, *J. Econ. Entomol.,* **54,** 796 (1961).

44. J. M. Franz, "Influence of Environment and Modern Trends in Crop Management of Microbial Control," in H. D. Burges and N. W. Hussey, Eds., *Microbial Control of Insects and Mites,* Academic Press, New York, 1971, p. 407.

45. W. Zwölfer, Studien zur Oekologie und Epidemiologie der Insekten. 1. Die Kieferneule *Panolis flammea* Schiff., *Z. Angew. Entomol.,* **17,** 475 (1931).

46. W. Zwölfer, Studien zur Oekologie, insbesondere zur Bevölkerungslehre der Nonne, *Lymantria monacha* L., *Z. ang. Entomol.,* **20,** 1 (1934).

47. G. Zimmermann, Effect of high temperatures and artificial sunlight on the viability of conidia of *Metarrhizium anisopliae, J. Invertebr. Pathol.,* **40,** 36 (1982).

48. R. J. Milner, J. T. Wood, and E. R. Williams, The development of milky disease under laboratory and field temperature regimes, *J. Invertebr. Pathol.,* **36,** 203 (1980).

49. T. F. Bird, Virus diseases of sawflies, *Can. Entomol.* **87,** 124 (1955).

50. Y. Tanada, Effect of high temperatures on the resistance of insects to infectious diseases, *J. Seric. Sci. Japan*, **36**, 333 (1967).

51. Y. Tanada and G. Y. Chang, Resistance of the alfalfa caterpillar, *Colias eurytheme*, at high temperatures to a cytoplasmic polyhedrosis virus and thermal inactivation point of the virus, *J. Invertebr. Pathol.*, **10**, 79 (1968).

52. H. Inoue and Y. Tanada, Thermal therapy of the flacherie virus disease in the silkworm, *Bombyx mori*, *J. Invertebr. Pathol.* **29**, 63 (1977).

53. M. Kobayashi, S. Inagaki, and S. Kawase, Effect of high temperature on the development of nuclear polyhedrosis virus in the silkworm, *Bombyx mori*, *J. Invertebr. Pathol.*, **38**, 386 (1981).

54. W. R. Kellen and D. F. Hoffmann, Thermoinactivation of a calicivirus of the navel orangeworm, *Amyelois transitella* (Lepidoptera: Pyralidae), and the effect of high temperature on larval resistance, *Environ. Entomol.*, **12**, 605 (1983).

55. V. P. Pospelov, The influence of temperature on the maturation and general health of *Locusta migratoria*, L., *Bull. Entomol. Res.*, **16**, 363 (1926).

56. G. R. Stairs, Effects of a wide range of temperature on the development of *Galleria mellonella* and its specific baculovirus, *Environ. Entomol.*, **7**, 297 (1978).

57. D. J. Witt and G. R. Stairs, Effects of different temperatures on *Tipula* iridescent virus infection in *Galleria mellonella* larvae, *J. Invertebr. Pathol.*, **28**, 151 (1976).

58. B. Hurpin, The influence of temperature and larval stage on certain diseases of *Melolontha melolontha*, *J. Invertebr. Pathol.*, **10**, 252 (1968).

59. B. Hurpin, Etude de diverses souches de maladie laiteuse sur les larves de *Melolontha melolontha* L. et sur celles de quelques espèces voisines, *Entomophaga*, **4**, 233 (1959).

60. C. M. Ignoffo, Effects of temperature on mortality of *Heliothis zea* larvae exposed to sublethal doses of a nuclear polyhedrosis virus, *J. Invertebr. Pathol.*, **8**, 290 (1966).

61. J. M. Franz, A. Krieg, and J. Reisch, Freilandversuche zur Bekämpfung des Eichenwicklers (*Tortrix viridana* L.) (Lep., Tortricidae) mit *Bacillus thuringiensis* im Forstamt Hanau, *Nachrichtenbl. Dtsch. Pflanzenschutzdienstes, Braunschweig*, **19**, 36 (1967).

62. J. Rozsypal, The sugar-beet pest, *Bothynoderes punctiventris* Germ., and its natural enemies (in Czech.), *Sbornik Chir. Pohyb. Ustroji*, C16, 92 pp. Abstr. *Rev. Appl. Entomol.*, A19, 427 (1931).

63. C. R. Flückiger, Unterschungen über drei Baculovirus-Isolate des Schalenwicklers, *Adoxophyes orana* F.v.R. (Lep., Tortricidae), dessen Phänologie und erste Feldversuche, als Grundlagen zur mikrobiologischen Bekämpfung dieses Obstschädlings, *Mitt. Schweiz. Entomol. Ges.*, **55**, 241 (1982).

64. L. A. Lacey, M. S. Mulla, and H. T. Dulmage, Some factors affecting the pathogenicity of *Bacillus thuringiensis* Berliner against blackflies, *Environ. Entomol.*, **7**, 583 (1978).

65. D. Molloy, R. Gaugler, and H. Jamnback, Factors influencing efficacy of *Bacillus thuringiensis* var. *israelensis* as a biological control agent of black fly larvae, *J. Econ. Entomol.*, **74**, 61 (1981).

66. S. P. Wraight, D. Molloy, H. Jamnback, and P. McCoy, Effects of temperature and instar on the efficacy of *Bacillus thuringiensis* var. *israelensis* and *Bacillus sphaericus* Strain 1593 against *Aedes stimulans* larvae, *J. Invertebr. Pathol.*, **38**, 78 (1981).

67. F. D'Herelle, Le Coccobacille des sauterelles, *Ann. Inst. Pasteur*, **28**, 1 (1914).

68. G. F. White, Hornworm septicemia, *J. Agric. Res.*, **26**, 477 (1923).

69. C. Vago, "Predispositions and Interrelations in Insect Diseases," in E. A. Steinhaus, Ed., *Insect Pathology: An Advanced Treatise*, Vol. 1, Academic Press, New York, 1963, p. 339.

70. C. Vago, Le problème de la flacherie en pathologie comparée, *11th Conf. Int. Tech. Seric. Murcie.* (1960).

71. R. L. Beard, Studies on the milky disease of Japanese beetle larvae, *Conn. Agric. Exp. Stn. Bull.,* **491,** 505 (1945).

72. S. R. Dutky, "The Milky Diseases," in E. A. Steinhaus, Ed., *Insect Pathology: An Advanced Treatise,* Vol. 2, Academic Press, New York, 1963, p. 75.

73. R. J. Milner, A new variety of milky disease, *Bacillus popilliae* var. *rhopaea,* from *Rhopaea verreauxi, Aust. J. Biol. Sci.,* **27,** 235 (1974).

74. R. E. McLaughlin, The role of certain gram-negative bacteria and temperature in larval mortality of the armyworm, *Pseudaletia unipuncta* (Hayworth), *J. Insect Pathol.,* **4,** 344 (1962).

75. A. Krieg, "Rickettsiae and Rickettsioses," in E. A. Steinhaus, Ed., *Insect Pathology: An Advanced Treatise,* Vol. 1, Academic Press, New York, 1963, p. 577.

76. O. F. Niklas, Zur Temperaturabhängigkeit der Vertikal-Bewegung rickettsiosekranker Maikäfer-Engerlinge (*Melolontha* spec.), *Anz. Schädlingskd.,* **30,** 113 (1957).

77. K. Aizawa, Multiplication mode of the silkworm jaundice virus. I. On the multiplication mode in connection with the latent period and LD_{50}-time curve, *Bull. Seric. Exp. Stn. (Tokyo),* **14,** 201 (1953) (Japanese with English summary).

78. K. Aizawa, "The Nature of Infections Caused by Nuclear-Polyhedrosis Viruses," in E. A. Steinhaus, Ed., *Insect Pathology: An Advanced Treatise,* Vol. 1, Academic Press, New York, 1963, p. 381.

79. A. Krieg, Untersuchungen über die Polyedrose von *Neodiprion sertifer* (Geoffr.), *Arch. Ges. Virusforsch.,* **6,** 163 (1955).

80. E. H. Glass, Laboratory and field tests with the granulosis of the red-banded leaf roller, *J. Econ. Entomol.,* **51,** 454 (1958).

81. K. Escherich and M. Miyajima, Studien über die Wipfelkrankheit der Nonne, *Natur. Z. Forst- Landwirtsch.,* **9,** 381 (1911).

82. S. Yokokawa and K. Yamaguchi, Induction of the cytoplasmic polyhedrosis in silkworm larvae by feeding of mulberry leaves with some agricultural chemicals, *J. Seric. Sci. Japan,* **29,** 133 (1960), (in Japanese).

83. D. W. Johnson, D. B. Boucias, C. S. Barfield, and G. E. Allen. A temperature-dependent developmental model for a nucleopolyhedrosis virus of the velvetbean caterpillar, *Anticarsia gemmatalis* (Lepidoptera: Noctuidae), *J. Invertebr. Pathol.,* **40,** 292 (1982).

84. P. J. McLeod, W. C. Yearian, and S. Y. Young, III, Inactivation of *Baculovirus heliothis* by ultraviolet irradiation, dew, and temperature, *J. Invertebr. Pathol.,* **30,** 237 (1977).

85. K. Suszuki and G. Sakizaki, On the resistance to low temperature in the 5th instar larva of *Bombyx mori, Syngyp-Shinpo,* **379,** 37 (1925), (in Japanese).

86. S. Sakai, Studies on the grasserie in the silkworm, *Bombyx mori, Sanshi-Gakuho,* **17,** 20 (1935), (in Japanese).

87. H. Suzuki, Grasserie in the silkworm, *Bombyx mori, Meeting Japan. Seric. Soc.,* **8,** 37 (1941).

88. N. Ishimori, On the temperature limits for grasserie virus induction and grasserie virus infection in the silkworm, *Bombyx mori* L., *J. Seric. Sci. Japan,* **20,** 51 (1951).

89. I. Kurisu, Studies on the virus diseases in the silkworm. I. On the mid-gut polyhedrosis, *Bull. Kumamoto Seric. Exp. Stn.,* **6,** 48 (1955), (in Japanese).

90. H. Ooba, The nature of artificially induced grasserie and flacherie in larvae of the silkworm, *Bombyx mori, J. Seric. Sci. Japan,* **25,** 211 (1956).

91. H. Aruga, Mechanism of the virus resistance in the silkworm, *Bombyx mori.* II. On the relation between the nuclear polyhedrosis and the cytoplasmic polyhedrosis, *J. Seric. Sci. Japan,* **26,** 279 (1957), (Japanese with English summary).

92. Y. Tanada, A polyhedrosis virus of the imported cabbageworm and its relation to a polyhedrosis virus of the alfalfa caterpillar, *Ann. Entomol. Soc. Am.,* **47,** 553 (1954).

93. E. A. Steinhaus, The duration of viability and infectivity of certain insect pathogens, *J. Insect Pathol.*, **2**, 225 (1960).

94. H. Aruga, N. Yoshitake, H. Watanabe, and T. Hukuhara, Studies on nuclear polyhedroses and their inductions in some lepidoptera, *Japan. J. Appl. Entomol. Zool.*, **4**, 51 (1960), (Japanese with English summary).

95. C. Vago, Phénomène de "latentia" dans une maladie à ultravirus des insectes, *Rev. Can. Biol.*, **10**, 299 (1951).

96. E. Kitajima, High temperature and grasserie, *Sangyo-Shinpo*, **393**, 148 (1926), (in Japanese).

97. T. Hukuhara and H. Aruga, Induction of polyhedroses by temperature treatment in the silkworm, *Bombyx mori* L., *J. Seric. Sci. Japan*, **28**, 235 (1959), (Japanese with English summary).

98. Y. Tanada, Description and characteristics of a granulosis virus of the imported cabbage-worm, *Proc. Hawaii. Entomol. Soc.*, **15**, 235 (1953).

99. C. G. Thompson, Thermal inhibition of certain polyhedrosis virus diseases, *J. Insect Pathol.*, **1**, 189 (1959).

100. Y. Tanada and A. M. Tanabe, Resistance of *Galleria mellonella* (Linnaeus) to the *Tipula iridescent* virus at high temperatures, *J. Invertebr. Pathol.*, **7**, 184 (1965).

101. P. Ferron, Etude en laboratoire des conditions écologiques favorisant la mycose à *Beauveria tenella* du ver blanc, *Entomophaga*, **12**, 257 (1967).

102. B. Schaerffenberg, *Beauveria bassiana* (Vuill.) Link als Parasit des Kartoffelkäfers (*Leptinotarsa decemlineata* Say), *Anz. Schädlingskd.*, **30**, 69 (1957).

103. J. W. Doberski, Comparative laboratory studies on three fungal pathogens of the elm bark beetle *Scolytus scolytus:* effect of temperature and humidity on infection by *Beauveria bassiana, Metarrhizium anisopliae*, and *Paecilomyces farinosus, J. Invertebr. Pathol.*, **37**, 195 (1981).

104. A. M. Boyce and H. S. Fawcett, A parasitic *Aspergillus* on mealy bugs, *J. Econ. Entomol.*, **40**, 702 (1947).

105. A. K. A. Mohamed, P. P. Sikorowski, and J. V. Bell, Susceptibility of *Heliothis zea* larvae to *Nomuraea rileyi* at various temperatures, *J. Invertebr. Pathol.*, **30**, 414 (1977).

106. A. L. Knight, Host range and temperature requirements of *Culicinomyces clavosporus, J. Invertebr. Pathol.*, **36**, 423 (1980).

107. L. Bailey, The natural mechanism of suppression of *Nosema apis* Zander in epizootically infected colonies of the honey bee, *Apis mellifera* Linnaeus, *J. Insect Pathol.*, **1**, 347 (1959).

108. G. Giodani, Amoeba disease of the honeybee, *Apis mellifera* Linnaeus, and an attempt at its chemical control, *J. Insect Pathol.*, **1**, 245 (1959).

109. J. P. Kramer, Observations on the seasonal incidence of microsporidiosis in European corn borer populations in Illinois, *Entomophaga*, **4**, 37 (1959).

110. R. Geiger, *Das Klima der bodennahen Luftschicht*, F. Vieweg & Sohn, Braunschweig, 1961.

111. A. L. Johnpulle, Temperatures lethal to the green muscardine fungus, *Metarrhizium anisopliae* (Metch.) Sorok, *Trop. Agric.*, **90**, 80 (1938).

112. C. M. Ignoffo, The effects of temperature and humidity on mortality of larvae of *Pectinophora gossypiella* (Saunders) injected with *Bacillus thuringiensis* Berliner, *J. Invertebr. Pathol.*, **4**, 63 (1962).

113. A. I. Sidorenko, Nonsporulating bacteria for the control of soil-inhabiting pest insects. *Fauna, Systematics and Ecology of Insects and Mites*, **10**, 130. Academy of Sciences, Novosibirk (1963), (in Russian).

114. A. Bonnefoi, M. Toucas, and H. Chaumont, Essais de thermorésistance de l'organisme responsable de la maladie laiteuse de la larve du hanneton (*Melolontha melolontha*), *Entomophaga*, **4**, 227 (1959).

115. A. Krieg, *Bacillus thuringiensis* Berliner, *Mitt. Biol. Bundesanstalt Land- Forstwirtsch. Berlin-Dahlem*, **103**, 3 (1961).

116. L. Bailey, The isolation and cultural characteristics of *Streptococcus pluton* and further observations on *Bacterium eurydice, J. Gen. Microbiol.*, **17**, 39 (1957).

117. B. Hurpin, Influence de certains facteurs du milieu sur la virulence de *Rickettsiella melolonthae* Krieg, *Proc. 12th Int. Congr. Entomol.*, London, p. 727 (1964).

118. W. A. L. David and B. O. C. Gardiner, The effect of heat, cold, and prolonged storage on a granulosis virus of *Pieris brassicae, J. Invertebr. Pathol.*, **9**, 555 (1967).

119. J. Brassel, Entwicklung von Methoden für die Produktion eines Granulosisvirus-Präparates zur mikrobiologischen Bekämpfung des Apfelwicklers, *Laspeyresia pomonella* (L.) (Lep., Tortricidae) und Schätzung der Produktionskosten, *Mitt. Schweiz. Entomol. Ges.*, **51**, 155 (1978).

120. S. Keller, Mikrobiologische Bekämpfung des Apfelwicklers (*Laspeyresia pomonella* (L.)) (=*Carpocapsa pomonella*) mit spezifischem Granulosisvirus, *Z. angew. Entomol.*, **73**, 137 (1973).

121. R. P. Jaques, "Persistence, Accumulation, and Denaturation of Nuclear Polyhedrosis and Granulosis Viruses," in M. Summers, R. Engler, L. A. Falcon, and P. Vail, Eds., *Baculoviruses for Insect Control: Safety Consideration,* Am. Soc. Microbiol., Washington, D. C., 1975, p. 90.

122. S. Watanabe, Studies on the grasserie virus of the silkworm, *Bombyx mori*. IV. Physical and chemical effects upon the virus, *Japan J. Exp. Med.*, **21**, 299 (1951).

123. A. Huger, "Granuloses of Insects," in E. A. Steinhaus, Ed., *Insect Pathology: An Advanced Treatise,* Vol. 1, Academic Press, New York, 1963, p. 531.

124. W. W. Martin, The dynamics of aquatic fungi parasitic in a stream population of the midge, *Chironomus attenuatus, J. Invertebr. Pathol.*, **44**, 36 (1984).

125. C. Teng, Studies on the biology of *Beauveria bassiana* (Bals.) Vuill. with reference to microbial control of insect pests (Chinese with English summary), *Acta Bot. Sin.*, **10**, 210 (1962).

126. G. G. Newman and G. R. Carner, Environmental factors affecting conidial sporulation and germination of *Entomophthora gammae, Environ. Entomol.*, **4**, 615 (1975).

127. G. G. Soares, Jr., and D. E. Pinnock, Effect of temperature on germination, growth, and infectivity of the mosquito pathogen *Tolypocladium cylindrosporum* (Deuteromycotina; Hyphomycetes), *J. Invertebr. Pathol.*, **43**, 242 (1984).

128. I. M. Hall and J. V. Bell, The effect of temperature on some entomophthoraceous fungi, *J. Insect Pathol.*, **2**, 247 (1960).

129. J. D. Vandenberg and R. S. Soper, Prevalence of Entomophtorales mycoses in populations of spruce budworm, *Choristoneura fumiferana, Environ. Entomol.*, **7**, 847 (1978).

130. E. Müller-Kögler, *Pilzkrankheiten bei Insekten,* Verlag Parey, Berlin, 1965.

131. R. W. Glaser, The green muscardine disease in silkworms and its control, *Ann. Entomol. Soc. Am.*, **19**, 180 (1926).

132. V. Vouk and Z. Klas, Ueber einige Kulturbedingungen des insektentötenden Pilzes *Metarrhizium anisopliae* (Metch.) Sorok., *Acta Bot., Inst. Bot. Univ. Zagreb*, **7**, 35 (1932).

133. J. D. Walstad, R. F. Anderson, and W. J. Stambaugh, Effects of environmental condition on two species of muscardine fungi (*Beauveria bassiana* and *Metarrhizium anisopliae*), *J. Invertebr. Pathol.*, **16**, 221 (1970).

134. G. C. Clerk and M. F. Madelin, The longevity of conidia of three insect-parasitizing hyphomycetes, *Trans. Br. Mycol. Soc.*, **48**, 193 (1965).

135. R. A. Daoust and D. W. Roberts, Studies on the prolonged storage of *Metarrhizium anisopliae* conidia: effect of temperature and relative humidity on conidial viability and virulence against mosquitoes, *J. Invertebr. Pathol.*, **41**, 143 (1983).

136. E. Schulz-Langner, Die Rolle des Ventiltrichters (Proventriculus) bei der Resistenz der Honigbiene gegen die Bösartige Faulbrut, *Z. Bienenforsch.*, **3**, 234 (1957).

137. G. L. Nordin, G. C. Brown, and J. A. Millstein, Epizootic phenology of *Erynia* disease of the alfalfa weevil, *Hypera postica* (Gyllenhal) (Coleoptera: Curculionidae), in central Kentucky, *Environ. Entomol.*, **12**, 1350 (1983).

138. G. E. Bucher, "Nonsporulating Bacterial Pathogens," in E. A. Steinhaus, Ed., *Insect Pathology: An Advanced Treatise*, Vol. 2, Academic Press, New York, 1963, p. 117.

139. R. T. White, Survival of type A milky disease of Japanese beetle larvae under adverse field conditions, *J. Econ. Entomol.*, **33**, 303 (1940).

140. E. A. Steinhaus, *Serratia marcescens* Bizio as an insect pathogen, *Hilgardia*, **28**, 351 (1959).

141. J. P. Harville, Ecology and population dynamics of the California oak moth *Phryganidia californica* Packard (Lepidoptera: Dioptidae), *Microentomology*, **20**, 83 (1955).

142. C. Vago and R. Caryol, Une virose à polyèdres de la noctuelle gamma, *Plusia gamma* L. (Lepidoptera), *Ann. Epiphyt.*, **4**, 421 (1955).

143. J. Szirmai, Biologische Abwehr mittels Virus zur Bekämpfung der *Hyphantria cunea* Drury, *Acta Microbiol. Acad. Sci. Hung.*, **4**, 31 (1957).

144. R. C. Wallis, Incidence of polyhedrosis of gypsy-moth larvae and the influence of relative humidity, *J. Econ. Entomol.*, **50**, 580 (1957).

145. S. Markovitch, Some climatic relations of armyworm outbreaks, *J. Tenn. Acad. Sci.*, **33**, 348 (1958).

146. C. G. Thompson and E. A. Steinhaus, Further tests using a polyhedrosis virus to control the alfalfa caterpillar, *Hilgardia*, **19**, 411 (1950).

147. D. Pimentel and M. Shapiro, The influence of environment on a virus-host relationship, *J. Insect Pathol.*, **4**, 77 (1962).

148. G. C. Brown and G. L. Nordin, An epizootic model of an insect-fungal pathogen system, *Bull. Math. Biol.*, **44**, 731 (1982).

149. P. L. Watson, R. J. Barney, J. V. Maddox, and E. J. Armbrust, Sporulation and mode of infection of *Entomophthora phytonomi*, a pathogen of the alfalfa weevil, *Environ. Entomol.*, **10**, 305 (1981).

150. A. E. Michelbacher, W. W. Middlekauff, and C. Hanson, Occurrence of a fungous disease in overwintering stages of the codling moth, *J. Econ. Entomol.*, **43**, 955 (1950).

151. C. C. Doane, *Beauveria bassiana* as a pathogen of *Scolytus multistriatus*, *Ann. Entomol. Soc. Am.*, **52**, 109 (1959).

152. E. A. Steinhaus, *Principles of Insect Pathology*, McGraw-Hill, New York, 1949.

153. M. F. Madelin, "Diseases Caused by Hyphomycetous Fungi," in E. A. Steinhaus, Ed., *Insect Pathology: An Advanced Treatise*, Vol. 2, Academic Press, New York, 1963, p. 233.

154. E. E. Schaefer, *The white fungus disease* (Beauveria bassiana) *among red locusts in South Africa, and some observations on the grey fungus disease* (Empusa grylli), Union. S. Africa Dep. Agric. Plant Ind. Ser. No. 18, Sci. Bull. 160, 28 pp. (1936).

155. V. Pospelov, "*Bothynoderes punctiventris* Germ., and Methods of Fighting it," in *An Agricultural Monography*, 2nd ed., Central Board of Land Administration and Agriculture, Dept. Agriculture, St. Petersburg, 1913 (in Russian). Abstr. in *Rev. Appl. Entomol.*, A2, 177–180.

156. P. Ferron, Influence of relative humidity on the development of fungal infection caused by *Beauveria bassiana* (fungi imperfecti monilales) in imagines of *Acanthoscelides obtectus* (Col.: Bruchidae), *Entomophaga*, **2**, 393 (1977).

157. W. A. Ramoska, The influence of relative humidity on *Beauveria bassiana* infectivity and replication in the chinch bug, *Blissus leucopterus*, *J. Invertebr. Pathol.*, **43**, 389 (1984).

158. J. A. Millstein, G. C. Brown, and G. L. Nordin, Microclimatic humidity influence on conidial discharge in *Erynia* sp. (Entomophthorales: Entomophthoraceae), an entomopathogenic fungus of the alfalfa weevil (Coleoptera: Curculionidae), *Environ. Entomol.*, **11**, 1166 (1982).

159. T. R. Gottwald and W. L. Tedders, Studies on conidia release by the entomogenous fungi *Beauveria bassiana* and *Metarrhizium anisopliae* (Deuteromycotina: Hyphomycetes) from adult pecan weevil (Coleoptera: Curculionidae) cadavers, *Environ. Entomol.*, **11**, 1274 (1982).

160. H. B. Girth, E. E. McCoy, and R. W. Glaser, Field experiments with a nematode parasite of the Japanese beetle, *N.J. Dep. Agric. Cir.*, **317**, 21 pp. (1940).

161. A. Couturier, Biologie d'un *Hexamermis* (Nematodes, Mermithidae), parasite des insectes défoliateurs de l'osier, *Ann. Épiphyt.*, **1**, 13 (1950).

162. H. E. Welch, A review of recent work on nematodes associated with insects with regard to their utilization as biological control agents, *Proc. 10th Int. Congr. Entomol. Montreal 1956*, **4**, 465 (1958).

163. S. R. Dutky, Insect microbiology, *Adv. Appl. Microbiol.*, **1**, 175 (1959).

164. D. Schvester, *Contribution a l'étude écologique des Coleoptères Scolytides. Essai d'analyse des facteurs de fluctuation des populations chez* Ruguloscolytus rugulosus *Muller 1818*, Ph.D. Thesis, Univ. Paris, Paris, 1957, 162 pp.

165. B. Wahl, Ueber die Polyederkrankheit der Nonne (*Lymantria monacha* L.), *Zentrabl. Forstwes.*, **36**, 193, 377; **37**, 247; **38**, 355 (1910–1912).

166. R. W. Glaser, Wilt of gypsy moth caterpillars, *J. Agric. Res.*, **4**, 101 (1915).

167. G. Wellenstein, *Die Nonne in Ostpreussen*, Paul Parey, Berlin, 1942.

168. J. L. Fetter-Lasko and R. K. Washino, In situ studies on seasonality and recycling pattern in California of *Lagenidium giganteum* Couch, an aquatic fungal pathogen of mosquitoes, *Environ. Entomol.*, **12**, 635 (1983).

169. H. E. Welch, *Hydromermis churchillensis* n.sp. (Nematoda: Mermithidae) a parasite of *Aedes communis* (DeG.) from Churchill, Manitoba, with observations of its incidence and bionomics, *Can. J. Zool.*, **38**, 465 (1960).

170. R. P. Jaques, The persistence of a nuclear polyhedrosis virus in the habitat of the host insect *Trichoplusia ni.* I. Polyhedra deposited on the foliage, *Can. Entomol.*, **99**, 785 (1967).

171. J. D. Podgwaite, K. S. Shields, R. T. Zerillo, and R. B. Bruen, Environmental persistence of the nucleopolyhedrosis virus of the gypsy moth, *Lymantria dispar*, *Environ. Entomol.*, **8**, 528 (1979).

172. W. A. L. David and B. O. C. Gardiner, Persistence of a granulosis virus of *Pieris brassicae* on cabbage leaves, *J. Invertebr. Pathol.*, **8**, 180 (1966).

173. L. L. J. Ossowski, The biological control of the wattle bagworm (*Kotochalia junodi* Heyl.) by a virus disease. I. Small-scale pilot experiments, *Ann. Appl. Biol.*, **45**, 81 (1957).

174. A. Burgerjon and P. Grison, Adhesiveness of preparations of *Smithiavirus pityocampae* Vago on pine foliage, *J. Invertebr. Pathol.*, **7**, 281 (1965).

175. H. R. Bullock, Persistence of *Heliothis* nuclear-polyhedrosis virus on cotton foliage, *J. Invertebr. Pathol.*, 9, 434 (1967).

176. J. M. Stephens, Survival of *Pseudomonas aeruginosa* (Schroeter) Migula suspended in various solutions and dried in air, *Can. J. Microbiol.*, **3**, 995 (1957).

177. V. V. Vlodavets, On the possibility of using *B. prodigiosum* as an experimental model for bacterial sprays, (In Russian), *Zh. Microbiol. Epidemiol. Immunol. SSSR*, **41**, 65 (1964).

178. J. Muspratt, Experimental infection of the larvae of *Anopheles gambiae* (Dipt., Culicidae) with a *Coelomomyces* fungus, *Nature*, **158**, 202 (1946).

179. C. Vago, D. Martouret, and F. Heitor, Conservation de virus et de bactéries entomopathogènes sous forme de comprimés, *Entomophaga*, **6**, 185 (1961).

180. J. Weiser, *Die Mikrosporidien als Parasiten der Insekten*, *Monogr. Angew. Entomol.*, **17**, 149 pp. (1961).

181. W. R. Kellen and J. J. Lipa, *Thelohania californica* n.sp., a microsporidian parasite of *Culex tarsalis* Coquillett, *J. Insect Pathol.*, **2**, 1 (1960).

182. M. Boczkowska, Contribution a l'étude de l'immunité chez les chenilles de *Galleria mellonella* L., contre les champignons entomophytes, *C. R. Soc. Biol.,* **119,** 39 (1935).

183. P. Voukassovitch, Contribution a l'étude d'un champignon entomophyte *Spicaria farinosa* (Fries) var. *verticilloides* Fron, *Ann. Épiphyt.,* **11,** 73 (1925).

184. C. E. Burnside, *Fungous Diseases of the Honey Bee,* U. S. Dep. Agric. Tech. Bull. 149, 43 pp. (1930).

185. A. Schmid, *Beitrag zur Mikrobiologischen Bekämpfung des Grauen Lärchenwicklers, Zeiraphera diniana (Gn.),* Ph.D. Thesis Nr. 5045, ETH Zurich, 1973.

186. R. Sieber and G. Benz, Juvenile hormone in larval diapause of the codling moth, *Laspeyresia pomonella* L. (Lepidoptera, Tortricidae), *Experientia,* **33,** 1598 (1977).

187. R. Sieber and G. Benz, The hormonal regulation of the larval diapause in the codling moth, *Laspeyresia pomonella* (Lep. Tortricidae), *J. Insect Physiol.,* **26,** 213 (1980).

188. F. Camponovo, *Biochemische Untersuchungen an zwei spezifischen Granulosisvirus-Stämmen des Apfelwicklers (Laspeyresia pomonella (L.) und Untersuchungen über die Entwicklung der Alterstoleranz gegen dieses Virus im letzten Larvenstadium,* Ph.D. Thesis Nr. 6535, ETH Zurich, 1980.

189. F. Camponovo and G. Benz, Age-dependent tolerance to baculovirus in last larval instars of the codling moth, *Cydia pomonella* L., induced either for pupation or for diapause, *Experientia,* **40,** 938 (1984).

190. M. Kobayashi and S. Kawase, Effect of alteration of endocrine mechanism on the development of nuclear polyhedrosis in the isolated pupal abdomens of the silkworm, *Bombyx mori, J. Invertebr. Pathol.,* **36,** 6 (1980).

191. E. A. Steinhaus and J. P. Dineen, Observations on the role of stress in a granulosis of the variegated cutworm, *J. Insect Pathol.,* **2,** 55 (1960).

192. H. Aruga and N. Yoshitake, Studies on the induction of nuclear and cytoplasmic polyhedroses by treating with x-rays and ultraviolet light in the silkworm, *Bombyx mori* L. *Japan J. Appl. Entomol. Zool.,* **5,** 46 1961), (Japanese with English summary).

193. O. Ege, Ein Beitrag zur Biologie einiger aphidivorer Entomophthoraceen, *Arch. Mikrobiol.,* **52,** 20 (1965).

194. G. E. Cantwell, Inactivation of biological insecticides by irradiation, *J. Invertebr. Pathol.,* **9,** 138 (1967).

195. R. P. Jaques, The inactivation of foliar deposits of viruses of *Trichoplusia ni* and *Pieris rapae* and tests on protectant additives, *Can. Entomol.,* **104,** 1985 (1972).

196. W. A. L. David, B. O. C. Gardiner, and M. Woolner, The effect of sunlight on a purified granulosis virus of *Pieris brassicae* applied to cabbage leaves, *J. Invertebr. Pathol.,* **11,** 496 (1968).

197. W. G. Yendol, R. A. Hamlen, and F. B. Lewis, Evaluation of *Bacillus thuringiensis* for gypsy moth suppression, *J. Econ. Entomol.,* **66,** 183 (1973).

198. G. Benz, Untersuchungen über die Pathogenität eines Granulosis-Virus des Grauen Lärchenwicklers *Zeiraphera diniana* (Guenée), *Agron. Glas., Zagreb,* **1962,** 566 (1962).

199. O. N. Morris, The effect of sunlight, ultraviolet and gamma radiations, and temperature on the infectivity of a nuclear polyhedrosis virus, *J. Invertebr. Pathol.,* **18,** 292 (1971).

200. W. A. Smirnoff, The effect of sunlight on the nuclear-polyhedrosis virus of *Neodiprion swainei* with measurement of the solar energy received, *J. Invertebr. Pathol.,* **19,** 179 (1972).

201. J. Brassel and G. Benz, Selection of a strain of the granulosis virus of the codling moth with improved resistance against artificial ultraviolet radiation and sunlight, *J. Invertebr. Pathol.,* **33,** 358 (1979).

202. W. C. Yearian and S. Y. Young, The persistence of *Heliothis* nuclear polyhedrosis virus on cotton plant parts, *Environ. Entomol.,* **3,** 1035 (1974).

203. E. C. Clark, Survival and transmission of a virus causing polyhedrosis in *Malacosoma fragile, Ecology, 37,* 728 (1956).

204. K. Aizawa, Inactivation of the silkworm jaundice virus by the ultraviolet irradiation, *J. Seric. Sci. Japan, 24,* 397 (1955).

205. G. Benz, Aspects of virus multiplication and average reduplication time for a granulosis virus of *Zeiraphera diniana* (Guenee), *Entomophaga, Mem. Hors. Serie 2,* 417 (1964).

206. C. M. Ignoffo, D. L. Hostetter, P. P. Sikorowski, G. Sutter, and W. M. Brooks, Inactivation of representative species of entomopathogenic viruses, a bacterium, fungus, and protozoan by an ultraviolet light source, *Environ. Entomol., 6,* 411 (1977).

207. C. Vago and M. C. Busnel, Photosensibilité du *Bacillus cereus* var. *alesti* au rayonnement ultraviolet, *Antonie van Leuwenhoek, 18,* 125 (1952).

208. A. Burgerjon and C. Yamvrias, Titrage biologique des préparations à base de *Bacillus thuringiensis* Berliner vis-à-vis de *Anagasta (Ephestia) kuhniella* Zell, *C. R. Acad. Sci. Paris, 249,* 2871 (1959).

209. G. E. Cantwell and B. A. Franklin, Inactivation by irradiation, of spores of *Bacillus thuringiensis* var. *thuringiensis, J. Invertebr. Pathol., 8,* 256 (1966).

210. D. E. Pinnock, R. J. Brand, and J. E. Millstead, The field persistence of *Bacillus thuringiensis* spores, *J. Invertebr. Pathol., 18,* 405 (1971).

211. A. Krieg, A. Gröner, J. Huber, and M. Matter, Ueber die Wirkung von mittel- und langwelligen ultravioletten Strahlen (UV-B und UV-A) auf insektenpathogene Bakterien und Viren und deren Beeinflussung durch UV-Schutzstoffe, *Nachrichtenbl. Dtsch. Pflanzenschutzdienstes (Braunschweig), 32,* 100 (1980).

212. C. M. Ignoffo, D. L. Hostetter, and R. E. Pinnell, Stability of *Bacillus thuringiensis* and *Baculovirus heliothis* on soybean foliage, *Environ. Entomol., 3,* 117 (1974).

213. M. Svestka, Ucinnost a perzistence syntetickeho pyrethroidu Permethrinu, *Lesnictvi, 24,* 267 (1978).

214. D. L. Hostetter, C. M. Ignoffo, and W. H. Kearby, Persistence of formulations of *Bacillus thuringiensis* spores and crystals on Eastern red cedar foliage in Missouri, *J. Kans. Entomol. Soc., 48,* 189 (1975).

215. C. C. Beegle, H. T. Dulmage, D. A. Wolfenbarger, and E. Martinez, Persistence of *Bacillus thuringiensis* Berliner insecticidal activity on cotton foliage, *Environ. Entomol., 10,* 400 (1981).

216. C. M. Leach, "A Practical Guide to the Effects of Visible and Ultraviolet Light on Fungi," in C. Booth, Ed., *Methods in Microbiology,* Vol. 4, Academic Press, New York, 1971, p. 609.

217. D. W. Roberts and A. S. Campbell, "Stability of Entomogenic Fungi", in C. M. Ignoffo and D. L. Hostetter, Eds., *Environmental Stability of Microbial Insecticides, Symposium 1974, Misc. Publ. Entomol. Soc. Am., 10,* 19 (1977).

218. C. Toumanoff, Actions des champignons entomophytes sur la pyrale du maïs (*Pyrausta nubilalis* Hübner), *Ann. Parasitol. Humaine Comparée, 11,* 129 (1933).

219. J. Weiser, "Sporozoan Infections," in E. A. Steinhaus, Ed., *Insect Pathology: An Advanced Treatise,* Vol. 2, Academic Press, New York, 1963, p. 291.

220. J. V. Maddox, The persistence of the Microsporida in the environment, *Misc. Publ. Entomol. Soc. Am., 9,* 99 (1973).

221. H. K. Kaya, Persistence of spores of *Pleistophora schubergi* (Cnidospora: Microsporida) in the field, and their application in microbial control, *J. Invertebr. Pathol., 26,* 329 (1975).

222. H. K. Kaya, Survival of spores of *Vairimorpha (= Nosema) necatrix* (Microsporida: Nosematidae) exposed to sunlight, ultra-violet radiation and high temperature, *J. Invertebr. Pathol., 30,* 192 (1977).

223. P. P. Sikorowski and J. H. Lashomb, Effect of sunlight on the infectivity of *Nosema heliothidis* spores isolated from *Heliothis zea, J. Invertebr. Pathol., 30,* 95 (1977).

224. G. E. Teetor and J. P. Kramer, Effect of ultraviolet radiation on the microsporidian *Octosporea muscaedomesticae* with reference to protectants provided by the host *Phormia regina, J. Invertebr. Pathol.,* **30,** 348 (1977).

225. E. A. Steinhaus, Polyhedrosis ("wilt disease") of the alfalfa caterpillar, *J. Econ. Entomol.,* **41,** 859 (1948).

226. R. P. Jaques, The persistence of a nuclear-polyhedrosis virus in soil, *J. Invertebr. Pathol.,* **6,** 251 (1964).

227. R. P. Jaques, The persistence of a nuclear-polyhedrosis virus in the habitat of the host insect, *Trichoplusia ni.* II. Polyhedra in soil, *Can. Entomol.,* **99,** 820 (1967).

228. R. P. Jaques, Leaching of the nuclear-polyhedrosis virus of *Trichoplusia ni* from soil, *J. Invertebr. Pathol.,* **13,** 256 (1969).

229. R. P. Jaques, Natural occurrence of viruses of the cabbage looper in field plots, *Can. Entomol.,* **102,** 36 (1970).

230. R. P. Jaques, Occurrence and accumulation of the granulosis virus of *Pieris rapae* in treated field plots, *J. Invertebr. Pathol.,* **23,** 351 (1974).

231. W. A. L. David and B. O. C. Gardiner, The persistence of a granulosis virus of *Pieris brassicae* in soil and in sand, *J. Invertebr. Pathol.,* **9,** 342 (1967).

232. D. G. Harcourt and L. M. Cass, Persistence of a granulosis virus of *Pieris rapae* in soil, *J. Invertebr. Pathol.,* **11,** 142 (1968).

233. R. P. Jaques and D. G. Harcourt, Viruses of *Trichoplusia ni* (Lepidoptera: Noctuidae) and *Pieris rapae* (Lepidoptera: Pieridae) in soils in fields of crucifers in Southern Ontario, *Can. Entomol.,* **103,** 1285 (1971).

234. E. D. Thomas, C. F. Reichelderfer, and A. M. Heimpel, Accumulation and persistence of a nuclear polyhedrosis virus of the cabbage looper in the field, *J. Invertebr. Pathol.,* **20,** 157 (1972).

235. Y. Tanada and E. M. Omi, Persistence of insect viruses in field populations of alfalfa insects, *J. Invertebr. Pathol.,* **23,** 360 (1974).

236. A. M. Crawford and J. Kalmakoff, A host-virus interaction in a pasture habitat. *Wiseana spp.* (Lepidoptera: Hepialidae) and its baculoviruses, *J. Invertebr. Pathol.,* **29,** 81 (1977).

237. C. G. Thompson, D. W. Scott, B. E. Wickman, Long-term persistence of the nuclear-polyhedrosis virus of the Douglas-fir tussock moth, *Orgyia pseudotsugata* (Lepidoptera: Lymantriidae), in forest soil. *Environ. Entomol.,* **10,** 254 (1981).

238. Y. Tanada, "Persistence of Entomogenous Viruses in the Insect Ecosystem," in *Entomolog. Essays to Commemorate the Retirement of Professor K. Yasumatsu,* Hokuryukan Publ., Tokyo, 1971 p. 367.

239. S. M. Saleh, R. F. Harris, and O. N. Allen, Recovery of *Bacillus thuringiensis* var. *thuringiensis* from field soils, *J. Invertebr. Pathol.,* **15,** 55 (1970).

240. S. M. Saleh, R. F. Harris, and O. N. Allen, Fate of *Bacillus thuringiensis* in soil: effect of soil pH and organic amendment, *Can. J. Microbiol.,* **16,** 677 (1970).

241. E. V. Kiselek, Survival of bacterial entomopathogens in tree crowns and in the soil around the trunk, *Vestn. Skh. Nauki (Moscow),* **5,** 68 (1974).

242. Y. Sekijima, Y. Akiba, K. Ono, K. Aizawa, and N. Fujiyoshi, Microbial ecological studies on *Bacillus thuringiensis.* I. Dynamics of *Bacillus thuringiensis* in soil of mulberry field, *Japan. J. Appl. Entomol. Zool.,* **21,** 35 (1977).

243. C. J. H. Pruett, H. D. Burges, and C. H. Wyborn, Effect of exposure to soil on potency and spore viability of *Bacillus thuringiensis, J. Invertebr. Pathol.,* **35,** 168 (1980).

244. A. W. West, H. D. Burges, R. J. White, and C. H. Wyborn, Persistence of *Bacillus thuringiensis* parasporal crystal insecticidal activity in soil, *J. Invertebr. Pathol.,* **44,** 128 (1984).

245. V. P. Pospelov, "Biological Methods of Controlling the Beet Weevil," in N. M. Kulagin and G. K. Pyatnitzkii, Eds., *The Beet Weevil and its Control,* Demyovo, Vsesoyuzn. Acad. S.-kh.

Nauk Lenina, Moscow 1940, p. 45 (in Russian); abstracted in *Rev. Appl. Entomol., A30,* 66–67.

246. G. K. Pyatnitzkii, "Agrotechnical Methods of Controlling the Beet Weevil," in N. M. Kulagin and G. K. Pyatnitzkii, Eds., *The Beet Weevil and its Control,* Demyovo, Vsesoyuzn. Akad. s.-kh. Nauk Lenina, Moscow 1940, p. 25 (in Russian); abstracted in *Rev. Appl. Entomol., A30* 64–65.

247. G. H. Bünzli and W. W. Büttiker, Fungous diseases of lamellicorn larvae in Southern Rhodesia, *Bull. Entomol. Res., 50,* 89 (1959).

248. A. J. Lingg and M. D. Donaldson, Biotic and abiotic factors affecting stability of *Beauveria bassiana* conidia in soil, *J. Invertebr. Pathol., 38,* 191 (1981).

249. J. Fargues, O. Reisinger, P. H. Robert, and C. Aubart, Biodegradation of entomopathogenic Hypomycetes: influence of clay coating on *Beauveria bassiana* blastospore survival in soil, *J. Invertebr. Pathol., 41,* 131 (1983).

250. K. M. Old and W. M. Robertson, Growth of bacteria within lysing fungal conidia in soil, *Trans. Br. Mycol. Soc., 54,* 337 (1970).

251. T. Naiki and T. Ui, Ultrastructure of sclerotia of *Rhizoctonia solani* Kuhn, invaded and decayed by soil microorganisms, *Soil. Biol. Biochem., 7,* 301 (1975).

252. K. M. Old, Giant soil amoebae cause perforation of conidia of *Cochliobolus sativus, Trans. Br. Mycol. Soc., 68,* 277 (1977).

253. M. Pussard, C. Alabouvette, and R. Pons, Etude préliminaire d'une amibe mycophage *Thecamoeba granifera* s.sp. *minor* (Thecamoebidae, Amoebidae), *Protistologica, 15,* 139 (1979).

254. R. K. Sprenkel and W. M. Brooks, Winter survival of the entomogeneous fungus *Nomuraea rileyi* in North Carolina, *J. Invertebr. Pathol., 29,* 262 (1977).

255. C. M. Ignoffo, C. Garcia, D. L. Hostetter, and R. E. Pinnell, Stability of conidia of an entomopathogenic fungus, *Nomuraea rileyi,* in and on soil, *Environ. Entomol., 7,* 724 (1978).

256. A. J. Walker, Fungal infections of mosquitoes, especially of *Anopheles costalis, Ann. Trop. Med. Parasitol., 32,* 231 (1938).

257. M. Laird, Parasites of Singapore mosquitoes, with particular reference to the significance of larval epibionts as an index of habit pollution, *Ecology, 40,* 206 (1959).

258. J. R. Linley and H. T. Nielson, Transmission of a mosquito iridescent virus in *Aedes taeniorhynchus* II. Experiments related to transmission in nature, *J. Invertebr. Pathol. 12,* 17 (1968).

259. J. Weiser, Protozoare Infektionen im Kampf gegen Insekten, *Z. Pflanzenkrankh. Pflanzenschutz, 63,* 625 (1956).

260. R. W. Glaser and C. C. Farrell, Field experiments with the Japanese beetle and its nematode parasite, *J. N. Y. Entomol. Soc., 43,* 345 (1935).

261. J. M. Hoy, The biology and host range of *Neoaplectana leucaniae,* a new species of insect parasitic nematode, *Parasitology, 44,* 392 (1954).

262. G. Benz, "Synergism of Microorganisms and Chemical Insecticides," in H. D. Burges and N. W. Hussey, Eds., *Microbial Control of Insects and Mites,* Academic Press, New York, 1971, p. 327.

263. R. P. Jaques and O. N. Morris, "Compatibility of Pathogens with Other Methods of Pest Control and with Different Crops," in H. D. Burges, Ed., *Microbial Control of Pests and Plant Diseases 1970–1980,* Academic Press, New York, 1981, p. 696.

264. K.-S. Chen, B. R. Funke, J. T. Schulz, R. B. Carlson, and F. I. Proshold, Effects of certain organophosphate and carbamate insecticides on *B. thuringiensis, J. Econ. Entomol., 67,* 471 (1974).

265. C. A. Creighton and T. L. McFadden, Complementary action of low rates of *Bacillus thuringiensis* and chlordimeform hydrochloride for control of caterpillars, *J. Econ. Entomol., 67,* 1902 (1974).

266. O. N. Morris, Susceptibility of some forest insects to mixtures of commercial *Bacillus thuringiensis:* chemical insecticides, *Can. Entomol.,* **104,** 1419 (1972).

267. O. N. Morris, Susceptibility of the spruce budworm and the marked tussock moth to *Bacillus thuringiensis:* chemical insecticide combination, *J. Invertebr. Pathol.,* **26,** 193 (1975).

268. J. T. Hamilton and F. I. Attia, Effects of mixtures of *Bacillus thuringiensis* and pesticides on *Plutella xylostella* and the parasite *Thyraeela collaris, J. Econ. Entomol.,* **70,** 146 (1977).

269. R. G. Luttrell, W. C. Yearian, and S. Y. Young, Laboratory and field studies on the efficacy of selected chemical insecticide-Elcar *(Baculovirus heliothis)* combinations against *Heliothis* spp., *J. Econ. Entomol.,* **27,** 57 (1979).

270. A. I. Mohamed, S. Y. Young, and W. C. Yearian, Effects of microbial: chemical pesticide mixtures on *Heliothis virescens, Environ. Entomol.,* **12,** 478 (1983).

271. A. I. Mohamed, S. Y. Young, and W. C. Yearian, Susceptibility of *Heliothis virescens* (F.) (Lepidoptera: Noctuidae) larvae to microbial agent-chemical pesticide mixtures on cotton foliage, *Environ. Entomol,* **12,** 1403 (1983).

272. N. A. Telenga, Intensification de la muscardine chez le charançon de la betterave a l'aide de l'hexachlorane, *Doklady Acad. Nauk. S.S.S.R.,* **109,** 665 (1956).

273. U. N. V. Ramaraje, H. C. Govindu, and K. S. S. Shastry, The effect of certain insecticides on the entomogenous fungi, *Beauveria bassiana* and *Metarrhizium anisopliae, J. Invertebr. Pathol.,* **9,** 398 (1967).

274. I. Olmert and R. G. Kenneth, Sensitivity of the entomopathogenic fungi, *Beauveria bassiana, Verticillium lecanii,* and *Verticillium* sp. to fungicides and insecticides, *Environ. Entomol.,* **3,** 33 (1974).

275. D. W. Johnson, L. P. Kish, and G. E. Allen, Field evaluation of selected pesticides on the natural development of the entomopathogen, *Nomuraea rileyi,* on the velvetbean caterpillar in soybeans, *Environ. Entomol.,* **5,** 964 (1976).

276. J. M. Livingston, W. C. Yearian, and S. Y. Young, Effect of insecticides, fungicides, and insecticide-fungicide combinations on development of lepidopterous larval populations in soybeans, *Environ. Entomol.,* **7,** 823 (1978).

277. T. L. Merriam and R. C. Axtell, Relative toxicity of certain pesticides to *Lagenidium giganteum* (Oomycetes: Lagenidiales), a fungal pathogen of mosquito larvae, *Environ. Entomol.,* **12,** 515 (1983).

278. R. Fritz, Action de quelques fongicides sur la croissance mycelienne de trois espèces d'entomophthorales, *Entomophaga,* **21,** 239 (1976).

279. W. L. Tedders, In vitro inhibition of the entomopathogenic fungi *Beauveria bassiana* and *Metarrhizium anisopliae* by six fungicides used in pecan culture, *Environ. Entomol.,* **10,** 346 (1981).

280. R. Loria, S. Galaini, and D. W. Roberts, Survival of inoculum of the entomopathogenic fungus *Beauveria bassiana* as influenced by fungicides, *Environ. Entomol.,* **12,** 1724 (1983).

281. D. F. Perry and J. P. Latge, The effect of eight fungicides on germination of *Conidiobolus obscurus* resting spores, *J. Invertebr. Pathol.,* **42,** 83 (1983).

282. E. Armstrong, Fumidil B and benomyl: chemical control of *Nosema kingi* in *Drosophila willistoni, J. Invertebr. Pathol.,* **27,** 363 (1976).

283. W. M. Brooks, J. D. Cranford, and L. W. Pearce, Benomyl: effectiveness against the microsporidian *Nosema heliothidis* in the corn earworm, *Heliothis zea, J. Invertebr. Pathol.,* **31,** 239 (1978).

284. R. P. Jaques, *The ecology of polyhedrosis of the cabbage looper, Trichoplusia ni (Hubner),* Ph.D. Thesis, Cornell Univ., Ithaca, New York, 1960.

285. U. Schnyder, *Untersuchungen einer Kernpolyedrose von Sterrha seriata Schrk. (= Ptychopoda seriata Schrk. = Acidalia virgularia Hb.) (Geometridae, Lepidoptera) und deren Beeinflussbarkeit durch Hunger, DDT, DNOC und Farnesylmethylaether,* Doctoral Thesis Nr. 3802 ETH, Zurich, 1967, 60 pp.

286. W. A. L. David, S. Ellaby, and G. Taylor, The effect of reducing the content of certain ingredients in a semisynthetic diet on the incidence of granulosis virus disease in *Pieris brassicae, J. Invertebr. Pathol.,* **20,** 332 (1972).

287. W. A. L. David and C. E. Taylor, The effect of sucrose content of diets on susceptibility to granulosis virus disease in *Pieris brassicae, J. Invertebr. Pathol.,* **30,** 117 (1977).

288. H. Aruga, "Induction of Virus Infections," in E. A. Steinhaus, Ed., *Insect Pathology: An Advanced Treatise,* Vol. 1, Academic Press, New York, 1963, p. 661.

289. M. Ripper, Bericht über die Tätigkeit der K. K. Landwirtschaftlichchemischen Versuchsstation in Görz im Jahre 1914, *Z. Landwirth. Ver. Oesterr.,* **18,** 12 (1915).

290. C. Vago, Facteurs alimentaires et activation des viroses latentes chez les insectes, *Compt. rend. 6th Congr. Int. Microbiol., Rome, 1953,* 556 (1953).

291. G. E. Bucher and P. Harris, Virus diseases and their interaction with food stress in *Calophasia lunula, J. Invertebr. Pathol.,* **10,** 235 (1968).

292. S. Matsumura, On "Kutho-disease" a kind of flacherie which attacks mainly young larvae of the silkworm *B. mori* in Japan, *Proc. 11th Congr. Zool. Padova,* p. 1 (1930).

293. O. I. Shvetzova, Yellow jaundice in the oak silkworm and methods of control in commercial production, *Mikrobiologiya,* **23,** 477 (1954).

294. H. Aruga, Relationship between insect virus disease and environmental factors, especially in the case of plant and *Bombyx mori. Nogyo Oyobi Engei,* **33,** 455 (1958) (Japanese).

295. J. D. Hare and T. G. Andreadis, Variation in the susceptibility of *Leptinotarsa decemlineata* (Coleoptera: Chrysomelidae) when reared on different host plants to the fungal pathogen, *Beauveria bassiana* in the field and laboratory, *Environ. Entomol.,* **12,** 1892 (1983).

296. G. Benz, Negative Rückkoppelung durch Raum- und Nahrungskonkurrenz sowie zyklische Veränderung der Nahrungsgrundlage als Regelprinzip in der Populationsdynamik des Grauen Lärchenwicklers, *Zeiraphera diniana* (Guenée), *Z. Angew. Entomol.,* **76,** 196 (1974).

297. G. Benz, Insect-induced resistance as a means of self-defense of plants, *Bull. SROP, 1977,* 155 (1977).

298. D. F. Rhoades, "Herbivore Population Dynamics and Plant Chemistry," in R. Denno and M. McClure, Eds., *Variable Plants and Herbivores in Natural and Managed Systems,* Academic Press, New York, 1983, p. 155.

299. R. E. Roome and R. A. Daoust, Survival of the nuclear polyhedrosis virus of *Heliothis armigera* on crops and in soil in Botswana, *J. Invertebr. Pathol.,* **27,** 7 (1976).

300. G. L. Andrews and P. P. Sikorowski, Effects of cotton leaf surfaces on the nuclear polyhedrosis virus of *Heliothis zea* and *Heliothis virescens* (Lepidoptera: Noctuidae), *J. Invertebr. Pathol.,* **22,** 290 (1973).

301. W. H. Stubblebine and J. H. Langenheim, Effects of *Hymenaea courbaril* leaf resin on the generalist herbivore *Spodoptera exigua* (beet armyworm), *J. Chem. Ecol.,* **3,** 633 (1977).

302. D. C. Kelly and T. Lescott, Baculovirus replication: protein synthesis in *Spodoptera frugiperda* cells infected with *Trichoplusia ni* nuclear polyhedrosis virus, *Microbiologica,* **4,** 35 (1981).

303. G. W. Felton and D. L. Dahlman, Allelochemical-induced stress: effects of L-canavanine on the pathogenicity of *Bacillus thuringiensis* in *Manduca sexta, J. Invertebr. Pathol.,* **44,** 187 (1984).

304. W. A. Smirnoff, Effects of some plant juices on the ugly-nest caterpillar *Archips cerasivoranus,* infected with microsporidia, *J. Invertebr. Pathol.,* **9,** 26 (1967).

305. A. B. Gukasian, Bacteriostatic and bacteriocidal influence of the foliage and its chemical components on the diseases of *Dendrolimus sibericus* Tschtn. larvae (in Russian), *Rep. Sib. Branch Acad. Sci. USSR,* **7,** 85 (1958).

306. D. J. Kushner and G. T. Harvey, Antibacterial substances in leaves: their possible role in insect resistance to disease, *J. Insect Pathol.,* **4,** 155 (1962).

307. W. A. Smirnoff, Tests of *Bacillus thuringiensis* var. *thuringiensis* Berliner and *B. cereus* Frankland and Frankland on larvae of *Choristoneura fumiferana* (Clemens), *Can. Entomol.,* **95,** 127 (1963).

308. W. A. Smirnoff, Inhibition of parasporal inclusion synthesis in crystalliferous sporeforming bacteria of the *"cereus"* group by an aqueous extract of *Viburnum cassinoides* Linnaeus (Caprifoliaceae) leaves, *J. Invertebr. Pathol.,* **7,** 71 (1965).

309. W. A. Smirnoff and P. M. Hutchinson, Bacteriostatic and bacteriocidal effects of extracts of foliage from various plant species on *Bacillus thuringiensis* var. *thuringiensis* Berliner, *J. Invertebr. Pathol.,* **7,** 273 (1965).

310. B. Maksymiuk, Occurrence and nature of antibacterial substances in plants affecting *Bacillus thuringiensis* and other entomogenous bacteria, *J. Invertebr. Pathol.,* **15,** 356 (1970).

311. W. A. Smirnoff, Effects of volatile substances released by foliage of various plants on the entomopathogenic *Bacillus cereus* group, *J. Invertebr. Pathol.,* **11,** 513 (1968).

312. W. A. Smirnoff, Effects of volatile substances released by the foliage of *Abies balsamea, J. Invertebr. Pathol.,* **19,** 32 (1972).

313. G. Benz, "Use of Viruses for Insect Suppression," in C. Papavizas, Ed., *Biological Control in Crop Production (BARC Symposium no., 5),* Allanheld, Osmun, Totowa, 1981, p. 259.

8

PATTERNS OVER PLACE AND TIME

JAROSLAV WEISER

Institute of Entomology
Academy of Sciences
Prague, Czechoslovakia

---

1. INTRODUCTION

Some features of a single infection in one host or a set of infections in a host population transcend the characteristics of a single case and have broad implications over place and time. The severity of a single infection in a single host or a group of hosts is characterized by virulence of the pathogen and the resistence of the host and is influenced by variations in dosages and the impact of temperature, moisture, type of food, and the infected instar. The overall impact of these factors to whole populations, however, is the opposite: biotic and abiotic factors together determine whether a disease remains an enzootic at varying but low incidence, largely unseen for long periods of time, or bursts out into an epizootic with dramatic insect mortality (1, 2). For every evaluation of broad situations in populations, minute details of interactions at lower levels must be known, which is what makes the generalization of patterns in populations influenced by pathogens so difficult to express in mathematical models (3).

    Data on diseases in populations are important for population ecology. "Natural mortality" as it figures in ecological data has many components with different causes and their impacts, including abiotic factors, diseases, and physiological mortality of old animals that have completed their vital functions (4). The "hidden impact" of diseases is evident from differences between "physiological" and "ecological" longevities in models of populations (4). "Realized mortality" results from decreasing impact of negative factors, biotic and abiotic, on a population. In each steady community that has "existed" for several seasons, there are infective pathogens in the environment. Once they are present, other factors become important to disease outbreaks, such as the activity and population density of the susceptible host, proper climatic conditions, and necessary contacts. Resistence of the host and virulence of the pathogen determine the dosage necessary for infection, the speed of development of the infection, and its severity. Disease prevalence is modulated by temperature, for which there is one optimum range for the host and another for the pathogen. In the range of coincidence or where the pathogen is favored, the result is disease development. In the range of temperatures unfavorable for the pathogen, the insects resist infections, which sometime remain hidden in or on the host and are activated in favorable conditions.

Host and pathogen population densities and short-term weather conditions are primarily responsible for oscillations of enzootics (Fig. 8.1), whereas host resistence and fertility together with climatic cycles are responsible for larger fluctuations of populations and the occurrence of cyclic epizootics. Natality, a distinct symptom of the fitness of a population, and actual recorded mortality are the main factors that indicate the "vital index." If the vital index is positive, it is a symptom of a growing population and an on-going outbreak. A vital index equal to 1 over a certain period characterizes a stagnation at the population peak; then comes a period of decline with a negative index leading to a period of hidden, inapparent survival of the pest, but not its extinction (5) (Fig. 8.2). Over long time periods, the growth curve of a population initially has a sigmoid configuration with an abrupt decrease from the peak to a minimum level lasting much longer than the peak. Only if conditions in nature change drastically and unfavorably during this period does a population run the risk of extinction (4, 6).

Diseases whose prevalence develops into a sigmoid curve (e.g., viral, fungal, bacterial infections) are the most influential factor in population fluctuations, and they appear simultaneously in all of the host populations over areas of

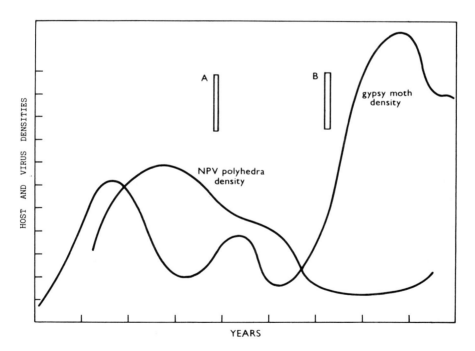

Figure 8.1. Model of gypsy moth density under the impact of a nuclear polyhedrosis virus. (A) Impact of an external factor (rain) reducing both host density and incidence of infection. (B) Impact of sunlight reducing only the pathogen density.

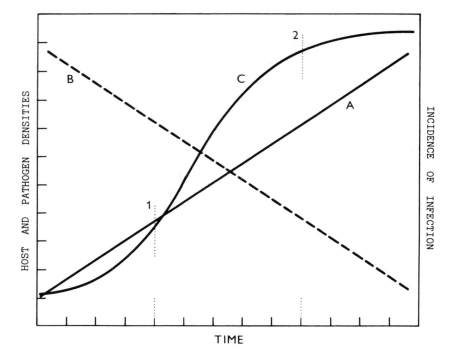

Figure 8.2. Interaction in a population model of the pathogen (A) and host (B) densities and the resulting incidence of infection (C). (1) First host contact with pathogen, (2) saturation of the biotope with pathogens.

hundreds of square kilometers (7–9). Other diseases remain enzootic and inapparent during the period as "background" infections with a constant incidence that is not very important to host population fluctuations (1). However, these same pathogens in a different host may cause spectacular epizootics and mortality. Both types of pathogens are density dependent. The enzootic diseases are more adapted to a parasitic life in a host and regularly infect 10 to 20% of the population. The acutely epizootic diseases are of two main types. Some are caused by pathogens such as nuclear polyhedrosis (NPV) or granulosis viruses (GV), which initiate infections and mortality wherever and whenever they are brought into contact with a susceptible host population. In others, the pathogens multiply in a dense host population in one situation but not when introduced artificially into a similar host population elsewhere and at another time. Host populations stressed by epizootics of *Entomophthora* or *Tarichium* are susceptible to panzootics over a large area of their distribution (9–11).

The foregoing discussion emphasizes the complications of analyzing local or area-wide patterns of disease; for example, the modeling of the interactions of hosts and diseases is difficult, and only in a few cases have simulated and real schemes of development been compared (12). Nevertheless, consideration of broad patterns in the ecology of entomopathogens may contribute to a better concept of the function of diseases in the dynamics of insect populations and the biological regulation of noxious insects.

2. PATTERNS OVER PLACE

During the long period of the formation of continents, insects colonized all accessible and suitable areas and brought the pathogens with them or acquired local pathogens with limited distributions. In addition to the widely distributed insect species with relatively stable geographical ranges, other species migrated or were introduced by man into new areas. The migrants of introduced species are of interest because we can follow the fates of their diseases during this process. There are also cases of zoogeographical analogues, that is, insects with close relatives which are well adapted to distant areas having the same climate. Such relatives have evolved with their own system of diseases. It would be interesting to summarize the impacts of distribution, isolation, migration, new environment, and eventually of climate on pathogens of certain insects.

2.1. Diseases of Insects with Global Distribution

The global distribution of a pest depends on the global distribution of its food plant and habitat. The distributions of some major pests depend on man-made habitats: field crops and orchards, stored products, mass-produced useful insects such as honeybees and silkworms, and major vectors of human diseases such as mosquitoes and blackflies. Published data indicate that the distribution of diseases in their hosts is known only when trained personnel are available — the distribution of diseases to some degree indicates the distribution of insect pathologists. The present discussion may, therefore, serve only as an introduction to a topic for which more information is necessary (13).

2.1.1. COMPLEX OF PESTS IN ORCHARDS AND FIELD CROPS. Lepidoptera which attack orchard trees in Europe comprise a group with many diseases in common. They are all susceptible to a similar group of pathogens, such as nonsporeforming bacteria, *Bacillus thuringiensis,* fungi, microsporidia, and nematodes. The Lepidoptera investigated in Europe include *Euproctis chrysorrhoea, Malacosoma neustrium, Lymantria dispar, Yponomeuta malinellus, Operophtera brumata, Hyphantria cunea, Aporia crataegi,* and a series of tortricids.

The viral infections in this group are host specific except for the cytoplasmic polyhedrosis virus (CPV). Infections are transmitted from contaminated leaves or on the ovipositors of entomophagous parasites (2, 14, 15). Epizootics with extensive mortality occur in *M. neustrium* and *L. dispar* with NPV, and in *H. cunea* with GV. Other mentioned host insects do not suffer extensive mortality from disease epizootics, even when microsporidia are present in more than 80%

of a population (16). *Lymantria dispar* and *E. chrysorrhoea* transfer pathogens from orchards to ornamental trees and deciduous forests. In some areas, after *L. dispar* outbreak ends with a NPV epizootics in oak forests, a population with high prevalence of latent virus persists in urban areas on *Robinia pseudoaccacia* which normally is not its preferred host tree. From there they again invade ornamentals outside urban areas.

Another important spreader of diseases from orchard habitats to ornamentals and deciduous forests is the winter moth, *O. brumata*. It transfers microsporidia of the browntail moth, *E. chrysorrhoea,* to oak tortricids, to *Orgyia antiqua,* and to a series of other hosts. Horizontal transfer of infections from field crops to nearby orchards is mediated by *A. crataegi* and its parasites, mainly *Apanteles glomeratus,* which it shares with the group of white butterflies. In this way *Pieris brassicae, P. rapae, P. napi, Plutella maculipennis,* and in some areas *Colias* sp. have in common, in addition to their specific NPVs and GVs, a group of microsporidia including *Nosema mesnili* and *Thelohania mesnili* and certain nonsporeforming bacteria transmitted by parasites (2, 14). In the cycle of transmission, braconid wasps play the role of vectors, though they too are susceptible to *Nosema mesnili* infection. For example, *A. glomeratus* larvae and adults are killed by the disease. The adult braconids transmit the disease on their ovipositors from caterpillar to caterpillar and do not become infected, though the larvae do in the infected host. Peroral infection of caterpillars and the infection by injection on the ovipositor both cause two types of infections affecting different host tissues. The virulence of the pathogen in both types of infections, peroral and injection by vector, is identical, and *N. mesnili* has similar virulence and routes of transmission throughout the entire distribution areas of *P. brassicae* and *P. rapae* (15).

Two Lepidoptera living in the same habitat have analogous groups of diseases, for example, the codling moth and the cabbage moth. *Cydia pomonella* has its specific group of pathogens which is similar but not identical with those of other orchard pests. It has a GV of limited zoogeographic distribution. In contrast to previously discussed insects, hibernating larvae are sometimes infected by *B. thuringiensis* and *Beauveria bassiana,* which kill a high percentage of larvae in their winter shelters. Chronic infection by *Nosema carpocapsae* is comparable to that by the microsporidia of white butterflies. In some limited foci the hibernating caterpillars of *C. pomonella* are infected with *Steinernema feltiae*. This infection is usually acquired by larvae which pass through soil containing invasive larvae of the nematode.

There is some interchange of insect diseases from the orchard-field crop complex through *Mamestra brassicae* to another complex, the noctuid moths.

2.1.2. DISEASES OF NOCTUID MOTHS. A second system of pathogens with relatively worldwide patterns infects larval and adult noctuids in field crops, for example, members of the genera *Agrotis, Feltia, Mamestra, Plusia, Trichoplusia, Heliothis, Spodoptera,* and many others. This system includes the same groups of worldwide pathogens with different species in different parts of the world. Each noctuid has a system of specific viruses which cause local epizootics in late instar caterpillars. Individual viruses coincide with individual distribution areas of host species with some overlapping and sometimes with great differences in infectivity. Although the *kurstaki* strain of *B. thuringiensis* was isolated from mass rearings of *Heliothis virescens,* this bacterium does not cause epizootics in this insect and appears only in individual cases when a caterpillar ingests a massive dose or is infected through a wound. The cannibalistic, aggressive habit of many of these species contributes to the transmission of some pathogens.

Infections with microsporidia are relatively common. Noctuids are hosts for many, if not all, species of *Vairimorpha. Pleistophora schubergi* has wide cross-infectivity for noctuids, even those strains with different host ranges. For example, the strains from the gypsy moth, *L. dispar,* and fir budworm, *Cacoecia murinana,* are not cross-infective to each other's host, but both strains infect *M. brassicae. M. brassicae* is a suitable host for microsporidia from mosquitoes, beetles, or snails (17). Microsporidia cause massive mortality only in rearings of field-collected noctuids during the first laboratory generation. In the field, infected caterpillars are victims of predators before the microsporidia are transmitted. Adults caught in light traps rarely are infected by microsporidia but have sporadic neogregarine infections.

Epizootics of nematodes, particularly Mermithidae and Steinernematidae, appear in distinct foci with a high concentration of invasive larvae in the soil. Within these foci, host larvae and pupae are killed at high prevalence rates (up to 70% of *Feltia segetum* or *Autographa gamma* in Central Europe). Suitable host material must be available for the focus to be maintained during subsequent seasons, but hosts can include white grubs or hibernating Colorado potato beetle *(Leptinotarsa decemlineata)* larvae and pupae as well as noctuids (18). Nematodes which attack noctuids kill more than 80% of *L. decemlineata* during the winter. Other nematodes such as *Pristionchus l'heritieri* complicate the picture in some foci when they invade caterpillars already killed by a developing population of *S. feltiae.* They introduce secondary bacteria and destroy the monoxenic condition in the host.

Fungal infections are more important to population dynamics of noctuids than to that of other insects in the same ecosystem. The Entomophthorales, in particular, often cause epizootics. For example, *Tarichium megaspermum* is a well-known pathogen of *Autographa gamma* on sugar beet and flax in Central Europe. During 1964–66 in Czechoslovakia the fungus killed 92% of last instar caterpillars throughout almost 100 km² in South Bohemia during the second week of August in each year. Such panzootics of entomophthoracean infections also are typical of other noctuids. Outbreaks of *Panolis flammea,* which attacks forests in Germany and Poland, were terminated by epizootics of *Entomophthora aulicae* in 22 thousand ha of pine forest in 1869 and again in 150 thousand ha in 1927 (7–9). There is no step-wise spread of the infection from one starting point or focus, but the environment is so seeded with resting spores of the fungus that a suitable sequence of climatic conditions can trigger an epizootics over the whole area at once (11). Similar phenomena were noted in America with *Trichoplusia ni* and *Heliothis* spp.

2.1.3. DISEASES OF APHIDS.   Another group of insects with world-wide distribution and identical diseases is the aphids. Due to their feeding on sterile saps they are not exposed to perorally transmitted infections, but contact infections by fungi play an important role in their population dynamics. Here we find one of the curious relationship of pathogen and host, where the pathogen replaces other control agents in destroying the host population. The entomophthoraceous fungi appear to have little selectivity during aphid population peaks, and they spread into young, unstressed colonies of susceptible hosts. Sometimes they appear in large, simultaneous epizootics or panzootics involving large areas.

One reason for such panzootics is that host contacts are not necessary for the spread of infective stages. For example, in England (19, 20) conidia of these fungi are common in dust samples from air streams over cultivated land. Therefore, certain critical abiotic conditions must help initiate epizootics. Studies of such phenomena in the USSR indicate that in the great plain area there is no significant difference in the frequency of epizootics in different climactic belts, namely, the boreal region, the mild central region, and the dry subtropics in the south. Wherever dense populations of aphids appear, epizootics are possible. The microclimate is the decisive factor, particularly high moisture and relative humidity for several days. Host populations in the boreal region have scarce, chronic infections; those in the central area have true epizootics and total destruction of susceptible hosts; and in the southern, dry zone outbreaks of aphids and disease result from irrigation or extraordinary rains. Similarly, studies in France (21, 22) have predicted epizootics based on a "key" of temperature and humidity. The minimum conditions for fungal epizootics are a temperature of at least 20°C and moisture from at least two rainy days. According to more detailed studies (23) of moisture alone, the critical factor is the dewpoint every morning and evening, during which the extensive distribution of conidia and adhesion of conidia to the host cuticle occur.

Other fungi, such as *Verticillium* spp., need more extended periods of high relative moisture, and these are achieved primarily in greenhouses and only occasionally in dense host populations in inundated forests or the moist subtropics (24, 25). Artificially high humidity breaks the host specificity of strains of *Verticillium lecanii* for aphids, scales, and the white fly so that it also attacks other insects such as thrips and hymenopterous parasites (*Encarsia* spp.).

2.1.4. DISEASES OF MOSQUITOES AND BLACKFLIES.   Mosquitoes and blackflies comprise a group of widely distributed insects with another typical system of pathogens. In contrast to phytophagous insects, there are no extensive outbreaks of diseases that reduce entire populations of mosquitoes in a whole country. Large fluctuations result only from abnormal abiotic, mainly climatic, conditions. Diseases of mosquitoes are generally endemic, not exceeding 10 to

15% prevalence in a population (Table 8.1). Their characteristic feature is that they are well adapted to the life cycle of the host, including egg diapause. Microsporidia in mosquitoes provide a good example. *Amblyospora* and *Parathelohania* are typical genera, the first infecting mainly *Aedes* and *Culex* spp., the second *Anopheles* spp. They apparently produce deadly infections in larvae and latent infections in adults. The visibly infected larvae are mainly males. In female adults the microsporidium invades the ovaries after the blood meal and is transmitted in eggs to the progeny. Similar, possibly sibling, species of microsporidia occur in different areas and in different mosquito hosts. The thick-walled spores from larval hosts do not directly infect young larvae of the same host and need a mediator.

In blackfly larvae a similar system of less specific pathogens causes enzootics. Instead of a mosaic of species of two genera, six microsporidian species are distributed in many host species all over the world (25). There is no good evidence for transovum transmission of these pathogens. Spores of these stream-inhabiting insects have a typical peculiarity: a system of surface structures, protrusions, hairs, ridges, mucous floatators, or membranes which affect their downstream distribution. If spores of these six microsporidia are released from one point in a stream, they become distributed differentially so that each would accumulate in its specific spot.

Only two groups of pathogens cause local mortality exceeding 10%. Entomophthoraceous fungi appear during the mass hatching of adults and occasionally kill large populations of mosquitoes and midges on plants and stones

TABLE 8.1
Life Cycle Characteristics and Percent Infection by Diseases of Mosquitoes and Blackflies in Temperate and Tropical Regions

	Temperate (boreal)[a]	Tropical[a]
Number of generations/year	1 (−3)	10−15
Life span:		
larvae	30−120 days	8−14 days
adults	1−5 months	1 month
Diseases:		
mosquito iridescent virus	+++	+
Coelomycidium (larvae)	5−15%	3−5%
Coelomomyces (larvae)	+	10−30%
Entomophthora (adults)	+++	−
Bacillus thuringiensis israelensis	+	+++
Microsporidia larval, dimorphic	15%	10−20%
in adults, dimorphic	2%	2%
monomorphic	+	20%
Mermithidae	++	++

[a]− = absent; + = rare; ++ = low prevalence rates; +++ = high prevalence rates.

around the breeding places. The second, more questionable group includes *B. thuringiensis* and *B. sphaericus*. Both have strains that are virulent in natural habitats, but there is no direct evidence that either bacterium causes natural epizootics (26, 27). My experience indicates that the field ecology of both bacilli is different from that in artificial applications. In natural habitats both *B. thuringiensis* and *B. sphaericus* infect individual mosquito larvae that encounter a dead insect and ingest an infectious dose. Their dead bodies serve as sources of infection for other larvae that feed on them. Spores and crystals are dispersed by the collecting activity of ciliates, rotifers, and other filtrators. Some material disappears in the sapropelic bottom. This loss is partially offset by their growth on proteinaceous matter, such as dead hosts or other living animals which have been wounded or infected with other pathogens (27, 28). By this process and by the drying of the habitat, the infectious material is concentrated and reintroduced into a host population. The geographic distribution of different strains of both bacilli depends on their introduction by migrating birds, other aquatic animals, ovipositing mosquitoes, or connections between bodies of water. For instance, both bacilli can be transported in the African savanna by adult *Simulium damnosum* blown by winds for hundreds of kilometers at the end of the dry spring period.

There is no definite evidence of a restricted zoogeographic distribution of most diseases of blackflies except for some cases. The system of endemic diseases in Europe, North America, Central Asia, Africa, Australia, and New Zealand is principally the same and includes *Thelohania bracteata, T. fibrata, T. varians, Vavraia multispora, Pleistophora simulii, P. debaisieuxi,* and the fungus *Coelomycidium simulii.* There are some pathogens with more specific distributions. *Caudospora* species infect mainly blackflies under boreal conditions or in high mountain regions. Species with surface structures are in mountain currents, for example, *Weiseria laurenti* in the Pyrenees and Sumava. A pathogen of local distribution is the microsporidium *Hirsutosporos austrosimulii* from blackflies in New Zealand, and a fungus with local distribution is *Simuliomyces lairdi* in Central Asia.

2.1.5. DISEASES IN MASS REARINGS.   Crowded insect populations are subject to high mortality from diseases that are distributed by contact and food contamination. Colony-forming insects, including ants, termites, wasps, and bees, have natural mechanisms to protect the colony, and only specific diseases for each host can become established. For example, *Nosema apis* is a pathogen only of the honeybee, wasps have only *Cordyceps* spp. as major pathogens, and only a few pathogens are known from termites (28–30) and ants (31). Artificially crowded populations in mass rearings of insects (silkworm, entomophagous insects, animals for sterile male releases) have no self-cleaning mechanisms and are notoriously susceptible to epizootics. Under these conditions microsporidia cause epizootics similar to those caused by viruses or bacteria in the field. This has been observed in mass rearings of silkworm, *Cactoblastis cactorum, Gnorimoschema operculella, Cydia pomonella,* and different species of *Heliothis.*

2.1.6. DISEASES OF STORED-PRODUCT INSECTS.   Diseases in stored-product insects (Table 8.2) have been studied by several authors (1, 2, 32–34). The endemic diseases, mainly caused by protozoa, do not effectively control populations of grain insects. In deficient diets of stored flour, the search for proteinaceous materials causes a much higher incidence of protozoan diseases in some host species than in others with the same ecology. An example is the distribution of *Farinocystis tribolii* in the three flour beetles: *Tribolium castaneum* (the most proteinophilic and cannibalistic), *T. confusum,* and *T. navale* (34), with prevalences of 80, 20, and 3%, respectively, in the same habitat. Increased susceptibility to diseases also can result from shortage of food. *Calandra granaria* usually does not have any infectious disease because it is hidden during its development in sterile grain kernels. But at the end of an infestation, when fragments and used kernels are the only food and shelter for the progeny, the weevil can acquire microsporidia or neogregarines of other grain insects present in the remains (35). In the same way, infections appear in mass rearings of the Angoumois grain moth, *Sitotroga cerealella,* used for production of egg parasites. Destructive epizootics are known in Lepidoptera infecting stored grain and seed, including *Ephestia kuhniella* and *Plodia interpunctella.* They are caused by viruses (mainly in America, but not in Europe) and Protozoa *(Mattesia dispora)* (33).

B. *thuringiensis* rarely causes natural epizootics (36, 37), though one such epizootic in flour mills of Thuringen introduced this pathogen into insect pathology. The author observed epizootics caused by B. *thuringiensis* in E. *kuhniella* on flour stored in sacks, in P. *interpunctella* on stored cacao beans and peanuts, and in G. *mellonella* on stored honey combs. B. *thuringiensis* was also found in small foci in hibernating codling moth larvae. To some extent parasitic Hymenoptera initially help spread the disease by puncturing infected hosts with their ovipositors (2, 32). The disease spread from dead larvae in which bacteria developed and which had been smeared on the ground among beans during transportation. Milling of dry, dead larvae caused, to some extent, contamination of food leading to further cases of disease.

Originally strains of B. *thuringiensis* were distributed in connection with sericulture [*alesti* and *anduze* (serovar H3a), and *sotto* (serovar H4a4b)] and with the distribution of stored products (most serovars) (Fig. 8.3). The B. *t.* strain *dendrolimus* is anomalous because it belongs to the *sotto* group (serovar H4a4b) but causes field epizootics in *Dendrolimus sibiricus* in the Siberian taiga. Species of *Dendrolimus* are common in the main sericulture area of China, and the endemic forest area is close to the Great Silk Route from China. Infected D. *sibiricus* caterpillars hibernate in soil, and dead bodies are deposited in one soil layer from which the bacteria are distributed the following spring by the hatching adults (37).

Distribution of grain from main production areas has contributed to the distribution of grain pests and of strains of B. *thuringiensis.* The first recorded appearance of the bacillus in Europe was connected with the introduction of E. *kuhniella* in European flour mills, probably from the USA (32, 36). There were some autochthonous infestations in old European flour mills. B. *t.* strains were

TABLE 8.2
Diseases in Pests of Stored Products[a]

Insect	Neogregarina	Coccidia	Microsporidia	Virus	Fungi	Bac. thuring.
Plodia	++	+	++	++[b]	•	++
Ephestia	+++	+	+++	+	•	++
Cadra	+	•	+	•	•	•
Sitotroga	+	−	+	−	•	+
Tenebrio	++	•	−	−	•[c]	•
Tribolium	++	++	++	−	•[c]	−
Laemophloeus	•	−	+	−	•[c]	−
Oryzaephilus	+	−	+	−	•[c]	−
Sitophilus	−	−	−	−	•[c]	−
Trogoderma	−	−	−	−	−	−
Bruchus	−	−	−	−	−	−

[a]+, ++ present in different quantity; • present but rare; − absent.
[b]Known only in America.
[c]Susceptible to application of conidia.

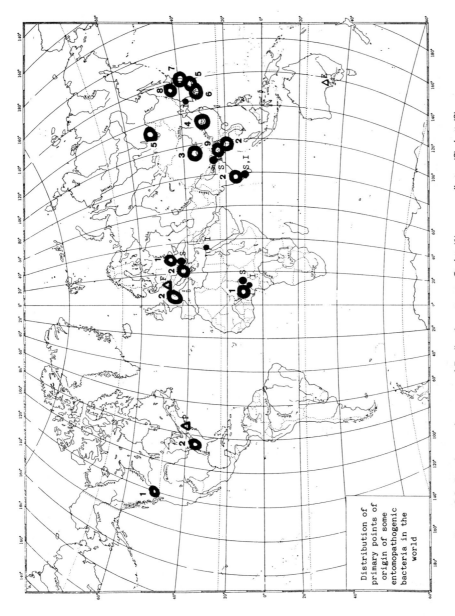

Figure 8.3. Distribution of serotypes of *Bacillus thuringiensis*. Circles: (1) serovar *galleriae*, (2) *alesti*, (3) *yunnanensis*, (4) *wuhanensis*, (5) *oo/ dendrolimus*, (6) *aizawai*, (7) *kyushuensis* and *tohokuensis*, (8) *kumamotoensis*, (9) *pakistani*. Dots: *B.t. israelensis* and *B. sphaericus*. Triangles: *B. popilliae*, *B. fribourgensis*, *B. euloomarahae*.

Distribution of
primary points of
origin of some
entomopathogenic
bacteria in the
world

227

sought in grain mills in the mountains of the Balkans where a minimal inter-connection with the distribution network of flour and food products ensured evidence of original conditions. The serovars, *thuringiensis, kurstaki,* and *morrisoni,* commonly infected the grain insects of Kosovo (38). These serovars appeared alternately in the investigated area without apparent segregation or interaction. Knowledge of the distribution of different strains of *B. thuringiensis,* based on identification of serovars, indicates that there are strains with regional distribution mainly in East and Southeast Asia (13, 28). The picture may become more obscured by the use of different bacterial preparations in agriculture in many countries. For example, serovars H1, H5a5b, and H4a4b are used in Entobakterin, Bitoxibacillin, and Dendrobacillin or Gomelin, respectively, produced in the USSR; H1 without exotoxin is in Bathurin from Czechoslovakia; H3a was the basis of Bactospeine in France; and *kurstaki* (H3a3b) is in all recent products in Europe and the USA. Only a few strains have some simple marker such as toxicity for *G. mellonella,* which is typical of the serovars *galleriae* (H5a5b) from Europe (USSR) and *aizawai* (H7) from East Asia. Detection of new serovars and new specific activities, including as yet unaffected groups of insects, is just a question of further testing.

2.2. Introduced Insects and Their Diseases

The spread of pathogens can be followed in insects that invade new areas. Examples include the Japanese beetle, the gypsy moth, the codling moth, and the Colorado potato beetle. Unfortunately, knowledge of their diseases in the old and new distribution areas is not complete.

2.2.1. JAPANESE BEETLE.   Introduction of the Japanese beetle, *Popillia japonica,* to the USA was the first case where pathogens and parasites were studied together for biological control. Dutky (39) found one rickettsia, two specific bacteria, one microsporidium, and two nematodes in this pest in populations in the USA. Several of these pathogens were propagated and distributed, and this effort initiated the practical microbial control of agricultural pests in general. This situation is unusual in that mortality agents were sought in the country where the insect was introduced but not in its original geographical range in East Asia. This is still true except for a search for the original bacterium symbiotic with *Neoaplectana glaseri* (40).

2.2.2. GYPSY MOTH.   The situation is different for insects introduced from Europe, where old data are available concerning pathogens. The gypsy moth, *L. dispar,* was introduced into the USA from Europe. In Central Europe it is a typical periodical pest with outbreaks repeated every 11 years (Fig. 8.4). Further south there are decreasing inter-gradation periods in the Mediterranean and Krymean regions (41) with outbreaks every three years, and in some localities in the marine belt, the insects were present continuously, without peaks of outbreaks. Several diseases have been reported from the gypsy moth in Europe and Asia (Table 8.3, Fig 8.5). These include an NPV; CPV; GV; a number of

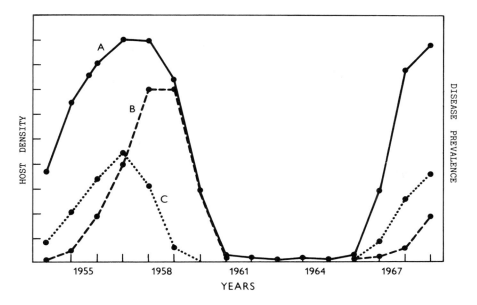

Figure 8.4. *Lymantria dispar* and its pathogens during the period of the insect outbreak in Slovakia, 1954–1968. (A) Host population, (B) prevalence of NPV, (C) prevalence of microsporidia.

microsporidia such as *Pleistophora schubergi, Nosema muscularis, N. lyman-triae, N. serbica,* and *Thelohania similis;* and some mermithids (16, 42). There are only limited data about the occurrence of fungal infections, and bacteria play only a limited role in the natural control of the gypsy moth, though data are available only from limited sources. The pathogens, except the NPV, do not cause high mortality levels, and they disappear during the inter-gradation period from the second to the sixth year after the last outbreak. After that, the microsporidia become more prevalent and infect 15 to 30% of the population before and during the peak period of the outbreak. As soon as the NPV appears in the population, the protozoa decline (hosts with mixed infections die first, often without sporulation). The population collapse is usually correlated with high prevalence of NPV. Occasionally the host population does not contact the virus during the peak phase, and the larvae die of starvation. In such situations the microsporidia appear but usually do not kill the caterpillars before pupation (42). Experimental evaluations of the inactivation of virions in polyhedra when exposed on vegetation to sunshine and to changes of weather indicate that the inactivation time corresponds to the time between outbreaks. Only hidden supplies of infective material remain in the area, and erratically behaving caterpillars of the growing population contact infective virions and introduce the disease again into the population.

Reports of naturally occurring disease in the gypsy moth in America reveal only the NPV and CPV. Microsporidia have not been observed yet. Interest-

TABLE 8.3

Percent Infections in Populations of Gypsy Moth in Deciduous Forests and Urban
Areas during Gradation Periods in Europe

Pathogen	Urban (%)	Field Populations (%)		
		Pregrad.	Gradation	Retrograd.
NPV	5–20	0–3	3–30	80–100
GV	(1)	[a]	2–5	[a]
CPV	1	2–5	2–5	20–30
Nosema serbica	1	1	3–30	3–15
N. lymantriae				
P. schubergi	1–2	1–2	5–20	30–50
Mermithids	[a]	[a]	2–5	2–5

[a]Present but rare.

ingly enough, *P. schubergi,* which infects *Choristoneura fumiferana* in Canada,
is not infectious for the gypsy moth (43).

2.2.3. CODLING MOTH.   Another example of an insect introduced from Europe to America, Africa, and Australia is the codling moth, *L. pomonella.* The
microsporidium *Nosema carpocapsae* is the only specific larval pathogen (be-
side nonspecific fungi, bacteria, and nematodes) in Europe, and it occurs over
the entire distribution area of the moth, though its prevalence varies greatly
from year to year and from area to area (18, 44) (Fig. 8.6). There is no evidence
of any difference in its virulence in material from different zoogeographic
regions. In its region of origin, the palaearctic, another microsporidium, *Pleistophora carpocapsae,* occurs in populations in the Kaspian region, and an NPV
was recorded from Central Asia. Both regions have wild apples (45). *N. carpocapsae* and Steinernematid nematodes are found in the nearctic region, but the
GV is the most commonly occurring pathogen. Perhaps because there are some
strain-related differences in *S. feltiae* between Eurasia and the USA, the only
recorded acquisition of the codling moth during its spread in America is the
GV, which does not occur in the palaearctic area. The virus does not naturally
infect European populations and does not spread when introduced in field
trials. This difference in susceptibility of distant populations of the codling
moth is evidently connected with the differentiation of local strains in mountain valleys in the Alps, where it is difficult to mate males from one valley with
females from another. In hibernating populations the pathogen list is identical
throughout the world and includes *Beauveria bassiana,* often in high prevalence, and *S. feltiae* (2, 46). A survey of codling moth in wild apples in its native
Central Asia probably would provide important information about natural
mortality agents and distinguish which pathogens it acquired during its introduction to different parts of the world.

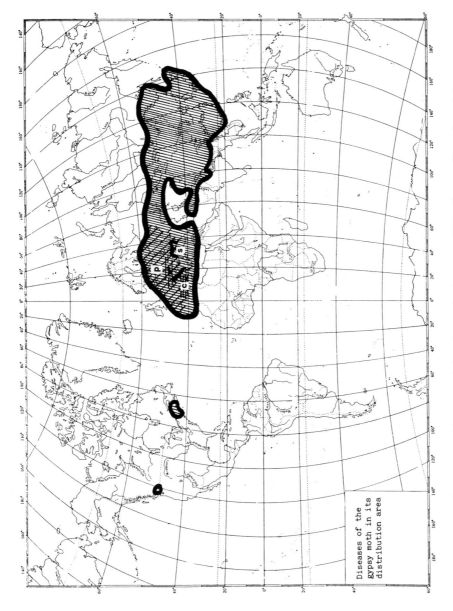

Figure 8.5. Diseases of the gypsy moth throughout its distribution area: NPV in all areas. p, *Pleistophora schubergi*; s, *Nosema lymantriae* and *N. serbica* in Europe; c, cytoplasmic polyhedrosis virus.

231

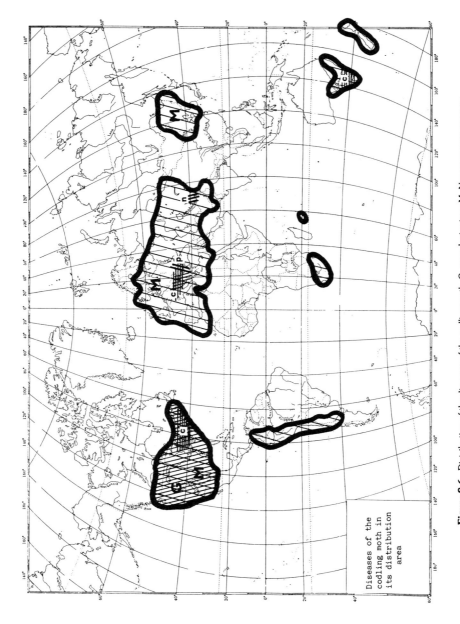

Figure 8.6. Distribution of the diseases of the codling moth. G, granulosis virus; M, *Nosema carpocapsae*; p, *Pleistophora carpocapsae*; c, *Steinernema feltiae*; n, nuclear polyhedrosis virus.

2.2.4. FALL WEBWORM. The fall webworm, *H. cunea,* was introduced to Europe in 1945 and is now found in Yugoslavia, Bulgaria, Romania, Hungary, Czechoslovakia, and the USSR. In America this insect occurs from the Great Lakes to southern Florida and from the Atlantic coast to British Columbia. In its nearctic region of origin, endemic pathogens include NPV, GV, the microsporidium *Vairimorpha necatrix,* two other *Vairimorpha* species, and a *Pleistophora* sp. (43). None of these pathogens is able to control this insect (47). The webworm had no pathogens when introduced into Europe. The microsporidum *Vairimorpha hyphantriae* appeared in 1951 – 53 in the northern limit zone on the Danube (Fig. 8.7). An NPV was recorded after 1954 in the original port of entry of the pest in Hungary. The GV was first recorded from central Yugoslavia in 1957, at which time it was absent from the rest of the distribution area of the webworm. The GV appeared in most localities in the invaded area after 1959, but its impact on populations was practically nil and it appeared only in laboratory rearings due to crowding and lack of sterilization of the environment by sunshine. In its new distribution area in Europe, *H. cunea* also has been found to be infected with a CPV, the microsporidium *P. schubergi,* various fungi, and local minor parasites and predators. There is no reason to believe that the European GV, NPV, or microsporidia became adapted to this host. But it is interesting that it took 5, 10, and 12 years, respectively, before the diseases, probably present in the introduced lot of insects, appeared as natural control agents.

2.2.5. COLORADO POTATO BEETLE. The Colorado potato beetle, *L. decemlineata,* is an oligophagous insect introduced from America to Europe. When this pest invaded potatoes in America it was actually an intruder which lost its natural diseases somewhere in its area of origin. Investigations of its mortality agents in the USA did not reveal any pathogens, only a predator *(Perillus bioculatus)* and parasite *(Doryphorophaga doryphorae).* Reported cases of mycoses or of nematodes were only occasional, unselective infections. When introduced into Europe, the beetle, under intense pressure from insecticides, could not establish any persistent populations to which any native pathogens could adapt. Nevertheless, in some areas in Europe diseases appeared with some regulating impact: a rickettsial organism in Yugoslavia (48) and a microsporidium in the USSR (49), neither of which regulated host populations. Certain microsporidia of chrysomelid and curculionid beetles from the same climatic zone were tested for their infectivity to the Colorado potato beetle. However, they did not become established in field populations of the beetle because their acute courses of infection did not allow density-dependent transmission (50), and they did not establish natural foci of infection.

2.3. Zoogeographical Analogues

In a few cases one can compare zoogeographical analogues and their array of limiting diseases from different areas. Analogous insect species have analogous

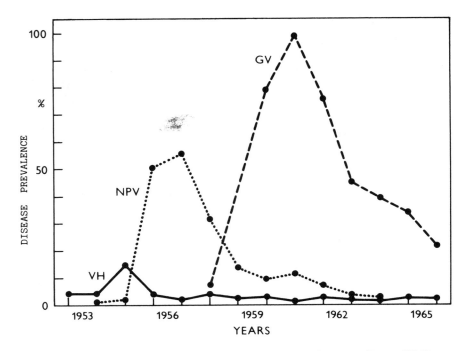

Figure 8.7. Onset of various diseases in the fall webworm after its introduction into Europe. VH, *Vairimor-pha hyphantriae;* NPV, nuclear polyhedrosis virus; GV, granulosis virus.

systems of pathogens which become more similar as ecological situations of the hosts become similar. When such analogues have been studied in detail, types of pathogens not occurring in hosts in specific regions can be identified and sought for introduction. Diseases of mosquitoes, blackflies, and other widely distributed pests already have been discussed. Further analogues include *Cacoecia murinana* and *Choristoneura fumiferana, Malacosoma neustrium* and *M. americanum* or *M. disstria, L. decemlineata* and *Polygramma undecimlineata,* and *Antheraea pernyi* and *A. mylitta* with *Bombyx mori.*

2.3.1. FIR BUDWORM. The fir budworm *C. murinana* from Europe has a true analogue in the spruce budworm *C. fumiferana* in America. Both insects have their viruses (NPV, GV, Pox) and a series of microsporidia, fungi, and nematodes. They differ slightly in the distribution and impact of these diseases on their populations, but it is evident that an identical system of factors provides population control. In Europe (18, 51, 52) the GV causes local epizootics which spread on the periphery but decline in their center, where uninfected (more resistant?) host populations appear. This is not the pattern that occurs with the GV of *C. fumiferana* in America. In Canada (53) microsporidia are more common than in Europe, and *Nosema fumiferanae* and *P. schubergi* are commonly present.

2.3.2. TENT CATERPILLARS AND THE LACKEY MOTH.   Tent caterpillars in America *(M. americanum* and *M. disstria)* and the lackey moth in Europe *(M. neustrium)* have an entirely different ecology, and therefore the occurrence of limiting pathogens is also different. The forest tent caterpillar *(M. disstria)* is a serious defoliator of deciduous forests over large areas of outbreak which usually end with a nuclear polyhedrosis epizootic. Other pathogens such as microsporidia (54) play only a minor role in the control of its populations. The NPV is less common in the lackey moth in Europe; foci of the virus in orchards are localized, though movement of females by wind and on cars from an NPV focus spreads virus to new areas. Infections by microsporidia are sporadically transmitted from the brown tail moth which occurs in the same habitats. Bacterial infections *(Enterobacter cloacae)* occasionally reduce lackey moth populations (2, 16).

2.3.3. POTATO BEETLES.   The Colorado potato beetle, which was discussed earlier, has an interesting geographical analogue, the eleven-banded potato beetle *P. undecimlineata.* This insect is common in Cuba on a shrub-like potato, *Solanum torvum.* It also occurs in some parts of the USA where it is narrowly sympatric with the Colorado potato beetle, though it does not attack potato crops. In Cuba, *S. torvum* is found in ruderal plant associations, and therefore *P. undecemlineata* is not treated with insecticides and forms natural populations. The insect has three efficient types of control agents: parasitic flies, microsporidia, and fungi (Table 8.4). The primary control agents are two tachinid flies, *Doryphorophaga australis* (in lowlands) and *Tachinophyton floridense* (at heights over 1000 m), which kill up to 80% of the insects in different localities. Unspecific bacterial septicemias resulting from attacks of tachinids increase the mortality up to 90% or more (55). Two microsporidia, *Pleistophora fidelis* and *Nosema polygrammae,* occur frequently but cause little mortality. They are transmitted in the feces of infected larvae or on the ovipositor of the tachinids (2, 50, 55). In Europe and USA, the prevalence of parasites and pathogens in Colorado potato beetle is nil or very low.

2.3.4. SILKWORMS.   The oak silkworm *A. pernyi* in Europe and the Tasar silkworm *A. mylitta* in India are reared for silk over wide territories. They are kept under similar conditions of crowding and distribution of rearing material (grain). They have analogous pathogens with similar effects on populations (56). Infections caused by NPV are severe but are well managed with surface sterilization of eggs. Microsporidia, which cause rearing complications, are introduced with food *(Pleistophora balbiani,* closely related to *P. schubergi,* in the midgut) or with infected eggs *(Nosema antheraeae* in different tissues). *Pleistophora balbiani* occurs mainly in Europe rearings, probably originating in some local Lepidoptera. *Nosema antheraeae* is transmitted through infected eggs and is present in two slightly different forms in both distribution areas, S. Ukraine and India. Its spread was enhanced during one period by inefficient thermal disinfection of eggs and pupae. Other diseases, including bacterial

TABLE 8.4

Comparison of Mortality Agents and Their Prevalence Rates in *Leptinotarsa decemlineata* in Europe and *Polygramma undecemlineata* in Cuba

Mortality Agent	*Leptinotarsa decemlineata*[a]	*Polygramma undecemlineata*[a]	
Endemic parasites	—	*Doryphorophaga*	16–94%
		Tachinophyton	28–45%
Endemic pathogens	—	*Pleistophora fidelis*	5–30%
		Nosema	
		polygrammae	5%
Introduced parasites	Mermithidae		10–60%
	Steinernematidae		5%
Introduced pathogens	*Beauveria*		10–30%
	Microsporidia		1%

[a]— = does not occur.

septicemia and *Beauveria* spp. infections, appear irregularly over the entire range of rearings. An analogous microsporidium *Nosema bombycis* in *B. mori*, does not infect *Antheraea* spp. even when present in the same region.

2.4. Regional Variability

Pathogens that have obligately intracellular development such as viruses or microsporidia are so adapted to the host that they are able to exist wherever the host is able to survive. They can be introduced into new climatic and zoogeographic zones inside the host and perform their roles (57, 58), though the speed and intensity of disease change with local conditions. Climatic conditions play an essential role in the transmission, activation, and infectious process of pathogens that enter the host through its cuticle (fungi, nematodes).

An example of differences caused by climate is found in the diseases of citrus pests or pests of field crops in California and Florida. Irrigation of plantations in California efficiently changes the dry subtropics into conditions suitable for the development of plants and pests that require high humidity (e.g., aphids, scale insects). However, the uncontrolled environment, particularly the high relative humidity, contributes to a wider distribution of fungal infections in Florida, for example, 18 to 64% infection by *Myiophagus ucrainicus* in *Lepidosaphes* scales or *Triplosporium fumosum* in *Pseudococcus citri* (59 – 61). These fungi occur in California only in intensively-irrigated plantations and at low prevalence rates. Another example is the appearance of *Nomuraea rileyi*, a white muscardine fungus, on caterpillars of different noctuids in Florida and the southeastern USA, especially in velvetbean caterpillars *Anticarsia gemmatalis* (12). The fungus is rare in California. Evaluation of environmental factors and the occurrence of this fungus in Florida revealed the complexity of its requirements.

Other fungi also provide examples of regional distribution. *Cordyceps* spp. (e.g., *C. sinensis* in China or *C. militaris* in Scandinavia) occur in the rainy zones of continents, often on mountain slopes exposed to excessive rains. Entomophthoraceae are more common in mild or cold climates of the North, and their number and frequency rapidly decrease in the more temperate zones of the Mediterranean climate and the African savanna. Other fungi such as *Massospora cicadina* occur in dry biotopes but only in a limited zoogeographical region. It is unknown why the Americas are the main area of distribution of this fungus and it does not infect the cicadas in Europe, Asia, or Africa (2, 13).

3. PATTERNS OVER TIME

Distributions of diseases in time have several underlying causes. Many insect diseases exhibit periodicity, occurring again and again during one seasonal period for many years. This is the case with some fungal diseases. *Coelomomyces chironomi* appears regularly in August even though host midges are present at other times. Infections of *Culex pipiens* by *Entomophthora culicis* or

E. curvispora usually appear in the first week of May, whereas *Entomophthora muscae* infections are typical in September or October. Their periodicity is probably related to climatic factors, or, in some cases, to a specific stage of development of the host. Periodicity is typical not only for fungi but also for some viruses. The mosquito iridescent virus appears only in the spring generation of *Aedes cantans* in Europe during the last two weeks of April and first week of May. In the subtropics (Florida) the virus appears in some mosquitoes in repeated appearances during the year (62).

Not only a suitable range of temperatures or a suitable stage of the host affects periodicity of insect diseases; the length of exposure to pathogens and the activity of the host also are important. For example, the winter generations of aquatic insects, such as blackflies, midges, caddisflies, or mayflies, have a higher prevalence of diseases than the summer generations. Also, fast development of host insects in the tropics causes problems in the artificial introduction of diseases, though infections and diseases caused by microsporidia develop more quickly in some hosts (flies, mosquitos). Microsporidioses in flies in hot climates develop to the same extent in 3 days as they do in a temperate climate (63) in 2 to 3 weeks. If tropical diseases, which complete their cycle in a week, can be studied in temperate hosts where the cycle lasts a month or two, the slower process permits more detailed and convenient observations.

Introduction of pathogens from temperate zones to a tropical climate eliminates periods of hibernation in low temperatures and subsequent epizootics because infectious pathogens remain more constantly available to hosts. This causes the shorter inter-gradation periods of the gypsy moth (see Section 2.2.2) (5, 41). Development of certain temperate insects takes several years (in the European chafer, *Melolontha melolontha,* 4 years), and the susceptible hosts are exposed to pathogens continuously for long time periods. Interruption of development during the winter in temperate climates is replaced in the tropical regions by periods of hot dry weather when the number of suitable hosts is reduced to a minimum. In temperate zones some insects find shelter in cool environments which reduce the speed of infections. The sawfly *Cephaleia abietis* hibernates in the deep sterile soil layer of forests where temperatures during the year do not rise over 15°C. Here the development of invasive larvae of *Steinernema kraussei,* which invade the larvae during their passage through upper layers of the soil, is slowed until the host is ready to pupate and leave the shelter. Visible development of the nematodes can be recorded only in 3% of the larvae during the summer. Invasive larvae remain in one soil layer and attack the sawfly whenever it crosses this layer on the way up or down.

In one example speed and intensity of contacts had a positive effect on the virulence of the transmitted pathogen. This was shown in sawfly viruses (64) where quick recycling of virus from the initially killed insects increased their infectivity and virulence. The speed of infection and mortality increased to an optimum after the seventh or eighth transfer. There is a lack of information concerning the same effect in other pathogens.

The time factor also plays an important role in the development of the host-pathogen system in nature. During the development of ecosystems (54) there are some stages in the succession of changes in the physical environment that enable extensive population growth of some organisms and subsequently of their diseases. A typical situation is the development of new agricultural land and large pastures and the subsequent invasion of different species of scarabaeid beetles. The concentration of white grubs of different species leads to a high prevalence of infectious diseases. For example, *Adelina sericesthis* became prevalent in grubs of *Melolontha hippocastani* in a new park lawn in Prague. Also, the beetle *Sericesthis pruinosa* similarly invaded lawns in Australia after which the *Sericesthis* iridescent virus, a coccidium, and bacterial infections caused by *Bacillus lentimorbus* and *B. euloomarahae* were recorded (65, 66). Outbreaks of *Popillia japonica* in golf courses in the USA during the 1940s resulted in a similar succession, with high rates of parasitism (39).

4. SUMMARY

The different factors and interactions influencing patterns of insect diseases over place and time are complex. In all groups of pests the array of pathogens that reduce their numbers fall into similar groups, each represented by regional or host-specific species. Insects, with worldwide distribution, studied because of their pest status, exhibit local variations in their representative diseases and in the severity of their impact. Natural periodicity of insect numbers also causes periodicities of their disease epizootics, and some features of this periodicity were presented. Pathogens that are mass produced and used efficiently as microbial insecticides do not often cause natural epizootics. Fungi, protozoa, and helminths require close contacts for their transmission, but viruses can cause epizootics in less dense populations. Epizootics over large areas result from climatic activation. Thus, some types of climate support the development of certain disease groups.

REFERENCES

1. H. D. Burges, Enzootic diseases of insects, *Ann. N. Y. Acad. Sci.,* **217,** 31 (1973).
2. J. Weiser, *Nemoci hmyzu* (Insect diseases), Academia, Praha, 1966.
3. D. E. Weidhaas, Simplified models of population dynamics of mosquitoes related to control technology, *J. Econ. Entomol.,* **67,** 622 (1974).
4. T. Park, Population ecology, *Encyclopedia Britannica* (1948).
5. F. Schwerdtfeger, *Die Waldkrankheiten,* P. Parey, Berlin, 1970.
6. E. P. Odum, The strategy of ecosystem development, *Science,* **164,** 262 (1969).
7. C. Bail, Pilzepidemien an der Forleule, *Noctua piniperda* L., *Z. Forst. Jagdwes.,* 243 (1869).
8. E. Muller-Kogler, Pilzbedingte Insektenkrankheiten, *Ber. Naturforsch. Wiss., Ver. Darmstadt,* 17 (1962).
9. L. Garbowski, Spostrzezenia nad owadomorkami, *Pr. Wydz. Chorob Roslin P. Inst. Nauk Rol. Bydgoszczy,* **4,** 3 (1927).

10. Y. Tanada, "Epizootiology of Insect Diseases," in P. DeBach, Ed., *Biological Control of Insect Pests and Weeds,* Chapman & Hall, London, 1964, p. 548.

11. J. Weiser, Notes on two new species of the genus *Tarichium* Cohn (Entomopht.), *Česká Mykol.,* **19,** 201 (1965).

12. L. P. Kish and G. E. Allen, *The biology and ecology of Nomuraea rileyi and a program for predicting its incidence on Anticarsia gemmatalis in soybeans,* Univ. Fla. IFAS Bull. 795 (1978).

13. J. Weiser, The milestones of invertebrate pathology, *J. Invertebr. Pathol.* **37,** 1 (1981).

14. H. Blunck, Mikrosporidien bei *Pieris brassicae,* ihren Parasiten und Hyperparasiten, *Z. Angew. Entomol.,* **36,** 316 (1954).

15. I. V. Issi, Mikrosporidios kapustnoj beljanki i drugich nasekomych i ego biologičeskoe značenie, Ph.D. Thesis, Univ. Leningrad, 1964.

16. J. Weiser, Mikrosporidien des Schwammspinners und der Goldafter, *Z. Angew. Entomol.,* **40,** 509 (1957).

17. J. Kramer, *Nosema necatrix* sp. n. and *Thelohania diazoma* sp. n. microsporidians from the armyworm *Pseudaletia unipuncta* (Haw.), *J. Invertebr. Pathol.,* **7,** 117 (1965).

18. J. Franz, "Influence of Environment and Modern Trends in Crop Management on Microbial Control," in H. D. Burges and N. W. Hussey, Eds., *Microbial Control of Insects and Mites,* Academic Press, New York, 1971, p. 407.

19. G. J. W. Dean and N. Wilding, *Entomophthora* infecting the cereal aphids *Metapolophium dirhodum* and *Sitobion avenae, J. Invertebr. Pathol.,* **18,** 169 (1971).

20. N. Wilding, Entomophthora conidia in the air-spora, *J. Gen. Microbiol.,* **62,** 149 (1970).

21. J. Misonier, Y. Robert, and G. Thoizon, Circonstances epidemiologiques semblant favoriser le developpement des mycoses a Entomophthorales chez trois aphides: *A. fabae, Capitophorous horni* et *Myzus persicae, Entomophaga,* **15,** 169 (1970).

22. Y. Robert, J. M. Rabasse, and P. Schultes, Facteurs de limitation des populations d'*Aphis fabae* dans l'ouest de la France, *Entomophaga,* **18,** 61 (1979).

23. R. Krejzová, Germination process in resting spores of some *Entomophthora* species and pathogenicity of spore material for lepidopterous larvae, *Z. Angew. Entomol.,* **85,** 42 (1978).

24. P. A Hall and H. D. Burges, Control of aphids in glasshouse with the fungus *Verticillium lecanii, Ann. Appl. Biol.,* **93,** 235 (1979).

25. E. I. Hazard and J. Weiser, Spores of *Thelohania* in adult female *Anopheles:* development and redescription of *Thelohania legeri* Hesse and *T. obesa* Kudo, *J. Protozool.,* **15,** 817 (1968).

26. J. Weiser, A mosquito virulent *Bacillus sphaericus* in adult *Simulium damnosum* from Northern Nigeria, *Zbl. Mikrobiol.,* **139,** 57 (1984).

27. J. Weiser and S. Prasertphon, Entomopathogenic sporeformers from soil samples of mosquito habitats in Northern Nigeria, *Zbl. Mikrobiol.,* **139,** 49 (1984).

28. H. T. Dulmage and K. Aizawa, "Distribution of *Bacillus thuringiensis* in Nature," in E. Kurstak, Ed., *Microbial and Viral Pesticides,* Marcel Decker, New York, 1982, p. 209.

29. C. Kalavati and C. C. Narasinhamurti, A. microsporidian parasite, *Duboscqia coptotermi* sp. n., from the gut of *Coptotermes heimi, J. Parasitol.,* **62,** 323 (1976).

30. R. Krejzová, Pathogenicity of a strain of *Paecilomyces fumosoroseus* isolated from *Zootermopsis* sp., *Vest. Cesk. Spol. Zool.,* **40,** 286 (1976).

31. C. Pérez, Sur *Duboscqia legeri,* microsporidie nouvelle parasite du *Termes lucifugus* et sur la classification des microsporidies, *C.R. Soc. Biol.,* **65,** 631 (1908).

32. O. Mattes, Parasitäre Krankheiten der Mehlmottenlarven und Versuche uber ihre Verwendbarkeit als biologische Bekämpfungsmittel, *Sitzungsber. Ges. Beförd. Naturwiss.,* Marburg, **62,** 381 (1927).

33. J. Weiser, Schizogregariny z hmyzu škodícího zásobám mouky, *Věst. Cesk. Spol. Zool.*, **16**, 199 (1953).

34. J. Weiser, Infektiosität der Kokzidien für Insekten, *Trans. Int. Colloq. Insect Pathol. Proc.*, Paris, 105 (1962).

35. K. Purrini, Zur Kenntnis der Mikrosporidienarten bei Vorratsschädlingen der Ordnung Lepidoptera, *Anz. Schaedlingskd., Pflanzplanzen-Umweltschutz.*, **49**, 42 (1977).

36. E. Berliner, Uber die Schlaffsucht der Mehlmottenraupen (Ephestia kuhniella Zel.) und ihren Erreger Bacillus thuringiensis n. sp., *Z. Angew, Entomol.*, **2**, 29 (1915).

37. E. V. Talalaev, Septicemija gusenic sibirskogo shelkopriada, *Mikrobiologija*, **25**, 99 (1956).

38. J. Vankova and K. Purrini, Natural epizootics by bacilli of the species *Bacillus thuringiensis* and *B. cereus, Z. Angew. Entomol.*, **88**, 216 (1979).

39. S. R. Dutky, Investigation of the diseases of the immature stages of the Japanese beetle, Ph.D. Thesis, Rutgers Univ., New Brunswick, New Jersey, 1937.

40. G. O. Poinar, The presence of *Achromobacter nematophilus* in the infective stage of a *Neoaplectana* sp. (Steinernematidae), *Nematologica*, **12**, 105 (1966).

41. J. V. Tchugunin, Očagovaja cikličnost razmnoženij nasekomych i mikrobiologičeskij metod borby s nimi, *Trans. 1st. Int. Conf. Insect Pathol. Biol. Control*, Prague, 81 (1958).

42. I. V. Issi, Vlijanie mikrosporidioza na plodovitost neparnogo šelkoprjada, *Lymantria dispar* v rjadu pokolenij, *Tr. Vses. Inst. Rast.*, *31*, 331 (1968).

43. J. Maddox, personal communication, 1983.

44. L. A. Falcon, Biological factors that affect the success of microbial insecticides: development of integrated control, *Ann. N. Y. Acad. Sci.*, **217**, 173 (1973).

45. P. A. Simchuk and I. V. Issi, *Pleistophora carpocapsae* sp. n., novij parazit jablonnoj plodožorki, *Parazitologija*, **9**, 293 (1975).

46. J. W. Doberski and H. T. Tribe, Isolation of entomogenous fungi from elm bark and soil with reference to ecology of *Beauveria bassiana* and *Metarhizium anisopliae, Trans. Br. Mycol. Soc.*, **74**, 95 (1980).

47. J. Weiser and J. Veber, Die Mikrosporidie *Thelohania hyphantriae* im Weissen Bärenspinner und anderen Mitgliedern seiner Biocoenose, *Z. Angew. Entomol.*, **40**, 55 (1957).

48. C. Sidor, personal communication, 1970.

49. J. Lipa, *Nosema leptinotarsae* sp. n., a microsporidian parasite of the Colorado potato beetle, *Leptinotarsa decemlineata* (Say.), *J. Invertebr. Pathol.*, **10**, 111 (1968).

50. Z. Hostounsky and J. Weiser, *Nosema gastroideae* sp. n. (Nosematidae, Microsporidia) infecting *Gastroidea polygoni* and *Leptinotarsa decemlineata, Acta Entomol., Bohemoslov.*, **70**, 345 (1973).

51. J. Weiser, Krankheiten des Tannentriebwicklers, *Cacoecia murinana* Hbn. in der Mittelsovakei, CSR, *Z. Pflanzenkr. Pflanzenschutz*, **63**, 193 (1956).

52. M. Čapek, R. Obrtel, and J. Weiser, Choroby, parasiti a episiti jedlového obaleče Ch. murinana v kalamitní oblasti středního Pohroní, *Lesn. Cas.*, **4**, 46 (1958).

53. G. G. Wilson, Effects of *Nosema disstriae* on the forest tent caterpillar *Malacosoma disstria, Proc. Entomol. Soc. Ont.*, **110**, 97 (1979).

54. W. A. Smirnoff, Adaptation of the microsporidian *Thelohania pristiphorae* from the tent caterpillars *Malacosoma disstria* and *Malacosoma americanum, J. Invertebr. Pathol.*, **11**, 321 (1968).

55. J. Weiser and Z. Hostounsky, *Nosema polygrammae* sp. n. and *Plistophora fidelis* sp. n. infecting *Polygramma undecimilineata* in Cuba, *Věst. Cesk. Spol. Zool.*, **39**, 104 (1975).

56. J. Veber, Contribution to the knowledge of diseases of the Chinese oak silkworm, *Antheraea pernyi* Guérin, *Proc. 1st Int. Conf. Protozool.*, Prague, 1962, p. 504.

57. I. V. Issi and V. P. Tchervickaja, O vlijanii temperatury na razvitie *Nos. mesnili* i *Plistophora schubergi, Zool. Zh.,* **48,** 1140 (1969).

58. E. A. Steinhaus, The importance of environmental factors in the insect-microbe ecosystem, *Bacteriol. Rev.,* **24,** 365 (1960).

59. M. H. Muma, Factors contributing to the natural control of citrus insects and mites in Florida, *J. Econ. Entomol.* **48,** 432 (1955).

60. A. G. Selhime and M. H. Muma, Biology of *Entomophthora floridana* attacking *Eutetranychus banksi, Fl. Entomol.,* **49,** 161 (1966).

61. A. T. Speare and W. W. Youthers, Is there an entomogenous fungus attacking the citrus rust mite in Florida?, *Science,* **60,** 41 (1924).

62. J. R. Linley and H. T. Nielsen, Transmission of a mosquito iridescent virus in *Aedes taeniorhynchus.* Experiment related to transmission in nature, *J. Invertebr. Pathol.,* **12,** 17 (1968).

63. J. Kramer, Generation time of the microsporidian *Octosporea muscaedomesticae* in adult *Phormia regina* (Dipt., Calliph.), *Z. Parasitenkd.,* **25,** 309 (1965).

64. W. A. Smirnoff, A virus disease of *Neodiprion swainei* Midl., *J. Insect Pathol.,* **3,** 29 (1961).

65. R. L. Beard, Two milky diseases of Australian Scarabaeidae, *Can. Entomol.,* **88,** 640 (1956).

66. J. Weiser and R. L. Beard, *Adelina sericesthis,* a new parasite of *Sericesthis pruinosa, J. Insect Pathol.,* **1,** 99 (1959).

DISEASE GROUPS

9

NONINFECTIOUS DISEASES

RANDY R. GAUGLER

Department of Entomology and Economic Zoology
Rutgers University
New Brunswick, New Jersey

1. INTRODUCTION

Disease may be broadly categorized as either infectious or noninfectious, denoting presence or absence of a transmissible living organism as the etiological factor. Insect pathologists have heavily emphasized infectious diseases in their research, reflecting an interest in the application of insect pathology for pest control purposes (i.e., microbial control). Noninfectious diseases have generally proven refractive to manipulation for control, resulting in a disproportionately small amount of study. Their neglect is compounded by the difficulty of diagnosis and by a lack of understanding of their nature. Although Steinhaus (1) argued that noninfectious diseases "are an integral part of insect pathology," many pathologists do not instinctively associate noninfectious etiological agents such as temperature, wind, wounds, and chemicals with disease, possibly because few insect pathology courses include a noninfectious disease component.

An understanding of disease dynamics in insect populations requires application of epizootiological principles. Tanada (2) recognized that epizootiology pertains to both infectious and noninfectious diseases, but the focus of epizootiological studies in insects has been limited almost exclusively to infectious diseases. This is in sharp contrast with medical epidemiology, which was dominated by microbiology until the middle of this century when rapid advances in disease control resulted in infectious diseases being replaced by noninfectious diseases as the foremost cause of human morbidity and mortality. Subsequent application of epidemiology to noninfectious diseases led to perhaps its greatest contribution: establishing smoking as the major cause of lung cancer. No such reorientation in research effort is anticipated in the epizootiology of insect diseases. Nevertheless, in view of the successes achieved in the epidemiology of noninfectious diseases, the study of these diseases from an epizootiological standpoint would appear to offer considerable promise.

The purpose of this chapter is to show that noninfectious as well as infectious diseases can be considered from an epizootiological perspective. The pathology of noninfectious diseases has been discussed elsewhere (3–7), and the reader is encouraged to consider these and other sources in obtaining a fuller understanding of the present topic.

2. CONCEPTS

Many of the general concepts of insect epizootiology, although developed for infectious diseases, are relevant to noninfectious diseases. This point can be illustrated with the Mt. St. Helens volcanic eruption as an example of a noninfectious disease epizootic. Here the etiological "agent" is volcanic ash, which is lethal to insects because the abrasive ash causes cuticular wounds resulting in excessive water loss (contributing causes of death are respiratory impairment due to blockage of spiracular valves and disruption of gut activity by ash boli)

(8). Those properties of abiotic etiological agents of significance to insect epizootics also have counterparts in biotic agents. The concepts of "pathogenicity" and "virulence," generally applied only to pathogens, are analogous to the single concept of toxicity, which can be measured and expressed as the minimum lethal dose in mg of ash. "Persistence" is readily applicable to both disease groups; the adverse effects of ash on insect populations decline with rainfall (9, 10). The ash is airborne, a "dispersal" route found in many infectious agents. In the case of ash, disease "transmission" is by direct integumental contact, and the "portal of entry" would be the actual site of cuticular injury. The disease "reservoir" would refer to the volcano itself. Additional terms borrowed from epidemiology and applied to epizootiology, including incidence, prevalence, environment, host, and host-related properties (i.e., susceptibility), are easily related to both infectious and noninfectious diseases.

Just as not all epidemiological terms and concepts are useful in insect epizootiology (see Chapter 1), not all epizootiological terms and concepts are appropriate to the study of noninfectious disease. Several unique problems are readily apparent. For example, "vertical transmission," very important in the occurrence and maintainence of many insect epizootics, is an inappropriate concept for volcanic ash and noninfectious diseases in general. The highly useful "basic reproductive rate" (11) has, by definition, no counterpart for abiotic agents. "Enzootic" is a term with limited usefulness because of the abrupt appearance and short persistence of most noninfectious diseases, but "epizootic wave" as an expression of disease incidence in time and space retains its utility.

Diseases not readily linked to an etiological agent are more abstruse, as parallels between the two disease groups become increasingly difficult and concepts previously discussed become less useful. Still, in examining freezing as a disease process, extreme cold might be reasonably viewed as an agent. The reservoir in this example would be air in the insect tissues, with airborne transmission, and a portal of entry at the site of tissue injury. With trauma as an example, physical force might be considered as the agent. Although strained, these examples illustrate that infectious and noninfectious diseases have characteristics in common useful in providing a conceptual framework for epizootiological studies of both groups.

Chronic noninfectious diseases, caused by lack or excess of any factor required for normal growth and development (e.g., most nutritional disorders), create even more formidable conceptual problems for the epizootiologist, most notably the absence of a specific etiological agent. Epidemiologists have tended to deal with this by deemphasizing the concept of an agent and stressing the principle of multiple etiology. Insect epizootiologists also have long been aware that disease outbreaks are seldom attributable to a single factor, but rather are the result of a complex of interacting factors (2, 12). The role of the abiotic environment in the expression of latent insect viroses (13) and Pasteur's early association of certain predisposing abiotic factors with flacherie in silkworms (1) are two noteworthy examples. Still, insect pathologists have tended to focus on the agent-host components of the epizootic triangle without fully analyzing the role of the environment in the initiation, rise, and decline of epizootics. Ultimate success in resolving the conceptual difficulties of investigating epizootics of noninfectious diseases may require that epizootiologists, like their epidemiological counterparts, regard an agent as an intrinsic element of the environment as a whole.

While our efforts at understanding epizootics of noninfectious diseases may be enhanced by considering similarities with infectious diseases, obvious and irreconcilable differences exist between the two disease groups. A rational approach will require that concepts originally derived from infectious diseases be adopted when possible and adapted where necessary. Significant contributions will not be realized without new disease models developed specifically for those properties unique to the noninfectious diseases of insects.

3. ETIOLOGY

3.1. Problems

A central objective in epizootiology is to suggest the cause of a disease outbreak. Noninfectious diseases sometimes are easily recognized on the basis of distinct symptomatology, as with many hormonal-induced pathologies, but more often they are indistinctive or even resemble infectious diseases, especially insect viroses.

The dearth of well-defined signs and symptoms is one of several factors complicating noninfectious disease etiology. Unlike the epidemiologist, the epizootiologist lacks an intelligent subject to assist diagnosis. Unlike most epizootologists, the investigator of noninfectious diseases is unable to fullfill Koch's postulates. Diagnostic methods such as serology, so critical to demonstrating the presence of specific infectious disease agents, are virtually unknown. The frequent absence of a discernable agent is the single greatest impediment to elucidating etiological questions. This is further complicated by the multiplicity of etiological factors, as the investigator must distinguish between and measure the contribution of each of a bewildering complex of interrelating factors. The impact of many environmental factors might be indirect; for example, caterpillars removed from the host plant by a strong wind might be uninjured, but starvation results if the insects are unable to quickly locate a suitable food source. Other factors can interact concurrently, and perhaps synergistically, as in the case of oxygen and temperature stress in aquatic habitats. As a consequence of these problems, researchers have tended to lump diseased insects lacking a microorganism or parasite into the noninfectious disease category without further differentiation (14, 15). Obviously, there is a pressing need for diagnostic tests and parallels of Koch's postulates designed specifically for nonreplicating disease agents of insects.

3.2. Methods of Study

The establishment of etiology represents the foremost challenge to the study of noninfectious diseases. Insect pathologists, preoccupied with microbial control and pathological processes, have in essence relinquished this area of investigation to ecologists. Interested in broader questions of the role of the abiotic environment in community structure, ecologists are currently debating the relative merits of the two methods of analysis: observational and experimental studies (16, 17).

Observational studies conducted with insects have been mostly prevalence (cross-sectional) studies (see Ch. 1). In considering catastrophic mortality (e.g., volcanic eruption, unseasonable frost, flooding, fire) a causal relationship often can be inferred by observation alone. Otherwise inferences must be made by statistical correlation of disease prevalence with insect density or other factors, as in Varley and Gradwells (18) use of "key factor analysis" to identify factors causing population changes in the winter moth *Operophtera brumata.*

Experimental studies permit the investigator to exert some control by manipulating pertinent variables to isolate causal relationships. Lejeune and Filuk (19) and Lejeune et al. (20), for example, submerged cocoons of the larch sawfly *Pristiphora erichsonii* in laboratory and field tests to implicate flooding as an important mortality factor.

In comparing the two approaches, direct experimentation is inherently more satisfying to the epizootiologist attempting to answer noninfectious etiological questions because causal relationships are more readily established. Observational data must be viewed with caution because correlation does not necessarily mean causation; the patterns observed may be chance associations where the variables under study are independently correlated with another factor. Alstad et al. (3) discussed problems encountered in attempts to describe effects in terms of relations that are correlative rather than causal. They supported their premise by providing many examples of correlation/causality confusion in the literature on air pollution and insects. Conversely, the experimental approach has been criticized as being artificial and short-term. The ideal strategy would be an incidence or cohort type of observational study resulting in life tables, a long and costly undertaking.

4. DISEASE DYNAMICS

Despite the heavy emphasis placed on microorganisms by insect pathologists, noninfectious diseases are presumed to be more important in regulating most insect populations. Ecologists recognize the powerful influence of the abiotic environment on insect population densities, but convincing evidence linking these factors to specific diseases is scarce and largely anecdotal.

The purpose of this section is to discuss the major categories of noninfectious diseases as direct agents of mortality; sublethal manifestations will not be included. Because the subject is large and diverse only a cursory examination is possible.

4.1. Mechanical Injuries

Like most other noninfectious diseases, physiological and structural changes associated with trauma in insects are well studied, but its impact on natural populations has rarely been investigated. Only the most obvious instances have been investigated, the foremost example being the Mt. St. Helens volcanic eruption.

Insect populations in the immediate vicinity of the 18 May 1980 eruption were obliterated, but adverse effects accompanying the heavy ash fall (1 cubic mile over Washington and neighboring states) were much more extensive. Observational studies indicated a wide range of susceptibilities to the ash deposits, with highly mobile insects (e.g., foragers, predators, and parasites) suffering disproportionately more mortality than relatively sedentary herbivores. For example, Frye (21) determined that populations of winged predators of pear psyllids were decimated, while psyllid population densities continued to increase as in previous years. Foraging honeybees were killed almost immediately by ash with some colonies being completely destroyed (22); beekeepers in the Columbia basin estimated losses of $2 million (23). Akre et al. (10) concluded that yellowjacket populations in western Washington had been "devastated," with *Dolichovespula* spp. being nearly eliminated, while ant colonies were not severely affected because only a small percentage of workers forage outside the nest at one time. Codling moth populations did not experience ash-related mortality because vulnerable stages were not present (24), and effects on insects associated with wheat were assessed as "not severe" (9). The scouring and smothering effects of ash in the stream environment virtually eliminated some species of caddisflies while other species were not significantly affected depending on feeding behavior and emergence patterns (25).

The Mt. St. Helens eruption represents one of the most conspicuous examples of a noninfectious disease epizootic. Unfortunately, there was no forewarning so that only simple field correlations and subjective observations are available. No effort was made to place the event in an epizootiological context, which would provide a more complete understanding of the phenomenon.

4.2. Physical Injuries

Extremes of temperature, moisture, wind, radiation, and atmospheric pressure, considered collectively as weather, are widely regarded to be the most significant abiotic factors regulating insect numbers. They influence rates of development, longevity, and reproduction, and they provide vital behavioral cues (e.g., for diapause and dispersal). The major physical factors—temperature, moisture, and wind—can also cause sudden, catastrophic reductions in insect populations (i.e., epizootics).

Insects usually are able to avoid high temperature, so that reports of heat injury are uncommon. By contrast, there are innumerable reports of mortality induced by low temperatures following a sudden frost or severe winter, including Berryman's (26) analysis of winter mortality in the fir engraver *Scolytus ventralis*. Investigating the effects of the coldest winter temperature in nearly 40 years, Berryman found that brood survival decreased in direct proportion to the temperature recorded at each sampling site. The greatest survival was recorded from populations residing beneath the snow line where there was protection from the cold, while brood above this level were virtually eliminated (0 and 1.6% survival) at the coldest sites. The role of low temperature as a cause of insect mortality is incontrovertible but, as with other noninfectious diseases, specifically when such mortality constitutes an epizootic remains obscure.

Rain appears to be a significant catastrophic mortality factor, particularly for small, sedentary stages of insects. Harcourt (27) reported rainfall to be important in the survival of the imported cabbageworm, *Pieris rapae,* causing an average mortality of 50% in the first two larval instars. These larvae were dislodged from the host plant and drowned in surface puddles or leaf axils. Later instars, which were not as easily displaced and when dislodged were usually able to regain the plant, sustained mortalities ranging from 0 to 12%. Most other evidence of mortality caused by rainfall is based on anecdotal accounts, summarized by Beirne (28), rather than hard quantitative data.

Lack of moisture also can limit insect abundance. Combining observational and experimental data, Birch and Andrewartha (29) reported that a severe summer drought killed 90% of diapausing eggs of *Austroicetes cruciata,* and they further concluded from an analysis of meteorological records that such catastrophic mortality might have occurred as many as seven times over the preceeding 50 years. Campbell (30) similarly implicated desiccation as a primary cause of sudden reductions in gypsy moth populations, but this finding is tenuous because of uncertain diagnosis.

Wind has not been thoroughly investigated as a mortality factor although there are many reports of migrating insects carried by wind currents being deposited in water or snow with lethal results (31, 32). Wellington (33) has observed that during outbreaks of spruce budworms, the waters near lake shores are sometimes covered by thousands of dispersing larvae caught by down drafts.

4.3. Chemical Injuries

Natural poisoning of insects by plant products are often difficult to verify and can be confused with viral diseases, but such poisonings represent peculiar instances of noninfectious disease caused by biotic agents (5). Exposure to man-made chemicals is much more strongly associated with adverse effects on insect populations.

Insect mortality from chemical insecticides is well understood, and discussion of this immense subject is best left to toxicological texts.

In reviewing the literature on the effects of atmospheric pollutants on insects, Alstad et al. (3) pointed to diverse studies indicating declines in insect population densities directly or indirectly attributable to airborne fluorides, sulfur, ozone, lead, and other emissions. The most convincing case was made for fluoride, which is known to cause reduced fecundity, fertility, feeding and growth rates, and death. Fluoride has been implicated in decline of silkworm, honey bee, European pine shoot moth, and bark beetles. Interestingly, fluoride contamination also has been correlated with outbreaks of some insect herbivores, possibly because of fluoride-induced plant stress.

Perhaps the best documented examples of noninfectious disease epizootics concern chemical toxins in aquatic environments, where striking changes in insect diversity and abundance have been demonstrated following pollution by heavy metals, phenols, detergents, and other toxic effluents (34). Research in this area recently has been focused on establishing the impact of acid rain on freshwater ecosystems through surveys and experimental manipulations (35, 36). These studies have shown that insect sensitivity varies greatly among taxa and even different life stages, but that, in general, species richness, diversity, and biomass are reduced with increasing acidity. It is difficult to envision an area of investigation where epizootiology could make a more significant contribution than acid rain.

4.4. Nutritional Diseases

Competition for limited food resources is generally regarded to be of significance in insect population dynamics. Many species have evolved means of reducing or avoiding malnutrition, as in the case of insects that disperse when populations exceed their food supply (32). Thus the precise role of nutrition in insect populations remains obscure. Clark (37) described a strong correlation between food depletion and high mortality among *Phaulacridium vittatum* nymphs, but this well documented example presumably does not constitute a true epizootic since acute food shortage appears to be a usual and predictable feature of the dynamics of this species. Doubtlessly, epizootics of starvation occur, but quantitative data are scarce. Few confirmed observations of nutrient deficiency have been made outside of the laboratory, probably because of difficulty in recognizing the generally nonspecific symptoms associated with malnutrition. Recent progress in the design of assays for the detection of starvation (38) might encourage further evaluation of this disorder.

4.5. Genetic Diseases

Genetic pathology is of little consequence in natural insect populations because deleterious genes in these short-lived organisms are repressed by natural selection, hence genetic diseases rarely reach epizootic proportions. Harmful mutations are more frequently observed in laboratory populations where inbreeding permits the buildup of homozygosity and the expression of recessive genes. Not surprisingly, most instances of mutation have been reported from honeybee and silkworm populations.

4.6. Neoplastic Diseases

The primary focus of epidemiological effort in recent years has been the investigation of tumors; however, neoplasms in insects are infrequently reported and, like hereditary disorders, generally are not a topic for epizootiological investigation.

5. CONCLUSIONS

Insect epizootiologists have focused their studies almost exclusively on infectious diseases, while community ecologists, seeking to explain patterns and causal processes in natural systems, persistently address questions regarding the abiotic environment, i.e., noninfectious disease, although ecologists do not use this term. Many investigators feel that the abiotic environment, especially meteorological factors, is the key to understanding community structure. Ecologists have not placed their findings in an epizootiological context by recording measurements such as incidence, relative risk, and case-fatality rates, since they do not share the same perspective on disease as pathologists. Irregardless, they have formed a foundation for the investigation of noninfectious disease epizootiology, albeit inadvertently. Further development will require a more comprehensive integration of insect pathology and insect ecology than presently exists.

A marriage of pathology and ecology may be more easily realized than is generally appreciated, as there is considerable common ground between these disciplines. Most significantly, the terms noninfectious disease, density-independent factors, abiotic agents, and physical environment are essentially equivalent. Similarly, catastrophic mortality as reported by an ecologist would be regarded as an epizootic from the pathologist's perspective.

Hopefully this discussion will stimulate research leading to broader application of ecological and pathological principles to noninfectious disease epizootiology. The available data is inadequate for such a synthesis, so even rudimentary studies would be valuable, but the most pressing need is for comprehensive studies with solidly documented causal relationships.

REFERENCES

1. E. A. Steinhaus, *Principles of Insect Pathology,* McGraw-Hill, New York, 1949.
2. Y. Tanada, "Epizootiology of Insect Diseases," in P. DeBach, Ed., *Biological Control of Insect Pests and Weeds,* Reinhold, New York, 1964, p. 548.
3. D. N. Alstad, G. F. Edmunds, Jr., and L. H. Edmunds, Effects of air pollutants on insect populations, *Annu. Rev. Entomol.,* **27,** 369 (1982).

4. E. A. Steinhaus and Y. Tanada, "Diseases of the Insect Integument," in T. Cheng (ed.) *Current Topics in Comparative Pathobiology,* Vol. 1, Academic Press, New York, 1971, p. 1.

5. L. Bailey, *Honey Bee Pathology,* Academic Press, London, 1981.

6. E. A. Steinhaus, Ed., *Insect Pathology: An Advanced Treatise,* Vol. 1, Academic Press, New York, 1963.

7. G. E. Cantwell, Ed., *Insect Diseases,* Vol. 2, Marcel Dekker, New York, 1974.

8. J. S. Edwards and L. M. Schwartz, Mount St. Helens ash: a natural insecticide, *Can. J. Zool.,* **59,** 714 (1980).

9. E. C. Klostermeyer, L. D. Corpus, and C. L. Campbell, Populations changes in arthropods in wheat following volcanic ash fall-out, *Melanderia,* **37,** 45 (1981).

10. R. D. Akre, L. D. Hansen, H. C. Reed, and L. D. Corpus, Effects of ash from Mt. St. Helens on ants and yellowjackets, *Melanderia,* **37,** 3 (1981).

11. R. M. Anderson and R. M. May, Infectious diseases and population cycles of forest insects, *Science,* **210,** 658 (1980).

12. E. A. Steinhaus, The importance of environmental factors in the insect-microbe ecosystem, *Bacteriol. Rev.,* **24,** 365 (1960).

13. H. Aruga, "Induction of Virus Infections", in E. A. Steinhaus, Ed., *Insect Pathology: An Advanced Treatise,* Vol. I, Academic Press, New York, 1963, p. 499.

14. R. W. Campbell and J. D. Podgwaite, The disease complex of the gypsy moth I. Major components, *J. Invertebr. Pathol.,* **18,** 101 (1971).

15. E. A. Steinhaus and G. A. Marsh, Report of diagnosis of diseased insects 1951–1961, *Hilgardia,* **33,** 349 (1962).

16. R. Lewin, Santa Rosalia was a goat, *Science,* **221,** 636 (1983).

17. R. Lewin, Predators and hurricanes change ecology, *Science,* **221,** 737 (1983).

18. G. C. Varley and G. R. Gradwell, "Population Models for the Winter Moth," in T. R. E. Southwood, Ed., *Insect Abundance, Symp. R. Entomol. Soc. London,* **4,** 132 (1968).

19. R. R. Lejeune and B. Filuk, The effect of water levels on larch sawfly populations, *Can. Entomol.,* **74,** 155 (1947).

20. R. R. Lejeune, W. H. Fell, and D. P. Burbidge, The effect of flooding on development and survival of the larch sawfly *Pristiphora erichsonii* (Tenthredinidae), *Ecology,* **36,** 63 (1955).

21. R. E. Frye, Impact of volcanic ash on pear psylla (Homoptera: Psyllidae) and associated predators, *Environ. Entomol.,* **12,** 222 (1983).

22. C. A. Johansen, J. D. Eves, D. F. Mayer, J. C. Bach, M. E. Nedrow, and C. W. Kious, Effects of ash from Mt. St. Helens on bees, *Melanderia,* **37,** 22 (1981).

23. C. Johansen, Mount Saint Helens blows the season for many Washington beekeepers, *Am. Bee J.,* **120,** 500 (1980).

24. J. F. Howell, Codling moth: the effects of volcanic ash from the eruption of Mount St. Helens on egg, larval, and adult survival, *Melanderia,* **37,** 52 (1981).

25. S. D. Smith, Preliminary report effects of Mt. St. Helens ashfall on lotic Trichoptera, *Melanderia,* **37,** 58 (1981).

26. A. A. Berryman, Overwintering populations of *Scolytus ventralis* (Coleoptera: Scolytidae) reduced by extreme cold temperatures, *Ann. Entomol. Soc. Am.,* **63,** 1194 (1970).

27. D. G. Harcourt, Major factors in survival of the immature stages of *Pieris rapae* (L.), *Can. Entomol.,* **98,** 653 (1966).

28. B. B. Beirne, Effects of precipitation on crop insects, *Can. Entomol.,* **102,** 1360 (1970).

29. L. C. Birch and H. G. Andrewartha, The influence of drought on the survival of eggs of *Austroicetes cruciata,* Sauss. (Orthoptera) in South Australia, *Bull. Entomol. Res.,* **35,** 243 (1944).

30. R. W. Campbell, The role of disease and desiccation in the population dynamics of the gypsy moth *Porthetria dispar* (L.) (Lepidoptera: Lymantriidae), *Can. Entomol.,* **95,** 426 (1963).

31. G. Thomas and F. Wood, Colorado beetle in Channel Islands, *EPPO Bull.,* **10,** 491 (1980).

32. C. G. Johnson, *Migration and Dispersal of Insects by Flight,* Methuen, London, 1969.

33. H. G. Wellington, "Insect Dispersal: A Biometeorological Perspective," in R. L. Rabb and G. G. Kennedy, Eds., *Movement of Highly Mobile Insects,* N. Carolina St. Univ., 1979, p. 104.

34. L. B. N. Hynes, *The Biology of Polluted Waters,* Liverpool Univ. Press, 1960.

35. T. M. Burton, R. M. Stanford, and J. W. Allan, "The Effects of Acidification on Stream Ecosystems," in F. D'Itri, Ed., *Acid Precipitation: Effects on Ecological Systems,* Ann Arbor Sci. Publ., 1982, p. 209.

36. R. J. Hall, J. M. Pratt, and G. E. Likens, Effects of experimental acidification on macroinvertebrate drift diversity in a mountain stream, *Water Air Soil Pollut.,* **18,** 273 (1982).

37. L. R. Clark, A population study of *Phaulacridium vittatum* Sjost. (Acrididae), *Aust. J. Zool.,* **15,** 799 (1967).

38. A. R. Greenway, L. E. Smart, J. Simpson, M. C. Smith, and J. H. Stevenson, Detection of starvation as the cause of death in honeybees, from thoracic glucose levels, *J. Apicult. Res.,* **20,** 180 (1981).

10

VIRAL DISEASES

HUGH F. EVANS

Research and Development Division
Forestry Commission
Farnham, Surrey, England

PHILIP F. ENTWISTLE

Natural Environment Research Council
Institute of Virology
Oxford, England

1. INTRODUCTION

A major task in this chapter is to attempt a synthesis of the steadily increasing knowledge of all groups of insect viruses yet be selective in describing those characteristics relevant to the theme of epizootiology. Initially we shall describe the groups of viruses pathogenic in insects. There has been a number of reviews in recent years, some dealing with insect viruses in general (1–3) and others with particular families (4–9) or groups of families (10,11). In the interests of setting the scene for an essentially ecological text we have also attempted to view viruses in the context of competition with each other and with other pathogens. The biological and ecological integration of viruses with hosts of varying life strategies is investigated, and we approach the inseparable but little studied area of the coevolution of host and virus. These last two ventures must currently be considered somewhat speculative. Their value, however, could be twofold: the speculations could prove to be correct (!), but more usefully the lines of reasoning adopted may contribute to the growth of a sounder and more all-embracing theory of the indivisible topic of host and pathogen ecology.

2. PRIMARY CHARACTERISTICS OF THE VIRUS GROUPS

The taxonomy of insect viruses has long been dominated by the observational approach constrained by the research tools available to early workers. This has resulted in a tendency to associate viruses with the host insect species in which they were discovered, thus ignoring the possibility that a single virus may have different hosts. Recent techniques, however, have provided the opportunities to identify viruses uniquely regardless of the host insect. Thus, although new examples of viruses previously undescribed in certain insect species are being reported frequently, the number of insect virus groups recognized by the International Committee on Taxonomy of Viruses (ICTV) has changed little since the late 1970s.

Table 10.1 summarizes the families or groups of viruses described in the 1982 classification of the ICTV (13) but also includes two newly described groups. This classification follows conventional lines in placing emphasis on type of nucleic acid, structure of the virus particle, and presence or absence of an inclusion body. These are useful descriptive tools for distinguishing the virus groups at a gross level. Further refinement is provided by the biochemical and biophysical characteristics in the context of virus replication. In this way the concepts of an evolutionary link between host and virus can be examined in later sections in the light of essential features not necessarily related to the taxonomy of the viruses.

A number of excellent reviews have been produced in recent years describing the general characteristics of insect viruses and these should be consulted for in-depth information (1–3,14). An outline of the essential properties of the nine virus families or groups named in Table 10.1 will be given below as a general background to the discussion that follows. The schemes of Harrap and Payne (1) have been used as a basis, with new information added where available. It is perhaps surprising that in spite of the advent of accurate techniques for virus taxonomy, the essential data on most of the virus groups have remained unchanged since Harrap and Payne's comprehensive review.

2.1. Baculoviridae (BV)

This family has received most attention, reflecting ease of detection and great potential as microbial control agents (15). The Baculoviridae (BVs) have double stranded DNA (dsDNA) genomes that are covalently closed. Molecular weights range between 50×10^6 and 110×10^6 daltons. Three subgroups are recognized (13) but the occluded viruses in subgroups A and B provide the main subjects of research in this family. However, it will undoubtedly pay to study further the characteristics of the other two subgroups of nonoccluded BVs in terms of their roles in the evolution of virus-host interactions (see Section 6.3).

Table 10.2 summarizes the major features of BVs and polydnaviruses. Although virion dimensions do not vary widely, the envelopment of the nucleocapsids can range from single to multiple with one or two envelopes present (16,

TABLE 10.1
Basic Characteristics of the Virus Groups (1, 11, 12)

Virus Group	Nucleic Acid	Segmentation	Virus Particle Symmetry	Inclusion Body Shape	Cryptogram
Baculoviridae	dsDNA	1	Baciliform	Polyhedral	D/2:80–100/8–15:VorVo/E:I/O
Subgroup A Nuclear polyhedrosis (NPV)					
B Granulosis (GV)				Cigar-shaped capsules	
C *Oryctes* virus NoBV					
D "Calyx" virus					
Reoviridae					
Cytoplasmic polyhedrosis (CPV)	dsRNA	10	Isometric	Polyhedral	R/2:Σ13–Σ18/25–30:50/S:I/O
Poxviridae					
Entomopoxivirinae (EPV)	dsDNA	1	Ovoid or brickshaped	Spheroid	D/2:140–240/5–6:X/*:I/O
Iridoviridae					
Iridescent virus (IV)	dsDNA	1	Isometric	—	D/2:130/15–20:S/S:I/I

Parvoviridae Densovirus (DNV)	ssDNA	1	Isometric	—	D/1:1.5–2.2/19–32:S/S:I/O
Other DNA viruses					
Polydnaviruses	dsDNA	1	Ovoid	—	
Ascoviruses	dsDNA	1	Allantoid	—	
Small RNA viruses					
Picornaviridae	ssRNA	1	Isometric	—	R/1:2.5/30:S/S:I, VI, O, R
Nudaurelia β group	ssRNA	1	Isometric	—	R/1:1.8/9.5–11.5:S/S:I/I
Nodamura virus group	ssRNA	2	Isometric	—	R/1:Σ1.5–Σ1.9/20–28:S/S:I/I
Caliciviruses	ssRNA	1	Isometric	—	
Rhabdoviridae Sigmavirus	ssRNA	(assumed)	Helical	—	*/*:/*Ve/E:I/C

TABLE 10.2

Biophysical Characteristics of Baculoviridae and Polydnaviruses (1, 12)

Subgrp.	I.B. Size	Vrn. Dimn.	Sed. Coef.	Buoy. Dens.
A NPV	0.3–15	$(40 - 140) \times (250 - 400)$	1228–1530	1.2–1.32
B GV	0.3–0.5	$(30 - 60) \times (260 - 360)$	1228–1530	
C Oryctes	—	120–220	1640	1.28
D Calyx	—	40×300		
Polydnaviruses	—	130×150		1.20

N.B. The following contractions are used for column headings in Tables 10.2–10.7, inclusive:

Buoy. Dens.	Buoyant density in CsCl
Dens.	Densisty of nucleic acid
G:C	Percentage G:C
I.B. M.W.	Inclusion body molecular weight $\times 10^{-3}$
I.B. size	Inclusion body size (μm)
I.B. type	Inclusion body type
M.W.	Molecular weight $\times 10^{-3}$
Nuc. Ac.	Nucleic acid
Nuc. Ac. Size	Size of nucleic acid $\times 10^{-6}$
Sed. Coef.	Sedimentation coefficient S_{20}^{w} (S)
Segs. RNA	Number RNA segments
Struct. Prots.	Structural proteins
Subgrp.	Subgroup
Vrn. Dimn.	Virion dimensions (nm)
Vrn. Prots.	Virion proteins
Vrs. grp.	Virus group

	Nuc. Ac.			Struct. Prots.	
			M.W.	Vrn.	Prots.
Size	Dens.	G:C	I.B.	No.	M.W.
58–119	1.701–1.715	37.4–62	25.2–31.4	11–25	12–160
60.4–119	1.6984–1.703	34.8–50	25.0–29.1	12–19	12–160
60–92	1.703	43–44		11–12	8.7–75.7
2.25	1.694			>18	
2–20		29–50		20–30	10–100

17). Baculoviruses in subgroups A (Fig.10.1A, B) and B (Fig,10.1C, D) are characterized by occlusion of virions in a paracrystalline protein matrix, the whole being known as an inclusion body (IB). In BVs the inclusion body is surrounded by a strongly glycosylated membrane. Convention has resulted in the IB protein being named after the type of BV: subgroup A (nuclear polyhedrosis viruses, NPVs) has polyhedral IBs made up of polyhedrin, subgroup B (granulosis viruses) (GV) has a protein called granulin.

The nucleocapsid is similar in all BV groups, whether occluded or not, and has a DNA core surrounded by a protein capsid. The virion is completed by an

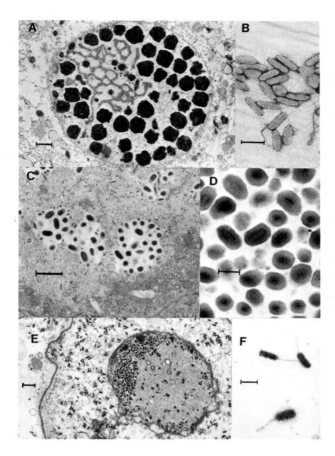

Figure 10.1. (A) Nucleus of *Spodoptera frugiperda* fat body cell infected with nuclear polyhedrosis virus. Many polyhedral inclusion bodies containing virions are present. The netlike virogenic stroma in the center of the nucleus is the site of virion assembly (bar = 1.4 μm). (B) Extracted and purified virions of the singly enveloped nuclear polyhedrosis virus of *Neodiprion sertifer* (bar = 250 nm). (C) Replication of a granulosis virus in the cytoplasm of a cell of *Pieris brassicae* (bar = 1 μm). (D) Purified, sectioned inclusion bodies of *P. brassicae* granulosis virus (bar = 250 nm). (E) Midgut nucleus of adult *Oryctes rhinoceros* infected with a subgroup C baculovirus (bar = 500 nm). (F) Purified virions of *O. rhinoceros* subgroup C baculovirus showing characteristic tail structure (bar = 250 nm).

outer lipoprotein envelope surrounding the nucleocapsid. In NPVs, the envelopes enclose single or multiple nucleocapsids, whereas in GVs and subgroup C BVs (Fig. 10.1 E, F) they contain single and rarely several nucleocapsids. Subgroup D BVs are nonoccluded and the nucleocapsid is surrounded by a single envelope whereas the polydnaviruses previously included in this subgroup possess two envelopes (17).

Precise taxonomic characterization is becoming increasingly dependent on restriction endonuclease (REN) cleavage of nucleotide sequences, thus providing a biochemical fingerprint of the test virus, initially of the DNA core (18). Subsequently, the technique has developed rapidly so that physical maps of the DNA have now been constructed (19). These maps form the initial blueprints for further analysis with techniques including DNA-DNA (20) and blot (21) hybridization. In this way the functions and relationships of particular regions of the genome have been elucidated. For example, Summers et al. (22) demonstrated that recombinants of *Autographa californica* multiply enveloped NPV (AcMNPV) and *Rachiplusia ou* MNPV (RoMNPV) held the coding sequence for polyhedrin at a constant position on the AcMNPV physical map. Hypotheses regarding the particular DNA fragments containing the polyhedrin gene followed, and the exact nucleotide sequences have now been determined for a number of BVs (23). Vlak and Rohrmann (24) reviewed the nature and role of polyhedrin including, on the basis of conservation of nucleotides, an evaluation of the evolution of certain BVs (see Section 6.3).

Less is known of the biology or biochemical nature of BV subgroups C and D. Recent work has indicated that the genome is made up of a series of covalently closed circles (25). Painstaking analysis of the DNA, including Southern blot hybridization and REN, confirmed that the genome included unique DNA sequences.

The size of the DNA genome of BVs is sufficient to code for a maximum of 100 proteins. Singh et al. (26), using two dimensional gel electrophoresis, demonstrated the presence of more than 80 polypeptides in AcMNPV and *Lymantria dispar* MNPV. It is likely that no more than 30 genes are involved in coding for structural components, leaving a large coding capacity for replication products (27) (see Section 3.2.2).

2.2. Nonoccluded DNA Viruses

Two groups of DNA containing viruses have been described recently and, although not yet recognized by the ICTV, have been ascribed names by the research workers involved

2.2.1. POLYDNAVIRUSES.   Polydnavirus is the name proposed by Stoltz et al. (12) for the nonoccluded viruses found in the ovaries of ichneumonid wasps. These viruses have many features in common with the nonoccluded BVs. The genome is packaged as a ds circular DNA core within a quasicylindrical nucleocapsid surrounded, unusually, by two envelopes. The few data on the biophysical characteristics of this group have been included, for convenience, in Table 10.2. Virus isolates are each known to contain an unique configuration of DNAs, for instance, no homology between *Hyposoter exiguae* and *Campoletis sonorensis* DNAs has been demonstrated (12). However, there was significant cross-reaction between antisera from *H. exiguae* and other ichneumonid genera in immunoblotting tests (28).

2.2.2. Ascoviruses. The second group of DNA viruses having many unusual features has been described by Federici (11). He suggested the name Ascovirus on the basis of virion-filled vesicles formed on disruption of the infected nucleus. Early studies indicate that this virus group possesses large enveloped virions (130 × 400 nm) which are allantoid in shape. Nucleic acid is linear dsDNA with molecular weight around 103×10^6 daltons.

2.3. Cytoplasmic Polyhedrosis Viruses (Reoviridae)

At the gross morphological level, the cytoplasmic polyhedrosis viruses (CPVs) (Fig. 10.2 B, C) bear a close resemblance to NPVs as evidenced by occlusion of virions in proteinaceous IBs (5). Indeed the biochemical composition of the polyhedral protein is remarkably similar to that of polyhedrin of NPVs, but the polyhedra themselves do not possess a polysaccharide membrane (29). In all other respects CPVs are clearly distinguishable from other insect virus groups.

Table 10.3 illustrates the major characteristics of the group. The double stranded RNA (dsRNA) and the presence of ten segments on the genome (30) place CPVs in close taxonomic association with the Reoviridae. The essential differences between the two groups lie in the presence of CPV virions embedded in IBs and the absence of a double capsid layer around these particles (13). The virions are isometric in shape and up to 10,000 may be occluded at random within IBs without disrupting the regular paracrystalline lattice (31). Many CPV virions remain nonoccluded even at a late stage of infection with up to 88% of *Malacosoma disstria* CPV virions remaining free within the cell cytoplasm (32).

The virions are composed of an electron-dense core surrounded by an outer envelope (33), but there is no good evidence for an outer capsid layer similar to those of the true reoviruses. Payne and Mertens (5) speculate that the polyhedrin of CPV serves the same purpose as in BVs. A further distinguishing feature of CPV virions is the presence of 12 spikes on the icosahedral particles (34).

Buoyant densities of CPV virions in CsCl are higher than those for true reoviruses (Table 10.3). Information on structural proteins is scant, although five polypeptides have been identified in *Bombyx mori* CPV (35). These are similar to reovirus cores in having at least two polypeptides with molecular weights over 100,000.

Most studies on CPVs have concentrated on the RNA genome which, as mentioned above, was found to be 10 segmented. This has formed the basis of a provisional classification of the CPVs (36). Electrophoretic profiles of RNA vary consistently between virus isolates reflecting the coding of each RNA segment for a specific protein. Payne et al. (37) identified the 12 virus types summarized in Table 10.3 with polyacrylamide gel electrophoresis. While reviewing the RNA profiles from 68 virus isolates from 49 species of Lepidoptera, Payne and Mertens (5) pointed out the discovery of five distinct types from a single species (*Spodoptera exempta*), thus graphically emphasizing the need for refined techniques in virus identification. CPV types 2, 3, 6, and 11 were the

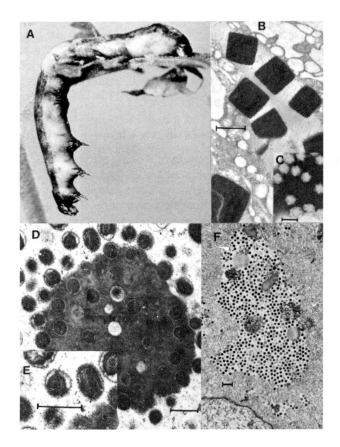

Figure 10.2. (A) Nuclear polyhedrosis virus-infected larva of *Spodoptera littoralis* in characteristic death posture. (B) Cytoplasmic polyhedrosis virus infection of a midgut cell of *Bombyx mori*. Cuboidal inclusion bodies and "spherical" virions are evident (bar = 1 μm). (C) Isolated and purified virions of *B. mori* cytoplasmic polyhedrosis virus showing typical "spikes" (bar = 100 nm). (D) Occlusion of virions in progress in an entomopoxvirus of *Melolontha melolontha* (bar = 500 nm). (E) Entomopoxvirus of *M. melolontha* showing the "brick"-like virions (bar = 500 nm). (F) Iridovirus (IV-22) infection in larval *Simulium* sp. (bar = 500 nm).

most frequently reported in their review, but this does not necessarily represent the true distribution in nature.

Emphasis has recently shifted to the molecular aspects of CPVs, but the limited techniques available, particularly the general paucity of cell culture systems, have restricted progress in comparison with similar work on BVs. The basic knowledge is that CPV dsRNA is not intrinsically infective (38). An RNA polymerase is necessary for transcription, and recent studies have concentrated on this enzyme. A detailed synopsis of the background and progress in the study of virus-associated enzymes can be found in Payne and Mertens (5). In brief

TABLE 10.3
Biophysical Characteristics of Cytoplasmic Polyhedrosis Viruses (Reoviridae) (1, 5)

Type	I.B. size	Vrn. Dimn.	Sed. Coef.	Buoy. Dens.	Nuc. Ac.			Struct. Prot.		
					Size	No. RNA Segs.	G:C	M.W. I.B.	Vrn. Prots.	
									No.	M.W.
1	0.2–10	68–69	371–440	1.37–1.44	14.63	10	40–43	27	5	31–144
2	"	"	"	"	14.36	"	36–37	30	4	32–129
3	"	"	"	"	15.09	"	"	28	6	36–146
4	"	"	"	"	15.53	"		25	3	40–143
5	"	"	"	"	14.82	"		26	5	28–142
6	"	"	"	"	15.32	"				
7	"	"	"	"	14.44	"				
8	"	"	"	"	15.23	"				
9	"	"	"	"	13.62	"				
10	"	"	"	"	15.58	"				
11	"	"	"	"	14.38	"				
12	"	"	"	"	14.73	"				
Unclassified										

summary, the enzyme activity acts to copy a single strand of the viral genome, but no modification of virions by external agencies such as heatshock, chelating agents, or chymotrypsin incubation is required for activation (39). The absence of an outer-capsid layer may indeed render such activation mechanisms unnecessary (40). Transcription of the genome segments by the polymerase results in monocistronic messenger RNAs that in turn are translated to virus-specific proteins. Miura (41) showed that ssRNAs resulting from polymerase activity (the plus strands) probably act as templates for the production of complementary minus strands to form the duplex genome RNAs. Other enzymes act in sequence to produce ss capped RNA which then acts as mRNA during transcription (42).

2.4. Entomopoxviruses

The entomopoxviruses (EPVs) (Fig. 10.2 D, E) are in the family Poxviridae (13), members of which include human and animal pathogens. At present the similarities remain at a structural and biochemical level, and no cross-infectivity to vertebrates has been demonstrated.

The EPVs are the least studied of the three groups of occluded insect viruses. They possess brick-shaped virions having distinctive forms depending on the virus species. Thus the so-called mulberry structure is commonly observed resulting from folding of the outer membrane. Virions also may exhibit lateral bodies and ropelike structures associated with the insect orders infected (Table 10.4). In the Orthoptera and Lepidoptera the virions are relatively small (350 × 250 nm) and possess a cylindrical core and two lateral bodies (43). EPV virions from Diptera are smaller and are characterized by a biconcave core and two lateral bodies (44). By contrast EPV virions in Coleoptera are larger (400 × 250 nm) and possess an offcenter core with a single lateral body in the concavity thus formed (45).

Occlusion of virus particles occurs within the paracrystalline protein matrix (spheroidin) of the inclusion body or spheroid. There also may be spindle-shaped or fusiform inclusions which appear to be free of virions (46). Like the IB protein of occluded BVs and CPVs, spheroidin is composed of a single polypeptide, although the molecular weight of $102\text{-}110 \times 10^3$ (47,48) is up to three times greater. There is no outer polysaccharide envelope to the intact spheroid. The virus particles are arranged randomly within the protein matrix, and the spheroid itself may be up to 24 μm on its longest edge (EPV of *Melolontha melolontha*) (49). Spindles, which are also paracrystalline protein, are commonly produced during EPV infection of Lepidoptera and Coleoptera, ranging in size from 0.5 to 25 μm (6). It is not clear what role the spindles play in the replicative cycle of EPV, particularly considering that they are not related antigenically or chemically to spheroidin (50).

Biophysical and biochemical studies of the virions have been few. EPV genomes are composed of ds DNA and make up about 5% of the virion (51). There are, however, differences in genome size depending on the insect order

TABLE 10.4
Biophysical Characteristics of Entomopoxviridae (1, 6)

Egs	I.B. Size	Spindle Present	Vrn. Dimn.	Buoy. Dens.	Nuc. Ac.			Struct. Prots.		
					Size	Dens.	G:C	M.W. I.B.	Vrn. Prots. No.	M.W.
Coleoptera, e.g., *Melolontha*	10–24	Yes	450 × 250	1.26–1.31	200.4	1.676	16.3	102–110	24–40	12–250
Lepidoptera, e.g., *Amsacta moorei*	1–4	Yes	350 × 250	"	134.7	1.678	18.5			
Diptera, e.g., *Chironomus luridus*	4–7	No	330 × 230 × 110	"	199.2	1.681	20.7–21.7			
Orthoptera, e.g., *Melanoplus sanguinipes*	2–11	No	320 × 350	"	123.2	1.678	18.6		39	

from which the EPVs were isolated and also on the technique used to determine the molecular weight. Thus the genome of *Choristoneura biennis* EPV (Lepidoptera) was approximately 137×10^6 daltons (52), and similar electron microscope measurements were found for EPVs isolated from other moths by Langridge and Roberts (53). Arif (6) later showed, using REN analysis, that *Choristoneura* spp. EPV genome was probably larger than this, 150×10^6 daltons. DNA from genomes of EPVs in Diptera and Coleoptera measured with an electron microscope were larger (200×10^6 daltons) than for Lepidoptera.

The few examples of virus structural proteins that have been studied with polyacrylamide gel electrophoresis show a range of polypeptides from 24 to 39 for different species (48). *C. biennis* EPV, however, has 40 polypeptides with a total molecular weight of 2.9×10^6 (54). These were assumed to represent 38% of the coding capacity of the genome but, as pointed out above, this figure may be lower (possibly 35%) if the mw determined by REN is the correct value. The functions of the various polypeptides are largely unknown.

A minimum of four enzymes are associated with the virions (nucleotide phosphohydrolase, acidic deoxyribonuclease, neutral deoxyribonuclease (55), and DNA-dependent RNA polymerase (56). Apart from demonstration of the presence of these enzymes and their presumed activities, nothing is known of their actual modes of action during replication.

2.5. Iridoviridae

This family of viruses was named on the basis of the characteristic iridescent green, blue, or purple seen in well-infected individuals. Although examples of "iridescent" viruses not exhibiting this sign have since been found, the generic name remains. The ICTV have included the iridescent viruses of arthropods in the same family, Iridoviridae, as a number of similar viruses affecting mammals, fish, and amphibians. To distinguish the arthropod group they suggested the subfamily name Invertebrate Icosahedral Cytoplasmic Deoxyribovirus (IICDV), although this has not yet been formally adopted (13).

An interim scheme for classification of the iridoviruses (IVs) was proposed in 1970 by Tinsley and Kelly (57). They noted that many of the virus "species" could be grouped by size and serological relationships into particular types, each of which was assigned a designation IV and type number. This scheme is now in use and a total of 32 types have been ratified, with a number of other possible examples awaiting designation (7).

Unlike the families already discussed, IVs do not possess inclusion bodies. IVs can be roughly divided into two size groups, those around 130 nm and those around 180 nm in diameter. The virion is packaged as an electron-dense DNA rich core surrounded by a single membrane, itself enveloped by an icosahedral capsid composed of up to 1562 subunits depending on IV type (58, 59).

Iridescence is produced by the physical conformation of virions into paracrystalline arrays (Fig. 10.2F). These interfere with the passage of light and reflect a particular wavelength depending on the distance between the particles.

Klug et al. (60) demonstrated that the array is made up of a face-centered cube with a spacing of 250 nm between individual particles.

Characteristic properties of the more intensively studied IVs are included in Table 10.5. The dsDNA core is thought to be a single linearly packaged duplex, although Wagner and Paschke (61) suggested that the R form of Type IV3 possessed a pair of duplex molecules. The size of nucleic acid varies between the IV types depending on the methods. For example, sedimentation gave consistently lower molecular weights than reassociation kinetics for IV2 and IV6 DNAs (62, 63). Base composition placed IV1, 2, and 6 in a group with G:C content around 30% whereas IV9 and IV18 had 40%. This grouping disguises the fact that IV1, 2, and 6 are distinct from each other according to homology and serological methods, whereas IV9, 18, and 3 show strong homology (61, 63).

Numbers of structural proteins vary from 9 to 25 depending on type. Types IV2, 6, 21, 22, 23, 25, 28, and 29 were separated by polypeptide profile according to SDS-PAGE (64, 65).

Serological techniques also have been employed in an attempt to separate various IV types. Harrap and Payne (1) summarized such research and concluded that IV6 and 24 were not antigenically related in any way, that IV1, 2, 9, 10, 16, 18, 21, 22, 23, 24, 25, and 28 were closely related, and that IV29 is linked to some of them.

2.6. Parvoviridae

The type genus of this family within the arthropods is *Densovirus*, also known as Densonucleosis Virus (DNV). Placing DNV in the Parvoviridae therefore relates this group to the vertebrate parvoviruses and dependoviruses, although neither of these have been recorded from arthropods (13).

DNVs are characterized by having a ssDNA core packaged in a virion with isometric symmetry. The virions range in size from 19 to 24 nm (Table 10.6) but most values are around 20 nm (1). Tijssen and Kurstak (66) showed that the capsid of DNV is icosahedral with 12 capsomeres of up to 3.5 nm arranged around a central area. More recently Nakagaki and Kawase (67), using electron micrograph rotation techniques, confirmed these results and suggested that *B. mori* DNV possessed virions with 60 molecules of molecular weight 5×10^{-3} forming 12 pentameric capsomeres arranged as an icosahedron.

The DNA is linear and is present as two complementary strands in different particles that are separately encapsidated (68). It has proved difficult to separate the ssDNA components; data on molecular weights of ss and dsDNA are both quoted in Table 10.5. Indeed the size of the genome is probably equivalent to the size of a total dsDNA formed on extraction from the virions (69). Kelly and Bud (70) demonstrated that the annealed strands of complementary ssDNA gave rise to dsDNA in linear or circular monomers or concatamers. Nakagaki and Kawase (71) confirmed that the molecular weight of the ssDNA within the virion was 1.7×10^6 daltons. *B. mori* DNV virus particles have a G:C content of

TABLE 10.5

Biophysical Characteristics of Iridoviridae (1, 7)

Type	Vnr. Dimn	Buoy. Dens.	Sed. Coef.	Nuc. Ac.			Struct. Prots.		
				Size	Dens.	G:C	No.	M.W. Range	Major polypepts.
1 (TIV) *Tipula paludosa*	130	1.33	2200	126	1.69	31.7	28	17.5–300	17.5, 41, 42, 85, 90
2 (SIV) *Sericesthis pruinosa*	130	1.33	2200	147	1.69	30.6	22	11.7–138.4	55
3 (MIV) *Aedes taeniorhynchus*	180	1.32	4440	243.7	1.7135	53.9	9	15.5–98	30, 54.5
6 (CIV) *Chilo suppresalis*	130	—	3300	152	1.689	29.6	19	10–113	65

TABLE 10.6
Biophysical Characteristics of Insect *Parvoviridae*: The Densoviruses (1, 8)

Type	Vrn. Dimn.	Buoy. Dens.	Sed. Coef.	Nuc. Ac. Size ssDNA	dsDNA	Dens.	G:C	Struct. Prots. No.	M.W.
DNV 1									
Galleria mellonella	19–24	1.40, 1.44	117–119	1.99	4.08	1.6937	35.1	4	42.6–107.3
DNV 2									
Junonia coenia	20–22		117–119	1.97	4.01	1.6936	37.0	4	41.9–109.6
DNV 3									
Agraulis vanillae	20		117–119	1.82				4	43–110
Bombyx mori	20–22	1.40, 1.45	117–119	1.7	3.4		42.7	4	50–77

43% and a DNA content of about 28%, (72), whereas Longworth et al. (73) demonstrated a DNA content of *Galleria mellonella* DNV at around 37%, perhaps reflecting differences in the putative coding capacity between different DNVs.

Genetic organization and mechanisms of transcription of DNVs are poorly understood. By analogy with the true parvoviruses and dependoviruses it is probable that viral DNA replication is controlled in the left third of the genome whereas the right two-thirds controls the production of viral structural proteins late in infection. However, DNV codes for four structural polypeptides (VP1 – VP4) compared with three for the other two parvovirus groups, the fourth polypeptide being the major component. The general scheme is assumed to be conversion of ssDNA to a linear duplex replicative form (RF) followed by viral capsid protein production and eventually progeny ssDNA packaged into preformed capsids (74). Unfortunately, corroborative evidence for this mechanism in DNV replication is lacking.

2.7. Small RNA Viruses

This generalized grouping covers a wide variety of RNA viruses that are less than 40 nm in size (1, 9, 10, 75, 76). Morphological features for distinguishing these viruses are generally lacking, and thus biophysical and biochemical characterizations are essential.

The basic features of the known small RNA viruses are summarized in Table 10.7. However, of the six recognized groups only three have been studied in sufficient detail to provide an appraisal of the mechanisms of viral synthesis.

2.7.1. PICORNAVIRUSES. Type viruses of this family are Cricket paralysis (CrPV), *Drosophila* C (DCV), and *Gonometa podocarpi* viruses, although it is doubtful if the latter is available in viable form for laboratory use (76). Picornaviruses have diameters between 27 and 32 nm with the majority measuring 30 nm. DCV and CrPV have been studied in detail and possess many features in common with mammalian picornaviruses, such as a single molecule of plus ssRNA with molecular weights between 2.5 and 3.0×10^6. The presence of three major (mw around 30×10^{-3}) and up to two minor (mw around 10×10^{-3}) structural polypeptides (1) plus resistance to acid confirms the identification of Picornaviridae.

Pullin et al. (77) and Clewley et al. (78) distinguished CrPV and different isolates of DCV by oligonucleotide mapping. However, there is little information about the structure of the RNA genome or the mechanisms of replication. Insect picornaviruses appear to replicate through dsRNA replicative intermediates. These code for poly-proteins of high molecular weight that are, in turn, proteolytically cleaved to give viral proteins (79). Based on present evidence, the replicative schemes of insect picornaviruses mirror those of their mammalian counterparts (13, 77, 79).

TABLE 10.7

Biophysical Properties of Small RNA Viruses of Insects (1, 75, 76)

Vrs. Grp.	Vrn. Dimn.	Buoy. Dens.	Sed. Coef.	Segs. RNA	Nuc. Ac. Size	Nuc. Ac. G:C	Struct. Prots. No.	Struct. Prots. M.W.
Picornaviridae								
Cricket paralysis virus (CrPV)	27	1.34	167	1	2.5–2.8	38.7	3 major 1 minor	43, 35, 34, 30
Drosophila C virus (DCV)	27	1.34	153	1	2.5–3.0		3 major 2 minor	31, 30, 28, 37, 8.5
Nudaurelia β virus	35	1.298	210	1	1.8–2.1	53	1	60–70
Nodamura virus	29	1.34	135	2	1.15–0.46	55,51	1 major 2 minor	40, 39, 43
Calicivirus:								
Cupped	38	1.32	185	1	2.5		2 major	55, 29, 70[a]
Smooth	28	variable	165	1	2.5		2 minor	23, 16, 29[a]
Rhabdoviridae								
Sigma virus	200 × 75				3.0		5	

[a]By rapid purification.

2.7.2. THE NUDAURELIA β GROUP OF VIRUSES.   The type species of this group is the ssRNA virus isolated from *Nudaurelia cytherea capensis*. Virions are 35 to 38 nm in diameter (Table 10.7) and are free of lipids. Nucleic acid is present as a single segment with molecular weight around 1.8×10^6 and G:C content of 53% (1). This accounts for approximately 12% of the total weight of the virion (80). A single structural polypeptide with molecular weight of 60×10^{-3} is present (9). The virion is packaged as an icosahedron which Finch et al. (81) showed was constructed of 240 repeats of the structural protein.

All attempts at growing members of this virus group in cell culture have failed, and the limited data on RNA synthesis have been obtained from cell-free systems using rabbit reticulocyte lysates (82). The demonstration of a product protein of molecular weight exceeding 100,000 assumed that a polymerase was present corresponding in size to this protein. Experiments of this type, involving pulse chasing of protein production, implied that capsid proteins were coded at the 5'-end of the genome while the 3'-end coded for polymerase (82). The only other feature so far demonstrated is the absence of a polyadenylated tract in the RNA (83) which was assumed by Moore et al. (76) to be a likely reason for the lack of *in vitro* replication of these viruses.

2.7.3. THE NODAMURA VIRUS GROUP.   Viruses in the family Nodaviridae (13) are characterized by the presence of ss RNA packaged as two segments of RNA within the same particle (84). Two examples have been studied extensively, the type Nodamura virus isolated from *Culex tritaeniorhynchus* mosquitoes in Japan (85) and black beetle virus (BBV) from *Heteronychus arator* in New Zealand (86). Unusually, Nodamura virus replicates in suckling mice and BHK cells in culture. These systems demonstrated the presence of two segments of RNA with molecular weights of 1.0×10^6 and 0.5×10^6 (87). Later studies showed that both segments were present in a single particle and that no polyadenylated tract was present (84).

Translation of BBV RNA has been studied in insect cell systems. Friesen and Rueckert (88) showed that a potential polymerase with molecular weight of 110,000, a major structural protein with related precursor, and two other low molecular weight proteins were present. It is probable that a protein of 407 amino acids demonstrated by Dasgupta et al. (89) is the precursor for the major structural polypeptide. The two RNAs in the virus particle code for polymerase (the larger RNA) and a capsid precursor (the smaller RNA) (88). Crump el al. (90), however, were unable to demonstrate sub-genomic RNA fragments thought by Friesen and Rueckert (91) to code for an open reading frame of 72 amino acids. Instead they showed that virion RNA was present as dsRNA replicative forms, the larger of which was a putative RNA-dependent RNA polymerase. Thus, although the availability of permissive insect cell lines for BBV has provided the most comprehensive information of small RNA virus translation, much is still to be learned of the full cycle of events.

2.7.4. CALICIVIRIDAE. An example of a likely calicivirus was found by Kellen and Hoffman (92) in the navel orangeworm *Amyelois transitella*. They demonstrated the presence of two types of virus particle, one of which was 38 nm in diameter and possessed the cuplike structure typical of caliciviruses. This had a single major capsid protein of molecular weight 78×10^{-3}. The cupped virus particles gave rise, by proteolytic cleavage, to smooth particles of 28 nm diameter possessing a capsid protein with molecular weight of 29×10^{-3} (75).

The genome contains ssRNA with a molecular weight of approximately 2.5 $\times 10^6$ (75). Although nothing is known of the replicative cycle, it is presumed that a VPG is present as in other caliciviruses (93) and that subgenomic RNA is a precursor of the single major capsid protein.

2.7.5. RHABDOVIRIDAE. The best known of the probable insect rhabdoviruses is sigma virus of the fruit fly *Drosophila melanogaster*. This has bullet-shaped virus particles measuring 200 nm long and 75 nm in diameter with associated surface spikes around 8 nm long (10). Nothing is known of replicative events, but some parallels with those of rabies virus have been drawn. In this respect there may be five structural proteins enclosing a single molecule of ssRNA having molecular weight of 3×10^6. Moore (10) hypothesized that, like rabies, the negative strand RNA is translated through a polymerase activity to messengers for structural proteins, some of which are involved in transcription. This virus is apparently symptomless apart from the unusual feature that infected adults are lethally sensitive to carbon dioxide.

2.7.6. OTHER SMALL RNA VIRUSES. The remaining small RNA viruses are unclassifed. Basic characteristics of some of these viruses are given in Table 10.7. They have been reviewed by Harrap and Payne (1) and Moore and Tinsley (9), and they will only be referred to when particular characteristics of their pathologies or replicative strategies add to the discussion on virus-host evolution.

3. THE MECHANISMS OF INFECTION

In the discussion above we have concentrated on the differences among virus groups so that, by using a range of methods, each can be identified uniquely to a specific level. This has necessitated the use of refined biochemical techniques to provide the fine details of characterization. In considering the mechanisms of viral infection we return initially to the gross morphological level by examining the common pathways of virus entry into the host insect. Discussion will then involve details of infection at the cellular and organ levels.

The primary route of virus ingress to the host is via the alimentary tract during the course of feeding (94, 95). This is more true of the larval than of the adult stages, and obviously any infection of the egg or pupal stages must result from some other route. Discussion will therefore concentrate on the adult and, more particularly, the larval stages of insects.

Examination of any text on the structure of insect alimentary tract (e.g., 96) reveals that it is a relatively simple structure. After ingestion, the food moves directly into the foregut, which, being ectodermal in origin, and covered with a cuticle, has no innate digestive function and provides a barrier to viral entry identical to the external epidermis (97). The hindgut is similarly an ectodermal invagination with cuticle and appears to play no part in the infection process. The midgut therefore is the only site available for entry of virus. Two characteristics of the midgut influence the passage of virus to host cells.

The first characteristic is the highly variable midgut pH. Table 10.8 gives a range of values for different insect orders and feeding specializations. In general there is a dichotomy between predatory and phytophagous insects such that the latter tend to have alkaline gut pHs often exceeding 9.0. Even among the plant-feeding insects there is a considerable range of observed values. Thus for example, Berenbaum (98) analyzed published midgut pH values for a large number of larval Lepidoptera, and showed a significant difference in pH values between insects feeding on plants high in tannin (mean pH 8.76) and those feeding on plants lacking tannin (mean pH 8.25).

The alkaline midgut results in dissolution of the polyhedral protein. Once in the midgut, virus degradation begins and it is therefore a matter of time before the virions in the lumen are destroyed (see Section 3.1). Effective passage of virus to gut cells is therefore a dynamic balance between dissolution in the alkaline gut contents and penetration of the gut epithelial cells. This generalization applies regardless of the virus group being considered.

The second characteristic of the midgut that has potential influence on virus entry is the presence of a peritrophic membrane (PM). This is a detached sheath secreted by the cells of the midgut and forms a lining throughout its lumen (96). There is some controversy as to the role of the PM but it is generally assumed to act as a barrier protecting the delicate gut cells against sharp items of food and other gut contents. In support of this the PM is absent or poorly formed in fluid-feeding insects (Hemiptera, most adult Lepidoptera, and some fluid-feeding Diptera) and completely absent in predatory Coleoptera (96). Richards and Richards (100) pointed out that as more information is gained, so the link between phylogeny or nutrition and the presence of the PM becomes more tenuous.

Two types of PM are recognized (96). In type I the PM is formed as an amorphous sheath composed of a concentric arrangement of microfibrils arranged as loose lamellae. This is produced directly from the surface of midgut cells as a secretion that condenses into a fibrous membrane, sometimes forming several layers. Type II PM is formed by viscous secretion from a group of cells at the anterior end of the midgut and differs from Type I in that it is a tube made up of cells arranged in a single layer and is of uniform circumference throughout its length. Type I PM is associated with most insect orders whereas Type II PM is typically found in larval and adult Diptera, the earwig, and possibly termites (96).

TABLE 10.8

Midgut pH Values in Relation to Insect Feeding Habits (96, 98, 99)

Feeding Habit	Example	Order	pH Range in Midgut	Comment
Predatory/saprophagous	*Lucilia*	Diptera	3.0–8.5	Ammonia production at posterior end of midgut
Grass	*Acheta domestica*	Orthoptera	4.3–7.2	Bacterial fermentation decreases pH
Scavengers	*Blatella*	Dictyoptera	4.4–6.3	Varies with diet, higher after protein digestion
Stored product	*Tribolium*	Coleoptera	5.2–6.0	
Flower	*Apis*	Hymenoptera	5.6–6.3 (6.8 in adult)	
Predatory	*Coccinella septempunctata*	Coleoptera	5.2–5.6	Predatory insects generally weakly acid pH
Phytophagous	—	Coleoptera	7.5–10.0	Plant feeders generally strongly alkaline pH but all lepidoptera are in this category
	Trichoplusia ni Galleria mellonella Tineola	Lepidoptera	7.5–10.0	
	—	Trichoptera	7.5–10.0	

Most information is available on Type II PM found in bloodsucking Diptera, and it is unfortunate that little work has been carried out on the role of Type I PM in Lepidoptera and Hymenoptera, the two orders for which most data on virus infection have been published. However, the data on chemical composition of the PM appear to apply to both types. Chitin is universally present although at a lower proportion (i.e., 13% in *B. mori* pupae) than in normal cuticle (100). Protein is the major component, ranging from 21 to 47%, while various mucopolysaccharides also have been reported (100).

The significance of the PM lies in its potential as a barrier to passage of virions from the gut lumen to the epithelial cells lining the gut. Permeability in its more general sense has been studied by various means. Passage of dyes takes place only if they are made up of particles that are not colloidal, for example, Congo red, thus implying an effective barrier to large particles (96). In a more detailed study Schildmacher (101) demonstrated that particles of colloidal gold greater than 20 nm in diameter would not pass through the PM of mosquito larvae whereas those of 2 to 4 nm in diameter pass more readily. Obviously for infection to take place virions must pass through the PM.

3.1 Primary Events After *per os* Ingestion

The previous discussion set the scene in describing the environment that ingested virus encounters once it enters the midgut. Two processes, occlusion body dissolution followed by passage of virus through the PM, are therefore common to all groups of occluded viruses; the latter step also applies to nonoccluded viruses.

Depending on pH there will be a gradual or rapid dissolution in the alkaline gut fluids. Enzymic action may also be involved but the predominant feature is the alkalinity of the gut. The proteinaceous IBs enable the occluded viruses to resist the gut fluids longer than nonoccluded viruses. However, the alkaline dissolution of IBs is also an essential step, for it is only the virions so released that are able to enter the midgut cells. Granados and Lawler (102) provided a comprehensive account of the early steps in polyhedral dissolution and passage of *AcMNPV* in *Trichoplusia ni* larvae. They demonstrated that polyhedra were dissolved to release enveloped nucleocapsids (ENC) very rapidly (within 15 minutes). Within two hours many ENCs were seen associated with the microvilli of midgut epithelial cells. Degradation of ENCs to naked nucleocapsids was rare and unusually AcMNPV ENCs were relatively resistant to the alkaline gut (pH 10.4) of *T. ni* (102). In contrast Aizawa (103) demonstrated a loss of 60% of released virions in *B. mori* NPV.

It is also conceivable that the PM plays an important role in whether virions in the midgut are degraded. Clearly if virions are unable to pass through the PM, or do so very slowly, they will be subject to the alkaline gut fluids for a longer period, hence increasing the probability of significant degradation. Granados and Lawler (102) noted in their study of gut events in AcMNPV infection of *T. ni* that ENCs "seemed to pass easily through the loose fibrous peritrophic membrane." It is possible that such rapid passage across the PM allows a relatively greater proportion of available inoculum to enter midgut cells before degradation takes place. Circumstantial evidence for partial degradation with time comes from Granados and Lawler's data for in vitro growth of midgut juice-treated, occluded virions in *T. ni* cell cultures. The titer of 3.8×10^7 IBs incubated in midgut juice dropped from 3.9×10^4 pfu/ml to 1.6×10^4 pfu/ml after 15 minute and 30 minute exposure, respectively, a drop of 59%. The PM of *B. mori* is thought to be intermediate between the extreme Type I and Type II examples, being formed by cells at the anterior end as well as by secretions of sheets from the remainder of the midgut cells (96). By analogy with the relatively impervious Type II PMs it is thus possible that the degradation of *B. mori* NPV virions noted by Aizawa (103) was a partial result of greater retention within the PM. Further research into the time course of virion transfer across the PM is clearly necessary before firm conclusions can be reached.

It appears that all occluded viruses (BVs, CPVs, and EPVs) are subject to the same process in the midgut lumen. Indeed the time course of events appears to be similar in all cases such that BV (102), CPV (104), and EPV IBs (105) were shown to dissolve within two hours.

Nonoccluded viruses, on the other hand, are more susceptible to degradation by gut juices. This is not surprising in view of the lack of a protein coat and the consequent direct action of digestive juices as soon as the virions enter the midgut. Paschke and Summers (106) reviewed early events in gut infection of a number of iridescent viruses in comparison with BVs. Stoltz and Summers (107) showed that Type IV3 mosquito iridescent virus (MIV) was almost completely degraded in the midgut of mosquitoes. Campbell (unpublished) in Paschke and Summers (108) demonstrated a range of responses such that degradation increased as MIV was moved along the gut. Virus at the anterior end of the midgut remained intact. The midgut fluids had a pH range of 6.6 to 7.1 which would lead to the expectation of minimal degradation since purified MIV virions were shown to be stable at pH 3 to pH 8 (109). No evidence was found for any other factor that could have produced the observed degradation independently of gut pH. Campbell also studied the PM and concluded that the structure was such (Type II) that it was unlikely that an MIV virion could pass through it by diffusion. This, in combination with the probable extensive degradation of the virion with time, may explain the low infectivity regardless of dosage of MIV when ingested.

Further evidence of the probable dual role of digestive juice and PM in determining the number of virions reaching gut cells to induce primary infection is provided by the so-called gut barrier. As Tinsley (97), Kelly (27), and others have pointed out, differences in susceptibility of many orders of magnitude are frequently observed when virions are injected directly into the hemocoel rather than introduced via the alimentary tract. Tinsley (97) cited many examples of induced infection in normally nonsusceptible hosts when virions were injected or the gut damaged. This was most marked for the arboviruses where insects could become vectors by puncturing the gut after administration of a normally noninfective virus, or by injection (110). Equally, *Nodamura* virus can replicate in hosts that are not susceptible *per os* if the virions are injected (10). In the case of occluded virus the enhanced susceptibility after injection occurs only if the IBs are dissolved to release free virions (BVs, 27; PVs, 111; EPVs, 112). Of course, it is not possible to distinguish the generalist barriers of gut juice and PM from the more specific barriers of the gut epithelial cells themselves. The mechanisms of viral entry to midgut cells and hence their possible role in defining specificity will be discussed in the next section.

A final topic relating to the PM is its changing status with larval age. Although the PM is not a permanent structure and is either produced continuously as a single tube (Type II) or discontinuously as a series of sheets (Type I), there are also differences in the thickness and permeability with larval age. For instance, Henson (113) showed that the Type I PM of first instar *Aglais (Vanessa) urticae* larvae was a very delicate structure difficult to remove, whereas it was much more substantial in older larvae. Likewise, the Type II PM of mosquitoes changes from a thin structure free of fibrous mats in the first instar to a thick, fibrous tube in third and fourth instar larvae (114). It is probable that the change to a more substantial PM with increasing age is universal in those insects possessing this structure. It is a distinct possibility, unfortunately not backed by suitable evidence, that the well-known phenomenon of decreasing susceptibility to *per os* infection with increasing age (115) is partially linked to a thickening PM barrier in certain insects. This is likely to be most marked for such viruses as MIV affecting mosquitos which have the Type II PM. This possible phenomenon would undoubtably repay definitive studies.

3.2. The Infection Process and General Pathology

3.2.1. INITIAL INFECTION IN MIDGUT CELLS.   Two possible processes are involved in the entry of virions to the cytoplasm of gut cells. Viropexis or entry by pinocytosis, the normal route of entry for vertebrate viruses, occurs in some cases while fusion of viral and cell membranes followed by passage of naked nucleocapsids into the cell cytoplasm is the alternative hypothesis (116). Evidence is good for fusion of enveloped nucleocapsids of BVs and EPVs to the membrane of gut microvilli (6, 102, 116). Viropexis as a means of viral passage was demonstrated for CPVs (5, 117) and Iridoviridae (118, 119) but no reliable data are available for the remaining virus groups. However, by analogy with related vertebrate viruses, viropexis is the likeliest route for parvoviruses and the various small RNA viruses.

Whatever the active process, the end result is the entry of a naked nucleocapsid into the columnar epithelial cell, infection rarely being observed in the goblet cells. While it is not possible to generalize, it does appear that the events within cells follow a reasonably common pattern. Naked nucleocapsids remain in the cytoplasm or migrate to the nuclei, depending on the virus group. During the so-called latent or eclipse period the nucleic acid core is exposed by some process of uncoating followed by the first stages of transcription to form progeny virus. A virogenic stroma is formed at this stage (13, 120), and it is within this structure that progeny virions are produced. Indeed, cytopathic effects are normally apparent only when the virogenic stroma begins to form. It is not always clear how the progeny virions are constructed (core first followed by envelopment or vice versa), and most observations have been of late stages of cellular infection when the virions are nearing maturity.

As an example of the replication of an insect virus at the cellular morphological level, the description of BV early infection by Granados (121) and Granados and Lawler (102) is a fine model. They proposed the following scheme for replication of AcMNPV in midgut cells.

1. After membrane fusion, naked nucleocapsids (NC) enter the cell cytoplasm and pass to the nucleus.
2. NCs align end-on to the nuclear pore and uncoat with the DNA core passing inside, or NCs pass in through the pore and uncoat inside.
3. Nucleus enlarges and a virogenic stroma (VS) develops.
4. Progeny NCs develop within VS; some are enveloped while other naked NCs leave nucleus and pass to cell basal membrane.
5. NCs at basal membrane bud through to the hemocoel and acquire a host-mediated envelope with peplomer structure (116, 122).
6. These nonoccluded virions act to pass infection between cells which they enter by viropexis or fusion (116).
7. Remaining NCs acquire an envelope and may become occluded in protein to form IB.

Although the mechanisms of BV progeny virus formation have not been universally agreed (see ref. 27), it is likely that the above scheme is correct at the morphological level. Similarly, those virus groups developing in the cytoplasm also produce virogenic stromata differing only in form and biochemical function.

CPV virion formation within the VS may take place through two possible mechanisms. The viral RNA core may form outside the VS and pass inside to become encapsidated (123), or the empty capsids are formed in sparse VS and then become filled with core material (33). Polyhedral occlusion takes place within the VS late in the replication cycle. Infection of other cells takes place through viropexis of nonoccluded NCs.

EPVs are unusual in that they have two forms of VS within the cytoplasm (see ref. 6). Type I VS is an amorphous electron-dense mass, whereas type II VS is made up of granular aggregates containing spherical vesicles. Virions are formed through development of membranes at the edges of the VS. Stoltz and Summers (124) showed that these were comprised of a trilaminar membrane surrounded by spicules. These gradually developed into a core and membrane plus the familiar lateral bodies.

The morphology of replication in nonoccluded viruses is relatively poorly studied. Iridoviruses appear to follow the standard pattern such that a paracrystalline VS forms in the cytoplasm. Virions are produced through initial dense core development followed by coating by the capsid protein (see ref. 7). Viropexis is the entry mechanism both into gut cells and into cells of other organs. Lee (125) gave a more detailed account of replication of type IV1 at the morphological and biochemical level and suggested that the capsid shell develops first followed by entry of the DNA core through a hole in this shell, a view at variance with that of Kelly and Tinsley (118). It remains to be seen which is the correct interpretation.

Small RNA viruses have received little attention at the morphological level. Taking one example, the Nudaurelia β virus group, there is some information on the likely course of events. Greenwood and Moore (126) studied Nudaurelia β virus infection in *T. ni* larvae with electron microscopy and enzyme-linked immunosorbent assay (ELISA). They showed by ELISA that multiplication took place in the cytoplasm of gut cells and, by electron microscopy, that virions were formed in crystalline arrays associated with vesicles. More recently Chao et al. (127) studied the cytopathology of an icosahedral virus of *Pseudoplusia includens* that in most characteristics resembled Nudaurelia β virus. They demonstrated by EM that a matrix of electron-dense material developed in the cytoplasm to form VS containing virions of two types, lucent and dense particles. The dense particles eventually aligned in paracrystalline arrays. Vesicles were also formed in the VS, confirming the findings of Greenwood and Moore (126) for Nudaurelia β virus.

The end points for primary cellular infection in the midgut are passage of progeny virus to infect other cells or production of mature virions accompanied by cell lysis and release of these virions. By the latter stages normal cell functions have been completely subverted with concomitant enlargement of the nucleus or whole cell. In the next section the secondary stages of infection are discussed and related to gross pathology of infection.

3.2.2. SECONDARY INFECTION AND GROSS PATHOLOGY.   Cell-to-cell passage of infective nonoccluded virus during the secondary stages of infection occurs by viropexis or occasionally fusion (116). In BV infection the nonoccluded virus (NOV) with peplomer structure derived from the basement membrane is serologically, biochemically, and biologically distinguishable from the nonoccluded virions released on dissolution of IBs in the gut (122). Using tissue culture assay Volkman et al. (122) showed that NOV was 2000 times more infective than IB-derived virions. Similar NOV have been demonstrated for EPVs (105).

Except for those viruses in which infection is confined to the midgut, infection of secondary organs must be accompanied by passage of infective virus through the basement membrane to the hemocoel and thence to the susceptible tissues. In doing so, virions are immediately exposed to the insect's internal defenses (130, 131). The main cellular defenses are phagocytosis by hemocytes of foreign bodies or encapsulation of larger bodies by combined action of many circulating hemocytes. It is not possible to state categorically that phagocytosis does not constitute a defense against virus infection, but there is certainly no evidence to support it (131). Indeed, the enhanced infectivity observed after injection of virions indicates that effective barriers to infection are unlikely in the hemocoel. The other possibility is a humoral defense reaction either natural or acquired by preexposure to virus. There is little evidence that such responses occur for viruses although there is ample evidence in the case of bacteria (131, 132). Aizawa (103) reported that injection of *Galleria mellonella* larvae with antiserum against its NPV reduced the level of infection relative to control larvae. A later report (133) identified a protein inhibitory to NPV in the hemolymph of *G. mellonella*. However, despite these reports it appears unlikely that humoral mechanisms play any significant role in protecting insects from attack by viruses. Indeed, in discussing the development of viral resistance in insect populations, Briese and Podgwaite (134) firmly believed that the only significant defenses are found in the gut lumen and gut epithelial cells themselves.

The sites of major replication for the various virus types can be grouped essentially into gut-only infection or virtually systemic infection. Pathological changes are accompanied by other recognizable characteristics, sometimes specific to a particular group. Thus behavioral and longevity changes are possible symptoms of virus attack. Harrap and Payne (1) and Payne and Kelly (2) outlined the features for identification of each group. We shall now describe the pathology and symptomatology of each group.

Baculoviruses may affect the gut only, as in the case of the NPVs of sawflies (Hymenoptera: Symphyta) (135) or also most other tissues including hemocytes, fat body, hypodermis, tracheal matrix, muscle cells, nerve ganglia, and pericardial cells (136). These near-systemic infections of occluded BVs are well described for Lepidoptera, whereas less is known of Neuroptera and Diptera. Massive quantities of IBs are produced (137) eventually giving the larvae a milky white appearance as most of the body tissues become infected, the end result being virtually a bag of virus (Fig. 10.2A). At this stage the hypodermis (usually the last organ to become infected) is extremely fragile, and the body wall frequently ruptures releasing the liquid body contents. Gut infection of sawflies provides a similar but localized picture so that the expanded midgut can be seen as a light structure beneath the hypodermis. In these cases disintegration of cells takes place accompanied by regurgitation of gut contents and diarrhea. Subgroup C BVs are represented by infections of coconut palm rhinoceros beetle *(Oryctes rhinoceros)* by *Oryctes* virus. Larval infection is most marked in the fat body (138) which gradually disintegrates while the abdomen expands through an enormous increase in the amount of hemolymph, resulting in a transparent appearance with possible evagination of the hind gut (138). Adult infection, on the the other hand, occurs mainly in midgut epithelial cells (139, 140) although replication has been reported in the ovarian sheath and spermatheca (141). In both larval and, particularly, adult gut infection, large quantities of infective virus accumulate in the gut lumen through disintegration of cells and are voided in the feces at the rate of up to 0.3 mg virus/day for an individual adult (142).

Far less is known of the so-called subgroup D BVs and polydnaviruses associated with parasitic Hymenoptera. These viruses, which replicate in the nuclei of calyx cells of female parasitoids, are injected into the hemolymph along with the parasitoid eggs. Once injected they fuse with hemocyte cell membranes in the conventional manner, and the nucleocapsid enters the cell and uncoats within the nucleus (143). The events within the hemocytes are not yet understood, but the "infected" cell then becomes incapable of encapsulating the parasitoid egg, preventing the host's normal defensive reaction to a large foreign body (144, 145).

Behavioral changes have often been associated with BV infection, especially for the occluded viruses of subgroups A and B. This phenomenon has been discussed by Entwistle and Evans (95) who reviewed the various possible effects of infection on both larval and adult stages. "Wipfelkrankheit" or "tree top disease,"the manifestation of enhanced activity and response to light in NPV-infected *Lymantria monacha* larvae, was one of the first recognized symptoms of BV disease, even though the causal agent was not known. Evans and Allaway (146) showed that NPV-infected *Mamestra brassicae* larvae were twice as active as healthy larvae and were effective disseminators of the pathogen. The mechanism behind the behavior changes is not known, but early infection of gut (sawflies), hemocytes, and nerve ganglia (Lepidoptera) may well mimic starvation so that the larvae continue to search for food even with a full gut.

Sublethal (chronic) infection and latency are associated with survival from challenge by virus. Although a detailed account is beyond the scope of this chapter, the two phenomena play a role in the replicative strategy of the BVs and other virus groups. In the case of occluded BVs there is evidence that sublethal infection in the larval stages may result in some form of carry-over effect into the adult stages and possibly into the next generation. Evans and Harrap (94) and Entwistle and Evans (95) have reviewed this topic citing lowered adult fecundity and fertility, changed sex ratios, and smaller size as typical effects in Lepidoptera. Evidence for latent or inapparent infections is circumstantial and depends on activation by some form of trigger such as low temperature (*B. mori* NPV, 147); feeding with heterologous virus (*S. littoralis* NPV, 148, *Adoxophyes orana* NPV, 149); or diet (GV of *Pieris rapae*, 150). Unfortunately, with very few exceptions, proof of the identity of activated virus infections has not been conclusive. It is hoped that techniques such as REN will allow more definitive studies of this phenomenon.

Associated with inapparent infections is the question of transovarian transmission by incorporation within the host genome (151). Latent infections in *B. mori* and *Galleria mellonella* were investigated by Skuratovakaya et al. (152). They observed that the genome copy number per cell was host-species related, and varied from two to six. By means of blot technique it was shown that in *B. mori* NPV DNA was integrated with cell DNA and that this was also true for a wild strain of *G. mellonella* though not for a laboratory strain.

Host range is the final characteristic of viral success. A wider host range may be expected to confer advantages to the virus and is a reflection of strategy in relation to host life-style in that more hosts are potentially available for continuing the infection process. This has been reviewed by Ignoffo (153), Evans and Harrap (94), Entwistle and Evans (95), Entwistle (154), and Evans (155) and will be discussed in relation to host r-K continuum in Section 5. Armed with techniques like REN and hybridization, more and more examples of cross-transmission are being demonstrated with certainty. The mechanisms allowing transmissibility between different hosts must lie in the initial infection process within the gut lumen, at the surface and within the epithelial cells themselves (106).

Replication of *Ascoviridae* takes place in the nuclei of host cells and culminates in nuclear and cytoplasmic hypertrophy (11). Fat body, hypodermis, and tracheal matrix are all infected, with release of virion-containing vesicles into the hemocoel producing a characteristic milky appearance. Disease in *T. ni* larvae could be induced by ingestion, but greater prevalence was achieved by inoculation (11). Symptoms included lethargy and prolonged development eventually resulting in death of larvae. A putative ascovirus of *Spodoptera frugiperda* was transmitted from infected to healthy host larvae by the braconid parasitoid *Cotesia marginiventris* (145). However, although the infected larvae did not die immediately, the parasitoid progeny failed to develop in larvae infected before or during parasitization.

Infection by *Cytoplasmic Polyhedrosis Viruses,* being restricted to the midgut, tends to be characterized by the accumulation of many IBs in the cytoplasm of infected cells which may result in a creamy-white appearance in the larval gut. Payne and Mertens (5) discussed CPV replication including pathological effects and reviewed early literature on this virus group.

Two consequences of CPV infection can be distinguished. First, virus administered to neonate larvae is likely to give rise to lethal infection with low LD_{50} values, for example, *Orgyia pseudotsugata* CPV $LD_{50} = 8$ IBs (156). In this respect CPVs resemble BVs in inducing lethal effects in neonate larvae after ingestion of low dosages of virus (115).

The second and more commonly observed effect is the development of chronic disease which may exhibit a number of characteristics but rarely death of the host. Notable among these is early reduction or cessation of feeding as the gut cells lose their structural integrity (31). Complete breakdown of the infected epithelial cells follows, and production of virus-contaminated feces accompanied by diarrhea are common (157). The reduced absorptive capacity of the gut gives rise to smaller larval, pupal, and ultimately adult stages, and it is surprising to find that molting takes place under such conditions. Indeed, the extended development time that results may be accompanied by an increased number of molts (158).

Sublethal effects are therefore the norm for CPVs, and this is reflected in continuing infection in the adult midgut. Deformed adults are common, and even if mating is successfully accomplished both fertility and fecundity of the abnormally small females may be reduced considerably (159). Continuation of infection into the adult results in transovum transmission of CPV (160), but this is a consequence of surface contamination only and transovarian transmission has not been demonstrated.

CPVs are credited with wide host ranges (161) and are one of the commonest causes of viral disease in laboratory cultures of insects. Unfortunately, in common with most examples of cross-transmission, there is rarely conclusive proof that prior contamination was eliminated or that progeny virus was the same as input virus. Allaway (162), however, successfully demonstrated cross-transmission of types 2 and 12 CPV between pierid and noctuid species. If wide host ranges are eventually proven then CPV replication strategy, particularly with respect to host "recognition," may provide valuable clues about virus-host coevolution.

Representatives of *Entomopoxviridae* have been found in Lepidoptera, Coleoptera, Diptera, and Orthoptera. Following initial gut infection, naked virions with peplomer structure bud through the basement membrane and pass to other organs of the body in a manner similar to BVs (105). Replication takes place mainly in the fat body (163) although hemocytes and hypodermis also are infected. Early signs of infection are color changes in the hemolymph which in *Elasmopalpus lignosellus* becomes light blue or white (164). As infection advances the fat body itself turns white and enlarges as virions are occluded in spheroids (165). Death from EPV infection usually results from rupture of the gut or cuticle releasing the large quantities of spheroids and spindles in the host body.

The major symptom of EPV infection is the extreme longevity of infected individuals. Mitchell and Smith (164) demonstrated that EPV-infected *E. lignosellus* larvae lived over twice as long (48 days compared with 20 days) as healthy larvae. Extended development of up to six months has been reported (EPV of *Orthnonius batesi*—Coleoptera: Scarabaeidae) (165). Activity patterns also vary with infection ranging from normal for *E. lignosellus* (164) to sluggish for *Melanoplus sanguinipes* (Orthoptera:Acrididae) (166). A full account of the pathology of EPV infection in the various insect orders affected is given by Granados (167).

As the name implies the single most striking feature of *Iridoviruses* is the characteristic iridescent color exhibited by larvae during the later stages of infection (168). Infection can be systemic although the major sites of replication are fat body, hemocytes, and hypodermis. For example, IV3 in mosquitoes affects fat body, hypodermis, hemocytes, nervous tissue, muscle, developing ovaries, tracheae, and imaginal disc (169). Iridescence is most obvious in fat body and hypodermis particularly late in infection when the paracrystalline arrays of virions are at their greatest concentrations before cellular breakdown occurs.

There is little change in behavior as infection proceeds, and larvae are able to continue with normal activity without apparent effect until just before death (7). However, *Tipula oleracea* larvae infected with IV-1 (170) and *Wiseana cervinata* larvae infected with IV-9 (171) change behavior to move upwards in their soil environment. This may have some adaptive significance for these iridescent viruses by maintaining a "pool" of virus near the soil surface. Infected larvae may also be extremely long lived as was the case for infected *Heliothis zea* larvae (89 days, compared to 13 days) (172). Infection of mature larvae can result in the passage of the virus to adults. Continuing degradation of the fat body and hypodermis would decrease the fecundity of infected females and therefore have an indirect sublethal effect on the population. Carter (173) demonstrated vertical transmission of IV-1 in field-collected *Tipula paludosa*. However, it is not known if this is transovum or transovarian transmission, although the balance of probabilities must support the former, bearing in mind the large quantities of virus likely to be present as contaminants in the adult fat body and hypodermis. The question of vertical passage was discussed by Hall (7) who, in reviewing IV-3 transmission cycles in *Aedes taeniorhynchus,* favored transovarian transfer from female to progeny based on data by Linley and Nielsen (174, 175). Such a mechanism, although a distinct possibility, must be considered circumstantial until evidence of transmission in the absence of contamination is available. A recent study by Sikorowski and Tyson (172) on the iridescent virus of *Heliothis zea* failed to demonstrate either transovum or transovarian transmission of the virus.

Natural prevalence may be extremely low (< 1% infection in mosquito IV-3) (176), or occasionally reach 50 to 70% (IV-13 infecting *Corothrella brakeleyi*) (177). Such variability links with viral strategy of extended development time thus optimizing inoculum availability in the face of intrinsically low host susceptibility.

Isolations of insect *Parvoviridae* are relatively recent and stem from the work of Meynadier et al. (178) on the DNV of *G. mellonella* (DNV 1). This particular DNV and the DNVs of *Bombyx mori* and *Junonia coenia* DNV 2 have received considerable attention, and most data on infection processes accrue from these studies.

G. mellonella DNV replicates in fat body, nerve cells, muscle cells, Malpighian tubules, gonads, and silk glands but not in the midgut (179). While transmission of DNV to *G. mellonella* larvae by the contaminated ovipositor of the parasitoid *Nemeritis canescens* has been demonstrated (180), it is not at all clear how this DNV is transmitted normally in the apparent absence of gut replication. Per os transmission via primary replication in the gut would certainly be expected in a virus that is reputed to be monospecific to *G. mellonella*. Once infected the susceptible tissues show characteristic accumulations of dense bodies in the hypertrophied cell nuclei which are destroyed in about six days, followed by death of larvae (181).

B. mori DNV contrasts with *G. mellonella* DNV in that it replicates only in the midgut (182). Flaccidity of the body is the major external sign, while the midgut changes to a pale yellow color as the epithelial cell nuclei become hypertrophied. No other tissues are affected.

Other DNV symptomatology points to a range of possible cytopathic effects. Suto et al. (183) studied the DNV of *Periplaneta fuliginosa* and observed paralysis of the hind legs and a swollen abdomen resulting from an enlarged fat body which was milky-white compared to the normal light brown color. *Agraulis vanillae* DNV induced a systemic infection of all tissues except the hypodermis, and mortality appeared to be concentrated at the pupal or early adult stage (184).

Behavioral changes have not been noted and the modes of vertical transmission remain obscure. Epizootics of DNV on silkworm farms indicate that foliage contaminated as a result of a probably identical midgut infection in the mulberry pyralid, *Glyphodes pyloalis* (185), and varying susceptibility with silkworm strains are the major regulators of infection.

Host range differences are marked between the better studied examples of DNV. As already mentioned, *G. mellonella* DNV 1 seems monospecific, whereas DNV 2 from *Junonia coenia* is cross-infective to other nymphalid butterflies and some noctuids but not to *G. mellonella* (186). *Aedes* DNV infects eight other species of mosquito in three genera (187), while *P. fuliginosa* DNV infects three other cockroach species (183).

Since little is known of the pathology of *Small RNA Viruses,* it is proposed that the groups discussed earlier be combined, with amplification only for features atypical of these viruses in general.

Replication of virus in the cell cytoplasm has been described in the previous section. It is probable that all the RNA viruses with the exception of *Rhabdoviridae* have a cycle of replication in the gut. Longworth (188) reported that when gut infection occurs, regurgitation and diarrhea are common, and this finding was confirmed by Moore and Tinsley (9). Infection may eventually result in a number of pathological conditions frequently related to disruption of the gut. Thus, reduced weight, flaccidity, increased longevity, and regurgitation/diarrhea are all possible (2).

Behavioral changes may be marked and in fact are sometimes characteristic of particular virus isolates but not necessarily linked to the virus group as a whole. Cricket paralysis virus (Picornaviridae) is a good example. This virus can cross-infect many species by injection but it is named after the change of behavior, lack of coordination, and eventual hind leg paralysis observed in infected *Teleogryllus oceanicus (189).* Another group of RNA viruses affects the brain and nervous system of bees causing a syndrome known as "chronic bee paralysis" (190) that seriously affects honeybee colonies. *Drosophila* X virus, a bisegmented genome virus of *Drosophila melanogaster* (191), and Sigma virus of *Drosophila* (Rhabdoviridae) (192) cause no overt pathological symptoms, but adult flies have enhanced sensitivity to CO_2, failing to recover when exposed to CO_2 anesthesia.

The wide host ranges combined with the inapparent nature of many small RNA virus infections are commonly accepted characteristics of these viruses (9). Demonstration of host ranges has been mainly through the injection of virus into potential hosts which, in view of the earlier discussion on gut barriers, may not provide a true picture of potential cross-infection in the wild. Indeed so little is known of the modes of transmission of these viruses that to establish host ranges before gaining an understanding of their epizootiology is premature. Scotti et al. (193) have provided one of the few appraisals of the ecological consequences of small RNA virus infection. They pointed out that careful sterilization of egg surfaces prevented transovum transmission of CrPV or DCV, but P, A, and iota viruses of *Drosophila* appeared to be transovarially transmitted. There was no clear mechanism for transmission in the field, but they felt that the *per os* route was likely in most cases.

If inapparent infections are the "norm" for small RNA viruses, then the probability of transovarian transmission must certainly be greater than for the other virus groups that have a more clear-cut epizootiology. However, as Moore and Tinsley (9) pointed out, the inapparency of many small RNA virus infections may simply be the result of the inability to detect the virus with existing techniques. There may well be subtle physiological changes and associated symptoms that can only be detected by comparison with healthy stock. Drosophila C virus (DCV) is a case in point. Virus extracted from apparently healthy insects produced overt symptoms after injection into "healthy" stock (194). However, the entire question of what is healthy or infected must cast doubt over some of these early results, particularly when injection is used as the means of introducing virus to the host. The use of serology has demonstrated that small RNA viruses such as CrPV are common in field populations. Reinganum et al., (195) found CrPV in 43% of sites examined. A survey for picornaviruses in *D. melanogaster* in laboratory and field populations by Plus et al. (196) detected positive reactions in serological tests in about a third of the samples. More detailed studies are required before inapparency can be demonstrated with certainty, although the ubiquitous nature of small RNA viruses would appear to add weight to the concept.

Transovarian transmission has been demonstrated for Sigma virus of *Drosophila* (192), the only proven example of this method of vertical transmission for insect viruses. Infections persist in female germinal tissue without overt signs, and in this respect it is also a reliable example of true inapparency (197). There is no doubt that such a mechanism provides the most secure means of vertical transmission for insect viruses, and it is possible that it is widespread in the field. The problems of distinguishing contamination and conventional transmission from transovarian transmission have yet to be solved fully, but the latest biochemical techniques should make this possible in the future.

4. JOINT INFECTIONS

The interactions of different pathogens in a single host species are probably of great epizootiological significance but have been little explored via ecological studies. There has, however, been a number of laboratory investigations which have shed light on the range and complexity of those events which almost certainly occur in the field. Krieg (198) and Sherman (199) reviewed this topic largely at a qualitative level, whereas Benz (200) dealt with both qualitative and quantitative pathogen-chemical insecticide interactions cogent to insect disease epizootiology. Entwistle and Evans (95) considered pathogen-pathogen and pathogen-chemical insecticide interactions, stressing the available methods of quantitative analysis.

Pathogen interactions can be viewed at several levels; pathogens may compete at the cellular level, at the level of individual hosts, and *ipso facto* in host populations. Layers of complexity arise out of host population age structure, climatic and microclimatic conditions, and the temporal gap between successive exposure to different pathogens. Further modulation through pathogen inoculum may also be encountered. It will probably be many years before any one pathogen interaction is properly explored at a host population level.

Joint infections of viruses may apparently occur in the same cell but not usually so. For instance, a dual infection of NPV and CPV in *Bombyx mori* was interpreted as interference at the cellular level as no cells had both viruses (201), as seemed also to be the case with NPV and GV in *Choristoneura fumiferana* (202). Nor did IV or NPV occur in the same cells in *Spodoptera frugiperda* cell cultures (203). However, joint CPV and NPV infections occurred in the same cells in *Spodoptera littoralis* cell cultures (204) as well as a picornavirus, IV, and BV in *Galleria mellonella* cells (205).

Interference at the whole insect level may occur in simultaneous infections. For example when two NPVs were fed to *Pseudaletia unipuncta* only one was symptomatically apparent (206). When NPV and GV were fed simultaneously to *Heliothis armigera* a dual infection resulted but produced a lower mortality than either virus separately (207). Multiple infection of IV and NPV in *S. frugiperda* cells resulted in a significant reduction of NPV replication (203). CPV inhibited nucleocapsid envelopment and NPV PIB formation in *S. littoralis* (204). More rarely, synergism may occur as was noted when NPV was fed to *P. unipuncta* infected with a GV in which a heat stable "synergistic" factor exists (208). Joint infections of a picornavirus isolated from apparently healthy *Aedes taeniorhynchus* (and also IV-3 infected individuals) resulted in higher IV-3 infection levels. No other "helper" viruses are known in insects (209). However, simultaneous GV and NPV infection in *Trichoplusia ni* resulted in neither interference nor enhancement (210, 211).

Interference is evident to a greater degree in lag phase exposures to successive pathogens. For instance, prior infection of *C. fumiferana* either with NPV or GV generally resulted in interference of the second virus (202), and if in *T. ni* NPV was fed 5 to 7 days after GV, NPV infection was interfered with (210, 211). As Kelly (203) remarked of IV and NPV in *S. frugiperda* cells, the inhibition phenomenon is affected by time and temperature. It is no doubt also affected by pathogen dosage. More studies embracing these features are necessary for the elucidation of rate and quantity aspects in interference.

Explanations of the phenomenon are lacking, but it is evident that interference is not necessarily due simply to preoccupation of cells with replication of the inhibiting virus. For instance, titers of Sindbis virus in BHK-21 cells were reduced by AcNPV though there was no autoradiographic evidence of AcNPV replication (199). In *B. mori,* u.v.-inactivated CPV protected against the challenge of live CPV even after the elapse of two host larval instars, though further experiments produced less lucid results (212, 213). Nevertheless, the impact of inactivated viruses, albeit in a contrary direction, cannot be dismissed as for instance when heat-inactivated *Aglais urticae* NPV potentiated "occult" NPV in *Lymantria dispar* (214) as did heat-inactivated *Mamestra brassicae* NPV for *Adoxophyes orana* NPV (149). However, NPV-GV interference in *H. armigera* and interference of NPV by IV in *S. frugiperda* cells was principally a function of live viruses (203, 207).

Multiple infections, at least by viruses that are not too distantly related such as the phylogenetically coherent NPVs, give rise to the interesting possibility of genetic recombination. When *G. mellonella* was jointly infected with GmNPV and AcNPV, and progeny NPV was serially passed four times by hemolymph transfer, REN analysis indicated such recombination (215).

Interactions of different pathogen species in a wild host population have been little investigated. It is, however, known that as the prevalence of microsporidiosis (Protozoa) increases with time in successive generations of *C. fumiferana,* the infection capacity of sprayed NPV decreases (J.C. Cunningham, personal communication); this also occurs with *Euproctis chrysorrhoea* (P. Sterling, personal communication). The antagonism of *Vairimorpha necatrix* (Protozoa: Microsporidia) to NPV in *Heliothis zea* has been formally demonstrated (216).

The development of adequate methods of quantitative analysis of interactions between pathogens is essential to both laboratory and field studies. In simple quantitative terms interactions have generally been accepted as either antagonistic, independent, additive, or synergistic. The use of some of these terms has, however, been inconsistent, and therefore Benz (200) proposed a series of quantitatively based definitions for the classification of simple mortality data:

1. *Independent synergism* (= independent action with zero correlation). A system of two components acting independently and not interfering with each other.
2. *Subadditive synergism.* A system of two components which together produce an effect greater than independent synergism but less than the algebraic sum of the two single effects. A weak potentiating effect is necessary to produce such a result.
3. *Supplemental synergism.* The two effective components together produce an effect greater than their algebraic sum.
4. *Potentiating synergism.* A system of a component causing an effect and a synergist whose sole influence is to increase the effect of the natural component. This type of synergism may be found when nonlethal concentrations of an insecticide are combined with a microorganism or where a helper virus increases infection levels of another virus, as for instance the

picornavirus helper of IV-3 in *A. taeniorhynchus* (209), a system also
described as heterologous interference (217).

5. *Coalitive action.* Each of the two components alone causes no measurable
effect but together produce a significant effect.

Figure 10.3 shows the application of Benz's analysis to the interaction of GV
and malathion, a chemical insecticide, in *Plodia interpunctella* (218), with an
outcome that is possibly additive.

A binomial test of McVay et al., (219) modified by Fuxa (216), can be used to
compare expected and observed percentage mortalities. Synergism or antagonism is indicated when the percentage observed mortality is significantly
greater, or less, than that expected.

In a purely graphic analysis, the Tammes-Bakuniak method, "isoboles" of
LC or LD_{50}s are plotted against concentrations or doses of each pathogen (Fig.
10.4a). It is generally accepted that synergism exists when the isobole crosses a
line drawn between the half LD_{50}s for each agent. In Figure 10.4b, the method is
shown applied to reveal the antagonistic interaction of an NPV and *V. necatrix,*
a protozoan, in *H. zea* (216).

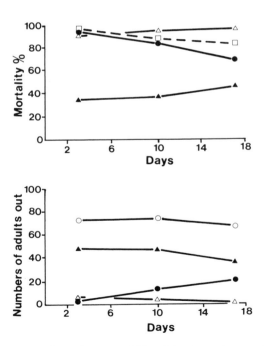

Figure 10.3. The interaction of granulosis virus (GV) and the chemical insecticide malathion in *Plodia interpunctella* analyzed after Benz (1971). ○ Control, ● GV (Pm); ▲ malathion (P_1); △ GV + malathion; □ independence á la Benz (Pm $+ 1 =$ Pm $+ P_1 (1 -$ Pm) (for days 10 and 18 this is a subadditive effect) (218).

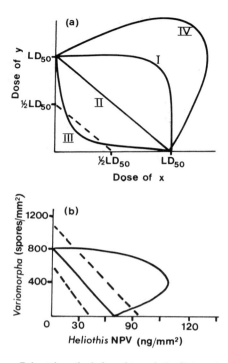

Figure 10.4. (a) The Tammes-Bakuniak method of graphic analysis of interactions of pesticides. I, independence; II, additivity; III, synergism; IV, antagonism. (b) Tammes-Bakuniak analysis indicating an antagonistic interaction of the microsporidian *Vairimorpha necatrix* and a nuclear polyhedrosis virus (NPV) in *Heliothis zea* L. Dotted lines indicate upper and lower 95% confidence limits (216).

5. INTEGRATION OF VIRUSES AND HOST LIFE STRATEGIES

Animals and plants exhibit a range of adaptations that fit different species to the occupancy and exploitation of different habitats. Some of these adaptations are morphological and physiological, but others concern aspects of population growth and behavior and can be synthesized into a comprehensive population dynamics synoptic model which relates the interaction of population growth rates and population density maxima to the constraints imposed by varying degrees of habitat stability. As habitat stability decreases, species tend to devote a greater proportion of their resources to elevating the reproductive rate (r). They tend to be small, for their taxon, and to be vagile because of the evanescent nature of their immediate habitat. At the other extreme of habitat stability with associated extended niche diversity, species tend to exploit specialist niches and, by longevity and a low reproductive rate, to achieve fairly stable population densities that are close to the carrying capacity (K) of the habitat. Many species are intermediate between these extremes, and the whole range is described as the r-K continuum (220).

The consistency of this model with insect population dynamics has been discussed (221–223), but the question of whether pathogens of insects have life strategy patterns that clearly fit them to hosts in different regions of the continuum or, indeed, if insect pathogens can themselves be legitimately classified in terms of the r-K synoptic model, has only recently been addressed (154).

5.1. Adaptation of Viruses to r-Selected Insect Hosts

The central problems for pathogens of r-strategist hosts (Table 10.9) are how, during the brief tenancy by the host of an ephemeral habitat, to produce enough inoculum to persist until the host returns at some future date, how to accompany the host from one refuge to another, or both.

NPVs are common in r-strategist Lepidoptera in which they are frequently epizootic, for example, in *Spodoptera exempta, Tiracola plagiata,* and *Trichoplusia ni.* NPVs of such hosts may have quite narrow host ranges, but this is not a prerequisite for adaptation to r-strategists. Indeed, the wider the host range the more likely the virus is to survive those periods when the r-strategist original host is unavailable. Additionally, species of *Spodoptera* may carry latent infections which can be induced by ingestion of heterologous NPVs (148); this could be a valuable adaptation to the life-style of a migrant host. The presence of NPV DNA sequences in the host genome has been demonstrated for *Bombyx mori* and *Galleria mellonella* (152), and it has been established that occult NPVs were induced by both live and heat-inactivated heterologous NPVs and even by a plant virus, tobacco mosaic virus (149, 214, 224). It is possible that CPVs can be similarly activated (225).

The mosquito *Aedes taeniorhynchus* routinely flies from one pool to another (226). Death of larvae from mosquito iridescent virus (MIV) genus *Chloriridovirus* (IV3) occurs most commonly in the fourth instar. Exposure of larvae to high concentrations of free MIV causes little infection (107) possibly because of rapid loss of infectivity after isolation from the host and the barriers to *per os* infection within the host. However, when MIV is acquired late in life through cannibalism of infected cadavers, there may be occult passage through the adult, and disease is not expressed until the following generation (174, 175). *Drosophila melanogaster* may also be regarded as an r-strategist as it has high fecundity (several hundred eggs laid) and considerable colonizing capacity in ephemeral habitats (227). Three picornalike viruses, P, A, and iota, are passed transovarially while one picornavirus, *Drosophila* C, is not. However, it is possible that *Drosophila* C can be transmitted via surface contamination of eggs by diseased females (228). Aphids and houseflies are regarded as r-strategists (223). A 27-nm single-stranded RNA virus of the aphid *Rhopalosiphum padi* (also detected in two other aphids) significantly decreases the longevity of infected hosts. It is transmitted transovarially, though to fully explain the infection rate it was felt another transmission route must exist (229). The vertical transmission route of a reovirus of the housefly, *Musca domestica,* is not fully understood, but shortly after oral infection of adults there can be an increase in

TABLE 10.9
Viruses in r-Selected Insect Hosts (after 154)

Host	Virus	Tropism	Vertical Passage	Host Range
Spodoptera exempta	MENPV	Mainly in tissues other than gut	Transovarial and environmental	Narrow
Tiracola plagiata	NPV		Environmental?	Narrow?
Trichoplusia ni	MENPV		Environmental?	Very wide
Aedes taeniorhynchus	Iridovirus	General	Transovarian	Wide in mosquitoes
Drosophila melanogaster	Picornalike (P,A & iota)		Transovarian	Narrow
	Picornavirus C	Tracheal cells and cerebral ganglia	Transovum?	Narrow?
Musca domestica	Reovirus	Gut?	Transovum?	Narrow?
Rhopalosiphum padi	ssRNA; 27 nm	?	Transovarian	Narrow?

mating and egg-laying activity. Though death may follow, some flies survive longer than others (16 as opposed to 3 to 4 days at 26°C and 70% RH) (230). Extended larval stages are often consequences of small RNA virus infection, but, although this may extend the available pool of inoculum, it has little significance for the individual insect. There is thus the possibility of adult carriage of virus to new breeding sites.

It thus seems likely that r-strategist hosts carry viruses from one ephemeral habitat to another but that at least in the Lepidoptera they also leave behind a long-term persistent reservoir, probably mainly in the soil where the persistence of NPVs can be measured in terms of years. Viruses in other host taxa seem to avoid the generation of such reservoirs and, it could be argued, have evolved further towards a homeostatic relationship with the host.

5.2. Adaptations of Viruses to K-Selected Insect Hosts

The problem of adjusting to the life-style of K-selected insects (Table 10.10) is to strike a balance between developing a mechanism of infection that is success-ful but not to the extent at which it creates an epizootic as this would lead to long-term population imbalance or even to local extinction in a host with innately low reproductive capacity.

Oryctes rhinoceros, the rhinoceros beetle, is a typical K-strategist. The BV subgroup C virus of *O. rhinoceros* (OrBV) may infect and kill any of the three larval instars or the adult but not the pupae. Infected adults soon become reproductively effete and commence to excrete large amounts of low persist-ence OrBV. The principal infection route appears to be inter-adult contact in breeding sites (decaying logs, compost heaps, etc.); contact of larvae with OrBV released from adults or from disintegrating infected larvae is of secondary importance. OrBV exists in an enzootic state in both western and eastern Malaysia and some other parts of SE Asia where it was first discovered by Hüger (138). Natural epizootics have not been observed.

The only known virus of codling moth, *Cydia pomonella,* is a granulosis virus (CpGV). The natural incidence of infection by this virus is not definitely known but is thought to be low as it has seldom been isolated. Passage via the egg has not been clearly substantiated (231), and in addition to some horizontal passage, vertical transmission is also probably effected by tits (Paridae) in Eu-rope, especially by *Parus major,* predating and scavenging living and dead infected larvae and distributing CpGV over the trees so that some neonate larvae acquire an infective dose before entering a fruit. As wild crab apple fruits are small it is also possible that larvae may become infected in moving from one fruit to another. The LD_{50} of CpGV decreases relatively little with host larval size (first instar five capsules; fifth instar 49 capsules) (232). However, the response of older larvae is markedly more heterogeneous so that while one capsule could cause death in 25% of both these larval instars, the doses to cause 70% mortality were 12 and 1578 capsules for first and fifth instars, respectively. This may be an adaptation to ensure infection and hence survival of CpGV in a

TABLE 10.10
Viruses in K-Selected Insect Hosts (after 154)

Host	Virus	Tropism	Vertical Passage Route	Host Range
Cydia pomonella	GV	General	Environmental	Narrow
Amyelois transitella	Calicivirus	Gut	Transovum or environmental?	Very narrow
Oryctes rhinoceros	BV subgroup C	Gut	Environmental	Very narrow
Merodon equestris	Not classified	Salivary gland	Transovum or transovarial	?
Glossina pallidipes	Linear DNA elongate BV-like particles	Salivary gland	Transovarial	Very narrow?
Dacus tryoni	Isometric	Gut	?	
?				

context generally unfavorable for transmission while minimizing the risk of a potentially catastrophic epizootic towards the end of the larval phase. The host range of CpGV seems very narrow so that the question of "maintenance" in "other" hosts may be irrelevant.

The phycetid moth *Amyelois transitella,* the navel orangeworm, is an omnivorous scavenger of wild legume seeds, dried berries, nuts, etc., and has also become a major pest of walnuts and almonds in California, USA. A calicivirus (AtCV) is lethal principally in neonate larvae but causes stunting of older larvae, pupae, and resultant adults. Larvae excrete active virus, but the mode of vertical passage is unknown though it seems likely it is via adult carriage of virus. No other hosts have been detected (92). AtCV appears to be closely adapted to its host, possibly because the life-style of *A. transitella* precludes easy access of other insects to its virus.

Probably the most closely host-adapted insect virus known, apart from *Drosophila σ* virus, is that of the tsetse fly, *Glossina pallidipes.* This is a ds linear DNA nonoccluded virus (233) that replicates in salivary glands and appears to be nonpathogenic. However, males with unusually enlarged salivary glands tend to be sterile. Transmission is thought to be from adult female offspring, but since the virus is infective when fed to, or injected into, adult flies it is also possible it could be acquired via vertebrate hosts (234) (there is no evidence for infection of vertebrates), by flies feeding close together and also possibly by abortive oviposition of parasitic Hymenoptera into puparia.

A nonoccluded BV-like virus of the large narcissus bulbfly *Merodon equestris* is also associated with hypertrophy of the salivary glands and atrophy of the gonads. This virus is passed from larva to adult and has been found only in an adult that has a band of black hairs on abdominal segment three, a "variety" which made up from 20 to 50% of different samples (235). In view of the biology of *M. equestris* (236) it seems likely this virus is passed vertically on or in the egg.

A cytoplasmic inclusion virus of the fruit fly *Dacus tryoni,* another K-strategist, causes a midgut infection, but apart from the fact that infection levels appear to be low little more is known (230).

As might be expected of viruses adapted to hosts of very specialized habitats, viruses of the five K-strategists discussed here have very limited host ranges. Some are probably voided over a considerable period of time, such as the gut-infecting virus of *Oryctes* and the salivary gland associated viruses of *Glossina* and *Merodon.* This is a valuable adaptive feature for viruses of hosts that meet at spatially very restricted loci and tend not to occur at innately high population densities. Natural epizootics are unknown, thus fulfilling one of the predicted features of viruses of K-strategist hosts.

5.3. r-K Intermediate Strategist Hosts of Insect Viruses

The majority of insect species may fall in this category. The regulation of one of them, *G. hercyniae,* has been analyzed in terms of the r-K synoptic model (221). Parasites may control *G. hercyniae* at low population densities but are inadequate to contain host populations that have exceeded a certain critical density. Introduction of an NPV with strong epizootic capacity accomplished population regulation. Like *G. hercyniae,* many r-K intermediates are forest pests displaying outbreak cycles of long periodicity which may be driven (237) by epizootics of BV diseases, such as *Zeiraphera diniana* (238), *Orgyia leucostigma* (239), *Lymantria dispar* (240), *L. ninayi* (241), and *L. fumida* (242). *Choristoneura fumiferana* however, seems little affected by its viruses which is probably a combined consequence of early feeding habits which isolate young larvae from contact with viruses and the mass migrational tendency of adults, effectively removing host populations from areas of virus accretion.

The reasonably strong stability of r-K intermediate strategist habitats favors environmental persistence of their viruses.

5.4. Synopsis

It could be argued that coevolution of hosts with their viruses has helped modify them to their position in the r-K continuum. With r-strategist Lepidoptera, virus pressure in the course of epizootics would select for shorter host life. As an indication of this process, where male larvae of Diprionidae and some Lepidoptera have a shorter development time than female larvae, under virus pressure the sex ratio is generally in favor of male survival. Smaller size, another feature of r-strategists, would be selected in parallel with this. Great fecundity and a capacity to escape by adult migration from areas of virus accumulation are features that would also have a selective advantage under pathogen pressure. Where the breeding habitat is particularly ephemeral, as with *Drosophila* and probably generally with *Musca,* epizootics will be unusual as also will be the case where horizontal transmission is weak (e.g., *Aedes* and *Rhopalosiphum*).

A central feature of viruses of K-selected insects is that the development of overly effective transmission would be disadvantageous because of the danger of disturbance of the steady host population state in the face of its low reproductive response rate. Precisely how virus transmission occurs will be dictated by the biology of individual hosts. K-strategists are often specialized so that close pathogen adaptations may develop which could lead to a narrow host range amongst other things.

In r-K intermediate hosts, studies have indicated that though occult passage of NPVs may occur it is not necessary to invoke the expression of latent infections to explain, as once was fashionable, the apparently sudden onset of epizootics. These can be accounted for by long-term viral persistence, as indicated by mathematical modeling of the epizootic process (237).

6. THE COEVOLUTION OF VIRUS AND HOST INSECT TAXA

Information on the prevalence and host range of the virus groups that have insect associations is of uneven quality. The apparent predominance of baculoviruses of subgroups A and B and of cytoplasmic polyhedrosis viruses (Reoviridae) is almost certainly because the occlusion body makes them more readily detectable than those not so occluded. For instance, it is likely that "only a small fraction of the naturally occurring nonoccluded baculoviruses have been reported, and many more remain to be detected and isolated" (243). However, the possibility should not be overlooked that, as referred to below, certain virus groups are especially associated with ecological situations, host insect taxa, and geographical areas less frequently probed than others. Despite these provisos there are probably sufficient data available, more so with some virus taxa than others, to permit generalizations about the taxonomic, physiological, and possibly geographic associations of at least some of the major insect virus groups.

6.1. Host Ranges of Insect Viruses

Table 10.11 suggests that Baculoviridae and Reoviridae are the most often observed viruses of insects and that occluded viruses constitute over 90% of all known infections, of which nearly 80% are concentrated in the Lepidoptera, 10% in the Diptera, and 3.5% in the Hymenoptera Symphyta. The distribution of nonoccluded viruses is quite different as 33% are in the Lepidoptera, 23% in the Diptera, 18% in the Coleoptera, and only 2.5% in the Hymenoptera. This account ignores the many viruses associated with the hive bee *Apis mellifera* as this host has been disproportionally investigated. These distributions are probably realistic for occluded viruses and are certainly true for nonoccluded viruses in so far as they show these to be the relatively more frequent in hosts other than the Lepidoptera.

Taking an overall view, in addition to the Baculoviridae and Reoviridae, present knowledge of the distribution of the Parvoviridae suggests these also to be strongly associated with Lepidoptera, whereas the Nudaurelia β group, tentatively given family status, are exclusively so. By contrast, the host associations of the Entomopoxviridae and Iridoviridae are more diffuse.

6.2. Ecological and Physiological Associations of Insect Viruses

The principal infection route of the occluded viruses is via gut tissue after liberation of virions from ingested IBs (see Section 3.1). In Baculoviridae, Reoviridae, and Entomopoxviridae alike, IB breakdown occurs under the alkaline conditions in the gut lumen (with a disulphide reducing agent apparently also necessary in the last family), and it therefore follows that occlusion body viruses will be successful only where the host gut presents the appropriate conditions. Gut pH is fairly strongly a function of host taxonomy rather than trophic physiology (Table 10.7), although there are influences of diet (98). The gut pH seems uniformly alkaline in larval Trichoptera and Lepidoptera (9.0 to 9.4) even where the food, as in *Galleria* and *Tineola,* is wax or hair rather than

TABLE 10.11

Number of Insects (or Related) Viruses Known in Major Taxa of Hosts

Host and Virus Taxa	Crustacea	Mites	Orthoptera	Dermaptera	Dictyoptera	Ephemeroptera	Odonata	Homoptera	Heteroptera	Neuroptera	Trichoptera	Lepidoptera	Coleoptera	Hymenoptera	Diptera	Totals	%
Baculoviridae																	
Subgroup A											1	337		25	21	384	47.4
B												103				103	12.7
C	3	2	2	1				3				1	6		2	20	2.5
Entomopoxviridae																	
Subgroup A													18			18	2.2
B			4									14				18	2.2
C															7	7	0.9
Iridoviridae																	
Iridovirus			1						1			4	6	1	2	15	1.9
Chloriridovirus															7	7	0.9
Parvoviridae					1	1	1					8			2	13	1.6
Reoviridae										2		166	2	1	32	203	25.0
Birnaviridae															1	1	0.1
Picornaviridae			1									2		1	1	5	0.6
Nodaviridae								1				1	2		1	5	0.6
Nudaurelia β												9				9	1.1
Caliciviridae												1				1	0.1
Retroviridae															1	1	0.1
Rhabdoviridae															1	1	0.1
	3	2	8	1	1	1	1	4	1	2	1	646	34	28	78	811	
%	0.4	0.3	1.0	0.1	0.1	0.1	0.1	0.5	0.1	0.3	0.1	79.7	4.2	3.5	9.6	100%	

plant material. On the other hand the guts of most other insects, including phytophagous Orthoptera, are acid to neutral, and thus these insects support no occluded viruses apart from some EPVs, the occlusion body dissolution characteristics of which differ somewhat from those of BVs and CPVs. As far as is known BV occlusion bodies are not disrupted under the gut conditions of adult Lepidoptera which may account for the virtual absence of overt infections. Martignoni (244) has demonstrated by injection of virions that they are not *per se* immune to infection.

It is tempting to argue that since the majority of hosts of BVs in subgroups A and B and of CPVs feed on plants above ground, rather than in the soil or aquatic habitats, the IB has evolved in response to associated pressures. The key selection pressure is possibly the rather poorer transmission mechanisms, particularly in BVs, necessitating virus persistence to ensure infection in following host generations. For virus to persist effectively it must (1) survive the injurious effects of solar ultra violet radiation; occluded viruses stand high in the hierarchy of uv tolerance by insect pathogenic microorganisms (245, 246). Nevertheless direct exposure can result in almost complete inactivation in as little as one day. However, much virus is protected by host cadaver components and is shaded by the plant or is in the soil. Persistence on plants can be of considerable duration. *Gilpinia hercyniae* NPV persists at least overwinter on spruce trees, *Picea* spp., where there is continuous movement of inoculum over the tree surface probably as a dynamic process of release from cadavers, erosion by rain, and inactivation by uv (247); (2) be present in infecting situations, in other words, in the sphere of host feeding; on structurally large plants this is often a consequence of pathologically modified host behavior resulting in death on branch tips, tree tops, etc., so that liberated virus widely contaminates plant tissues (94). On plants of lesser stature, virus can be returned by rain splash from soil and the movement of small animals. Inclusion bodies can persist physically on foliage for considerable periods of time. The nature of the forces of acquisition and retention by foliage have recently been analysed in depth for NPVs and are complex (248).

The IB also protects virions from proteolytic digestion in the presence of neutral to acid conditions. In NPVs this capacity seems particularly to be vested in the polyhedral envelope (249). As a result NPVs and GVs tolerate passage through the gut of both vertebrate and invertebrate predators which thus act as dispersal agents (99). However, no such envelope is detectable in the IBs of CPVs. Strong glycosylation is a feature of the NPV polyhedral envelope, but experiments with labeled lectins failed to yield evidence for glycosylation at the periphery of CPV IBs (Christian and Enwistle, unpublished). Nevertheless these IBs were able to tolerate passage through the gut of a bird, *Sturnus vulgaris,* the starling, and to induce infection in their insect host (Carter and Entwistle, unpublished). This same capacity to tolerate the presence of many enzymes may also be connected with the capacity of occluded viruses to tolerate the microbiologically complex milieu of the decaying host cadaver (250) and the soil (94).

6.3. Insect Virus Evolution

It is interesting to look for taxonomic and ecological associations in insect-associated viruses as a possible guide to some aspects of their evolution. We may postulate, for instance, that where a discrete virus group appears to have a unique association with a particular host taxon the evolution of the virus group in that biophysical form is probably not older than the host taxon itself. For example, the Nudaurelia β viruses are known only from the Lepidoptera whose emergence was in the Jurassic and whose major phase of radiation was coincident with that of the angiosperms in the Cretaceous. In contrast, the Nodaviridae, Parvoviridae, Picornaviridae, and subgroup C baculoviruses have hosts in a diversity of insect orders, some of them ancient, such as the Dictyoptera, Odonata, and Homoptera, (Table 10.12), possibly indicative of ancient lineage not necessarily accompanied by any comparatively more recent efflorescence. In an arguably intermediate position are the occluded groups of the Baculoviridae probably of fairly recent lineage but with very extensive radiation in Lepidoptera in and after the Cretaceous.

TABLE 10.12
Distribution of Iridoviruses among Host Taxa

	Aquatic			In Soils			On Plants		
	Ch	Ir	Unc[a]	Ch	Ir	Unc	Ch	Ir	Unc
Diptera	8	1	15	—	3		—	1	—
Hymenoptera	—	—	—	—	—				
						1	—	1	—
Coleoptera	—	—	—	—	6	1	—	—	—
Lepidoptera	—	—	—	—	2	—	—	1	—
Heteroptera	—	1	—	—	—				
						—	—	—	—
Ephemeroptera	—	1	—	—	—				
						—	—	—	—
Cladocera	—	1	—	—	—				
						—	—	—	—
Isopoda	—	—	—	—	—				
						3	—	—	—
Mollusca	—	—	1	—	—				
						—	—	—	—
Polychaeta	—	1	—	—	—				
	8	5	16	0	11	5	0	3	0
			29			16			3

[a]Ch, *Chloriridovirus;* Ir, *Iridovirus;* Unc, genus not classified.

The Iridoviridae are especially interesting because the state of present knowledge suggests strongly that they are not only an ancient family but also that evolution has led to a dichotomy based on ecologically different routes. The genus *Chloriridovirus* (180 nm diameter virions) appears to have evolved exclusively in association with aquatic Diptera (Chaoboridae, Ceratopogonidae, Culicidae, and Simuliidae). The genus *Iridovirus* (130 nm diameter virions), however, though it has some associations with overtly aquatic insects and other arthropods, seems to occur principally in soil-dwelling insects and some more clearly terrestrial hosts (Table 10.10). On the basis of its much wider range of associated host orders *Iridovirus* is probably to be judged the older genus of which *Chloriridovirus* is a more specialized derivative. Vertical and horizontal transmission of the *Iridovirus* (IV-1) of *Tipula paludosa* are simply by larval cannibalism alone; in addition to horizontal passage by cannibalism, the *Chloriridovirus* (IV-3) of *Aedes taeniorhynchus* has transovarian vertical route. This evolved complexity may explain why the genome of the latter virus is the larger of the two though it has been suggested the genomes of both genera contain repeat sequences and that the greater size of *Chloriridovirus* (based on analysis of IV-3) is due to a greater multiplicity of such repeats (61).

Among insect-associated viruses only the baculoviruses have been investigated at a molecular level with a view to illuminating their evolutionary pathways. For instance Rohrmann et al (251) employed homology studies of the structurally highly conserved N-terminal polyhedrin and granulin sequences and antigenic comparisons of these proteins to construct a preliminary phylogenetic tree. They compared SNPVs of *Tipula paludosa* (Diptera), *Neodiprion sertifer* (Hymenoptera, Symphyta), *Bombyx mori, Orgyia pseudotsugata,* GV of *Pieris brassicae,* and MNPV of *O. pseudotsugata* (Lepidoptera). By use of REN analysis, DNA hybridization, and Southern blot hybridization, Smith and Summers (252) studied DNA homology of 18 baculoviruses including representatives of all three subgroups. It seems likely subgroup C baculoviruses represent an ancestral type from which the occluded groups are secondarily derived perhaps consequent upon the selection pressures, discussed above, encountered by arthropods invading the harsher terrestrial sphere. Unlike Rohrmann et al. (251), Smith and Summers (252) did not include a comparison of occluded viruses of representatives of older insect groups such as *N. sertifer* and *T. paludosa* but restricted their study to BVs of lepidopterous hosts.

The homologous DNA sequences among the three BV subgroups provide good evidence for a common ancestry. Comparison of nucleotide changes necessary to account for differences in amino acid sequences (using a method developed by Fitch (253)) showed that IB body proteins were in general related, but alignment of sequences indicated many areas of amino acid change and some areas of substitution or deletion. Further details and discussion are to be found in Vlak and Rohrmann (24).

A cardinal feature of BV evolution is that they are consistently associated with phytophagous or detritus-feeding insects. The Diptera and Hymenoptera have been in existence for over 230 million years, and the latter are known to have been phytophagous for the whole of this period. Phytophagy probably arose from detritus feeding, as is still a feature of such present day dipterous groups as the Tipulidae and Culicidae both of which support BV infections. In contrast, the emergence and massive radiation of the Lepidoptera was a phenomenon associated with Angiosperm radiation in the Cretaceous, about 60 to 130 million years ago. This evolutionary pattern is in general reflected in the molecular phylogeny of occluded BVs. The SNPVs are well separated from the MNPVs, and, compatible with the ancestry of their hosts, the NPVs of Diptera and Hymenoptera are remote from those of Lepidoptera. However, the findings of Rohrmann et al (251) indicate a greater relatedness of lepidopteran to hymenopteran NPVs than the GVs (Fig. 10.5), which seems unlikely in view of the apparently exclusive association of the latter with the Lepidoptera. Diptera and Lepidoptera are mutually more closely related than they are to Hymenoptera (254), and it is suggested by Rohrmann et al. that on the basis of molecular phylogeny, correlation of NPV and host phylogeny would require Hymenoptera to have been cross-infected by an ancestral lepidopteran NPV, as suggested by molecular phylogeny. It is not possible to predict if new evidence is likely to emerge to modify these tentative conclusions, but it is worth remembering that in Hymenoptera Symphyta the NPVs are monotropic, attacking only gut tissue, whereas in all other orders they are polytropic. Associated with the monotropic link with the midgut is the capacity, unavailable in Lepidoptera, for NPVs to transmit vertically by contamination of the egg surface. The Diprionidae, a family of Hymenoptera Symphyta exclusively associated with conifers

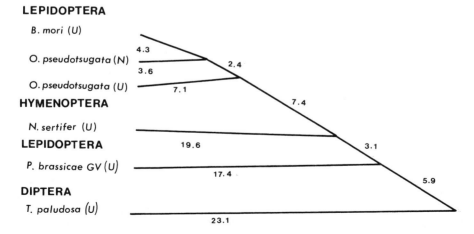

Figure 10.5. The molecular phylogeny of six *Baculovirus* polyhedrins based on N-terminal amino acid sequences. The phylogeny was determined using the method of Fitch (253) with gaps created to align homologous sequences and given a value of 4 nucleotide substitutions per gap. The numbers represent the average phyletic distance in nucleotide substitution between points of branching (251).

(Gymnosperms), is one of the most ancient extant groups of phytophagous Hymenoptera. Many of the species are known to have NPVs, and these apparently have more limited host ranges than do those of Lepidoptera, suggesting a long period of coevolution. NPVs are little known in the more recently evolved sawfly families. Considerations such as these suggest the possibility of the Diprionidae being an ancestral host of NPVs which early diverged into a Dipteran-Lepidopteran stock. In view of the restriction of GVs to Lepidoptera, this virus group probably diverged much later in Lepidoptera of the Cretaceous period or possibly in a pre-Cretaceous ancestral Trichopteran-Lepidopteran stock (251).

6.4. Geographical Distribution of Insect Virus Groups

Though NPV infections are easy to recognize, in contrast to the nonoccluded viruses, it is interesting to note that while expertise is concentrated in temperate and Mediterranean type climatic areas, fewer nonoccluded viruses have been found here than in the tropics. This is especially true of the Nudaurelia β group. It is also notable that epizootics of nonoccluded virus diseases, apart from Cricket paralysis virus, are almost unknown outside the tropics. In the tropics small RNA virus disease epizootics have occurred in *Nudaurelia cytherea* (South Africa), *Gonometa podocarpi* (East Africa), *Lymantria ninayi* (Papua, New Guinea), *Darna trima,* and *Thosea exigua* (island of Borneo). Thus it is possible that such viruses are relatively more successful in hotter climates and have evolved K-strategies to cope with the more constant availability of their hosts.

7. CONCLUSIONS

The diversity among the viruses pathogenic in insects is becoming increasingly apparent, and it is almost a certainty that further virus families and genera remain to be encountered. However, there has often been a dichotomy in the study of viruses, such that the observational or epizootiological approach was divorced from the biochemical with the literature on the two aspects having little cross-reference. This was true especially for the occluded viruses such as baculoviruses, cytoplasmic polyhedrosis viruses, and entomopoxviruses which could be recognized by symptomatology and examination under the light microscope alone. An extensive body of literature thus developed where viruses were being described on the basis of simple criteria with each being associated with a single insect species. Consequently, the occluded viruses make up more than 90% of the records in the Famulus catalogue of Martignoni and Iwai (255).

Fortunately, the recognition of a need for a sound taxonomic approach using all techniques available, including detailed biophysics and biochemistry, is now the accepted ideal, and more research laboratories are adopting a multidisciplinary attitude to the study of viral diseases.

Consequently, the revelations made in recent years, and likely to be made in the future, center around the less easily diagnosed nonoccluded viruses, for example, Nodaviridae, Picornaviridae, and the Nudaurelia β and Ascovirus groups. As the scale of representation of such groups becomes filled out, the role viruses play in the direct regulation of insect population densities (e.g., overt mortality consequent of some picornavirus and iridovirus infections) and by less direct routes (e.g., virus-induced sterility in *Glossina* and *Merodon* (234, 235)) will be clarified.

Already it is recognized that baculovirus infections can drive the population outbreak cycles of many forest Lepidoptera (237), and it is also known that in physically stable pastures (i.e., permanent leys) accumulated undisturbed populations of virus can strongly regulate grass-feeding Lepidoptera (256). These phenomena are documented for temperate areas, but it is increasingly apparent that in warmer climatic zones nonoccluded viruses frequently fulfill similar ecological functions. Clearly, a proper global assessment of the role of viruses in host population ecology will depend on a greater concentration of studies outside the temperate sphere.

In this chapter reference was made to the complexities of interference and other aspects of interplay between different pathogens in single insect hosts. This has necessarily been done more at the allusive rather than at a properly documented level from which the principles of interactions can be clearly derived, and this reflects a paucity of adequately comprehensive studies. Most insects, when observed for a period of time, can be shown to naturally suffer from a variety of pathogens often in conjunct space and time. In such instances neither host nor pathogen ecology can adequately be described without such complexity being recognized. A further variable, not discussed here, is the effects of ingestion of food inevitably of variable quality and of a wide range of secondary plant metabolites.

Insects have diversified, that is evolved to fit, into niches definable only by widely different patterns of microclimate, food resource, physical structure, the influence of natural energy pressures, and temporal stability. It seems inevitable that such has led, through evolutionary time, to appropriate adaptations ensuring the "success" of pathogens and which must have resulted in frequent coevolutionary adjustments of both host and pathogen. We have attempted to probe this, particularly through inspecting the relevant characteristics of pathogen biology and ecology in the conceptual framework afforded by the r-K synoptic model of Southwood and Comins (220). Some conclusions can be drawn. For instance, viruses of r-selected host strategists may be strongly epizootic while the low reproductive, and hence recovery, rates of K-selected hosts would make such a pathogen strategy counter-productive through increased likelihood of host extinction, albeit locally. We actually state that epizootics do not occur in K-strategist hosts, but this implies a catastrophic influence. However, it seems more logical to allow it the freedom of being relative so that an epizootic could be defined in terms not just of the immediate proportion of hosts killed but also of the capacity of the host population for recovery. These ideas seem worth further exploration, but this will be difficult until more is known of virus biology and ecology. Indeed, the scatter of examples cited in this text indicates where, in the context of the position of hosts in the r-K continuum, information on viruses is most lacking.

Finally, and tentatively, we attempted to explore the coevolution of host and virus groups. This exercise is rendered difficult by the weakness of the insect palaeontological "data base," and therefore of host evolutionary patterns. Knowledge of the trophic habits of earlier host forms is poor, and there is a sparcity of studies on virus evolution, especially at a molecular level. However, while our views may be regarded as premature or tenuous, we regard this subject as valuable if only because it seems it cannot be dissociated from evaluation of coadaptation in the context of the r-K continuum with which it may jointly and eventually give rise to a unifying theory of host-virus ecology. The biochemical tools for studying virus evolution at the genome level are now available and, as Vlak and Rohrmann (24) pointed out, the complex relationships of proteins coding within the genome reflect the evolutionary links between virus and host. Additionally, as one of us (154) has attempted to explain, a field theory based on coevolution between virus and host may provide a context in which to evaluate the different methods by which viruses may be deployed as pest control agents of host occupying different positions in the r-K continuum.

ACKNOWLEDGMENTS

Our thanks to Mr. J.S. Robertson for his helpful comments on the manuscript and to Mrs. J. Bald and Miss J. Murdock for all their work in typing and word processing. We also thank Miss M.K. Arnold for all the electronmicrographs (except Figs. 2B and E) which are used with the kind permission of Dr. M. Bergoin.

REFERENCES

1. K. A. Harrap and C. C. Payne, The structural properties and identification of insect viruses, *Adv. Virus Res.,* **25,** 273 (1979).

2. C. C. Payne and D. C. Kelly, "Identification of Insect and Mite Viruses," in H. D. Burges, Ed., *Microbial Control of Pests and Plant Diseases 1970–1980,* Academic Press, London, New York, 1981, p. 61.

3. E. W. Davidson, Ed., *Pathogenesis of Invertebrate Microbial Diseases,* Allanheld, Osmun, Totowa, New Jersey, 1981.

4. P. Faulkner, "Baculovirus," in E. W. Davidson, Ed., *Pathogenesis of Invertebrate Microbial Diseases,* Allanheld, Osmun, Totowa, New Jersey, 1981, p. 3.

5. C. C. Payne and P. P. C. Mertens, "Cytoplasmic Polyhedrosis Viruses," in W. K. Joklik, Ed., *The Reoviridae,* Plenum Press, New York, London, 1983, p. 425.

6. B. M. Arif, The Entomopoxviruses, *Adv. Virus Res.,* **29,** 195 (1984).

7. D. W. Hall, "Pathobiology of Invertebrate Icosahedral Cytoplasmic Deoxyriboviruses (Irido-

viridae)," in K. Maramorosch and K. E. Sherman, Eds., *Viral Insecticides for Biological Control*, Academic Press, New York, London, 1985, p. 163.

8. S. Kawase, "Pathology Associated with Densovirus," in K. Maramorosch and K. E. Sherman, Eds., *Viral Insecticides for Biological Control*, Academic Press, New York, London, 1985, p. 197.

9. N. F. Moore and T. W. Tinsley, The small RNA-viruses of insects, *Arch. Virol.*, **72**, 229 (1982).

10. N. F. Moore, "The Replication Schemes of Insect Viruses in Host Cells," in K. Maramorosch and K. E. Sherman, Eds., *Viral Insecticides for Biological Control*, Academic Press, New York, London, 1985, p. 635.

11. B. A. Federici, Enveloped double-stranded DNA insect virus with novel structure and cyto-pathology, *Proc. Natl. Acad. Sci. USA*, **80**, 7664 (1983).

12. D. B. Stoltz, P. Knell, M. D. Summers, and S. B. Vinson, Polydnaviridae — a proposed family of insect viruses with segmented, double-stranded, circular DNA genomes, *Intervirology*, **21**, 1 (1984).

13. R. E. F. Matthews, Classification and nomenclature of viruses. Fourth Report of the International Committee on Taxonomy of Viruses, *Intervirology*, **17**, 200 pp (1982).

14. K. Maramorosch and K. E. Sherman, *Viral Insecticides for Biological Control*, Academic Press, New York, 1985.

15. WHO, The use of viruses for the control of insect pests and disease vectors. Report of a joint FAO/WHO Meeting on insect viruses, *W. H. O. Tech. Rep. Ser.* No. 531, Geneva (1973).

16. C. C. Payne, D. Compson, and S. M. deLooze, Properties of the nucleocapsids of a virus isolated from *Oryctes rhinoceros, Virology*, **77**, 269 (1977).

17. D. B. Stoltz, "Viruses of Parasitoid Hymenoptera," in *Proc. 3rd. Int. Congr. Invertebr. Pathol.*, Brighton, U.K., p. 1160 (1982).

18. D. Nathans and H. O. Smith, Restriction endonucleases in the analysis and restructuring of DNA molecules, *Annu. Rev. Biochem.*, **44**, 273 (1975).

19. J. M. Vlak and G. E. Smith, Orientation of the genome of *Autographa californica* nuclear pollyhedrosis virus — a proposal, *J. Virol.*, **41**, 1118 (1982).

20. D. C. Kelly, The DNA contained by nuclear polyhedrosis viruses isolated from four *Spodoptera* sp. (Lepidoptera, Noctuidae): genome size and homology assessed by DNA reassociation kinetics, *Virology*, **76**, 468 (1977).

21. E. M. Southern, Detection of specific sequences among DNA fragments separated by gel electrophoresis, *J. Mol. Biol.*, **98**, 503 (1975).

22. M. D. Summers, G. E. Smith, J. D. Knell, and J. P. Burand, Physical maps of *Autographa californica* and *Rachiplusia ou* nuclear polyhedrosis virus recombinants, *J. Virol.*, **34**, 693 (1980).

23. B. J. L. Hooft van Iddekinge, G. E. Smith, and M. D. Summers, Nucleotide sequence of the polyhedrin gene of *Autographa californica* nuclear polyhedrosis virus, *Virology*, **131**, 561 (1983).

24. J. M. Vlak and G. F. Rohrmann, "The Nature of Polyhedrin," in K. Maramorosh and K. E. Sherman, Eds., *Viral Insecticides for Biological Control*, Academic Press, New York, 1985, p. 489.

25. P. J. Krell, M. D. Summers, and B. S. Vinson, Virus with a multipartite superhelical DNA genome from the ichneumonid parasitoid *Campoletis sonorensis, J. Virol.*, **43**, 859 (1983).

26. S. P. Singh, R. T. Gudauskas, and J. D. Harper, High resolution two-dimensional gel electrophoresis of structural proteins of baculoviruses of *Autographa californica* and *Porthetria (Lymantria) dispar, Virology*, **125**, 370 (1983).

27. D. C. Kelly, Baculovirus replication, *J. Gen. Virol.*, **63**, 1 (1982).

28. D. Cook and D. B. Stolz, Comparative serology of viruses isolated from ichneumonid parasitoids, *Virology,* **130,** 215 (1983).

29. K. A. Harrap, The structure of nuclear polyhedrosis viruses, I. The inclusion body, *Virology,* **50,** 114 (1972).

30. J. Kalmakoff, L. J. Lewandowski, and D. R. Black, Comparison of the ribonucleic acid subunits of reovirus, cytoplasmic polyhedrosis virus and wound tumor virus, *J. Virol.,* **4,** 851 (1969).

31. H. J. Arnott, K. M. Smith, and S. L. Fullilove, Ultrastructure of a cytoplasmic polyhedrosis virus affecting the monarch butterfly *Danaus plexippus, J. Ultrastruct. Res.,* **24,** 479 (1968).

32. Y. Hayashi, Occluded and free virions in midgut cells of *Malacosoma disstria* infected with cytoplasmic polyhedrosis virus (CPV), *J. Invertebr. Pathol.,* **16,** 442 (1970).

33. T. G. Andreadis, A new cytoplasmic polyhedrosis virus from the saltmarsh mosquito, *Aedes cantator* (Diptera:Culicidae), *J. Invertebr. Pathol.,* **37,** 160 (1981).

34. K. Yazaki and K. I. Miura, Relation of the structure of cytoplasmic polyhedrosis virus and the synthesis of its messenger RNA, *Virology,* **105,** 467 (1980).

35. P. P. C. Mertens, A study of the transcription and translation (in vitro) of the genomes of cytoplasmic polyhedrosis viruses types 1 and 2, Ph.D. Thesis, Univ. Oxford (1979).

36. C. C. Payne and C. F. Rivers, A provisional classification of cytoplasmic polyhedrosis viruses based on the sizes of the RNA genome segments, *J. Gen. Virol.* **33,** 71 (1976).

37. C. C. Payne, M. Piasecka-Serafin, and B. Pilley, The properties of two recent isolates of cytoplasmic polyhedrosis viruses, *Intervirology,* **8,** 155 (1977).

38. S. Kawase, "Chemical Nature of the Cytoplasmic-Polyhedrosis Virus," in H. Aruga and Y. Tanada, Eds., *The Cytoplasmic-Polyhedrosis Virus of the Silkworn,* Univ. Tokyo Press, Tokyo, 1971, p. 37.

39. L. J. Lewandowski, J. Kalmakoff, and Y. Tanada, Characterization of a ribonucleic acid polymerase activity associated with purified cytoplasmic polyhedrosis virus of the silkworm, *Bombyx mori, J. Virol.,* **4,** 857 (1969).

40. A. A. Van Dijk and H. Huismans, The in vitro activation and further characterization of the bluetongue virus-associated transcriptase, *Virology,* **104,** 347 (1980).

41. K. I. Miura, The cap structure of eukaryotic messenger RNA as a mark of a strand carrying protein information, *Adv. Biophys.,* **14,** 205 (1981).

42. R. E. Smith and Y. Furuichi, A unique class of compound, guanosine-nucleoside tetraphosphate G (5′) pppp (5′)N, synthesized during the in vitro transcription of cytoplasmic polyhedrosis virus of *Bombyx mori, J. Biol. Chem.,* **257,** 485 (1982).

43. R. R. Granados and D. W. Roberts, Electron microscopy of a poxlike virus infecting an invertebrate host, *Virology,* **40,** 230 (1970).

44. J. Weiser, A pox-like virus in the midge *Camptochironomus tentans, Acta Virol.,* **13,** 549 (1969).

45. M. Bergoin, G. Devauchelle, and C. Vago, Electron microscopy study of *Melolontha* poxvirus: the fine structure of occluded virions, *Virology,* **43,** 453 (1971).

46. F. T. Bird, The development of spindle inclusions of *Choristoneura fumiferana* Lepidoptera:Tortricidae) infected with entomopox virus, *J. Invertebr. Pathol.,* **23,** 325 (1974).

47. S. L. Bilimoria and B. M. Arif, Subunit protein and alkaline protease of Entomopoxvirus spheroids, *Virology,* **96,** 596 (1979).

48. W. H. R. Langridge and D. W. Roberts, Structural proteins of *Amsacta moorei, Euxoa auxiliaris* and *Melanoplus sanguinipes* entomopoxviruses, *J. Invertebr. Pathol.,* **39,** 346 (1982).

49. B. Hurpin and P. Robert, Sur la spécificité de *Vagoiavirus melolonthae, Entomophaga,* **12,** 175 (1967).

50. M. Bergoin and S. Dales, "Comparative Observations on Poxviruses of Invertebrates and Vertebrates," in K. Maramorosch and E. Kurstak, Eds., *Comparative Virology,* Academic Press, New York, 1971, p. 169.

51. W. J. McCarthy, R. R. Granados, and D. W. Roberts, Isolation and characterization of Entomopox virus from virus-containing inclusions of *Amsacta moorei* (Lepidoptera:Arctiidae), *Virology, 59,* 59 (1974).

52. B. M. Arif, Isolation of an entomopoxvirus and characterization of its DNA, *Virology, 69,* 626 (1976).

53. W. H. R. Langridge and D. W. Roberts, Molecular weight of DNA from four entomopoxviruses determined by electron microscopy, *J. Virol., 21,* 301 (1977).

54. S. L. Bilimoria and B. M. Arif, Structural polypeptides of *Choristoneura biennis* Entomopoxvirus, *Virology, 104,* 253 (1980).

55. B. G. T. Pogo, S. Dales, M. Bergoin, and D. W. Roberts, Enzymes associated with an insect Poxvirus, *Virology, 43,* 306 (1971).

56. W. J. McCarthy, C. F. Neser, and D. W. Roberts, RNA polymerase activity of *Amsacta moorei* Entomopox virions, *Intervirology, 5,* 69 (1975).

57. T. W. Tinsley and D. C. Kelly, An interim nomenclature system for the iridescent group of insect viruses, *J. Invertebr. Pathol., 16,* 470 (1970).

58. N. G. Wrigley, An electron microscope study of the structure of *Tipula* iridescent virus, *J. Gen. Virol., 6,* 169 (1970).

59. V. F. Nanyakov, Fine structure of the iridescent virus Type 1 capsid, *J. Gen. Virol., 36,* 73 (1977).

60. A. Klug, R. E. Franklin, and S. P. F. Humphreys-Owen, The crystal structure of *Tipula* iridescent virus as determined by Bragg reflection of visible light, *Biochem. Biophys. Acta., 32,* 203 (1959).

61. G. W. Wagner and J. D. Paschke, A comparison of the DNA of the "R" and "T" strains of mosquito iridescent virus, *Virology, 81,* 298 (1977).

62. A. J. D. Bellett and R. B. Inman, Some properties of deoxyribonucleic acid preparations from *Chilo, Sericesthis* and *Tipula* iridescent viruses, *J. Mol. Biol., 25,* 425, (1967).

63. D. C. Kelly and R. J. Avery, The DNA content of four small iridescent viruses: genome size, redundancy and homology determined by renaturation kinetics, *Virology, 57,* 425 (1974).

64. G. P. Carey, T. Lescott, J. S. Robertson, L. K. Spencer, and D. C. Kelly, Three African isolates of small iridescent viruses: Type 21 from *Heliothis armigera* (Lepidoptera:Noctuidae), Type 23 from *Heteronychus arator* (Coleoptera:Scarabaeidae), and Type 28 from *Lethocerus columbiae* (Hemiptera Heteroptera:Belostomatidae), *Virology, 85,* 307 (1978).

65. D. C. Kelly, M. D. Ayres, T. Lescott, J. S. Robertson, and G. M. Happ, A small iridescent virus (Type 29) isolated from *Tenebrio molitor:* a comparison of its proteins and antigens with six other iridescent viruses, *J. Gen. Virol., 42,* 95 (1979).

66. P. Tijssen and E. Kurstak, Studies on the structure of the two infectious types densonucleosis virus, *Intervirology, 11,* 261, (1979).

67. M. Nakagaki and S. Kawase, Capsid structure of the densonucleosis virus of the silkworm, *Bombyx mori, J. Sericult. Sci. Japan, 51,* 420 (1982).

68. A. H. Barwise and I. O. Walker, Studies on the DNA of a virus from *Galleria mellonella, FEBS Lett., 6,* 13 (1970).

69. D. C. Kelly, A. H. Barwise, and I. O. Walker, DNA contained by two densonucleosis viruses, *J. Virol., 21,* 396 (1977).

70. D. C. Kelly and H. M. Bud, Densonucleosis virus DNA: analysis of fine structure by electron microscopy and agarose gel electrophoresis, *J. Gen. Virol., 40,* 33 (1978).

71. M. Nakagaki and S. Kawase, DNA of a new parvo-like virus isolated from the silkworm, *Bombyx mori, J. Invertebr. Pathol., 35,* 124 (1980).

72. S. Kawase and S. K. Kang, On the nucleic acid of a newly isolated virus from the flacherie-diseased silkworm larvae, *J. Sericult. Sci. Japan,* **45,** 87 (1976).

73. J. F. Longworth, T. W. Tinsley, A. H. Barwise, and I. O. Walker. Purification of a nonoccluded virus of *Galleria mellonella, J. Gen. Virol.,* **3,** 167 (1978).

74. G. Siegl, R. C. Bates, K. I. Berns, B. J. Carter, D. C. Kelly, E. Kurstak, and P. Tattersall, Characteristics and taxonomy of Parvoviridae, *Intervirology,* **23,** 61 (1985).

75. B. Hillman, T. J. Morris, W. R. Kellen, and D. Hoffman, An invertebrate calici-like virus: evidence for partial virion disintegration in host excreta, *J. Gen. Virol.,* **60,** 115 (1982).

76. N. F. Moore, B. Reavy, and L.A. King, General characteristics, gene organization and expression of small RNA viruses of insects, *J. Gen. Virol.,* **66,** 647 (1985).

77. J. S. K. Pullin, N. F. Moore, J. P. Clewley, and R. J. Avery, Comparison of the genomes of two insect picornaviruses, cricket paralysis virus and Drosophila C virus, by ribonuclease T, oligonucleotide fingerprinting, *FEMS Microbiol. Lett.,* **15,** 215 (1982).

78. J. P. Clewley, J. S. K. Pullin, R. J. Avery, and N. F. Moore, Oligonucleotide fingerprinting of the RNA species obtained from six *Drosophila* C virus isolates, *J. Gen. Virol.,* **64,** 503 (1983).

79. B. Reavy and N. F. Moore, Cell-free translation of cricket paralysis virus RNA: analysis of the synthesis and processing of virus-specified proteins, *J. Gen. Virol.,* **55,** 429 (1981).

80. J. K. Struthers and D. A. Hendry, Studies on the protein and nucleic acid components of *Nudaurelia capensis β virus, J. Gen. Virol.,* **22,** 355 (1974).

81. J. T. Finch, R. A. Crowther, D. A. Hendry, and J. K. Struthers, The structure of *Nudaurelia capensis β* virus: the first example of a capsid with icosahedral surface symmetry T=4, *J. Gen. Virol.,* **24,** 191 (1974).

82. B. Reavy and N. F. Moore, In vitro translation of the genome of a small RNA virus from *Trichoplusia ni, J. Invertebr. Pathol.,* **44,** 244 (1984).

83. L. A. King and N. F. Moore, The RNAs of two viruses of the *Nudaurelia β* family share little homology and have no terminal poly (A) tracts, *FEMS. Microbiol. Lett.,* **26,** 41 (1985).

84. J. F. E. Newman and F. Brown, Further physiochemical characterization of Nodamura virus. Evidence that the divided genome occurs in a single component, *J. Gen. Virol.,* **38,** 83 (1978).

85. W. F. Scherer, J. E. Verna and G. W. Richter, Nodamura virus, an ether- and chloroform-resistant arbovirus from Japan, *Am. J. Trop. Med. Hyg.,* **17,** 120 (1968).

86. J. F. Longworth and G. P. Carey, A small RNA virus with a divided genome from *Heteronychus arator* (F.) (Coleoptera:Scarabaeidae), *J. Gen. Virol.,* **18,** 119 (1976).

87. J. F. E. Newman and F. Brown, Evidence for a divided genome in Nodamura virus, an arthropod-borne picornavirus, *J. Gen. Virol.,* **21,** 371 (1973).

88. P. D. Friesen and R. R. Rueckert, Synthesis of black beetle virus proteins in cultured *Drosophila* cells: differential expression of RNAs 1 and 2, *J. Gen. Virol.,* **37,** 876 (1981).

89. R. Dasgupta, A. Ghosh, B. Dasmahapatra, L. A. Guarino, and P. Kaesberg, Primary and secondary structure of black beetle virus RNA2, the genomic messenger for BBV coat protein precursor, *Nucl. Ac. Res.,* **12,** 7215 (1984).

90. W. A. L. Crump, B. Reavy, and N. F. Moore, Intracellular RNA expressed in black beetle virus-infected *Drosophila* cells, *J. Gen. Virol.,* **64,** 717 (1983).

91. P. D. Friesen and R. R. Rueckert, Black beetle virus: messenger for protein B is a subgenomic viral RNA, *J. Virol.,* **42,** 986 (1982).

92. W. R. Kellen and D. F. Hoffman, A pathogenic non-occluded virus in hemocytes of the navel orangeworm, *Amyelois transitella* (Pyralidae:Lepidoptera), *J. Invertebr. Pathol.,* **38,** 52 (1981).

93. F. L. Schaffer, D. W. Ehresmann, M. K. Fretz, and M. E. Soergel, A protein VpG, covalently linked to 36S calicivirus RNA, *J. Gen. Virol.,* **47,** 215 (1980).

94. H. F. Evans and K. A. Harrap, "Persistence of Insect Viruses," in B. W. J. Mahy, A. C.

Minson, and G. Darby, Eds., *Virus Persistence,* 33rd Symp. Soc. Gen. Miocrobiol. Cambridge Univ. Press, London, 1982, p. 57.

95. P. F. Entwistle and H. F. Evans, "Viral Control," in L. I. Gilbert and G. A. Kerkut, Eds., *Comprehensive Insect Physiology, Biochemistry and Pharmacology,* Vol. 12, Pergamon Press, Oxford, 1985, pp. 347.

96. V. B. Wigglesworth, *The Principes of Insect Physiology,* 7th Ed. Chapman & Hall, London, 1972.

97. T. W. Tinsley, "Factors Affecting Virus Infection of Insect Gut Tissue," In K. Maramorosch and R. E. Shope, Eds., *Invertebrate Immunity,* Academic Press, New York, 1975, p. 55.

98. M. Berenbaum, Adaptive significance of midgut pH in larval Lepidoptera, *Am. Nat.,* **115,** 138 (1979).

99. P. F. Entwistle, Passive carriage of baculoviruses in forests, *Proc. 3rd Int. Colloq. Invertebr. Pathol., Brighton, UK,* p. 344, (1982).

100. A. G. Richards and P. A. Richards, The peritrophic membranes of insects, *Annu. Rev. Entomol.,* **22,** 219 (1977).

101. H. Schildmacher, Darmkanal und Verdauung bei Stechmuckenlarven, *Biol. Zentralbl.,* **69,** 390 (1950).

102. R. R. Granados and K. A. Lawler, In vivo pathway of *Autographa californica* baculovirus invasion and infection, *Virology,* **108,** 297 (1981).

103. K. Aizawa, Infection of the greater wax moth, *Galleria mellonella* Linn., with the nuclear polyhedrosis of the silkworm, *J. Insect Pathol.,* **4,** 122 (1962).

104. C. C. Payne, Properties and Replication of Some Occluded Viruses, Ph.D. Thesis, Univ. Oxford, (1971).

105. R. R. Granados, Entry of an insect poxvirus by fusion of the virus envelope with the host cell membrane, *Virology,* **52,** 305 (1973).

106. J. D. Paschke and M. D. Summers, "Early Events in the Infection of the Arthropod Gut by Pathogenic Insect Viruses," in K. Maramorosch and R. E. Shope, Eds., *Invertebrate Immunity,* Academic Press, New York, 1975, p. 75.

107. D. B. Stoltz and M. D. Summers, Pathway of infection of mosquito iridescent virus, I. Preliminary observations on the fate of ingested virus, *J. Virol.,* **8,** 900 (1971).

108. W. R. Campbell, Unpublished information cited in J. D. Paschke and M. D. Summers, ref. 106.

109. G. W. Wagner, J. D. Paschke, W. R. Campbell, and S. R. Webb, Biochemical and biophysical properties of two strains of mosquito iridescent virus, *Virology,* **52,** 72 (1973).

110. J. L. Hardy, E. J. Houk, L. D. Kramer, and W. C. Reeves, Intrinsic factors affecting vector competence of mosquitoes for arboviruses, *Annu. Rev. Entomol.,* **28,** 229 (1983).

111. R. M. Faust and G. E. Cantwell, Inducement of cytoplasmic polyhedrosis by intrahemocoelic injection of freed viral particles into the silkworm *Bombyx mori, J. Invertebr. Pathol.,* **11,** 119 (1968).

112. G. Devauchelle, M. Bergoin, and C. Vago, Etude ultrastructurale du cycle de réplication d'un Entomopoxvirus dans les hémocytes de son hôte, *J. Ultrastruct. Res.,* **37,** 301 (1971).

113. H. Henson, The structure and postembryonic development of *Vanessa urticae* 1. The larval alimentary canal, *Quart. J. Microsc. Sci.,* **74,** 321 (1931).

114. A. G. Richards and P. A. Richards, The origin and composition of the peritrophic membrane of the mosquito *Aedes aegypti, J. Insect Physiol.,* **17,** 2253 (1971).

115. H. F. Evans, Quantitative assessment of the relationships between dosage and response of the nuclear polyhedrosis virus of *Mamestra brassicae, J. Invertebr. Pathol.,* **37,** 101 (1981).

116. Y. Tanada and R. T. Hess, "The Cytopathology of Baculovirus Infections in Insects," in R. C. King and H. Akai, Eds., *Insect Ultrastructure,* Vol. 2, Plenum Publ. Corp., New York, 1984.

117. J. M. Quiot and S. Belloncik, Characterisation d'une polyédrose cytoplasmique chez le lépidoptère *Euxoa scandens,* Riley (Noctuidae, Agrotinae): études in vivo et in vitro, *Arch. Virol.,* **55,** 145 (1977).

118. D. C. Kelly and T. W. Tinsley, Iridescent virus replication: a microscope study of *Aedes aegypti* and *Antherea eucalypti* cells in culture infected with iridescent virus types 2 and 6, *Microbios,* **9,** 75 (1974).

119. W. B. Mathieson and P. E. Lee, Cytology and auto-radiography of Tipula iridescent virus infection of insect suspension cell cultures, *J. Ultrastruct. Res.,* **74,** 59 (1981).

120. K. Maramorosch, *The Atlas of Insect and Plant Viruses,* Academic Press, New York, 1977.

121. R. R. Granados, Infectivity and mode of action of baculoviruses, *Biotechnol. Bioeng.,* **22,** 65 (1980).

122. L. E. Volkman, M. D. Summers, and C. H. Hsieh, Occluded and non-occluded nuclear polyhedrosis virus grown in *Trichoplusia ni:* comparative neutralization, infectivity and in vitro growth studies, *J. Virol.,* **19,** 820 (1976).

123. M. Kobayashi, "Replication Cycle of Cytoplasmic-Polyhedrosis Virus as Observed with the Electron Microscope," in H. Aruga and Y. Tanada, Eds., *The Cytoplasmic-Polyhedrosis Virus of the Silkworm,* Univ. Tokyo Press, Tokyo, 1971, p. 103.

124. D. B. Stoltz and M. D. Summers, Observations on the morphogenesis and structure of a hemocytic poxvirus in the midge *Chironomus attenuatus, J. Ultrastruct. Res.,* **40,** 581 (1972).

125. P. E. Lee, "The Replication of Iridoviruses in Host Cells," in K. Maramorosch and K. E. Sherman, Eds., *Viral Insecticides for Biological Control,* Academic Press, New York, London, 1985, p. 545.

126. L. K. Greenwood and N. F. Moore, Determination of the location of an infection in *Trichoplusia ni* larvae by a small RNA-containing virus using enzyme-linked immunosorbent assay and electron microscopy, *Microbiologica,* **7,** 97 (1984).

127. Ya-Chan Chao, S. Y. Young, and K. S. Kim, Cytopathology of the soybean looper, *Pseudoplusia includens,* infected with the *Pseudoplusia includens* icosahedral virus, *J. Invertebr. Pathol.,* **45,** 16 (1985).

128. G. A. De Zoeten, G. Gaard, and F. B. Diez, Nuclear vesiculation associated with pea enation mosaic virus-infected plant tissue, *Virology,* **48,** 638 (1972).

129. R. E. F. Matthews, Induction of diseases by viruses with special reference to turnip yellow mosaic virus, *Annu. Rev. Pl. Pathol.,* **11,** 147 (1973).

130. K. Maramorosch and R. E. Shope, *Invertebrate Immunity,* Academic Press, New York, London, 1975.

131. H. G. Boman, "Insect Responses to Microbial Infections," in H. D. Burges, Ed., *Microbial Control of Pests and Plant Diseases 1970–1980,* Academic Press, London, New York, 1981, p. 768.

132. J. S. Chadwick, "Hemolymph Changes with Infection or Induced Immunity in Insects and Ticks," in K. Maramorosch and R. E. Shope, Eds., *Invertebrate Immunity,* Academic Press, New York, London, 1975, p. 241.

133. T. Kawarabata and A. Aizawa, Proc. Joint U.S.-Japan Semin. on Microbial Control of Insect Pests, Fukuoka, April 1971, p. 143, cited in J. S. Chadwick, ref. 132 (1967).

134. D. T. Briese and J. D. Podgwaite, "Development of Viral Resistance in Insect Populations," in K. Maramorosch and K. E. Sherman, Eds., *Viral Insecticides for Biological Control,* Academic Press, New York, London, 1985, p. 361.

135. J. C. Cunningham and P. F. Entwistle, "Control of Sawflies by Baculovirus", in H. D. Burges, Ed., *Microbial Control of Pests and Plant Diseases 1970–1980,* Academic Press, London, 1981, p. 379.

136. G. Benz, A nuclear polyhedrosis of *Malacosoma alpiola* (Stand.), *J. Insect Pathol.,* **5,** 215 (1963).

137. H. F. Evans, C. J. Lomer, and D. C. Kelly, Growth of nuclear polyhedrosis virus in larvae of the cabbage moth, *Mamestra brassicae* L., *Arch. Virol.,* **70,** 207 (1981).

138. A. M. Hüger, A virus disease of the Indian rhinoceros beetle *Oryctes rhinoceros* caused by a new type of insect virus *Rhabdionvirus oryctes* gen. n., sp. n., *J. Invertebr. Pathol.,* **8,** 38 (1966).

139. A. M. Hüger, Grundlagen zur biologischen Bekämpfung des Indisohen Nashornkäfers *Oryctes rhinoceros:* histopathologie der Virose bei Käfern, *Z. Angew. Entomol.,* **72,** 309 (1973).

140. C. C. Payne, The isolation and characterization of a virus from *Oryctes rhinoceros, J. Gen. Virol.,* **25,** 105 (1974).

141. P. Monsarrat, G. Meynadier, G. Croizier, and C. Vago, Recherches cytopathologiques sur une maladie virale du Coléoptère *Oryctes rhinoceros* L., *C. R. Acad. Sci. Ser. D.,* **276,** 2077 (1973).

142. P. Monsarrat and J. C. Veyrunes, Evidence of *Oryctes* virus in adult feces and new data for virus characterization, *J. Invertebr. Pathol.,* **27,** 387 (1976).

143. W. N. Norton, S. B. Vinson, and D. B. Stoltz, Nuclear secretory particles associated with the calyx cells of the ichneumonid parasitoid *Campoletis sonorensis* (Cameron), *Cell Tissue Res.,* **162,** 195 (1975).

144. K. M. Edson, S. B. Vinson, D. B. Stoltz, and M. D. Summers, Virus in a parasitoid wasp: suppression of the cellular immune response in the parasitoid's host, *Science,* **211,** 5824 (1981).

145. J. J. Hamm, D. A. Nordlung, and O. G. Marti, Effects of a nonoccluded virus of *Spodoptera frugiperda* (Lepidoptera:Noctuidae) on the development of a parasitoid, *Cotesia marginiventris* (Hymenoptera:Braconidae), *Environ. Entomol.,* **14,** 258 (1985).

146. H. F. Evans and G. P. Allaway, Dynamics of baculovirus growth and dispersal in *Mamestra brassicae* L. (Lepidoptera:Noctuidae) larval populations introduced into small cabbage plots, *Appl. Environ. Miocrobiol.,* **45,** 493, (1983).

147. M. Himeno, F. Matsubara, and K. Hayashiya, The occult virus of nuclear polyhedrosis of the silkworm larvae, *J. Invertebr. Pathol.,* **22,** 292 (1973).

148. D. J. McKinley, J. A. Brown, C. C. Payne, and K. A. Harrap, Cross-infectivity and activation studies with four baculoviruses, *Entomophaga,* **26,** 79 (1981).

149. M. Jurkovicova, Activation of latent virus infections in larvae of *Adoxophyes orana* (Lepidoptera:Noctuidae) by foreign polyhedra, *J. Invertebr. Pathol.,* **34,** 213 (1979).

150. K. D. Biever and J. D. Wilkinson, A stress-induced granulosis virus of *Pieris rapae, Environ. Entomol.,* **7,** 572 (1978).

151. M. E. Martignoni and J. E. Milstead, Trans-ovum transmission of the nuclear polyhedrosis virus of *Colias eurytheme* Boiduval through contamination of the female genitalia, *J. Insect Pathol.,* **4,** 113 (1962).

152. L. Skuratovakaya, L. Strokovskaya, L. Alexeenko, N. Miriutha, and I. Kok, Replication of baculovirus genome at productive and latent infections: structural aspects. *Abstr. 6th Int. Congr. Virol.,* p. 334, Sendai, Japan.

153. C. M. Ignoffo, "Evaluation of in vivo Specificity of Insect Viruses," in M. Summers, R. Engler, L. A. Falcon, and P. Vail, Eds., *Baculoviruses for Insect Pest Control: Safety Considerations,* Am. Soc. Microbiol., Washington, D.C., 1975, p. 52.

154. P. F. Entwistle, "Epizootiology and Strategies of Microbial Control," in J. M. Franz and M. Lindaur, Eds., *Biological Plant and Health Protection,* Fischer Verlag, Stuttgart (In Press).

155. H. F. Evans, "Ecology and Epizootiology of Baculoviruses," in R. A. Granados and B. A. Federici, Eds., *The Biology of Baculoviruses,* CRC Press (In Press).

156. M. E. Martignoni, P. J. Iwai, K. M. Hughes, and R. B. Addison, A cytoplasmic polyhedrosis of *Hemerocampa pseudotsugata, J. Invertebr. Pathol.,* **13,** 15 (1969).

157. C. C. Payne, "Cytoplasmic Polyhedrosis Viruses," in E. W. Davidson, Ed., *Pathogenesis of Invertebrate Microbial Diseases,* Allanheld Osmun, Totowa, New Jersey, 1981, p. 61.

158. A. Magnoler, Effects of a cytoplasmic polyhedrosis on larval and postlarval stages of the gypsy moth, *Porthetria dispar, J. Invertebr. Pathol.,* **23,** 263 (1974).

159. M. M. Neilson, Effects of a cytoplasmic polyhedrosis on adult Lepidoptera, *J. Invertebr. Pathol.,* **7,** 306 (1965).

160. N. Bellemare and S. Belloncik, Etudes au laboratoire des effets d'une polyédrose cytoplasmique sur le ver gris blanc *Euxoa scandens* (Lepidoptère:Noctuidae Agrotinae), *Ann. Soc. Entomol. Que.,* **26,** 28 (1981).

161. S. Tanaka, "Cross Transmission of Cytoplasmic-Polyhedrosis Viruses," in H. Aruga and Y. Tanada, Eds., *The Cytoplasmic-Polyhedrosis Virus of the Silkworm,* Univ. Tokyo Press, Tokyo, 1971, p. 201.

162. G. P. Allaway, *Infectivity of some occluded insect viruses,* Ph. D. Thesis, Univ. London, (1982).

163. E. Kurstak and S. Garzon, "Entomopoxviruses (Poxviruses of Invertebrates)," in K. Maramorosch, Ed., *The Atlas of Insect and Plant Viruses,* Academic Press, New York, London, 1977, p. 29.

164. F. L. Mitchell and J. W. Smith, Pathology and bioassays of the lesser cornstalk borer *(Elasmopalpus lignosellus)* Entomopoxvirus, *J. Invertebr. Pathol.,* **45,** 75 (1985).

165. R. H. Goodwin and R. J. Roberts, Diagnosis and infectivity of entomopoxviruses from three Australian scarab beetle larvae (Coleoptera:Scarabaeidae), *J. Invertebr. Pathol.,* **25,** 47 (1975).

166. J. E. Henry, B. P. Nelson, and J. W. Jutila, Pathology and development of the grasshopper inclusion body virus in *Melanoplus sanguinipes, J. Virol.,* **3,** 605 (1969).

167. R. R. Granados, "Entomopoxvirus Infections in Insects" in E. W. Davidson, Ed., *Pathogenesis of Invertebrate Microbial Diseases,* Allanheld Osmun, Totowa, New Jersey, 1981, p. 101.

168. N. Xeros, Development of the *Tipula* iridescent virus, *J. Insect Pathol.,* **6,** 261 (1964).

169. D. W. Hall and D. W. Anthony, Pathology of a mosquito iridescent virus (MIV) infecting *Aedes taeniorhynchus, J. Invertebr. Pathol.,* **18,** 61 (1971).

170. J. B. Carter, The mode of transmission of *Tipula* iridescent virus. I. Source of infection, *J. Invertebr. Pathol.,* **21,** 123 (1973).

171. M. Fowler and J. S. Robertson, Iridescent virus infection in field populations of *Wiseana cervinata* (Lepidoptera: Hepialidae) and *Witlesia* sp. (Lepidoptera:Pyralidae) in New Zealand, *J. Invertebr. Pathol.,* **19,** 154 (1972).

172. P. P. Sikorowski and G. E. Tyson, Per os transmission of iridescent virus of *Heliothis zea* (Lepidoptera:Noctuidae), *J. Invertebr. Pathol.,* **44,** 97 (1984).

173. J. B. Carter, Personal communication cited in D. W. Hall, ref. 169.

174. J. R. Linley and H. T. Neilson, Transmission of a mosquito iridescent virus in *Aedes taeniorhynchus.* I. Laboratory experiments, *J. Invertebr. Pathol.,* **12,** 7 (1968a).

175. J. R. Linley and H. T. Nielsen, Transmission of a mosquito iridescent virus in *Aedes taeniorhynchus.* II. Experiments related to transmission in nature, *J. Invertebr. Pathol.,* **12,** 178 (1968b).

176. D. W. Hall and D. W. Anthony, An "R" type *Iridovirus* from *Aedes vexans* (Meigen), *Mosq. News,* **36,** 536 (1976).

177. H. C. Chapman, T. B. Clark, D. W. Anthony, and F. E. Glenn, An iridescent virus from larvae of *Corothrella brakeleyi* (Diptera: Chaeoboridae) in Louisiana, *J. Invertebr. Pathol.,* **18,** 284 (1971).

178. G. Meynadier, C. Vago, G. Plantevin, and P. Atger, Virose d'un type habituel chez le Lépidoptère, *Galleria mellonella* L., *Rev. Zool. Agr. Appl.,* **63,** 207 (1964).

179. S. Garzon and E. Kurstak, Ultrastructural studies on the morphogenesis of the densonucleosis virus (parvovirus), *Virology,* **70,** 517 (1976).

180. E. Kurstak and C. Vago, Transmission du virus de la densonucléose par le parasitisme d'un Hyménoptère, *Rev. Can. Biol.,* **26,** 311 (1967).

181. A. Amargier, C. Vago, and G. Meynadier, Etude histopathologique d'un nouveau type de vrose mis en évidence chez le lépidoptère *Galleria mellonella, Arch. Ges. Virusforsch,* **15,** 659 (1965).

182. H. Watanabe, S. Maeda, M. Matsui, and T. Shimizu, Histopathology of the midgut epithelium of the silkworm *Bombyx mori,* infected with a newly isolated virus from the flacherie diseased larvae, *J. Sericult. Sci. Japan,* **45,** 29 (1976).

183. C. Suto, F. Kawamoto, and N. Kumada, A new virus isolated from the cockroach, *Periplaneta fuliginosa* (Serville), *Microbiol. Immunol.,* **23,** 207 (1979).

184. D. C. Kelly, M. D. Ayres, L. K. Spencer and C. F. Rivers, Densonucleosis virus 3: a recent insect parvovirus isolated from *Agraulis vanillae* (Lepidoptera:Nymphalidae), *Microbiologica,* **3,** 455 (1980).

185. H. Watanabe and J. Shimizu, Epizootiological studies on the occurrence of densonucleosis in the silkworm, *Bombyx mori,* reared at sericultural farms, *J. Sericult. Sci. Japan,* **49,** 485 (1980).

186. C. F. Rivers and J. F. Longworth, A non-occluded virus of *Junonia coenia* (Nymphalidae:Lepidoptera), *J. Invertebr. Pathol.,* **20,** 369 (1972).

187. N. N. Lebinets, D. B. Tsarichkova, L. V. Karpenko, A. G. Kononko, and L. P. Buchatskij, Studies of the *Aedes aegypti* L. densonucleosis virus effect on pre-imaginal stages of different species of blood-sucking mosquitoes, *Mikrobiol. Zh.,* **40,** 352 (1978).

188. J. F. Longworth, Small isometric viruses of invertebrates, *Adv. Virus Res.,* **23,** 103 (1978).

189. C. Reinganum, G. T. O'Loughlin, and T. W. Hogan, A non-occluded virus of the field cricket *Teleogryllus oceanicus* and *T. commodus* (Orthoptera:Gryllidae), *J. Invertebr. Pathol.,* **16,** 214 (1970).

190. L. Bailey, Viruses attacking the honey bee, *Adv. Virus Res.,* **20,** 271 (1976).

191. D. Teninges, A. Ohanessian, C. Richard-Moland, and D. Contamine, Isolation and biological properties of Drosophila X virus, *J. Gen. Virol.,* **42,** 241 (1979).

192. D. Teninges, D. Contamine and G. Brun, "*Drosophila* Sigma Virus," in D. H. L. Bishop, Ed., *Rhabdoviruses* Vol. III, CRC Press, Boca Raton, Florida, 1980, p. 113.

193. P.D. Scotti, J. F. Longworth, N. Plus, G. Croizier, and C. Reinganum, The biology and ecology of strains of an insect small RNA virus complex, *Adv. Virus Res.,* **26,** 117 (1981).

194. N. Plus, G. Croizier, C. Reinganum and P. D. Scotti, Cricket paralysis virus and *Drosophila* C virus: serological analysis and comparison of capsid polypeptides and host ranges, *J. Invertebr. Pathol.,* **31,** 296 (1978).

195. C. Reinganum, S. J. Gagen, S. B. Sexton, and H. P. Vellacott, A survey for pathogens of the black field cricket, *Teleogryllus commodus,* in the Western District of Victoria, Australia, *J. Invertebr. Pathol.,* **38,** 153 (1981).

196. N. Plus, G. Croizier, F. X. Jousset, and J. David, Picornaviruses of laboratory and wild *Drosophila melanogaster:* geographical distribution and serotypic composition, *Ann. Microbiol. (Inst. Pasteur),* **126 A,** 107 (1975).

197. P. L'Heritier, Drosophila viruses and their role as evolutionary factors, *Evol. Biol.,* **4,** 185 (1970).

198. A. Krieg, "Interactions Between Pathogens," in H. D. Burges and N. W. Hussey, Eds., *Microbial Control of Insects and Mites,* Academic Press, London, New York, 1971, p. 459.

199. K. E. Sherman, "Multiple Virus Interactions," in K. Maramorosch and K. E. Sherman, Eds., *Viral Insecticides for Biological Control,* Academic Press, London, New York, 1985, p. 735.

200. G. Benz, "Synergism of Micro-Organisms and Chemical Insecticides," H. D. Burges and N. W. Hussey, Eds., *Microbial Control of Insects and Mites,* Academic Press, London, New York, 1971, p. 327.

201. S. Tanaka and H. Aruga, Interference between the midgut nuclear polyhedrosis virus and the cytoplasmic polyhedrosis virus in the silkworm, *Bombyx mori, J. Sericult. Sci. Japan,* **36,** 169 (1967).

202. F. T. Bird, Polyhedrosis and granulosis viruses causing single and double infections in the spruce budworm, *Choristoneura fumiferana* Clemens, *J. Insect Pathol.*, **1**, 406 (1959).

203. D. C. Kelly, Suppression of baculovirus and iridescent virus replication in dually infected cells, *Microbiologica*, **3**, 177 (1980).

204. J. M. Quiot, C. Vago, and M. Tchoukchry, Experimental study of the interaction of two invertebrate viruses in lepidopteran cell culture, *C. R. Acad. Sci.*, **290**, 199 (1980).

205. F. Odier, Mise en évidence et étude d'un complex d'unmaladies a parvovirus, baculovirus et iridovirus, *Entomophaga*, **22**, 397 (1977).

206. K. S. Ritter and Y. Tanada, Interference between two nuclear polyhedrosis viruses of the armyworm, *Pseudoletia unipuncta, Entomopohaga*, **23**, 349 (1978).

207. V. H. Whitlock, Simultaneous treatments of *Heliothis armigera* with a nuclear polyhedrosis and a granulosis virus, *J. Invertebr. Pathol.*, **29**, 297 (1977).

208. Y. Tanada, H. Inoue, R. T. Hess, and E. M. Omi, Site of action of a synergistic factor of a granulosis virus of the armyworm, *Pseudaletia unipuncta, J. Invertebr. Pathol.*, **34**, 249 (1980).

209. G. W. Wagner, S. R. Webb, J. D. Paschke, and W. R. Campbell, A picornavirus isolated from *Aedes taeniorhynchus* and its interaction with mosquito iridescent virus, *J. Invertebr. Pathol.*, **24**, 380 (1974).

210. R. E. Lowe and J. D. Paschke, Simultaneous infection with the nucleopolyhedrosis and granulosis viruses of *Trichoplusia ni, J. Invertebr. Pathol.*, **12**, 86 (1968).

211. R. E. Lowe and J. D. Paschke, Pathology of a double viral infection of *Trichoplusia ni, J. Invertebr. Pathol.*, **12**, 438 (1968).

212. H. Aruga and H. Watanabe, Interference between uv-inactivated and active cytoplasmic-polyhedrosis viruses in the silkworm, *Bombyx mori*, II. Silkworm strain and time interval of inoculation, *J. Sericult. Sci. Japan*, **39**, 382 (1970).

213. H. Aruga and H. Watanabe, Interference between uv-inactivated and active cytoplasmic polyhedrosis viruses in the silkworm, *Bombyx mori* L. *J. Sericult. Sci. Japan*, **40**, 176 (1971).

214. J. C. Cunningham and J. F. Longworth, The activation of occult nuclear-polyhedrosis viruses by foreign nuclear polyhedra, *J. Invertebr. Pathol.*, **10**, 361 (1968).

215. G. Croizier, D. Godse, and J. Vlak, Selection de types viraux daus les infections double à Baculovirus chez larves de Lepidoptère, *C. R. Acad. Sci.*, **290**, 579 (1980).

216. J. R. Fuxa, Interactions of the microsporidium *Vairimorpha necatrix* with a bacterium, virus, and fungus in *Heliothis zea, J. Invertebr. Pathol.*, **33**, 316 (1979).

217. F. A. Murphy, "Cellular Resistance to Arbovirus Infection," in L. Bulla and T. C. Cheng, Eds., *Pathobiology of Invertebrate Vectors of Disease*, New York Acad. Sci., New York, 1975, p. 197.

218. D. K. Hunter, S. J. Collier, and D. F. Hoffman, Compatability of malathion and the granulosis virus of the Indian meal moth, *J. Invertebr. Pathol.*, **25**, 389 (1975).

219. J. R. McVay, R. T. Gudauskas, and J. D. Harper, Effects of *Bacillus thuringiensis*-nuclear polyhedrosis virus mixtures on *Trichoplusia ni* larvae, *J. Invertebr. Pathol.*, **29**, 367 (1977).

220. T. R. E. Southwood and H. N. Comins, A synoptic population model, *J. Anim. Ecol.*, **45**, 949 (1976).

221. T. R. E. Southwood, "The Relevance of Population Dynamic Theory to Pest Status," in J. M. Cherrett and G. R. Sager, Eds., *Origins of Pest, Parasite, Disease and Weed Problems*, Blackwell, Oxford, U.K., 1977, p. 35.

222. T. R. E. Southwood, "Bionomic Strategies and Population Parameters," in R. M. May, Ed., *Theoretical Ecology*, Blackwell, Oxford, U.K., 2nd Ed., 1981, p. 30.

223. R. G. Conway, "Man Versus Pests," in R. M. May, Ed., *Theoretical Ecology*, Blackwell, Oxford, U.K., 2nd Ed., 1981, p. 356.

224. D. C. Kelly, T. Lescott, M. D. Ayres, D. Carey, A. Coutts and K. A. Harrap, Induction of a

non-occluded baculovirus persistently infecting *Heliothis zea* cells by *Heliothis armigera* and *Trichoplusia ni* nuclear polyhedrosis viruses, *Virology,* **112,** 174 (1981).

225. T. D. C. Grace, The development of a cytoplasmic polyhedrosis in insect cells grown in vitro, *Virology,* **18,** 33 (1962).

226. E. T. Nielsen, The initial stage of migration in salt-marsh mosquitoes, *Bull. Entomol. Res.,* **49,** 305 (1958).

227. M.-Demerec (Ed.) *Biology of Drosophila,* Hafner Publ. Co., New York, London, 1965.

228. F.-X. Jousset and N. Plus, Etude de la transmission horizontale et de la transmission verticale des picornavirus de *Drosophila melanogaster* et de *Drosophila immigrans, Ann. Microbiol. (Inst. Pasteur),* **126B,** 231 (1975).

229. C. J. D'Arcy, P. A. Burnett, and A. D. Hewings, Detection, biological effects and transmission of a virus of the aphid *Rhopalosiphum padi, Virology,* **114,** 268 (1981).

230. A. Y. Moussa, A new virus disease in the housefly, *Musca domestica* (Diptera), *J. Invertebr. Pathol.,* **31,** 204 (1978).

231. L. K. Etzel and L. A. Falcon, Studies of transovum and transstadial transmission of a granulosis virus of the codling moth, *J. Invertebr. Pathol.,* **27,** 13 (1976).

232. R. F. Sheppard and G. R. Stairs, Dosage-mortality and time-mortality studies of a granulosis virus in a laboratory strain of the codling moth, *Laspeyresia pomonella, J. Invertebr. Pathol.,* **29,** 216 (1977).

233. M. Odindo, C. C. Payne, and N. Crook, A novel virus isolated from the tsetse fly *Glossina pallidipes, Abstr. 6th Int. Congr. Virol.,* p. 336, Sendai, Japan, (1984).

234. M. O. Odindo, D. M. Sabwa, P. A. Amutalla, and W. A. Otieno, Preliminary tests on the transmission of virus-like particles to the tsetse *Glosia pallidipes, Insect Sci. Application,* **2,** 219 (1981).

235. A. Amargier, J.-P. Lyon, C. Vago, G. Meynadier, and J.-C. Veyrunes, Mise en évidence et purification d'un virus dans la prolifération monstrueuse glandulaire d'Insectes. Etude sur *Merodon equestris* F. (Diptère, Syrphidae), *C. R. Acad. Sci. Paris,* **289D,** 481 (1979).

236. W. E. H. Hodson, The large narcissus bulbfly, *Merodon equestris* Fab. (Syrphidae), *Bull. Entomol. Res.* **23,** 429 (1932).

237. R. M. Anderson and R. M. May, The population dynamics of microparasites and their invertebrate hosts, *Proc. R. Soc. Phil. Trans,* **291,** 251 (1981).

238. M. E. Martignoni, Contributo alla conoscenza di una granulosi di Eucosma griseana (Hübner) (Tortricidae, Lepidoptera) quale fattore limitante il pullulamento dell'insetto nella Engadina alta, *Mitt. Schweiz Anst. Forstl. Versuchswesen,* **32,** 371 (1957).

239. D. E. Elgee, Persistence of a virus of the white-marked tussock moth on Balsam fir foliage, *Bi-monthly Res. Notes, Can. For. Serv.,* (1975).

240. G. Mihalache and D. Pirvescu, Epizootille verotice in padurile infestate de defoliatorul *Lymantria dispar* L., *Silv. Exploat. aduritor,* **92,** 146 (1977).

241. H. Roberts, *Lymantria ninayi* B. Br. (Lep. Fam. Lymantriidae) a potential danger to Pinus afforestation in the highlands of Papua New Guinea, *Trop. For. Res. Note SR 37, Papua New Guinea,* 12 pp. (1979).

242. K, Katagiri, Review on microbial control of insect pests in forests in Japan, *Entomophaga,* **14,** 203 (1969).

243. A. M. Crawford and R. R. Granados, Nonoccluded baculoviruses, *Proc. 3rd Int. Colloq. Invertebr. Pathol., Brighton, UK,* 1982, p. 154.

244. M. E. Martignoni, Progressive nucleopolyhedrosis in adults of *Peridroma saucia* (Hübner), *J. Insect Pathol.,* **6,** 368 (1964).

245. C. M. Ignoffo, D. L. Hostetter, P. P. Sikorowski, G. Sutter, and W. M. Brooks, Inactivation of representative species of entomopathic viruses, a bacterium, fungus and a protozoan by an ultraviolet light source, *Environ. Entomol.* **6,** 411 (1977).

246. A. Krieg, A. Groner, J. Huber, and M. Matter, Uber die Wirkung von mittel-und langwellen ultravioletter Strahlen (UV-B and UV-A) auf insektenpathogene Bakteria und Viren und deren Becenflussing durch UV-Schultzstoffe, *Nachrichtenbl. Dtsch. Pflanzenshutzdienstes (Braunschweig), 32,* 100 (1980).

247. H. F. Evans and P. F. Entwistle, "Epizootiology of the Nuclear Polyhedrosis Virus of European Spruce Sawfly with Emphasis on Persistence of Virus Outside the Host," in E. Kurstak, Ed., *Microbial and Viral Pesticides,* Marcel Dekker, New York, 1982, p. 449.

248. D. A. Small, Aspects of the attachment of a nuclear polyhedrosis virus from the cabbage looper *(Trichoplusia ni)* to the leaf surface of cabbage *(Brassica oleracea),* Ph.D. Thesis, Univ. Oxford, UK, (1985).

249. I. Gipson and H. A. Scott, An electron microscope study of the effects of various fixatives and thin section enzyme treatments on a nuclear polyhedrosis virus, *J. Invertebr. Pathol., 26,* 171 (1975).

250. N. Dubois, Effectiveness of chemically decontaminated *Neodiprion sertifer* polyhedral inclusion body suspensions, *J. Econ. Entomol., 69,* 93 (1976).

251. G. F. Rohrmann, M. N. Pearson, J. J. Bailey, R. R. Becker, and G. S. Beaudreau, N-terminal polyhedrin sequences and occluded *Baculovirus* evolution, *J. Mol. Evol., 17,* 329 (1981).

252. G. E. Smith and M. D. Summers, DNA homology among subgroup A, B and C baculoviruses, Virology, **123,** 393 (1982).

253. W. M. Fitch,"The Phyletic Interpretation of Macromolecular Sequence Information: Simple Methods," in M. D. Hecht, P. C. Gordy, and B. M. Hecht, Eds., *Major Patterns in Vertebrate Evolution,* Plenum Press, New York, 1977, p. 169.

254. H. B. Boudreaux, *Arthropod Phylogeny with Special Reference to Insects,* John Wiley & Sons, New York, 1979.

255. M. E. Martignoni and P. J. Iwai, "A Catalogue of Viral Diseases of Insects, Mites and Ticks," in H. D. Burges, Ed., *Microbial Control of Pests and Plant Diseases 1970–1980,* Academic Press, London, New York, 1981, p. 897.

256. J. Kalmakoff and A. M. Crawford, "Enzootic Virus Control of *Wiseana* spp.," in E. Kurstak, Ed., *Microbial and Viral Pesticides,* Marcel Dekker, New York, Basel, 1982, p. 435.

11

DISEASES CAUSED BY BACTERIA AND OTHER PROKARYOTES

ALOYSIUS KRIEG

Biologische Bundesanstalt für Land-und Forstwirtschaft
Darmstadt, Federal Republic of Germany

---

1. INTRODUCTION

The microecology of arthropod-associated bacteria embraces diverse relationships. Bacteria may serve as food for insects (e.g., for mosquito larvae) or are often associated with the insect's digestive functions. Characteristic gut microflora have been established in many insects. Closer relationships occur in mutualistic symbiosis between specialized bacteria and their hosts, particularly in blood- or plantsap-sucking arthropods. Moreover, certain arthropods act as vectors of circulative arthropod-born pathogens (for plants and animals). In some cases, such vector-transmitted prokaryotes are moderately pathogenic for the vector itself. Finally, facultative and obligate pathogenic bacteria are well known as etiological agents causing severe intoxications or infectious diseases of arthropods that may result in epizootics.

The association of arthropods with prokaryotes is often considered the result of a long coevolutionary process during which special relationships developed. This is evident with many mutualistic symbiotes and also obligate pathogenic prokaryotes, including arthropod-borne pathogens. However, there is no indication of evolutionary adaptation of the facultative pathogens or toxigenic bacteria for their potential insect hosts. Nevertheless, these bacteria have conquered the ecological niche "insect" as an additional substrate for their development. Furthermore, our knowledge is very poor regarding the epizootiology of nearly all nonobligate pathogenic bacteria of insects.

In the following paragraphs, the important insect-pathogenic bacteria and arthropod-borne prokaryotes are surveyed with respect to the relevant etiological and epizootiological features. Although extensive diagnostic studies are a prerequisite for epizootiological evaluations and models, special details concerning the microbiology and taxonomy of these pathogens are not considered here, and the readers should refer to pertinent literature (1, 2).

2. Gut Microflora and Pathogens

Since most insect pathogenic bacteria enter their host *per os* and *via* the alimentary tract, they interact with the gut microflora. Studies with axenically reared insects can demonstrate the role of the gut microflora on the ingested pathogens and their survival in nonaxenic host insects.

Although the external environment (substrate) governs the bacteria that are available, several internal factors determine their establishment within the intestine. The limiting internal microecological factors are: (mostly high) pH, oxygen deficit, special nutritional conditions, and several secreted substances including immune factors (e.g., lysozyme and other bactericides). In addition, the bacteria also can be inhibited by foreign metabolites (e.g., antibiotics) ingested with food.

Normally, insects have only small numbers of bacteria within their gut lumens, but abnormal conditions (e.g., enhanced temperatures, reduced feeding, changes in diet, uptake of antibiotics) can favor the multiplication of certain gut bacteria and sometimes induce disturbance or even diarrhea.

In many arthropods (e.g., locusts and many lepidopterous larvae) sterile hatching larvae quickly acquire a more or less individual gut flora by ingesting contaminated food. In other cases (e.g., flies reared in the laboratory), a characteristic cyclic gut flora is established because the eggs become contaminated externally by fecal bacteria. In holometabolic insects, however, a cycle of gut flora is often prevented by an autosterilization process of the intestine during pupation. This is typical of mosquitoes, in which the larval gut flora is not transmitted to adults.

Many blood-sucking arthropods have sterile intestines, or, if they become contaminated, they are often able to rid themselves of ingested bacteria. Only potential pathogens can induce a fatal enteral infection after a blood meal, especially at high temperatures, as in the case with *Yersinia pestis* in *Xenopsylla cheopis* (3).

Typical gut-inhabitants of many insects are *Streptococcus faecalis* (4) and several species of Enterobacteriaceae. Only as a consequence of toxic effects, trauma, or parasitism (e.g., by nematodes) do bacteria of the gut microflora invade the hemocoel, but all apathogenic species (e.g., *Enterobacter aerogenes, Escherichia coli,* or *Bacillus megaterium*) are eliminated sufficiently by the immune mechanisms of the hemolymph (see Section 5.3.2). However, the so-called facultative pathogens (e.g., *Proteus* spp., *Serratia* spp., or *Bacillus cereus*), which are largely immunoresistant, can induce an advanced bacteremia within the hemocoel followed by a septicemia. It is often difficult to differentiate between such secondary invaders and true primary pathogens. Furthermore, at *post mortem* the apathogenic saprophytes may multiply within the cadaver. Therefore Koch's postulates must be fulfilled for the identification of a bacterium as a primary pathogen.

3. Toxigenic and Pathogenic Eubacteria

Most insect-pathogenic Eubacteria are able to grow under saprophytic conditions and are, therefore, facultative pathogens. Their pathogenicity is predominantly based on the production of host-specific toxins, and these bacteria can be designated as toxigenic saprophytes (e.g., *Bacillus thuringiensis*). However, with obligate pathogens, which multiply only within their hosts and cannot be isolated and cultured on artificial media (e.g., *Bacillus popilliae*), special in vivo indicator systems are necessary for their demonstration and cloning.

3.1. Pseudomonadaceae

The Pseudomonadaceae are gram-negative, strictly aerobic, rod-shaped bacteria, motile by polar flagella (2).

3.1.1. *PSEUDOMONAS AERUGINOSA.* *P. aeruginosa,* a facultative pathogen, has caused outbreaks in laboratory rearings of grasshoppers (*Melanoplus vivittatus* (5, 6)) and bugs (*Oncopeltus fasciatus* (7)). Furthermore, it has been isolated from flies (*Stomoxys calcitrans, Musca domestica*), coleopterous larvae (*Melolontha* spp., *Diabrotica* spp.), and lepidopterous larvae (*Agrotis* spp.). Strains of *P. aeruginosa*, however, are only moderately virulent. When grasshoppers are fed a high dose, the bacteria normally do not multiply in the gut and most of them are rapidly eliminated. However, when the bacteria accidentally penetrate the gut wall into the hemocoel, an intensive bacterial growth occurs followed by fatal septicemia. The low slopes of the dosage-mortality regression lines indicate that the insect populations are heterogenous with respect to their susceptibility. Outbreaks in insectary-reared grasshoppers are assumed to originate from imported egg pods, the foam of which is contaminated at a low percentage with small numbers of *P. aeruginosa*. The bacteria infect the susceptible hatching nymphs, which die rapidly and contaminate their environment with high numbers of bacteria. From such a focus, the disease can become epizootic (5, 6). The bacteria also can be acquired from contaminated food, by cannibalism, or by contact with decaying insects. In nature, however, epizootics caused by *P. aeruginosa* occur rarely, if at all. The contamination of egg pods in nature is suggested to be soil-borne because *P. aeruginosa* is readily isolated from soil and sewage.

Beard (8) tested the epizootic significance of *P. aeruginosa* in the field in Canada (Alberta). The application of a bacterial culture to egg beds and as a bait to nymphs and adults resulted in poor suppression of the grasshopper population, and only a few caged nymphs died from infections.

In the rearing of the milkweed bug (*O. fasciatus*), the preconditions for spontaneous laboratory epizootics are unknown, especially the source of *P. aeruginosa*. Temperature and high humidity may favor the outbreak. The infection is usually fatal (7). Adult bees (*Apis mellifera*) also are susceptible to *P. aeruginosa* and develop fatal septicemia after ingesting the pathogen (9). In flies (e.g., *Musca domestica*), the bacterium can persist in pupae and adults from larvae that were fed *P. aeruginosa*. Therefore, flies may be a reservoir host (10).

In vertebrates, *P. aeruginosa* was first isolated from infections of wounds, and from respiratory and urinary tracts in man, and later from necrotizing enteritis and abortion in swine and cattle. It was also found to be a plant pathogen inducing diseases such as internal brown rot of onions in the field and in storage. Since *P. aeruginosa* has been detected repeatedly on plants and vegetables, it may be imported into hospitals and insectaries with plant material.

P. aeruginosa grows on simple artificial media, such as nutrient broth, and normally produces a blue to greenish-brown pigment, though colorless strains are known. The optimum temperature for growth is 37°C, and it can also grow at 41°C but not at 4°C. Biochemical methods are sufficient for recognition. Strains are identified by serotyping (0- and H- antigens) or bacteriocin-typing, but the results of phagotyping are often inconsistent.

3.1.2. OTHER PSEUDOMONADS. Fluorescent pseudomonads, such as *P. fluorescens, P. chlororaphis, P. aureofaciens,* and *P. putida*, are occasionally found as potential pathogen of insects. For example, *P. fluorescens* has been investigated in soil insects such as the larvae of *Agrotis segetum, Amphimallon* sp., *Melolontha* spp., *Phyllopertha* sp., and in *Leptinotarsa decemlineata*. In addition, pathogenic strains have been reported from mosquito larvae (*Aedes, Anopheles, Culex, Culiseta*) (11). *P. chlororaphis* has been isolated from phytophagous insects such as the larvae of *Aporia crataegi, Cacoecia crataegana*, and *Euproctis chrysorrhoea*. *P. putida* has been found virulent for *Dacus oleae* adults after ingestion (12).

In contrast to *P. aeruginosa,* fluorescent pseudomonads do not grow at 41°C, but usually at 4°C; their temperature optimum lies between 25 and 30°C. They are usually isolated from water, soil, and spoiled food.

3.2. Enterobacteriaceae

The Enterobacteriaceae are gram-negative, facultatively anaerobic, rod-shaped bacteria, motile by peritrichous flagella (2).

3.2.1. *SERRATIA* SPP. Red-pigmented strains of this genus (usually *S. marcescens*) have been recovered frequently from diseased insects (6, 13). These bacteria have been isolated from Orthoptera (*Schistocerca gregaria, Periplaneta americana*), Coleoptera (*Melolontha melolontha, Tenebrio molitor*), Hymenoptera (*Neodiprion lecontii*), Lepidoptera (Bombyx mori, *Malacosoma* spp., *Carpocapsa pomonella*), and Diptera (*Drosophila* sp., *Ceratitis capitata,*

Dacus dorsalis, Musca domestica) Many insect species are susceptible to ingested high doses of red-pigmented strains, for example, *S. gregaria, N. lecontii, Pristiphora erichsonii, Pieris brassicae, Lymantria dispar, Trichoplusia ni, Ostrinia nubilalis,* and *Galleria mellonella.*

Pathogenic nonpigmented strains of *Serratia* have been isolated from *S. gregaria, Melanoplus bivittatus, Melolontha* spp., *Leptinotarsa decemlineata, Bombyx mori,* and several noctuids.

In 1979, Grimont et al. (14) attempted a reclassification of insect-associated strains of *Serratia* spp. in light of new taxonomic knowledge. Several nonpigmented strains, previously not recognized as *Serratia* spp., have now been classified as *Serratia liquefaciens.* Such strains are *Bacillus melolonthae-liquefaciens, Bacillus noctuarum, Paracolobactrum rhynocoli,* and *Enterobacter cloacae* type B.

Strains of *Serratia* spp. grow on ordinary media, such as nutrient agar, with an optimum temperature of between 25 and 30°C. No growth occurs at 37°C. Strain differentiation is possible by serotyping (0-antigens), isoesterase pattern, bacteriocin-reaction, and phagotyping.

Some biotypes of *S. marcescens* have been recognized in clinical material. This species is now found more frequently in human pathology with wound infections, which are often followed by septicemia in patients with delayed immune defense.

S. marcescens is known as an insect pathogen of only moderate virulence. Steinhaus (13) reported the inconsistency of pathological reations within a particular host population. Stephens (15) stated that *S. marcescens* was comparable in its virulence to that of *P. aeruginosa* in locusts.

Serratia has been reported as a potential pathogen in natural populations of members of the Acridiidae. Usually it induces outbreaks only in laboratory rearings of grasshoppers and other insects rather than epizootics in the field. After ingestion, *Serratia* spp. can persist and even grow in the guts of several insects. The bacterium is found in the feces. A fatal septicemia is induced only after accidental penetration through the gut wall and invasion into the hemocoel. Infections may originate from contaminated eggs, since *S. marcescens* has been isolated from eggs of *Ceratitis capitata* and *Heliothis zea* (16). Cannibalism may be an important factor in transmission (17). Fecal contaminations, however, provide a means whereby the disease becomes enzootic.

In a small field experiment of a swarm of desert locust nymphs in Kenya (Africa), Stevenson (17) applied *S. marcescens* in a bait and as a spray on plants. The examination of dead locusts (8 days after application) indicated a specific infection, but the swarm was not significantly reduced.

Recently *S. liquefaciens* and *S. marinorubra* have been isolated from the larval gut contents of healthy and diseased larvae of *Costelytra zealandica* (Coleoptera) collected from several locations throughout New Zealand. Virulent strains of both bacteria are suggested as etiological agents of the honey disease of grass grubs. In contrast to other *Serratia*-induced infections, this disease does not cause rapid death, but rather a general wasting of the larvae.

The disease is presumed to cause a breakdown in the population peaks of *C. zealandica* grubs in pastures. A high incidence of the disease was found in the Canterbury region of New Zealand. In a small field experiment, 2 months after the application of a bacterial culture a number of larvae showed symptoms of the disease, and this was followed by a reduction in population density (18).

From several regions of the USA, Burnside (19) described fatal outbreaks in honeybee colonies due to septicemia in adults. He isolated "*Bacillus* (syn. *Pseudomonas) apisepticus*" as the etiological agent which was identified later as *S. marcescens.* A similar bee disease was found in Europe, especially in Switzerland (20). A typical symptom was weakening of activity, resulting in the inability of a great number of bees to fly in front of hives. It was not possible to infect bees perorally, but only after an intensive contact from spraying or dipping the bees into a bacterial suspension. Wille and Pinter (20) suggested, therefore, a tracheal infection. Since the infection was often correlated with parasitism by mites, the latter may act as a vector of the disease.

The fact that *Serratia* affects parasite-host systems is of ecological importance. Bucher recorded that *Itoplectis conquisitor* transmitted *S. marcescens* to pupae of *Galleria mellonella* when the ovipositor of the parasite became contaminated. Another hymenopterous parasite, *Exeristes comstockii,* suffered overt mortality in the adult stage after parasitizing larval populations of *G. mellonella* enzootically infected with *S. marcescens.* Later, similar results were obtained with the tachinid parasite *Lixophaga diatraea* when its host, *Diatraea saccharalis,* was fed on an artificial diet contaminated with the bacterium (21). In another case, a laboratory stock of the hymenopterous parasite *Ernestia consobrina,* with *Mamestra brassicae* as host, collapsed as a result of a natural infection in the system with *S. liquefaciens* (22).

3.2.2. OTHER ENTEROBACTERIA. *Coccobacillus acridiorum* (syn. *Enterobacter cloacae* type A), a potential pathogen of locusts (23), was first described in 1911 by d'Herelle from Yucatan (Mexico) as a virulent pathogen of *Schistocerca pellens.* Subsequent infection experiments, however, were mostly inconsistent. Later workers isolated this bacterium from the guts of healthy locusts (6). Only when injected into the hemocoel does it induce a reproducible septicemia. Therefore, this bacterium is often considered to be secondary invader rather than a primary pathogen for grasshoppers. Similar isolates from locusts, previously described as *E. cloacae* type B, have been reclassified as *Serratia liquefaciens* (14). Reexamination of type A, however, is still incomplete.

Several species of *Proteus,* such as *P. vulgaris, P. mirabilis,* and *P. rettgeri,* also have been described as potential pathogens of grasshoppers (24). Since their infectivity is limited, they do not cause epizootics in nature.

3.3. Streptococcaceae

The Streptococcaceae are gram-positive, aerobic to anaerobic, coccoid bacteria in pairs or chains and are nonmotile or rarely motile (2).

3.3.1. *STREPTOCOCCUS FAECALIS.* *Streptococcus faecalis* is a widely distributed inhabitant of the alimentary tracts of many insects (4) and is predominant in the gut microflora of *Galleria mellonella* (5). Occasionally it occurs as a potential pathogen in several insects. The bacterium has low virulence when experimentally injected into the hemocoel of *G. mellonella.*

S. faecalis grows on simple artificial media, such as nutrient broth, under microaerophilic conditions at 10 to 45°C. It belongs to the serological Lancefield group D (group specific precipitogens). The species is characterized by biochemical features including isoesterase pattern. Bacteriocines or bacteriophages have been used for strain differentiation.

During epizootics of *Lymantria dispar* in North America, a motile yellow variant of *S. faecalis* was isolated as the causative agent (25, 26). The bacterium is ingested with food and multiplies in the gut lumen causing diarrhea. The larvae die as a result of desiccation and become mummified. The anal discharge contains sufficient streptococci to infect other larvae. Furthermore, the bacteria can grow in decaying larvae and contaminate the environment. In a field experiment, Doane (27) reported the control of *L. dispar* with the application of this pathogen.

3.3.2. *STREPTOCOCCUS PLUTON.* Bailey et al. (28) stated that *Streptococcus pluton,* not *S. faecalis,* causes European foulbrood of the honeybee. Sometimes other bacteria, such as *Lactobacillus eurydicae* or *Bacillus alvei-circulans,* are found as secondary invaders. Because fecal streptococci are rarely found within the alimentary tracts of worker bees, it is suggested that strains of *S. faecalis* isolated from beehives have come from the feces of wax moth larvae (*G. mellonella*).

S. pluton grows only in special media (containing glucose or fructose) under microaerophilic conditions at 20 to 45°C with an optimum at 35°C. Serological and other characters have not been reported.

S. pluton is ingested with contaminated food and multiplies in the larval gut inducing characteristic symptoms of the disease. Most larvae die on the bottoms of their open cells. Young larvae are highly susceptible but the older ones are more tolerant and may survive. Because survivors discharge their gut contents including bacteria before pupating, the discharged material is the most infective source of European foulbrood within the beehive. Epizootics usually occur in the spring and may reappear in the fall, but outbreaks often seem to be self-limited becaused the diseased larvae are removed from the hives by worker bees (29).

Differences in the susceptibility of various races of honeybees have been reported against *S. pluton.* The replacement of the German race of bees by the supposedly more resistant Italian race may be responsible for the virtual elimination of European foulbrood in North America (30).

3.4. Bacillaceae

The Bacillaceae are rod-shaped, spore-forming, aerobic or anaerobic bacteria which are usually gram-positive and motile by peritrichous flagella (2, 31).

3.4.1. *BACILLUS LARVAE.* American foulbrood of honeybee is caused by *Bacillus larvae* first described by White (32). It is distributed over many parts of the world where European races of *Apis mellifera* are cultured, especially North America and Europe, but also Australia, New Zealand, and Hawaii. Other races of honebee in Africa and Asia seem to be more resistant.

   B. larvae requires thiamine and several amino acids for growth and sporulation. The so-called J-medium (containing yeast extract, trypton, and glucose) is very useful (31). Development occurs between 25 and 40°C with an optimum at 35°C. Serological methods have been used for species characterization (0-antigen) (33) and for strain differentiation (precipitogens) (34).

   When *B. larvae* is ingested by young bee larvae, the spores germinate in the gut lumen followed by a rapid vegetative growth. After infecting limited regions of the midgut epithelial cells, the bacterium invades the hemocoel and induces bacteremia. Defense by phagocytic hemocytes is weak, and the fat body, tracheal matrix, and hypodermal cells become infected (35, 36). Heat-labile exotoxins produced by vegetative cells of *B. larvae* kill larvae, prepupae, and adults and also are toxic for several other insects after injection (37).

   Only the bee larvae are susceptible to *B. larvae.* After infection, they usually do not die before the brood have been sealed. Older larvae often complete their development to the pupal stage before death. Adult workers may transmit the spores from the remains of dead larvae to young susceptible broods. When fed spores, the worker bees eject them in their feces for about 2 months (38). Spores of *B. larvae* are also found in the guts of worker bees that had direct contact with spore-fed queens. Such spore passage from queens to attendant workers may be of some ecological significance (39). The addition of pollen to the larval diet may reduce the mortality from American foulbrood. Since drone larvae are fed more pollen, they are usually less susceptible than the queen or worker larvae.

   Resistance to *B. larvae* is associated with the behavior of worker bees (rapid removal of infected brood and contaminated material; type of feeding), the age of the colony, and inherited factors of resistance. In some infection experiments, bee colonies bred for resistance showed lower mortality compared with colonies bred for susceptibility (40). Differences in the LD_{50} of the larvae, however, were usually small (41), with a resistance factor of 1.9. Since the slope of the dosage-mortality regression line was lower in the more "resistant" line, the observed effect might be due to increased variability in the system rather than to enhanced resistance.

   Shimanuki et al. (42) studied the effect of serial passages of *B. larvae* through two lines of *Apis mellifera* with different degrees of susceptibility. After three passages through larvae (16 or 24 hours old), some increase in virulence was obtained in both host lines.

3.4.2. *BACILLUS POPILLIAE.*   Dutky (43) recognized *B. popilliae* (*B.p.*) as the causative agent of the milky disease in larvae of *Popillia japonica* found in the USA (New Jersey). He described two types of bacteria: *B.p.* var. *popilliae,* which produced a parasporal body during sporulation, and *B.p.* var. *lentimorbus,* which did not. Similar pathogens from scarabaeid larvae were found in Europe (44), Australia (45), and New Zealand (46).

B.p. needs thiamine, biotin, and several amino acids, but shows only vegetative growth on the J-medium mentioned above. Production of virulent spores is guaranteed only within susceptible hosts as the result of an infection. Each variety of *B.p.* infects only a few closely related host species (after application *per os* to the larval stage). For the mass production of spores in vivo, the spore inoculum is usually injected intrahemocoelically into larvae or adults. More constant results and a shorter incubation period are obtained by this method than with spore-fed larvae. *B.p.* only develops between 15 and 37°C.

The classification of milky disease pathogens is based on morphological features (form and size of spore and parasporal body) and the host range. Serological studies (based on precipitogens) suggest that several types are distinct but represent a common group. Further studies, however, are needed to confirm the present classification of the four varieties of *B.p.* (var. *popilliae, melolonthae, rhopaea,* and *lentimorbus*) and a probable separate species: *B. euloomarahae* (47).

After ingestion, the milky disease spores germinate within the gut of susceptible larvae. The vegetative bacteria induce localized infections in the midgut epithelium. In spite of the replacement of some infected cells, an invasion of the bacteria into the hemocoel occurs through focal necroses of the gut wall (48, 49). Because no striking immune defense occurs, a massive bacteremia takes place. As a consequence of the prolific production of spores in the hemolymph, the diseased larvae become whitish in appearance before death. Whether the parasporal crystals of *B.p.* are toxic to scarabaeid larvae has not yet been adequately determined.

The continuation of vegetative growth and the production of virulent spores are only possible in living and not in dead hosts. Accordingly, for the perpetuation of milky disease in nature it is advantageous to select strains of moderate virulence that produce a maximum number of spores. Otherwise if strains with high virulence are selected, the hosts die before sporulation. Such strains are not useful either for their perpetuation in nature or for control experiments (50).

According to field studies in the Eastern part of the USA, *B.p.* var. *popilliae* was still present in the soil 30 years after its original application (51). Milky disease became established, and a progressive increase of infections and a decline of the beetle populations were demonstrated in *P. japonica* as well as in *Anomala orientalis.* However, in 1975, a serious reduction in the effectiveness of milky disease was reported (52) in Connecticut. At sites with a former infection rate of 40 to 100%, a maximum of only 20% was found. This decline was associated with loss of virulence and reduction in spore yield (about ¼ of the previous yield). A less virulent strain apparently became dominant in Connecticut after 1960 (53). Meanwhile, the beetle populations also increased in milky disease areas in New York and Ohio. Besides a reduced virulence of the bacteria, an increased tolerance or resistance also may have developed in host populations. Such a resistance is indicated because even the most virulent spore preparations resulted in fewer infections than anticipated (52).

In contrast to the successful control of *P. japonica* and *A. orientalis* by *B.p.* var. *popilliae* in the USA, the use of milky disease pathogens for microbial control in other cases has been hampered, perhaps by ecological limitations (54). For example, *B.p.* var. *popilliae* is not a dominant factor in controlling populations of *Amphimallon majalis* in New York (55). Also *B.p.* var. *melolonthae* seems to be of minor importance in the suppression of *Melolontha melolonthae* in France (56, 57).

The narrow host range of the milky disease pathogens indicates that they have no effect on nontarget organisms including vertebrates. *B.p.* spores remain dormant until ingestion by scarabaeid larvae and therefore present no risk for use in pest control.

3.4.3. *BACILLUS SPHAERICUS*.

B. sphaericus is found in soil and soil-aquatic systems and has a worldwide distribution. But only strains isolated from mosquito larvae (in California, Middle America, India, Ceylon, Indonesia, Philippines, and Japan) are pathogenic or toxic for Nematocera. The first insecticidal strain (K) isolated from *Culiseta incidens* in California was described in 1965 by Kellen et al. (58); however, the activity of this strain was low compared with others, for example, the strain SS II-1 from India. But spores of strains 1593 from *Culex fatigans* (Indonesia) and 2297 from *Culex quinquefasciatus* (Ceylon) are approximately four logs more toxic than SS II-1. The host range of *B. sphaericus* includes the genera *Culex, Anopheles,* and *Aedes.*

The effectiveness of some strains, such as K, is caused by an infection, but most of the others are predominantly toxigenic. Toxin production begins in the vegetative stage but is greatest with the onset of sporulation (e.g., SS II-1). In most insecticidal strains, such as 1539 and 2297, the spore is accompanied by a parasporal, probably insecticidal, crystal. No crystals are found in nontoxic strains or in those that are not toxic in the spore stage. The crystals are assumed to be solubilized in the gut, and the toxic subunits penetrate the peritrophic membrane. The swelling of the mid-gut is the first sign of intoxication. Ingested spores of moderately toxic strains (such as K) may germinate and grow in the gut lumen, but bacterial growth is not necessary for highly toxic strains (59).

B. sphaericus grows and sporulates on ordinary media, such as nutrient broth, with a temperature optimum between 25 to 35°C and a maximum at 40 to 45°C. At present, nontoxic and insecticidal strains cannot be differentiated by biochemical features. The toxic strains belong to only four groups of H-serotypes (representatives are K, SS II-1, 1593, and 2297). Another useful technique for strain identification is phagotyping. High correlation between serotypes and phagotypes has been reported (60).

The larvicidal activity of *B. sphaericus* in laboratory and field tests depends on spore concentration, larval instar, and temperature. The persistence of spores is influenced by water quality; for example, it is higher in sewage habitats than in fresh water (61). Rapid spore settling and, consequently, the low probability of subsequent generations of larvae coming into contact with the bacteria may limit the cycling potential of this bacillus. Multiplication of *B. sphaericus* after larval death suggests that the cadavers may be involved in the spread of the bacterium, since *Culex* larvae can become infected when dead bacillus-carrying larvae are brought into their aquatic environment.

Bacillus sphaericus is a useful larvicide for the control of several mosquito species which are well-known vectors of medically important diseases (62). The lack of effects on nontarget organisms including vertebrates is important for safety considerations.

3.4.4. *BACILLUS THURINGIENSIS.*   *B. thuringiensis* (*B.t.*) is the most important and best known insect pathogen for the biological control of insect pests. It is, therefore, produced on a commercial scale and applied in many countries. *B.t.* is closely related to *B. cereus* but differs from it by producing an insecticidal parasporal crystal. Consequently, variants of *B.t.* lacking crystals cannot be differentiated from *B. cereus.* It is well documented that prolonged subculturing of *B.t.* strains can result in an irreversible loss of crystals and pathogenicity. Recent investigations have shown that genes coding for crystal production are located within plasmids as well as in the bacterial chromosome. Furthermore, conjugal transfer of toxin-coding plasmids is possible within the *B. cereus/B. thuringiensis* group (63).

Strains of the *B.t./B. cereus* group grow and sporulate under aerobic conditions on ordinary media, such as nutrient broth, at an optimal temperature of 30°C (maximum of 40 to 45°C). As in the case of *B. cereus,* the *B.t.* varieties are characterized by serological (H-antigens) and biochemical properties including an isoesterase pattern. At present, about 30 varieties belonging to 20 H-serotypes are known (64). Strains can be identified by specialized phages, bacteriocines, or antibiotic pattern.

During its vegetative phase *B.t.* (as well as *B. cereus*) can produce several minor, relatively unspecific toxins, such as α-exotoxin and β-exotoxin. Only the so-called δ-endotoxin, which is produced during sporulation and is localized in the parasporal crystal, is strongly related to the host-specific pathogenicity of *B.t.*

After the ingestion of spores and parasporal crystals of *B.t.* by susceptible larval insects, the crystals are solubilized and activated by alkaline gut proteases. The toxic subunits induce a dosage-dependent feeding stoppage and typical lesions in the larval mid-gut. The toxin attacks the membrane of the gut epithelium cells resulting in a rapid swelling, cellular vacuolization, and blebbing. The gut cells then separate from each other and from the basal membrane, sloughing into the lumen of the gut, and eventually lyse. This breakdown of the mid-gut epithelium is followed by spore germination within the gut lumen. Massive multiplication of the bacterial cells may or may not occur in the gut lumen. The bacteria may invade the hemocoel before larval death and induce a fatal septicemia (65–67). Finally, *B.t.* can be enriched by saprophytic growth within the dead larvae and possibly sporulate.

Since distinct strains of *B.t.* are pathogenic only for insects belonging to a special taxonomic order or group, they have been separated into pathotypes (67). Strains of *pathotype A* are toxic or pathogenic only for Lepidoptera with var. *thuringiensis* as the prototype (68). *Pathotype B* is effective only against Nematocera (Diptera) with the prototype var. *israelensis* (69). The recently isolated var. *tenebrionis* (67) is representative of *pathotype C* which is pathogenic only against some Coleoptera, especially members of the family Chrysomelidae. Further analysis of the host range of strains belonging to pathotype A has demonstrated clusters of strains with a pathological preference for special groups of Lepidoptera. Such clusters may represent subtypes of pathotype A with molecular variants in their crystal protein.

The preferred habitat of *B. cereus* and of other members of the genus *Bacillus* is the soil, in which their spores persist in a dormant stage. Thus, the insect pathogenic bacilli may belong to the so-called soil-borne bacteria. Actually, *B.t.* var. *israelensis* (strain A 60) was found initially in soil samples of a semipermanent breeding place of mosquitoes in the Negev desert (Israel) (69), but later it was isolated from mosquito larvae in different climatic regions of the world.

A survey of the natural occurrence and distribution demonstrated that *B.t.* was only rarely present in agricultural soils in the USA. Most of the isolates were not identical with the specific strains of *B.t.* which had been used for the control of lepidopterous pests in these areas. The authors (70) stated that the presence of *B.t.* in soil did not indicate any enhanced value in epizootiology. In their study on the fate of *B.t.* in the soil, West and Burges (71) demonstrated not only a gradual decrease of viable spores but also a rapid inactivation of the crystal toxin. Although the authors concluded that the potential of *B.t.* in the soil to reinfect subsequent generations of host species was greatly reduced, it is possible that spores would be dispersed by nontarget soil arthropods or vermes as carriers. For example, *B.t.* was isolated not only from dead nymphs of cicadas but also from dead earthworms.

In Japan *B.t.* occasionally produced serious losses in silkworm rearings. Ishiwata (72) first isolated *B.t.* var. *sotto* from the so-called sotto disease. Subsequently, several varieties of *B.t.* were found in the litter of sericultural farms or from the soil surrounding them (73). The epizootiology of *B.t.* infections of *Bombyx mori,* however, is not adequately understood. The same is true for the epizootiology of *B.t.* in populations of the Siberian silkworm (*Dendrolimus sibiricus*) (74) and of the European corn borer (*Ostrinia nubilalis*) (75).

The presence of *B.t.* in the stored-product environment is ecologically important. In 1911, Berliner (68) first isolated *B.t.* from the flour moth, *Ephestia kühniella,* after an outbreak in a mill in Thüringen (Germany). Later, similar epizootics were reported from mills in France (76) and Yugoslavia (77). Recently, *B.t.* was isolated from grain and grain products from Egypt and Germany (78). Furthermore, the incidence of *B.t.* was reported in stored-product insects of different origin and from outbreaks in insectaries in England (79, 80). The assumption was that these epizootics were initiated by a high incidence of *B.t.* derived from diseased and decayed larvae of the storage moth. The distribution of the pathogen within the substrate resulted largely from the migratory behavior of infected larvae. Such larvae often accumulated at the surface of the substrate. Since moths laid their eggs at the surface, where the susceptible hatching larvae first forage, the larvae could be contaminated rapidly. Their cadavers provided more inoculum to infect the less susceptible, older larvae (80).

Varieties of *B.t.* that are often found in stored-product insects, such as var. *thuringiensis,* are distributed worldwide. Others, such as var. *sotto* from *Bombyx mori* and var. *dendrolimus* from *Dendrolimus sibiricus,* have been found only in East Asia or Siberia, respectively. Today, however, isolates of var. *thuringiensis,* var. *galleriae,* and var. *kurstaki* are no longer of special ecological significance because they are used extensively in microbial control.

Larval parasites can act as vectors of *B.t.,* for example *Nemeritis canescens* (Ichneumonidae, Hymenoptera) in populations of *E. kühniella* (81, 82). Such transmission provides not only host regulation but also uniformity in the prevalence of the pathogen within the biotope. *B.t.* is harmless not only to this hymenopterous parasite but also to others like *Apanteles glomeratus* (83). Sometimes *B.t.* can cause the death of parasite larvae indirectly as a result of the premature death of the host (84). Egg parasites, such as *Trichogramma evanescens,* however, are never influenced by *B.t.* in any stage of their development (85), and are not vectors of *B.t.*

Although *B.t.* is very pathogenic to many insects, its role as a limiting factor in population dynamics is not very important. This may be caused mostly by an insufficient density and distribution of the pathogen within the environment of potential hosts. Since *B.t.* lacks transmission and spreading facilities, it must be applied like a chemical insecticide in the microbial control of pests.

The HD-1 strain of *B.t.* var. *kurstaki* (pathotype A), isolated by Dulmage (86) from *Pectinophora gossypiella,* is now used worldwide for the control of lepidopterous pests (87, 88). Target insects that are sufficiently controlled by this strain include: *Pieris* spp.; *Plutella xylostella* and *Trichoplusia ni* on cabbage; *Ostrinia nubilalis* on corn; *Tortrix viridana, Lymantria dispar, Choristoneura fumiferana, Malacosoma* spp., *Thaumetopoea* spp., and *Dendrolimus sibiricus* in forests; *Hyphantria cunea, Euproctis chrysorrhoea,* and *Hyponomeuta* spp. in orchards; *Ephestia* spp. and *Plodia interpunctella* in stored products; and *Galleria mellonella* in bee combs.

In the field the longevity of *B.t.* on the surface of leaves is limited by sunlight. For example, within 1 day on soybean leaves there was a 90% reduction in spore viability and a 65% reduction in insecticidal activity against *Trichoplusia ni* (89). Similar results have been obtained by other authors.

In connection with the application of commercial preparations of *B.t.* var. *kurstaki* at recommended dosages, safety tests and environmental monitoring indicated no special risks to human health or the environment (87). Also, nontarget insects, such as the honeybee and other pollinators, were not harmed.

Control experiments with A-60 strain of *B.t.* var. *israelensis* (pathotype B) have been successful in many parts of the world against mosquito larvae of medical importance (88), mostly in the genera *Aedes, Culex,* and *Anopheles.* The efficacy of *B.t.* var *israelensis* was limited by great water depth, abundant food supply in polluted water, and high larval density (90). In the case of Simuliidae (vectors of onchocerciasis), effective control was often handicapped by the high flow rate of streams in the breeding sites.

B.t. var. *israelensis* does not harm fish, amphibians, or aquatic arthropods, including predatory dipterous larvae such as *Chaoborus, Mochlonyx,* and *Toxorhynchites.*

A marked tolerance usually occurs with increasing age of the larvae (91). For example, in tests with *Aedes vexans,* the LD_{50} of last instar larvae is about 50-fold higher than of the first (92), and in *Ephestia cautella* the LD_{50} of larvae is 100-fold higher in the last instar as compared to the first (93).

Within one species, there may be inherent differences among different populations in susceptibility to *B.t.* (94). For example, in populations of *Aedes caspius* and *Culex pipiens* three- to sevenfold differences in the LD_{50} have been observed (95). Similarly, in populations of *Ephestia cautella* and *Plodia interpunctella,* differences in the LD_{50} are about six- to tenfold (96).

Attempts have been made with lepidopterous insects to select for increased resistance to *B.t.* var. *kurstaki.* Laboratory experiments with *Plutella xylostella* in which a high selection pressure was maintained over 10 or even 30 generations (97, 98) indicated no change in sensitivity of the selected lines. This finding was supplemented with a marked reduction in viability and fertility as a result of the application of sublethal doses of *B.t.* over many generations (98).

3.4.5. OTHER BACILLI.   Strains of *B. cereus* are facultative pathogens of the larvae of *Carpocapsa pomonella* (Lepidoptera) (99), *Pristiphora erichsonii* (Hymenoptera) (100), and *Musca domestica* (Diptera) (101). In these and other insects, *B. cereus* induces a fatal septicemia if spores or vegetative cells accidentally penetrate the gut wall or enter into the hemocoel as a result of a traumatic process. Furthermore, several strains of *B. alvei, B. brevis, B. circulans,* and intermediates are described as potential pathogens of mosquito larvae *Culex tarsalis* (62).

3.4.6. *CLOSTRIDIUM* SPP.   In 1954, Bucher (102) isolated *Clostridium* spp. from diseased larvae of *Malacosoma pluviale* in Western Canada. When spores were fed to host larvae, they germinated within the gut and the bacteria multiplied under the microaerobic conditions. At the time of sporulation, the first symptoms of the disease occurred, suggesting possible toxic substances. Feeding was reduced and the larvae tended to empty their guts. Finally, the larvae became inactive, shrunken, and sluggish before death. *Clostridium* spp. were never found in the hemolymph. Several species of *Malacosoma* were more or less susceptible. Infected young larvae showed the usual severe damage, whereas older ones could survive. Nothing is known about the epizootiology and natural distribution of these pathogens.

These insect pathogenic clostridia are gram-negative bacilli which grow but do not sporulate on special media under microaerobic conditions. There are two species based on biochemical characters: *C. malacosomae* and *C. brevifaciens* (102).

4. OBLIGATELY PARASITIC PROKARYOTES

In contrast to facultative pathogens, the obligate pathogens multiply only within viable hosts or host cells in spite of the fact that culturing on special media is possible in a few cases. Not all obligate pathogens induce an acute disease. Attenuated infections or a carrier stage can also result from infection. In such states, obligate pathogens may remain within a host which acts as a vector or a reservoir. The life cycle of obligate pathogens is often complex and can involve vertical transmission or transmission to alternate hosts. The involvement of vertebrates or plants as hosts is a special ecological adaptation.

4.1. Spiroplasmas and Mycoplasma-Like Organisms

The cells of spiroplasmas are gram-positive, are bound only by membrane, and lack a typical bacterial cell wall. They are pleomorphic and vary in shape from nearly spherical forms on the surface of semisolid media to helical filaments in liquid cultures, and they are motile by rotation or undulation. Cells of variants that lack spirals look like cells of mycoplasmas. Similar to mycoplasmas, the growth of spiroplasmas requires special media with cholesterol or even other sterols. A relatively narrow optimum temperature at 30°C is characteristic for most spiroplasmas (103).

The first spiroplasmas were found in connection with insect-transmitted yellows and witches broom diseases of plants (earlier considered to be caused by arthropod-borne viruses) (104). When vector insects ingest sieve-tube sap from infected plants, the spiroplasmas primarily infect the gut epithelium cells. Later, they are found in the hemolymph and in several tissues including ovaries. Finally, the agent is transmitted with the saliva. Some plant pathogenic spiroplasmas exhibit pathogenicity to their vector insects. Often the same agent (e.g., corn stunt spiroplasma) that is harmless for one vector (*Dalbulus maidis*) is detrimental to another (*Dalbulus elimatus*). However, most spiroplasmas induce severe diseases in plants but not in their insect hosts. This may be due to long adaptation during the coevolution of the spiroplasma and its special vector, resulting even in a vertical transmission of the agent *via ovo*. The relationship of spiroplasmas to plants may be a more recent development.

Whereas spiroplasmas are transmitted only by a limited number of vector insects (Cicadelloidea, Fulgoroidea, and Psyllidae), they are usually transferable to many susceptible plant species of several families. The incidence of these spiroplasmas is dependent on the population density of the vector insect, the percentage of carriers, the frequency of sucking, the life-span of the vector, and probably the rate of vertical transmission.

The first vertebrate-pathogenic spiroplasma was isolated from rabbit ticks (*Haemophysalis leporis palustris*) (105). Some tick-born spiroplasmas have been cultivated on special media at 37°C.

Although the sex-ratio factor of *Drosophila* spp. was identified as a *Spiroplasma* in 1974, this spirochetelike organism had been detected earlier (106). The maternally inherited agent eliminates males in the embryonic stage or later, but does not affect females. In adult flies, the sex-ratio spiroplasma is found in large numbers within the hemolymph. It is transmissible to other flies only experimentally by injection. Until now, this *Spiroplasma* has not been cultivated on media suitable for other spiroplasmas. The sex-ratio spiroplasma usually carries a lysogenic virus which plays a major role in the epizootiology of this disease. This bacteriophage multiplies extensively within the spiroplasma and destroys it. In this way, the normal sex ratio of *Drosophila* flies is restored (104).

A spiroplasma pathogenic for honeybees was first discovered in North America and later in Hawaii. It is transmitted by feeding from bee to bee. After penetrating the gut wall and entering the hemocoel, the agent induces a fatal septicemia (107). Since nectar has been suggested as a natural source of the agent, many flowers have been examined, and numerous spiroplasmas have been isolated, especially from the nectar of tulip trees (*Liriodendron* spp.). None of these isolates, however, is pathogenic for bees. All the flower-born spiroplasmas appear to be associated with flower-visiting insects (103).

The host range of the bee spiroplasma includes *Drosophila* spp. and other insects. Spiroplasmas from various sources also multiply in larvae of *Galleria mellonella* after injection. In contrast to the honeybee spiroplasma which has only moderate virulence (similar to *Spiroplasma citri*), a relatively high mortality has been obtained with three flower-born spiroplasmas in *G. mellonella* (108).

The bee spiroplasma is pathogenic for chicken embryos. Its cultivation is possible in a special artificial medium.

Several serogroups have been created for the identification of spiroplasmas (103): group I_1—*Spiroplasma citri*; I_2—honeybee spiroplasma; I_3—corn-stunt spiroplasma; I_4—tick-born spiroplasma 277-F; group II—*Drosophila* sex-ratio spiroplasma; groups III and IV—flower-born spiroplasmas; group V—tick-born spiroplasma SMCA.

4.2. Rickettsias and Chlamydia-Like Organisms

These prokaryotes have small coccoid to rod-shaped or disk-like gram-negative cells, which are usually not cultivable on artificial media. Rickettsias have an evident association with arthropods (109–111).

Members of the tribe Rickettsieae are etiological agents of vertebrate diseases transmitted by ectoparasites. The prototype of the typhus fever group is *Rickettsia prowazeckii*, which causes epidemic typhus in man. The rickettsia grows profusely in the cytoplasm of the gut epithelium cells of its vector, the human body louse (*Pediculus humanus*), inducing death, usually within two weeks. Other lice or fleas from small rodents act as reservoirs. The causative agent of murine or endemic typhus fever of man is *R. typhi*. Humans act only as incidental hosts, and the primary reservoirs are rats. The rat louse, *Polyplax spinulosus*, and the rat flea, *Xenopsylla cheopis*, are the main vectors. The human body louse and the human flea *Pulex irritans* are both alternative hosts of *R. typhi*. In contrast to the early death of lice, the longevity of fleas is not reduced by these rickettsiae. It is concluded, therefore, that the host relationship of rat-*R. typhi*-flea may be the ancestor of the younger cycle man-*R. prowazeckii*-louse.

Rickettsiae of the spotted fever group are transmitted to man by the bite of ticks. The best known species is *Rickettsia rickettsii*, the etiological agent of Rocky mountain fever. Another species, *R. sibirica,* is distributed throughout Northern and Central Asia (North Asia tick typhus). Both rickettsiae are transmitted from wild rodents or dogs by *Dermacentor* ticks in which they multiply intracellularly within the cytoplasm and nuclei of several tissues. They are also found in the gonads of ticks and are transmitted by the females to at least some of the offspring. Tick hosts are not only vectors but also reservoirs. The fact that rickettsiae induce severe pathogenic effects only in vertebrates but not in ticks may be due to a longer period of coevolution of rickettsiae and ticks.

The louse-transmitted *R. prowazekii* shows not only growth within several ticks but sometimes also a vertical transmission after injection. Most species of tick-born rickettsiae are able to multiply within the gut epithelium of *Pediculus humanus* after rectal application.

Many species of the genus *Rickettsia* share common O-antigens with special *Proteus* strains (Weil-Felix-reaction), for example *R. prowazeckii* and *R. sibirica* with *Proteus* OX-19, and *R. rickettsii* with *Proteus* OX-2.

The epizootiology of all arthropod-transmitted rickettsioses of vertebrates is very complex and dependent on the ecology of all the hosts involved. Important parameters include the population density of the relevant vectors, the percentage of carriers, the frequency of blood meals, and the life-span of the vectors. In the case of tick-transmitted rickettsiae, the rate of vertical transmission is important.

Fluctuations in the virulence of rickettsiae have been reported by several authors. An interference phenomenon with *R. rickettsii* prevents the occurrence of mutant strains of rickettsiae and limits definite strains to different areas of the tick biotope (see Section 5.2.3).

All rickettsiae of the tribe Wolbachieae are associated only with arthropod hosts. The species which induce no or only a mild form of rickettsiosis are in the genus *Wolbachia*. The type species is *Wolbachia pipientis,* which was first described in *Culex pipiens* from North America and China. The only organs consistently infected by *W. pipientis* are the gonads. The spermatogonial cells often are destroyed, but such detrimental effects in the ovary are not usual. It has been suggested that *W. pipientis* is only transmitted vertically *via ovo*. Yen and Barr (112) demonstrated that *W. pipientis* causes sexual incompatibility between geographic populations of *C. pipiens* (e.g., in Germany and California). Incompatibility results from rickettsia-infected sperms from carrier males of noncompatible strains not being able to fertilize healthy females. Therapy with antibiotics (e.g., addition of tetracycline to the rearing water of larvae or pupae) eliminates this kind of incompatibility. Healthy males can fertilize heterologous rickettsia-infected or noninfected females (112, 113). Similar types of *Wolbachia* have been found in other mosquitoes of the genera *Culex, Culiseta,* and *Aedes. Wolbachia*-like, ovo-transmitted microorganisms which may induce sexual incompatibility have also been reported from Lepidoptera, such as *Ephestia cautella*. They contribute to the genetic isolation of natural populations as shown in the crossings between moths from Iran and USA (114).

Typical *Wolbachia*-like rickettsiae have been isolated also from several ticks (e.g., *Argas arboreus* in Egypt; *Dermacentor andersoni, Rhipicephalus sanguineus,* and *Amblyomma americanum* in the Americas; *Haemophysalis inermis* and *Hyalomma asiaticum* in Eurasia). They multiply within vacuoles of host cells, especially of Malpighian tubules and ovarial tissues. These agents do not induce striking pathological symptoms in their tick-host and are usually transmitted vertically *via ovo*. Their prototype is *Wolbachia persica* isolated from *A. arboreus* (115), which is successfully transferable to *Pediculus humanus* and *Tenebrio molitor*. A profuse growth in the yolk sac of chicken embryo also has been reported.

In 1956, Philip created the genus *Rickettsiella* for rickettsialike organisms that are highly pathogenic for their insect hosts. The type species, *R. popilliae,* was detected in 1950 by Dutky and Gooden (116) from white grubs of the beetle *Popillia japonica* in the eastern USA. A closely related type, *R. melolonthae,* was found in diseased grubs of *Melolontha melolontha* in Europe (117, 118).

When ingested, the coccoid to kidney-shaped small rickettsiae penetrate the gut wall of the insect host and replicate in cells of the fat body and other tissues of the hemocoel including gonads. Within the cytoplasmic vacuoles, the small rickettsiae undergo transformation to bacterialike forms which multiply by binary fission. Characteristic protein crystals are formed within pleomorphic cells of *Rickettsiella*. The bacterialike forms often enlarge to giant cells or even to a rickettsiogenic stroma which reverts to small rickettsia at the end of reproduction (119). Finally, the infected host cells undergo lysis, and masses of rickettsiae, crystals, and rickettsia-filled vacuoles are released into the hemolymph. At an advanced stage of infection, a blue-greyish discoloration occurs in the larvae. When larvae become infected with a sublethal dose, they form diseased pupae and adults.

Since *R. melolonthae* multiplies only with susceptible hosts, the prevalence of rickettsiosis is dependent on the population density of scarabaeid larvae in the biotope. Only at high levels of host density can the epizootics regulate the host population. Vertical transmission *via ovo* is important for the persistence of *R. melolonthae* during periods of low host density (120, 121).

All larval instars of *Melolontha* spp. are very susceptible to *R. melolonthae*, and the diseased grubs usually die within several months after infection at 15 to 25°C. When the temperature decreases in the biotope, healthy larvae move down in the ground, but rickettsia-infected ones move upwards before death. This change in behavior may be of epizootiological significance, because in this way the upper layers of soil become contaminated by decaying larvae. Since the adult females lay their eggs in the upper regions of the soil, susceptible young larvae can become infected. *R. melolonthae* is very resistant outside its hosts and may retain its activity within the soil for several years.

Since *R. melolonthae* has a relatively wide host range, host alternation is possible. Ingestion of the agent by larvae of many species of Melolonthidae, Hopliidae, and Rutelidae induces a typical rickettsiosis. Natural infections caused by *Rickettsiella* also have been reported from diseased larvae of other Coleoptera belonging to Cetonidae, Lucanidae, Tenebrionidae, and Carabidae. Furthermore, a closely related or even identical *Rickettsiella* has been found causing epizootics in larvae of *Tipula paludosa* (Diptera). Because these larvae are often living in the same soil substrate (pastures) as the larvae of *Melolontha melolontha,* the pathogen may alternate from one host to the other.

After the injection of *Rickettsiella melolonthae,* other species of Coleoptera, Diptera (*Tipula paludosa*), and even Lepidoptera (*Galleria mellonella*) can be infected (118). Moreover, small rodents such as mice can acquire a mild rickettsiosis after the injection or inhalation of *R. melolonthae.*

Some types of *Rickettsiella* have been described from Orthoptera such as *Blatta orientalis, Schistocerca gregaria,* and *Gryllus bimaculatus.* Some are probably identical because their host ranges overlap. For example, *R. grylli* is infectious *per os* for species of Gryllidae, Tettigoniidae, and Acridiidae. In *G. bimaculatus,* since the gonads of both sexes are often heavily infected, many diseased insects become sterile. Antibiotics may restore the function of ovaries, but not that of testes (122). As in the case of *R. melolonthae, R. grylli* has low virulence for mice.

Immunological test with different types of *Rickettsiella* revealed two O-serogroups (123): group I, *Rickettsiella* from *Melolontha, Cetonia,* and *Tipula;* group II, *Rickettsiella* from *Gryllus* and from *Armadillidum* spp. (terrestrial Isopoda).

Chlamydiaceae are usually pathogens of vertebrates without any association with arthropods. However, some chlamydialike pathogens with disklike "elementary bodies" were recently described from diseased midge larvae (Chironomidae) in Europe (124) and North America (125). *"Rickettsiella chironomi,"* first detected in 1942 by Weiser within diseased larvae of *Camptochironomus tentans* from Germany, belongs to this group (124–126).

After ingestion, the chlamydialike pathogen invades the hemocoel of *C. tentans* and attacks predominantly the fat body. During its reproductive cycle within the cytoplasm of host cells, the morphogenesis of the agent is similar to that of other Chlamydiaceae (124, 125). In an advanced stage of the disease, the midge larvae show a whitish discoloration and die usually one or two weeks after infection. The source of infectious material for epizootics is diseased larvae, from which the microorganisms are released into the surrounding water after larval death. This pathogen appears to be involved in the population dynamics of midge larvae in their aquatic biotope.

Recently, similar chlamydialike microorganisms have been isolated from Arachnida, such as the scorpion *Buthus occidentalis* and the spider *Pisaura mirabilis.* Morel (126, 127) proposed the new genus *Porochlamydia* for those microorganisms. No common O-antigens with *Rickettsiella* have been found (123).

5. EPIZOOTIOLOGICAL PARAMETERS

The causation and development of an epizootic wave is dependent on a sufficient density of the host population (see Ch. 4), a sufficient concentration of the pathogen in the epicenter (see Ch. 5), and an efficient means of transmission (see Ch. 6). In addition, environmental factors (see Ch. 7) can decisively influence an epizootic.

5.1. Pathogen Transfer to Hosts

An important condition for the continuation of the chain of infections is the efficiency of pathogen transfer. The most common route for entomopathogenic bacteria is *horizontal transmission* between contemporaries based on direct or indirect contact with infected hosts of the same or another species. Usually the bacteria are taken up by the host feeding on infected organisms or contaminated substrates. The dispersal of pathogens occurs passively by mechanical means, such as wind or rain, or by organisms, such as infected hosts or nonsusceptible healthy carriers, including parasites and predators.

Sometimes alternative hosts favor the persistence and chance of transmission of a pathogen. This is possible if the host range is not too narrow. For example, different species of Scarabaeidae may act as hosts of *Rickettsiella melolonthae.* Alternate hosts that are largely pathogen-tolerant are useful reservoirs. Extreme forms of alternative hosts are known for many spiroplasmas and rickettsiae, which are transmitted from their arthropod hosts (as vector) to vertebrates and plants. In such systems the transmission critically depends on the interacting pattern of the population dynamics of all host species involved.

In the case of less than minimal host density, or of low infectivity of obligate pathogens, persistence is guaranteed only by *vertical transmission.* In this way, the pathogen is transferred from parent to offspring (similar to that of genetic factors) and is limited, therefore, to one host species. Vertical transmission is well known for obligately pathogenic prokaryotes with low virulence which persist in the enzootic state and are characterized by an equilibrium between host and pathogen (such as *Rickettsoides* and *Wolbachia*). The more virulent the pathogen, the less is the chance for an equilibrium. Accordingly, between epizootics, the highly virulent pathogens (such as *Bacillus popilliae* and *Bacillus thuringiensis*) usually persist outside their host. Persistence in the environment, however, requires some capacity to survive (see Section 5.2.2).

Sometimes, obligate pathogens are transferred in more than one way. For example, *Rickettsia rickettsii* as well as *Rickettsiella melolonthae* are transmitted horizontally by feeding and, in addition, vertically *via ovo*.

As shown by cybernetic concepts in population dynamics (128), only host density-dependent factors can act as regulators, which form with the controlled populations a complex of interacting feedback systems. Strong density dependence usually occurs with obligate pathogens, which are transmitted horizontally and multiply and persist only within their hosts.

Obligate pathogens with a low transmission rate may persist only in populations of high density or at low density within hosts having a long life span, such as ticks. However, pathogens with a high transmission rate may also persist within populations of low density.

Certain bacteria, such as food poisoning agents which multiply preferentially outside of their hosts on suitable nutrients, are more or less accidentally present in the biotope and act as density-independent control agents. But other saprophytic pathogens act as density-dependent agents to a large degree because they multiply predominantly within living hosts or cadavers. Furthermore, they induce epizootics mostly at a high host density because this favors horizontal pathogen transfer and enhances the predisposition of the host insects for pathogens by stress.

5.2. The Pathogen

5.2.1. PATHOGENICITY AND VIRULENCE. The epizootic efficiency of a pathogen is based on its infectivity, transmission rate, and virulence (see Ch. 5). For obligate pathogens (such as *Rickettsiella melolonthae*), virulence is based on the invasive mechanism, the reproduction rate, and the immunotolerance of the agent. Sometimes toxins may also be involved.

Since facultative pathogens (such as *Bacillus cereus*) lack a special invasive mechanism, they enter the hemocoels of their host accidentally. Their virulence is identical with their potency to induce a septicemia after injection. This is based on the production of relatively unspecific toxins, cytolysins, or other metabolites, which act also as immunoinhibitors. Only the so-called toxigenic bacteria (such as *Bacillus thuringiensis)* produce, in addition, cell-specific toxins, which are effective after ingestion. Usually a multivariable complex of factors is responsible for the various kinds of pathotypes and for differences in the virulence of bacterial strains.

In *B. thuringiensis*, the production of a parasporal crystal is a useful marker for toxicity. Acrystalliferous variants have no greater effect on insects than *B.*

cereus. Several pathotypes of *B. thuringiensis* are known according to their host specificity. The pathotype of each isolate is genetically fixed, and the virulence is relatively stable in clones. This situation is changed only by genetic processes like mutation, recombination, and other forms of gene transfer.

Bacillus larvae has only one pathotype, but in *B. popilliae* each variety possesses a limited but sometimes overlapping host range. Modifications in virulence by selection have been reported with both obligate pathogens (42, 50). This may be based on the cloning of heterogenous isolates or may result from a phenotypic adaptation. However, the experimental enhancement of virulence with other insect pathogens by serial host passages often seems to be unsuccessful, as shown with the facultative pathogens *Pseudomonas aeruginosa* (7) and *Serratia marcescens* (13).

5.2.2. PERSISTENCE.   Several obligate pathogens survive only within living insects such as hosts or reservoirs. A relatively high degree of persistence of pathogens has been recorded within host cadavers which protect the pathogen from the negative influences of the environment. This was shown, for example, with *Rickettsiella melolonthae.* Mummified cadavers are, therefore, often an ecologically important source of insect pathogenic bacteria.

Only spores of Bacillaceae persist without difficulty in the environment, even at high temperatures. For example, dried spores of *B. popilliae* and *B. thuringiensis* retain their activity for several years and survive prolonged time periods in the soil. This contrasts with their fate on exposed surfaces (e.g., with spores on leaves) where they become inactivated within a relatively short time (89).

5.2.3. ANTAGONISM.   Antagonists of pathogens may play a role as limiting factors under epizootic or enzootic conditions. The most important types are competing microorganisms and so-called hyperparasites.

Interference is possible between different strains or varieties of *obligate pathogens* with mixed infections. Kunkel (129) reported cross-protection between two strains of the aster yellows spiroplasma within their vector *Macrosteles fascifrons.* Leafhoppers which first acquired the Eastern strain were unable to transmit the California strain after feeding later on plants infected with this strain and *vice versa.* But with another spiroplasma, the corn-stunt agent, only unilateral interference was reported (130).

Some rickettsiae show cross-protection. According to Burgdorfer et al. (131), a natural infection of the vector *Dermacentor andersoni* with avirulent rickettsiae prevents all later infections of the tick with the pathogenic *Rickettsia rickettsii.* This corresponds to the sharp border of spotted fever appearance in the Rocky Mountains. The ticks are vectors of *R. rickettsii* only west of the Bitterroot Valley, while the eastern ticks are carriers of the avirulent *R. montana* which prevents the establishment of the virulent type.

Double infections with different strains of *Bacillus popilliae* may have an antagonistic effect. Beard (132) reported that a strain of var. *popilliae* inhibited the growth of an isolate of var. *lentimorbus* in larvae of *Popillia japonica*. Both pathogens did not occur simulaneously in the same host.

Competition also is known between *nonobligate pathogens*. Several factors, such as virulence, relative dose, and succession of infections, govern which pathogen will become predominant in mixed infections. Grasshoppers fed different combinations of *Pseudomonas aeruginosa* and *Serratia marcescens* generally die from an infection produced by only one of the microorganisms. The pathogen taking precedence is usually the one applied in greater numbers. However, when both bacteria are fed in equal numbers, all deaths are caused by *S. marcescens,* which is known to be somewhat more virulent (15).

In other cases, however, other forms of competition depend on the activity of bacteriocines or antibiotics. Such molecular antagonists are produced by many bacteria and are able to inhibit or kill sensitive microorganisms. In experiments with larvae of *Philosamia cynthia*, Pendleton (133) demonstrated that after the feeding of two strains of *B.t.*, the dominance of one strain was caused by antibiotic antagonism rather than virulence.

Bacteriocines or antibiotics are considered to play a role in stabilizing the gut microflora and in suppressing ingested bacteria. For example, *Streptococcus faecalis*, which predominates in the gut microflora of *Galleria mellonella,* produces a bacteriocine which inhibits ingested gram-positive bacteria. In addition some gram-negative bacteria, such as *P. aeruginosa* and *S. marcescens*, are also suppressed (134), but the mechanism of this antagonistic effect is not sufficiently understood.

More or less host-specific viruses, *Bacteriophages*, may act as antagonists of Eubacteria and other prokaryotes. Vegetative bacterial cells are very sensitive to bacteriophages, but spores are completely resistant. Depending on the virulence of the phage and on the tolerance of the bacterial host, infections may be lytic or not. Bacterial cells latently infected by phages often show cross-protection to closely related phages.

Reports of the ecological significance of bacteriophages in insect pathology are rare. Raun et al. (135) studied the influence of phages on *Bacillus thuringiensis* infection of *Ostrinia nubilalis*. The application of virulent phages resulted only in a retarded course of the disease and not in a decrease of the final mortality. Only in *Drosophila spp.* was it demonstrated that specific phages were able to control in vivo a *Spiroplasma* and to restore the normal sex-ratio in infected flies (104).

5.3. The Host

5.3.1. RESISTANCE AND TOLERANCE.   Since pathogens reduce the longevity and/or the reproductive rate of their hosts, they act as selection factors for enhanced tolerance or even resistance of the host. Such selection may be strengthened if the contact with the pathogen is permanent or prolonged. This may result in an ecological equilibrium between host and pathogen to produce an enzootic state. Selection is relatively high if pathogen-induced regulative cycles of the host are short. In the case of long cycles or episodes, however, the selective pressure is limited to the epizootic wave.

Reports of acquired resistance in connection with bacterial pathogens are rare. For example, Rothenbuhler and Thompson (40) reported a small increase of tolerance or resistance in bee larvae after selection experiments with *Bacillus larvae*. Furthermore, Dunbar and Beard (52) suggested some natural increase of resistance in field populations of *Popillia japonica* after a long period of contact with *Bacillus popilliae*. Selection experiments with *Bacillus thuringiensis*, however, did not reveal any enhancement of resistance within populations of *Plutella xylostella* (97, 98).

Nevertheless, in different populations of one host species, natural differences in susceptibility may preexist (94). However, evaluations on the basis of LD_{50} mostly do not exceed the factor 10 as demonstrated for populations of *Ephestia cautella* and *Plodia interpunctella* with *B.t.* var. *kurstaki* (96) and for populations of *Aedes caspicus* with *B.t.* var. *israelensis* (95).

Low slopes in dosage-mortality regression lines are typical for bacterial pathogens, indicating that a large increase in dosage induces only a relatively small enhancement in effectiveness. This is due to large variability in individual host responses. Furthermore, different instars usually show a shift of the regression line, demonstrating in general that larvae become less susceptible with increasing age (91–93). In many holometabolic insects, only the larval stage is highly susceptible to pathogenic bacteria. Other developmental stages, like eggs and pupae, are unable to take up such pathogens. Adults usually show a high degree of tolerance to pathogenic bacteria, such as for *Bacillus larvae*. They often act as reservoirs of pathogens and aid their dispersal. Pupae, adults, and even embryos, however, may die as the result of a chronic infection acquired within a previous developmental stage (e.g., *Rickettsiella* infections of *Melolontha* spp.). The same can occur after an experimental injection of a pathogen.

5.3.2. IMMUNITY.   The exoskeleton and peritrophic membrane of the insect gut are effective barriers against invading bacteria. Antibiotic substances (bactericides) and enzymes (lysozyme), which are present within the gut lumen, may act as limiting factors of ingested bacteria. Moreover, the gut milieu, (high pH and proteinases) often activates bacterial protoxins, for example, the crystal toxin of *B. thuringiensis*. In the hemocoel, humoral and cellular factors are involved in immune defense. Active hemocytes are responsible for clearing the hemolymph of invaded or injected bacteria by phagocytosis or nodule formation (136–138). Among inducible immunoproteins of the hemolymph, lysozyme is bacteriolytic for many gram-positive bacteria (138, 139), and some bactericidal proteins (like complement) are effective against gram-negative bacteria (140, 141). In contrast to vertebrates, however, an acquired immunity based on immunoglobulins (antibodies) cannot be found in arthropods. In general, the immune system of arthropods is relatively unspecific with respect to the target bacteria, nonetheless it is self-protecting for the host.

The immune factors of arthropods are primarily responsible for the natural resistance against bacteria which are known to be nonpathogenic. Pathogenic bacteria, such as *Bacillus popilliae* or *Serratia marcescens*, however, may overcome the immune defense by passive resistance or special inhibitors. Finally, pathogenic bacteria often grow so fast in the hemolymph that immune reactions do not destroy them. In spite of experiments in which an adaptive protection of immune factors can be demonstrated from the injection of vaccines (apathogenic bacteria or inactivated pathogens), the epizootic relevance of such limited acquired immunity against bacterial pathogens is still controversial. Biotests have often shown that successive peroral applications of virulent pathogens caused a superinfection within the insect rather than a delayed or abortive infection as a result of an immune response induced by a foregoing infective process.

5.4. Environmental Factors

Several environmental factors have important influence on the interrelationship between insect hosts and bacteria. Some stressors, such as malnutrition and extreme climatic conditions, are known as predisposing causes of several diseases. Seasonal fluctuations in the quality of diet and differences in weather conditions, therefore, may be responsible for (often periodic) fluctuations of LD_{50} values within a host population throughout the year.

Secondary metabolites in the chemical defense of plants may influence not only frass behavior of insects (and, therefore, the ingestion of pathogens) but also their predisposition to disease. Furthermore, the foliage may contain bactericidal substances (142, 143). Such secondary plant substances may protect insects against bacterial diseases. On the other hand, some plant substances are known as attractants which can indirectly stimulate the uptake of pathogens. Thus, foliage of different quality may modify the pathogen-induced rate of mortality as shown with *B. thuringiensis* (144).

For aquatic insects, like mosquito larvae, water quality influences the efficiency of pathogens such as *B. sphaericus* or *B.t.* var. *israelensis*. For example, small slit and clay particles (in contrast to sand) can reduce the pathogen-induced rate of mortality (145). Even though eutrophic conditions favor the effect of *Bacillus sphaericus* (61), they also reduce the effect of *B.t.* var. *israelensis* (90). The value of soil parameters and of climatic factors on the persistence of pathogens in the environment has already been discussed.

Of all the climatic factors, only temperature is important in the intensity of pathogen ingestion, the development of a bacterial infection, and its pathogenesis for insects. This is due to the fact that physiological processes are usually determined by the rule of van t'Hoff. In the effect of *B. thuringiensis* on *Ephestia kühniella*, the logarithm of the reciprocal of the time necessary to kill half the larvae ($1/Lt_{50}$) is proportional to the temperature (146). *B.t.* var. *kurstaki* has, therefore, little activity against caterpillars below 15°C, but at 25 to 30°C the toxic and pathogenic effects accelerate. Similarly, the efficiency of *Bacillus sphaericus* against mosquito larvae declines noticeably with decreasing temperature. In contrast, the effect of *B.t.* var. *israelensis* is not so temperature-sensitive (147) and, therefore, mosquito larvae living in cold water are sufficienty controlled at 5.5 to 8.5°, as shown for *Aedes cantans* and *Aedes rusticus* (90).

The influence of environmental conditions on the pathogenic process has to be taken into consideration in addition to other important epizootic parameters, such as virulence, dosage, timing, transmission (including application techniques), and persistance, if bacteria are to be used for the biological control of insect pests or evaluated in epizootiology.

REFERENCES

1. G. E. Bucher, "Identification of Bacteria Found in insects," in H. D. Burges, Ed., *Microbial Control of Pests and Plant Diseases*, Academic Press, London, 1981, p. 7.

2. M. P. Starr, H. Stolp, H. G. Trüper, A. Balows, and H. G. Schlegel, Eds., *The Prokaryotes*, Springer Verlag, Berlin-New York, 1981.

3. L. Kartman, F. M. Prince, and S. F. Quan, New knowledge on the ecology of sylvatic plague, *Ann. N. Y. Acad. Sci.*, **70**, 668 (1958).

4. J. D. Martin and J. O. Mundt, Enterococci in insects, *Appl. Microbiol.*, **24**, 575 (1972).

5. G. E. Bucher and J. M. Stephens, A disease of grasshoppers caused by the bacterium *Pseudomonas aeruginosa* (Schroeter) Migula, *Can. J. Microbiol.*, **3**, 611 (1957).

6. G. E. Bucher, "Nonsporulating Bacterial Pathogens," in E. A. Steinhaus, Ed., *Insect Pathology: An Advanced Treatise*, Vol. 2, Academic Press, New York, 1963, p. 117.

7. A. Dorn, Studies on the fat body of *Oncopeltus fasciatus* invaded by *Pseudomonas aeruginosa*, *J. Invertebr. Pathol.*, **29**, 347 (1977).

8. R. Beard, Field experiments with *Pseudomonas aeruginosa* (Schroeter) Migula to control grasshoppers, *Can. Entomol.*, **90**, 89 (1958).

9. G. B. Landerkin and H. Katznelson, Organisms associated with septicemia in the honey bee, *Apis mellifera*, *Can. J. Microbiol.*, **5**, 169 (1959).

10. A. Bacot, On the persistence of bacilli in the gut of an insect during metamorphosis, *Trans. Roy. Entomol. Soc. London*, pt. II., 497 (1911).

11. N. D. Mikhnovskaya and T. N. Povazhnaya, (*Pseudomonas fluorescens* bacteria as pathogens of mosquito larvae) (in Russian), *Med. Parazitol. (Moscow)* **44**, 690 (1975).

12. G. E. Haniotakis and N. Avtzis, Mortality in *Dacus oleae* (Gmelin) through infection with *Pseudomonas putida*, *Ann. Zool. Écol. Anim.*, **9**, 299 (1977).

13. E. A. Steinhaus, *Serratia marcescens* Bizio as an insect pathogen, *Hilgardia*, **28**, 351 (1959).

14. P. A. D. Grimont, F. Grimont, and O. Lysenko, Species and biotype identification of *Serratia* strains associated with insects, *Curr. Microbiol.*, **2**, 139 (1979).

15. J. M. Stephens, Note on effects of feeding grasshoppers two pathogenic species of bacteria simultaneously. *Can. J. Microbiol.*, **5**, 313 (1959).

16. J. V. Bell, *Serratia marcescens* found in eggs of *Heliothis zea*: tests against *Trichoplusia ni*, *J. Invertebr. Pathol.*, **13**, 151 (1968).

17. J. P. Stevenson, Epizootiology of a disease of the Desert locust, *Schistocerca gregaria* (Forskal), caused by non-chromogenic strains of *Serratia marcescens* Bizio, *J. Insect Pathol.*, **1**, 232 (1959).

18. G. Stucki, T. A. Jackson, and M. J. Noonan, Isolation and characterisation of *Serratia* strains pathogenic for larvae of the New Zealand grass grub *Costelytra zealandica*, *N. Z. J. Science*, **27**, 255 (1984).

19. C. E. Burnside, Septicemic condition of adult bees, *J. Econ. Entomol.*, **21**, 379 (1928).

20. H. Wille and L. Pinter, Untersuchungen über bakterielle Septikämie der erwachsenen Honigbiene in der Schweiz, *Bull. Apicol.*, **4**, 141 (1961).

21. E. G. King, J. V. Bell, and D. F. Martin, Control of the bacterium *Serratia marcescens* in an insect host-parasite rearing program, *J. Invertebr. Pathol.*, **26**, 35 (1975).

22. A. M. Huger and A. Krieg, Über eine fatale Bakteriose in einem Parasit/Wirt-System, *Jahresber. Biol. Bundesanst. Land- Forstwirtsch. Berlin Braunschweig*, 1983, p. 80.

23. G. E. Bucher, The bacterium *Coccobacillus acridiorum* d'Herelle: its taxonomic position and status as a pathogen of locusts and grasshoppers, *J. Insect Pathol.*, **1**, 331 (1959).

24. G. E. Bucher, Potential bacterial pathogens of insects and their characteristics, *J. Insect Pathol.*, **2**, 172 (1960).

25. B. J. Consenza and F. B. Lewis, Occurrence of motile, pigmented streptococci in lepidopterous and hymenopterous larvae, *J. Invertebr. Pathol.*, **7**, 86 (1965).

26. C. C. Doane, Primary pathogens and their role in the development of an epizootic in the gypsy moth, *J. Invertebr. Pathol.*, **15**, 21 (1970).

27. C. D. Doane, Field application of a *Streptococcus* causing brachyosis in larvae of *Porthetria dispar*, *J. Invertebr. Pathol.*, **17**, 303 (1971).

28. L. Bailey, The pathogenicity for honeybee larvae of microorganisms associated with European foulbrood, *J. Insect Pathol.*, **5**, 198 (1963).

29. L. Bailey, The epizootiology of European foulbrood of the larval honeybee, *Apis mellifera* Linnaeus, *J. Insect Pathol.*, **2**, 67 (1960).

30. W. C. Rothenbuhler, "Semi-domesticated insects: honeybee breeding," in M. A. Hoy and J. J. McKelvey, Eds., *Genetics in Relation to Insect Management*, Rockefeller Found., 1979, p. 84.

31. R. E. Gordon, W. C. Haynes, and C. H. N. Pang, *The Genus Bacillus*, (Agric. Handbook 427) Agric. Res. Serv. USDA Washington, 1973.

32. G. F. White, The cause of American foulbrood, *U. S. Dep. Agric. Bur. Entomol. Circ.*, **94**, 1 (1906).

33. W. Fritsch, Untersuchungen an 56 Stämmen des *Bac. larvae* White über deren antigene Verhältnisse mit Hilfe der Agglutination, *Arch. Bienenkunde, 1957*, 22 (1957).

34. Y. S. Peng and K. Y. Peng, A study on the possible utilization of immunodiffusion and immunofluorescence techniques as the diagnostic methods for American foulbrood of honeybees (*Apis mellifera*), *J. Invertebr. Pathol.*, **33**, 284 (1973).

35. J. F. Bamrick, Resistance to American foulbrood in honeybees. V. Comparative pathogenesis in resistant and susceptible larvae, *J. Insect Pathol.*, **6**, 284 (1964).

36. E. W. Davidson, Ultrastructure of American foulbrood disease pathogenesis in larvae of the worker honeybee, *Apis mellifera*, *J. Invertebr. Pathol.*, **21**, 53 (1973).

37. N. G. Patel and L. K. Cutcomb, The toxicity of enzyme fractions of *Bacillus larvae*, *J. Econ. Entomol.*, **54**, 773 (1961).

38. W. T. Wilson, Resistance to American foulbrood in honeybees. XII. Persistence of viable *Bacillus larvae* spores in faeces of adults permitted flight, *J. Invertebr. Pathol.*, **20**, 165 (1972).

39. R. A. Bitner, W. T. Wilson, and J. D. Hitchcock, Passage of *Bacillus larvae* spores from adult queen honeybees to attendant workers (*Apis mellifera*), *Ann. Entomol. Soc., Am.*, **65**, 899 (1972).

40. W. C. Rothenbuhler and V. C. Thompson, Resistance to American foulbrood in honeybees. I. Differential survival of larvae of different genetic lines, *J. Econ. Entomol.*, **49**, 470 (1956).

41. T. R. Hoage and W. C. Rothenbuhler, Larval honeybee response to various doses of *Bacillus larvae* spores, *J. Econ. Entomol.*, **59**, 42 (1966).

42. H. Shimanuki, P. A. Hartman, and W. C. Rothenbuhler, The effect of serial passage of *Bacillus larvae* White in the honeybee, *J. Invertebr. Pathol.*, **7**, 75 (1965).

43. S. R. Dutky, Two new spore-forming bacteria causing milky diseases of Japanese beetle larvae, *J. Agric. Res.*, **61**, 57 (1940).

44. H. Wille, *Bacillus fribourgensis* n. sp., Erreger einer "milky disease" im Engerling von *Melolontha melolontha* L., *Mitt. Schweiz. Entomol. Ges.*, **29**, 271 (1956).

45. R. L. Beard, Two milky diseases of Australian Scarabaeidae, *Can. Entomol.*, **88**, 640 (1956).

46. L. J. Dunbleton, Bacterial and nematode parasites of soil insects, *N. Z. J. Sci. Technol.*, A **27**, 76 (1945).

47. R. J. Milner, "Identification of the *Bacillus popilliae* Group of Insect Pathogens," in H. D. Burges, Ed., *Microbial Control of Pests and Plant Diseases,* Academic Press, London, 1981, p. 45.

48. C. M. Splittstoesser, H. Tashiro, S. L. Lin, K. H. Steinkraus, and J. B. Fiori, Histopathology of the European chafer, *Amphimallon majalis* infected with *Bacillus popilliae, J. Invertebr. Pathol.*, **22**, 161 (1973).

49. C. M. Splittstoesser, C. Y. Kawanishi, and H. Tashiro, Infection of the European chafer, *Amphimallon majalis* by *Bacillus popilliae:* light and electron microscope observations, *J. Invertebr. Pathol.*, **31**, 84 (1978).

50. S. R. Dutky, "The Milky Diseases," in E. A. Steinhaus, Ed., *Insect Pathology: An Advanced Treatise,* Vol. 2, Academic Press, New York, 1963, p. 75.

51. P. O. Hutton and P. P. Burbutis, Milky disease and Japanese beetle in Delaware, *J. Econ. Entomol.*, **67**, 247 (1974).

52. D. M. Dunbar and R. L. Beard, Present status of milky disease of Japanese and Oriental beetle in Connecticut, *J. Econ. Entomol.*, **68**, 453 (1975).

53. L. A. Bulla, R. N. Costilow, and E. S. Sharpe, Biology of *Bacillus popilliae, Adv. Appl. Microbiol.*, **24**, 1 (1978).

54. M. G. Klein, "Advances in the Use of *Bacillus popilliae* for Pest Control," in H. D. Burges, Ed., *Microbial Control of Pests and Plant Diseases,* Academic Press, London, 1981, p. 183.

55. H. Tashiro, G. G. Gyrisco, F. L. Gambrell, B. J. Fiori, and H. Breitfeld, N. Y. Agric. Exp. Stn. Geneve Bull., **828** (1969); cited after Klein (54).

56. B. Hurpin and P. Robert. Experiments on simultaneous infections of the common cockchafer *Melolontha melolontha, J. Invertebr. Pathol.*, **11**, 203 (1968).

57. B. Hurpin and P. Robert, Comparison of the activity of certain pathogens of the cockchafer *Melolontha melolontha* in plots of natural meadowland, J. Invertebr. Pathol., **19**, 291 (1972).

58. W. R. Kellen, T. B. Clark, J. E. Lindegren, B. C. Ho, M. H. Rogoff, and S. Singer, *Bacillus sphaericus* Neide as a pathogen of mosquitoes, *J. Invertebr. Pathol.*, **7**, 442 (1965).

59. E. W. Davidson, "Bacterial Diseases of Insects Caused by Toxin-Producing Bacilli other than *Bacillus thuringiensis*," in E. W. Davidson, Ed., *Pathogenesis of Invertebrate Microbial Diseases,* Allanheld, Osmun Publ., New Jersey, 1981, p. 269.

60. A. Yousten, H. de Barjac, J. Hedrick, V. Cosmao-Dumanoir, and P. Myers, Comparison between bacteriophage typing and serotyping for the differentiation of *Bacillus sphaericus* strains, *Ann. Microbiol. Inst. Pasteur,* **131 B**, 297 (1980).

61. J. A. Hornby, B. C. Hertlein, R. Levy, and T. W. Miller, Persistent activity of mosquito larvicidal *Bacillus sphaericus* 1593 in fresh water and sewage, *WHO/VBC/* 81.830., Geneva, 1981.

62. S. Singer, "Potential of *Bacillus sphaericus* and Related Spore-Forming Bacteria for Pest Control," in H. D. Burges, Ed., *Microbial Control of Pests and Plant Diseases,* Academic Press, London, 1981, p. 283.

63. B. C. Carlton, J. M. Gonzalez, and B. J. Brown, Assignment of delta-endotoxin genes of *Bacillus thuringiensis* to specific plasmids by curing and plasmid transfer analyses, *Abstr. 3rd Int. Colloq. Invertebr. Pathol., Microbial Contr.,* Univ. Sussex, Brighton, 1982, p. 68.

64. H. de Barjac, "Identification of H-serotypes of *Bacillus thuringiensis,"* in H. D. Burges, Ed., *Microbial Control of Pests and Plant Diseases,* Academic Press, London, 1981, p. 35.

65. A. M. Heimpel and T. A. Angus, The site of action of crystalliferous bacteria in Lepidoptera larvae, *J. Insect Pathol.,* **1,** 152 (1959).

66. J. F. Charles and H. de Barjac, Histopathologie de l'action de la delta-endotoxine de *Bacillus thuringiensis* var. *israelensis* sur les larves de l'*Aedes aegypti* (Dipt., Culicidae), *Entomophaga,* **26,** 203 (1981).

67. A. Krieg, A. M. Huger, G. A. Langenbruch, and W. Schnetter, *Bacillus thuringiensis* var. *tenebrionis* ein neuer gegenüber Larven von Coleopteren wirksamer Pathotyp, *Z. Angew. Entomol.,* **96,** 500 (1983).

68. E. Berliner, Über die Schlaffsucht der Mehlmottenraupe (*Ephestia kühniella* Zell.) und ihren Erreger *Bacillus thuringiensis* n. sp., *Z. Angew. Entomol.,* **2,** 29 (1915).

69. L. J. Goldberg and J. Margalit, Bacterial spore demonstrating rapid larvicidal activity against *Anopheles sergentii, Uranotaenia unguiculata, Culex univittatus, Aedes aegypti* and *Culex pipiens, Mosq. News,* **37,** 355 (1977).

70. A. J. deLucca, J. G. Simonson, and A. D. Larson, *Bacillus thuringiensis* in soils of United States, *Can. J. Microbiol.,* **27,** 865 (1981).

71. A. W. West and H. D. Burges, Ecology of *Bacillus thuringiensis* in soil, *Proc. 3rd Int. Colloq. Invertebr. Pathol., Microbial Control,* Univ. Sussex, Brighton, 1982, p. 319.

72. S. Ishiwata, (On a kind of severe flacheria [sotto disease]) (in Japanese), *Dainihon Sanshi Kaiho,* **9,** 1 (1901).

73. M. Ohba and K. Aizawa, Serological identification of *Bacillus thuringiensis* and related bacteria isolated in Japan, *J. Invertebr. Pathol.,* **32,** 303 (1978).

74. A. B. Gukasyan, *Bacteriological control methods of Siberian Silkworm* (in Russian), *Nauka,* Moskva, 1970.

75. S. Metalnikov and V. Chorine, Maladies microbiennes chez les chenilles de *Pyrausta nubilalis* Hb., *Ann. Inst. Pasteur,* **43,** 136 (1929).

76. E. Kurstak, Données sur l'epizootie bactérienne naturelle provoquée par un bacillus du type *Bacillus thuringiensis* sur *Ephestia kühniella* Zeller, *Entomophaga Mem. Hors. Sér.,* **2,** 245 (1964).

77. J. Vaňková and K. Purrini, Natural epizootics caused by bacilli of the species *Bacillus thuringiensis* and *Bacillus cereus, Z. Angew. Entomol.,* **88,** 216 (1979).

78. A. Krieg, Über das natürliche Vorkommen von *Bacillus thuringiensis* in Getreide und Getreideprodukten im Hinblick auf eine biologische Bekämpfung von Mehlmotten (Phycitidae) im Vorratsschutz, *Nachrichtenbl. Dtsch. Pflanzenschutzdienst (Braunschweig),* **34,** 153 (1982).

79. J. R. Norris, The classification of *Bacillus thuringiensis, J. Appl. Bacteriol.,* **27,** 439 (1964).

80. H. D. Burges and J. A. Hurst, Ecology of *Bacillus thuringiensis* in storage moths, *J. Invertbr. Pathol.,* **30,** 131 (1977).

81. E. S. Kurstak, Le rôle de *Nemeritis canescens* Gravenhorst dans l'infection à *Bacillus thuringiensis* Berliner chez *Ephestia kühniella* Zeller, *Ann. Epiphyt.* (Paris), **17,** 335, 451 (1966).

82. R. Flanders and I. M. C. Hall, Manipulated bacterial epizootics in *Anagasta* populations, *J. Invertebr. Pathol.,* **7,** 368 (1965).

83. O. Mück, S. Hassan, A. M. Huger, and A. Krieg, Zur Wirkung von *Bacillus thuringiensis* Berliner auf die parasitischen Hymenopteren *Apanteles glomeratus* L. (Braconidae) und *Pimpla turionella* (L.) (Ichneumonidae), *Z. Angew. Entomol.,* **92,** 303 (1981).

84. M. Tamashiro, Effect of insect pathogens on some biological control agents in Hawaii, *Proc. U.S.-Japan Sem. Microbial Control Insect Pests,* Fukuoka 1967, p. 147.

85. S. A. Hassan and A. Krieg, Über die schonende Wirkung von *Bacillus thuringiensis*-Präparaten auf den Parasiten *Trichogramma cacoeciae* (Hym.: Trichogrammatidae), *Z. Pflanzenkr. Pflanzenschutz.,* **82,** 515 (1975).

86. H. T. Dulmage, Insecticidal activity of HD-1, a new isolate of *Bacillus thuringiensis* var. *alesti, J. Invertebr. Pathol.,* **15**, 232 (1970).

87. A. Krieg, Bekämpfung von Insekten im Pflanzenschutz mit *Bacillus thuringiensis*-Präparaten und deren Einfluß auf die Umwelt, *Anz. Schädlingskde.,* **56**, 41 (1983).

88. J. M. Franz and A. Krieg, *Biologische Schädlingsbekämpfung,* 3rd ed., Verlag P. Parey, Berlin, 1982.

89. C. M. Ignoffo, D. L. Hostetter, and R. E. Pinnel, Stability of *Bacillus thuringiensis* and *Baculovirus heliothis* on soybean foliage, *Environ. Entomol.,* **3**, 117 (1974).

90. W. Schnetter, S. Engler-Fritz, C. Aly, and N. Becker, Anwendung von *Bacillus thuringiensis* var. *israelensis* Präparaten gegen Stechmücken am Oberrhein, *Mitt. Dtsch. Ges. Allg. Angew. Entomol.,* **4**, 18 (1983).

91. A. M. Afify and M. M. Matter, Zunehmende Toleranz (LT-Werte) von *Anagasta kühniella* Z. gegen *Bacillus thuringiensis* mit dem Alter der Larvalentwicklung, *Anz. Schädlingskde.,***43**, 97 (1970).

92. W. Schnetter, S. Engler, J. Morawcsik, and N. Becker, Wirksamkeit von *Bacillus thuringiensis* var. *israelensis* gegen Stechmücken und nontarget-Organismen, *Mitt. Dtsch. Ges. Allg. Angew. Entomol.,* **2**, 195 (1981).

93. W. H. McGaughey, Effect of larval age on the susceptibility of almond moth and Indianmeal moth to *Bacillus thuringiensis, J. Econ. Entomol.,* **71**, 923 (1978).

94. K. Aizawa, T. Kawarabata, and F. Sato, Response of the silkworm *Bombyx mori* to *Bacillus thuringiensis* Berliner, *J. Sericult. Sci. Japan,* **31**, 253 (1962).

95. C. Sinégre, B. Gaven, J. L. Jullien, and O. Crespo, Activité du sérotype H-14 de *Bacillus thuringiensis* vis-à-vis des principales espèces de moustiques anthropophiles du littoral mediterranean francais, *WHO/VBC/79.743.,* Geneva, 1979.

96. R. A. Kinsinger and W. H. McGaughey, Susceptibility of populations of Indianmeal moth and almond moth to *Bacillus thuringiensis, J. Econ. Entomol.,* **72**, 346 (1979).

97. M. Devriendet and D. Martouret, Absence de resistance a *Bacillus thuringiensis* chez la taigne des crucifères, *Plutella maculipennis* (Lep.: Hyponomeutidae), *Entomophaga,* **21**, 189 (1976).

98. G. A. Langenbruch and A. Krieg, Prüfung der Möglichkeit zur Entsehung einer Resistenz gegenüber Bakterienpräparaten zur Schädlingsbekämpfung, *Jahresber. Biol. Bundesanst. Land- Forstwirtsch. Berlin Braunschweig,* 1976, p. 81.

99. J. M. Stephens, Disease in codling moth larvae produced by several strains of *Bacillus cereus, Can. J. Zool.,* **30**, 30 (1952).

100. A. M. Heimpel, A strain of *Bacillus cereus* Fr. and Fr. pathogenic for the larch sawfly, *Pristiphora erichsonii* (Htg), *Can. Entomol.,* **86**, 73 (1954).

101. K. Aizawa and N. Fujiyoshi, Selection and breeding of bacteria for control of insect pests in the sericultural countries, *Proc. U.S.-Japan Sem. Microbial Control Insect Pests,* Fukuoka, 1967, p. 79.

102. G. E. Bucher, Artificial culture of *Clostridium brevifaciens* n. sp. and *C. malacosomae* n. sp., the causes of brachytosis of tent caterpillar, *Can. J. Microbiol.,* **7**, 641 (1961).

103. J. G. Tully and R. F. Whitcomb, "The Genus *Spiroplasma*," in M. P. Starr, H. Stolp, H. G. Trüper, A. Balows, and H. G. Schlegel, Eds., *The Prokaryotes,* Springer-Verlag, Berlin-New York, 1981, p. 2271.

104. K. Maramorosch, Spiroplasmas: agents of animal and plant diseases, *Bioscience,* **31**, 374 (1981).

105. L. P. Brinton and W. Burgdorfer, Cellular and subcellular organization of the 211 F agent, a spiroplasma from the rabbit tick *(Haemophysalis leporispalustris)* (Acari: Ixodidae), *Int. J. Syst. Bacteriol.,* **26**, 534 (1976).

106. D. F. Poulson and B. Sakaguchi, Nature of "sex ratio" agent in *Drosophila, Science,* **133**, 1489 (1961).

107. T. B. Clark, *Spiroplasma* sp., a new pathogen in honey bees, *J. Invertebr. Pathol.*, **29**, 112 (1977).

108. R. V. Dowell, H. G. Basham, and R. E. McCoy, Influence of five spiroplasma strains on the growth rate and survival of *Galleria mellonella* (Lepidoptera: Pyralidae) larvae, *J. Invertebr. Pathol.*, **37**, 231 (1981).

109. A. Krieg, "Rickettsiae and Rickettsioses," in E. A. Steinhaus, Ed., *Insect Pathology: An Advanced Treatise,* Vol. 1, Academic Press, New York, 1963, p. 577.

110. E. Weiss, "The Family Rickettsiaceae: Human Pathogens," in M. P. Starr, H. Stolp, H. G. Trüper, A. Balows, and H. G. Schlegel, Eds., *The Prokaryotes,* Springer-Verlag, Berlin-New York, 1981, p. 2137.

111. E. Weiss and G. A. Dash, "The Family Rickettsiaceae: Pathogens of Domestic Animals and Invertebrates; Non-Pathogenic Arthropod Symbiotes," in M. P. Starr, H. Stolp, H. G. Trüper, A. Balows, and H. G. Schlegel, Eds., *The Prokaryotes,* Springer-Verlag, Berlin-New York, 1981, p. 2161.

112. J. H. Yen and A. R. Barr, The etiological agent of cytoplasmic incompatibility in *Culex pipiens, J. Invertebr. Pathol.,* **22**, 242 (1973).

113. P. E. M. Fine, On the dynamics of symbiote-dependent cytoplasmic incompatibility in culicine mosquitoes, *J. Invertebr. Pathol.,* **31**, 10 (1978).

114. W. R. Kellen and D. F. Hoffman, *Wolbachia* sp. (Rickettsiales: Rickettsiaceae) a symbiont of the almond moth, *Ephestia cautella:* ultrastructure and influence on host fertility, *J. Invertebr. Pathol.,* **37**, 273 (1981).

115. E. C. Suitor and E. Weiss, Isolation of a rickettsialike microorganism *(Wolbachia persica* n. sp.) from *Argaas persicus* (Oken), *J. Infect. Dis.,* **108**, 95 (1961).

116. S. R. Dutky and E. L. Gooden, *Coxiella popilliae,* n. sp., a rickettsia causing blue diesase of Japanese beetle larvae, *J. Bacteriol.,* **63**, 743 (1952).

117. H. Wille and M. E. Martignoni, Vorläufige Mitteilung über einen neuen Krankheitstypus beim Engerling von *Melolontha vulgaris* F., *Schweiz Z. Allg. Pathol., Bakteriol.,* **15**, 470 (1952).

118. B. Hurpin, Spécifité de *Rickettsiella melolonthae* et pathogenie pour le vertebrates, *Ann. Soc. Entomol. France, N.S.,* **7**, 439 (1971).

119. A. M. Huger and A. Krieg, New aspects of the mode of reproduction of *Rickettsiella* organisms in insects, *J. Invertebr. Pathol.,* **9**, 442 (1967).

120. A. Krieg, Ultrahistologische Untersuchungen im Zusammenhang mit dem Problem einer Übertragung von Rickettsien bei Scarabaeiden, *Naturwissenschaften,* **53**, 484 (1966).

121. O. F. Niklas, Transovariale Weitergabe von *Rickettsiella melolonthae* (Krieg) Philip über zwei Zuchtgenerationen von *Amphimallon solstitiale* (L.) (Col.: Melolonthidae), *Entomophaga,* **14**, 225 (1969).

122. R. Martoja, Sur quelques aspects de la biologie des Orthoptères en relation avec la présence de concentrations microbiennes, *Ann. Soc. Entomol. France, N.S.,* **2**, 753 (1966).

123. G. Croizier, G. Meynadier, G. Morel, and M. Capponi, Comparison immunologique de quelques *Wolbachiae* et recherche de communauté antigénique avec d'autres membres l'ordre des Rickettsiales, *Bull. Soc. Pathol. Exot.,* **68**, 133 (1975).

124. P. Götz, *"Rickettsiella chironomi":* an unusual bacterial pathogen which reproduces by multiple cell division, *J. Invertebr. Pathol.,* **20**, 22 (1972).

125. B. A. Federici, Reproduction and morphogenesis of *Rickettsiella chironomi,* an unusual intracellular prokaryotic parasite of midge larvae, *J. Bacteriol.,* **143**, 995 (1980).

126. G. Morel, Two Chlamydiales (Rickettsias) in an arachnida: the spider *Pisaura mirabilis* Cl., *Experientia,* **34**, 344 (1978).

127. G. Morel, Surface projection of a chlamydia-like parasite of midge larvae, *J. Bacteriol.,* **144**, 1174 (1980).

128. H. Wilbert, Cybernetic concepts in population dynamics, *Acta Biotheoretica,* **19,** 54 (1970).

129. L. O. Kunkel, Cross protection between strains of yellows-type viruses, *Adv. Virus Res.,* **3,** 251 (1955).

130. K. Maramorosch, Cross-protection between two strains of corn stunt virus in an insect vector, *Virology,* **6,** 448 (1958).

131. W. Burgdorfer, S. F. Hayes, and A. J. Mavros, "Nonpathogenic Rickettsiae in *Dermacentor andersoni*: Limiting Factor for the Distribution of *Rickettsia rickettsii,*" in W. Burgdorfer and R. L. Anacker, Eds., *Rickettsia and Rickettsial Diseases,* Academic Press, New York, 1981, p. 585.

132. R. L. Beard, Competition between two entomogenous bacteria, *Science,* **103,** 371 (1946).

133. I. R. Pendleton, Ecological significance of antibiotics of some varieties of *Bacillus thuringiensis, J. Invertebr. Pathol.,* **13,** 235 (1959).

134. J. Jarosz, Gut flora of *Galleria mellonella* suppressing ingested bacteria, *J. Invertebr. Pathol.,* **34,** 192 (1979).

135. E. S. Raun, G. R. Sutter, and M. A. Revelo, Ecological factors affecting the pathogenicity of *Bacillus thuringiensis* var. *thuringiensis* to the European corn borer and Fall armyworm, *J. Invertebr. Pathol.,* **8,** 365 (1966).

136. N. A. Ratcliffe and S. J. Gagen, Studies on the in vivo cellular reactions and fate of injected bacteria in *Galleria mellonella* and *Pieris brassicae* larvae, *J. Invertebr. Pathol.,* **28,** 17 (1976).

137. N. A. Ratcliffe and S. J. Gagen, Cellular defense reactions of insect hemocytes in vivo: nodule formation and development in *Galleria mellonella* and *Pieris brassicae* larvae, *J. Invertebr. Pathol.,* **28,** 373 (1976).

138. A. Krieg, Zur potentiellen Pathogenität einiger Sporenbildner (Genus: *Bacillus*) gegen Larven von *Galleria mellonella* und deren Ursachen, *Z. Angew. Entomol.,* **93,** 355 (1982).

139. W. Mohrig and B. Messner, Immunreaktionen bei Insekten. I. Lysozym als grundlegender antibakterieller Faktor im humoralen Abwehrmechnismus der Insekten, *Biol. Zentralbl.,* **87,** 439 (1968).

140. J. M. Stephens and J. H. Marshall, Some properties of an immune factor isolated from the blood of actively immunized wax moth larvae, *Can. J. Microbiol.,* **8,** 719 (1962).

141. I. Fye, A. Pye, T. Rasmuson, G. Boman, and I. A. Boman, Insect immunity. II. Simultaneous induction of antibacterial activity and selective synthesis of some hemolymph proteins in diapausing pupae of *Hyalophora cecropia* and *Samia cynthia, Infect. Immun.,* **12,** 1426 (1975).

142. D. J. Kushner and G. T. Harvey, Antibacterial substances in leaves: their possible role in insect resistance to disease, *J. Insect Pathol.,* **4,** 155 (1962).

143. W. A. Smirnoff and P. M. Hutchison, Bacteriostatic and bactericidal effects of extracts of foliage from various plant species on *Bacillus thuringiensis* var. *thuringiensis* Berliner, *J. Invertebr. Pathol.,* **7,** 273 (1964).

144. U. Skatulla, Unterschiedliche Wirkungen von *Bacillus thuringiensis* (B.) auf *Orgyia antiqua* in Abhängigkeit von der Fraßpflanze, *Anz. Schädlingskd.,* **46,** 46 (1973).

145. W. A. Ramoska, S. Watts, and R. E. Rodriguez, Influence of suspended particulates on the activity of *Bacillus thuringiensis* serotyp H-14 against mosquito larvae, *J. Econ. Entomol.,* **75,** 1 (1982).

146. M. M. Matter, *Effect of Bacillus thuringiensis Berliner on the development and fecundity of Anagasta kühniella (Zeller), under different environmental conditions,* M.Sc. Thesis, Dept. Entomol. Fac. Sci. Univ., Cairo, 1969.

147. S. P. Wraight, D. Molloy, H. Jamnback, and P. McCoy, Effects of temperature and instar on the efficacy of *Bacillus thuringiensis* var. *israelensis* and *Bacillus sphaericus* strain 1593 against *Aedes stimulans* larvae, *J. Invertebr. Pathol.,* **38,** 78 (1981).

12

FUNGAL DISEASES

RAYMOND I. CARRUTHERS
RICHARD S. SOPER

United States Department of Agriculture
Agricultural Research Service
Plant Protection Research Unit
Boyce Thompson Institute
Cornell University
Ithaca, New York

---

1. INTRODUCTION

The first microorganism to be recognized as an agent of disease was the fungus, *Beauveria bassiana,* which Bassi demonstrated to be the causal agent of the white muscardine disease of the silkworm, *Bombyx mori.* Bassi de Lodi's "germ theory of disease" was developed from this host/pathogen life system. Since then fungi have been shown to cause many insect diseases. Although these microbial pathogens share much in common with viruses, bacteria, and protozoa, there are many unique differences. Perhaps the most fundamental is the route of infection. Other microbial pathogens infect their host primarily via the digestive tract, whereas fungi usually infect by direct penetration of the insect body wall. As a consequence, phytophagous insects with sucking mouthparts are susceptible to few infectious diseases other than those caused by fungi. For some unknown reason, the beetles seem to be relatively more susceptible to fungal diseases than other insect groups. This is not to say that insects in other orders are immune to fungal pathogens, as fungi are known from practically all insect taxa. In fact, entomopathogenic fungi fill a general role in the regulation of many different insect groups but are of unique significance in respect to aphids, leafhoppers, planthoppers, and their allies.

Epizootics of insect fungal diseases are relatively common in nature and can be spectacular. The sight of many hectares of hardwood tree boles covered with dead and dying forest tent caterpillars *Malacosoma disstria* infected by the pathogen *Erynia crustosa* is extremely impressive. Likewise, entire pastures covered with millions of grasshopper cadavers killed by *Entomophaga grylli* is an awesome sight for any biologist (Fig. 12.1). Such fungal-induced epizootics have been noted by man for over two millennia with some even attaining ritualistic status in early civilizations (2).

There are 750 species of entomopathogenic fungi, exclusive of the 115 genera in the order Laboulbeniales, which are grouped into 85 genera found throughout the classes of fungi (3). However, the majority of species are contained in the Hyphomycetes, Zygomycetes, order Entomophthorales, and Ascomycetes, in particular the genus *Cordyceps*. These pathogens are found in insect hosts in all ecological niches. There are those that attack the aquatic stages of insects, subterranean grubs, insects on herbaceous plants, and forest insects; in fact very few habitats and insect groups are without fungal pathogens.

Entomopathogenic fungi and their population level interactions with insect hosts have only recently been studied at the level of sophistication employed in human or plant epidemiology. Even though reports on fungal epizootiology are found scattered throughout the literature, in only a few cases has there been a systematic approach to describing and understanding the factors that control insect/fungal disease abundance and distribution. In this chapter we do not present a literature review and synthesis of all available epizootiological literature. Instead, we first attempt to summarize the status of the field of fungal epizootiology and the techniques currently employed in this area of research. This is followed by summaries of several specific host/pathogen life systems which are chosen because of their relevance as models illustrating the current level of understanding of fungal disease epizootics and their place within their associated ecosystems. Lastly, we present, by way of examples, the direction of research that we feel to be essential in furthering the understanding of host pathogen dynamics.

2. STATUS OF EPIZOOTIOLOGICAL TECHNIQUES ASSOCIATED WITH THE EVALUATION OF THE DYNAMICS OF INSECT/FUNGAL DISEASE

2.1. Current State of Fungal Epizootiology

The scientific evaluation of the dynamics of insect/fungal diseases and the understanding of the role such diseases play in the regulation of both host and pathogen populations is a relatively new science. Studies considering theoreti-

Figure 12.1. (A) A natural epizootic of *Entomophaga grylli* in an Arizona population of the clear-winged grasshopper *Camnula pellucida*. (B) Nearly 100% of the grasshoppers present in this pasture were infected by *E. grylli* and died attached to the top of the foliage in the typical pattern of an Entomophthorales mycosis.

cal or applied aspects of these host-pathogen systems are rare when compared with the numerous investigations of insect competition, predation, and parasitism (4–6). Epidemiological studies and the tools used to aid in the evaluation of disease dynamics have advanced much further in other disciplines including plant (7–9) and human (10–13) pathology.

Interest in using insect pathogens as control agents within Integrated Pest Management (IPM) programs has generated research aimed at the development of fungal pathogens as microbial insecticides (14). Much of this research has been focused on the isolation, development, and production of highly pathogenic microbes with genetic characteristics favoring long-term storage and use. This orientation has allowed pathogens to be applied to target hosts as replacements for insecticides. Although some successes have been noted (15–18), the majority of attempts to use fungal pathogens as a substitute for insecticides have resulted in unpredictable and less than adequate responses (19–23).

Applying insect pathogens as an inundative release presents no direct problems in itself and may actually prove to be a desirable technique for current and future IPM programs. Problems arise, however, because pathogens have been treated as direct substitutes for chemical insecticides and are applied without an adequate understanding of their population biology or their interactions within the local environment. The realization of this lack of understanding as a limiting problem has caused the field of epizootiology to be a major research priority among insect pathologists (17, 19, 21).

2.1.1. THEORETICAL BASIS FOR POPULATION STUDIES.   Insect populations have long been one of the major focal points for research and development of theoretical and applied population studies (24–26). In fact, much of the early epidemiological work on plant pathogens was based on theoretical and applied population studies of the dynamics of insect populations (27). The basic population characteristics representing the dynamics and stability of these systems are, in fact, common to all populations independent of their taxonomic classification and therefore can be applied equally well to fungal pathogens and their insect hosts. These population characteristics can be classified in several ways, but three main categories (temporal, spatial, and genetic) interact in complex ways to regulate plant and animal populations.

Temporal insect population characteristics are those that fluctuate in reference with time. Examples include birth rate, death rate, and migration rates. These are classically aggregated into a common population parameter, the intrinsic rate of increase (r_m) (4). Populations may be subdivided into smaller units of interest (e.g., age classes, susceptibility classes, etc.) with time-dependent rates used to describe the change of population numbers between these units. Similarly, temporal characteristics may describe interactions between populations (e.g., attack rates, feeding rates, infections rates, etc.).

Spatial characteristics of insect populations may include the area occupied by the population, the aggregation patterns of the organisms, and their movement within that area. Spatial characteristics are of major importance in evaluating the dynamics of host-pathogen systems because interactions between species are highly spatially dependent (28). The role of genetics in population dynamics, regulation, and the evolution of life histories is still poorly understood, although efforts are under way to expand this general area of research (29).

The epizootiological evaluation of insect host/fungal pathogen systems has just begun to explore these population characteristics and how they affect the regulation of both host and pathogen populations. Most of the effort to date has been centered on the temporal aspects of host/pathogen dynamics (see Sections 3.1–3.1.5) with only a few studies taking spatial dynamics into consideration (30–33, Brown, unpublished). Population genetic factors, other than pathogen isolate characterization and evaluation of pathogenicity via laboratory bioassay, have essentially been ignored in studies of insect host/pathogen systems.

Evaluation of basic population parameters via the application of analytical models has lead to substantial theory in the general areas of population dynamics, regulation, and epidemiology. Application of analytical modeling techniques in the area of insect pathology (34–39) has suggested that pathogens may indeed play a major role in regulating host numbers. In the most thorough treatment to date, Anderson and May (37) present a host-pathogen model as described by Figure 12.2 and the following two equations:

$$dX/dt = a(X + Y) - bX - \beta XY + \gamma Y \tag{1}$$

$$dY/dt = \beta XY - (\alpha + b + \gamma)Y \tag{2}$$

Where

X = susceptible host population β = infection rate
Y = infecteds γ = recovery rate
a = birth rate α = Rate of disease-induced mortality
b = death rate

Through evaluation and successive alterations of this basic model, Anderson and May demonstrated the effects of disease-induced host mortality, diminished host reproduction, latent periods of infection, free-living infective stages, stress-related pathogenicity, and density-dependent constraints on theoretical host/pathogen dynamics. Their major findings included evidence for pathogen-induced regulation of host populations, an inverse relationship between pathogenicity and the ability to regulate host populations, theoretical host threshold levels necessary for the maintenance of a pathogen population, and the ability of pathogens to produce stable constant or stable cyclic oscillation patterns in the host population.

As complexity was added in this modeling system (e.g., seasonal changes in population rate processes in place of average values), the population responses were highly altered from those in the less complex models initially used to evaluate host/pathogen dynamics. Such changes in model response point to one of the major pitfalls of using simplified analytical models as a basis for general population theory. That is, simple models are abstract representations of actual biological systems, and thus their evaluation must be interpreted as such and no more. Although theoretical modeling will no doubt aid in the development of our understanding of disease dynamics, a complete under-

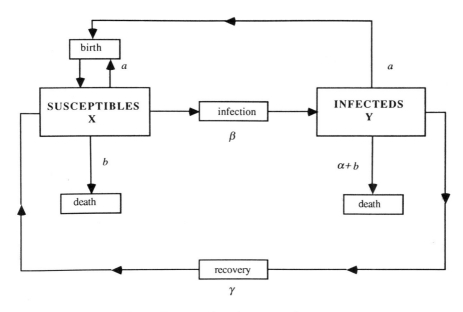

Figure 12.2. A conceptual host-pathogen model developed and evaluated by Anderson and May (37) using equations 1 and 2. Although the host population is divided into two strata (susceptibles and infecteds), the pathogen is not explicitly considered in this model.

standing of these systems will only be possible after actual biological populations have been examined in detail and their interactions with the biotic and abiotic environment characterized.

2.1.2. EMPIRICAL EPIZOOTIOLOGICAL STUDIES.   Epizootiological studies of the dynamics of specific host/pathogen systems have followed two general approaches, being either descriptive (see observational epizootiology, Ch. 1) or experimental. Only a limited number of studies have integrated both of these approaches in developing an overall understanding of epizootiological systems.

Descriptive epizootiology is based on the collection and analysis of monitoring data describing the host and pathogen populations. Although temporal population dynamics have been the focus of the majority of epizootiological studies of fungi, some attention has been directed toward the spatial aspects (30–33). Temporal epizootiological studies usually include the sampling of age-specific host density, an assessment of disease prevalence (typically as percent infected hosts by age class), an assessment of pathogen inoculum levels where feasible, and the recording of relevant abiotic environmental variables. These data are then either presented graphically (40–43) or statistically to identify possible correlations between the variables examined. Multiple regression analysis is the most frequent method employed to relate both biotic and abiotic factors to disease prevalence. These techniques indicate that host population levels, pathogen inoculum levels, and moisture levels in the environment are the most common variables statistically identified as key factors affecting the prevalence of fungal diseases (30, 31, 44, 45). Although gross correlations between such variables are helpful in determining which factors may be important in respect to changes in population numbers and disease prevalence, these techniques provide little understanding of the cause and effect relationships that govern the interactions and dynamics of host/pathogen systems (45). Regression analysis may in fact not identify key factors regulating disease dynamics if the factors being examined were not limiting over the duration of the study period (see 45). Furthermore, regression models based on gross population and environmental data are too general to be of much value in predicting the dynamics of natural or man-manipulated disease systems (31).

Experimental epizootiological studies typically focus on more specific aspects of host/fungal systems in a controlled or more thoroughly monitored environment. Numerous experimental studies have evaluated such things as pathogen-induced mortality between differentially treated host populations (42, 46, 47), the effect of mycosis on host fecundity (32, 48), or pathogen-specific population parameters including sporulation, germination, and infection rates (49–52). Although the orientations of experimental epizootiological studies are diverse, the objective of most of these studies seems to be either to differentiate between natural situations or experimental treatments (46, 53–55) or to characterize the response of a particular system component to controlling environmental stimuli (49, 56–58).

As in the case of descriptive epizootiological studies, research with the sole objective of separating experimental treatments usually provides a minimal amount of forward-projecting information. Even though research of this type provides scientifically enlighting data, it typically cannot be used under even slightly different conditions and consequently is of restricted predictive value. On the other hand, experimental studies producing results in the form of quantitative stimulus-response functions (see Section 4.2) for specific system components can be evaluated over the entire experimental range investigated. Experimental studies of this type provide basic information from which detailed systems models can be constructed. Linkage of specific system components and quantification of their individual or combined responses in model form can be used predictively given adequate model detail and verification.

2.1.3. PREDICTIVE ABILITIES.   Predictive (or forward projecting) models have been developed and used extensively in biology (59, 60), particularly in the area of IPM (61–66). These applications range from the prediction of basic patterns of phenology or the timing of critical population events to highly sophisticated simulation models that aid decision makers in managing entire cropping systems. The application of predictive models in the area of insect/fungal epizootiology has been limited primarily to the prediction of basic phenological events. That is not to say that several very good insect/pathogen models have not been developed, but that most of the modeling activities to date have been more theoretical or explorative in nature rather than predictive (see Sections 3 and 4 and Ch. 3).

One of the earliest modeling efforts in the area of insect/fungal epizootiology (67) has been used and evaluated in a predictive mode more than any other model published to date. Although simple in structure, this model incorporates substantial biological detail to predict future levels of *Nomuraea rileyi* infection of the velvet bean caterpiller *Anticarsia gemmatalis* (see Section 3.1.5). Although this model proved to be somewhat reliable, it has served more as a seed in the area of insect/fungal pathogen life system models than as a predictive tool. This and other modeling efforts have clearly pointed out that epizootiological systems are complex and that prediction of their dynamics will require detailed experimentation and analysis of specific host/pathogen life systems.

3. EPIZOOTICS IN RELATION TO ECOSYSTEMS: EXAMPLES OF SPECIFIC HOST PATHOGEN LIFE SYSTEMS

Disease dynamics and the development of epizootics are controlled by many diverse factors including the environment, as well as density and dispersion of both the host and pathogen populations. An overriding consideration, however, is the ecosystem in which host and pathogen interactions occur. Ecosystems provide the matrix within which these organisms develop, interact, and evolve with one another and with other abiotic and biotic factors. For that reason, we have chosen to present examples of these studies organized by general ecosystem type. Although the ecosystems presented within any given category may be quite diverse, we feel that several common factors link many of these examples. The over-all commonality of each ecosystem type will be discussed prior to the specific examples within each general category, and similarities across ecosystems will be discussed at the end of this section.

3.1. Agricultural Crops

3.1.1. AGROECOSYSTEMS AS INFECTION COURTS. Agricultural ecosystems vary tremendously between individual crop species and between different types of production schemes used to grow a particular crop, but they possess several inherent similarities that can affect interactions between hosts, pathogens, and the environment. Many of these effects stem from the structure of the agroecosystems themselves or from energy-related inputs that are necessary to maintain their overall stability.

Modern industrialized agriculture has been developed through genetic and mechanistic engineering, with the goal of maximizing material through-put for human use (68). This has been accomplished largely by increasing biological and physical homogeneity, an approach that runs counter to natural forces that tend toward higher levels of diversity (69–72). This increased homogeneity has been maintained primarily through large inputs of energy in the form of labor and, more recently, through increased use of fossil fuels and their derivative chemicals (fertilizers and pesticides) (68, 73).

Decreases in biological and physical diversity, particularly with respect to the crop itself and the resulting microenvironment, are known to induce increased pest population levels and subsequent crop damage (27, 74, 75). Likewise, increased insect pest levels associated with agroecosystems (76–78) and the more uniform microenvironmental conditions within crop canopies often stimulate the development of disease epizootics (45). This homogeneity not only induces high population densities of the host insect, but many times it also alters pest distribution in time and space, creating patterns that can affect disease development in the field (32, 33).

On the other hand, agricultural ecosystems are often characterized by sudden alterations (e.g., harvesting, plowing, spraying, etc.) which catastrophically disrupt the ecosystem as a whole, including possible host/pathogen interactions that might lead toward long-term biological control of some pest species. Pesticide applications, particularly fungicides, provide a classic example of such a catastrophic alteration which highly limits the development and spread of fungal diseases while producing little or no direct effect on associated insect host populations (79).

Although it is true that agricultural ecosystems are less diverse than natural ecosystems and that catastrophic events periodically disrupt the balance of associated populations, it has been shown that the structure of these systems is still quite complex (68) and that many have evolved to their own points or cyclic patterns of equilibrium (71, 80). This is clearly true for several agroecosystems that include fungal pathogens of insects as common and consistent members of the biotic community. Despite major disruptions in their associated agroecosystems, several of the pathogens discussed in the following sections have consistently maintained or expanded their populations while in some cases dominating as primary mortality factors of their host.

As managers of agroecosystems we need to become more aware of the structure and interactions that regulate these life systems, so that we can modify them to our own advantage. Careful management based on knowledge of the underlying agroecosystem has been and can be used to manipulate host/pathogen interactions to enhance disease development. In the long run, a more comprehensive understanding of host and pathogen dynamics will allow us to use these systems more effectively as components of multifaceted pest management schemes.

3.1.2. ENTOMOPHTHORALES AFFECTING APHIDS AND LEAFHOPPERS. The aphids and leafhoppers are a major concern in plant protection because of their ability to spread plant viruses and, in some cases, because of damage caused by feeding-induced production of plant toxins. There are many groups of predators and insect parasitoids which attack these plant pests but as far as entomopathogens are concerned, these insects are susceptible only to the fungi. The other major groups, viruses, bacteria, and protozoa, are transmitted per os, and since the aphids and leafhoppers have sucking mouthparts and feed on plant sap they can not normally ingest pathogens. Fungi, on the other hand, infect by direct penetration of the germ tubes produced by spores. Among the entomopathogenic fungi, the Entomophthorales contain many species which frequently cause epizootics in aphid and leafhopper populations (81, 82). Despite the general observations that these pathogens can cause considerable mortality in aphid and leafhopper populations, there is a controversy over what limits their effectiveness and therefore how they may be manipulated for crop protection.

Many researchers have cited moisture as being the all-important environmental factor that influences the course of epizootics in aphid populations. Wilding (77) reported that infections of the pea aphid *Acyrthosiphon pisum* were positively correlated to the average rainfall recorded 12 days prior to disease observations. Others have also attempted to quantify this relationship. Missonier et al. (83) determined that for an enzootic of *Entomophthora* sp. to be sustained in an aphid population there must be a minimum of 90% relative humidity (RH) for at least 8 hours per day. To increase disease levels to epizootic proportions, the RH must exceed 90% for 10 hr/day and there must be 5 hours of rain per day for 3 consecutive days.

Voronina (84) used the "hydrothermal coefficient" of Selyaninov (see 21, 85) to define ecological zones in the Soviet Union according to their ability to support epizootics of *Entomophthora* sp. in pea aphid populations (Fig. 12.3). She calculated this coefficient (GTK) as sum of precipitation (mm) X 10.0/sum of ave. temperature ($^\circ$ C) > 10°. In those areas where the GTK exceeds 1.4, *Entomophthora* sp. regularly control pea aphid populations below economic levels. When the GTK is between 1.0 and 1.4, fungi are still an important aphid mortality factor but some crop damage can occur. In very dry areas, where the GTK is below 1.0, the pea aphid causes crop losses almost every year. This approach to predicting areas where epizootics will occur based on a measurement of moisture may have some value, but there are notable exceptions. The spotted alfalfa aphid *Therioaphis trifolii maculata* populations in the Bet Shan Valley, Israel, are decimated each year with epizootics caused by *Erynia radicans* (86). This is a hot arid area that even supports date palms. Nevertheless, moisture is not a limiting factor for epizootics even though the GTK would be well below 1.0. Moisture is important for the transmission of mycoses caused by species of the Entomophthorales, as pointed out by numerous authors (50, 84). In the arid climate of the Bet Shan, moisture is provided by both heavy dewfall and irrigation. If the hydrothermal coefficient were modified to substitute hours of leaf wetness or rainfall, its predictive value may be improved.

Temperature is another abiotic factor generally referred to as limiting disease prevalence caused by Entomophthorales. This effect is complicated because the developmental rates of both pathogen and host are regulated by temperature. This is most apparent in multivoltine insects, such as aphids and leafhoppers. If the temperature favors a quick generation time for the host but is above or below the optimum for the pathogen, the insect population may still build to

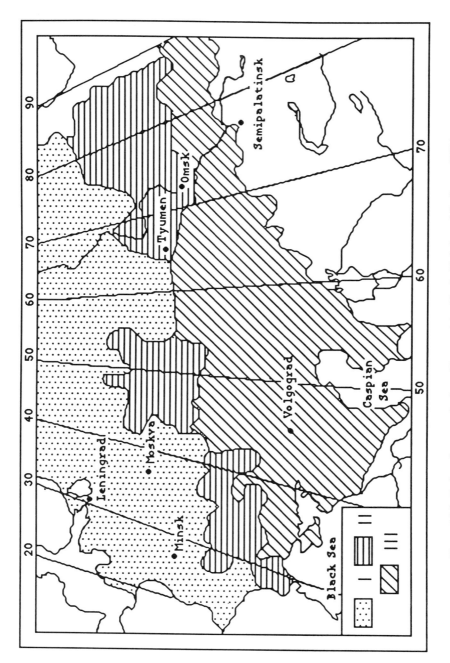

Figure 12.3. Ecological zones in the Soviet Union as defined by the Hydothermal Coefficient, GTK (see text). Areas identified by Zone I are characterized by high moisture conditions leading to consistant control of the pea aphid by *Entomophthora* sp. In Zone II, these pathogens are still important but damaging aphid populations occur. Zone III is characterized by dry conditions which prohibit pathogen development.

crop-damaging levels. Conversely, if temperatures favor a brief incubation period for the pathogen but retard insect development, epizootics can result. Provided that the air is moist, temperature affects three stages of development in the Entomophthorales during current season transmission. First, conidial formation and discharge, then conidial germination and infection, and finally vegetative growth within the insect. All of these stages may have different thresholds of minimum and maximum temperatures for activity as well as different optima. Most species can form and discharge conidia at 5° C and then germinate. However, it may take over 16 hours for these events to occur, whereas at an optimum of around 20° C, all of this may happen in 4 hours (87). Once infection has taken place, the median lethal time (LT_{50}) will depend upon how long the temperature is near the optimum for vegetative growth. All of this is further complicated by the fact that differences commonly exist among pathotypes of the same fungus in their response to these abiotic influences. During a final selection process, Milner and Soper (88) chose an isolate of *Erynia radicans,* based on its ecological derivation, for introduction against the spotted alfalfa aphid in Australia. Two isolates gave the lowest median lethal dose (LD_{50}) values of all strains tested in the laboratory. One was obtained from an epizootic on this aphid near Ithaca, New York, and the other was from a similar epizootic in the Bet Shan Valley, Israel. The climatic conditions in the intended release site were more similar to those in Israel, providing a deciding factor in selecting the Israeli strain for introduction. This strain has since become established and spread several hundred kilometers from Armidale, New South Wales to Brisbane, Queensland, frequently causing epizootics in spotted alfalfa aphid populations (89).

3.1.3. ENTOMOPHTHORALES MYCOSES OF BEETLES. Although numerous Coleoptera are known to be susceptible to fungi, very few pathogens from the Entomophthorales are known to infect this order of insects. A notable exception to this generalization is found in the mycosis of immature weevils from the Genus *Hypera* induced by the pathogen *Erynia* (= *Entomophthora* = *Zoophthora*) *phytonomi.* This mycosis was first described by Arthur (90, 91) from *E. phytonomi* infections of the cloverleaf weevil *H. punctata.* Since its first description, this pathogen has been noted as an important natural control of the cloverleaf weevil in North America (92) and *H. variabilis* in Israel (93).

After the introduction of the alfalfa weevil *H. postica* to North America, 17 years of alfalfa insect-parasite surveys were conducted (94 – 97). These revealed only moderate levels of *E. phytonomi* prevalence in cloverleaf weevil populations, while no primary pathogens of the introduced alfalfa weevil were identified. In the summer of 1973, a disease similar to that found in cloverleaf weevil populations was cited as killing large numbers of larvae and pupae of the alfalfa weevil in southern Ontario, Canada (46). Subsequently, this disease was found to cause substantial alfalfa weevil mortality in other North American alfalfa production areas (98 – 103). Although originally thought to be *E. phytonomi,* studies of the morphology of conidia and resting spores have suggested otherwise (93, 104). The taxonomy is still unresolved, but the pathogen causing widespread epizootics in alfalfa weevil populations is now thought to be a different but closely allied species within *Erynia* (D. M. MacLeod, unpublished data). We will refer to this pathogen or pathogen complex only as *Erynia* sp.

It is unclear whether the pathogen affecting the alfalfa weevil was imported into North America in conjunction with alfalfa weevil introductions, or parasite introductions, or if the pathogen switched from another host species onto the introduced alfalfa weevil. Although the origin of the pathogen is still unknown (99), the fungus has become well established as a major natural biological control agent of the alfalfa weevil throughout much of its range.

Field-oriented studies have been restricted primarily to observational data documenting either the level of disease prevalence (99, 102, 105) or the levels of larval and pupal mortality induced by *Erynia* sp. (46). Observed prevalence levels during epizootics vary between sites and seasons, but levels ranging from 30 to 70% are not atypical at the time of peak larval occurrence. Disease prevalence approaching 100% has been cited in the late phases of host population development. Disease levels of this magnitude produced mortality between 65 and 90% in weevil larvae and an additional 40 to 50% in pupae (46). Laboratory studies suggest that infections of young larvae result in conidial production whereas infections of later instars usually produce resting spores (106).

No attempts have been made to correlate these descriptive data to relevant abiotic conditions, although Harcourt et al. (46) speculated that the epizootics observed in Ontario, Canada in 1973 were the result of "2 weeks of warm, showery weather." Millstein et al. (57) demonstrated, however, that conidial production and discharge depended on environmental conditions; i.e., *Erynia* sp. produced conidia only when moisture conditions in the microenvironment of the alfalfa canopy exceeded 91% RH for at least 3 hours. No effect of temperature (above freezing) was detected on the timing or duration of conidial discharge. This microclimate-pathogen relationship was used in the development of an analytical population model (107). This model provided the quantitative base from which detailed pest management strategies were developed to maximize the effect of this pathogen in controlling alfalfa weevil larvae (107).

The proposed strategy developed by Brown and Nordin (107) involves an earlier-than-normal harvest of first cutting alfalfa. This early harvest concentrates the susceptible weevil larvae under high density conditions in windrows of hay that are left several days in the field for drying. The microenvironmental conditions in the windrows are warm and extremely humid, ideal for disease development and spread. Although this strategy necessitates an earlier-than-normal harvest date and thus lower first cutting yields, it is economically attractive since second cutting alfalfa will produce higher yields and insecticide costs will be reduced. Brown and Nordin (107) estimate a $40 per acre per year savings for alfalfa producers implementing this strategy. This host/pathogen system seems to provide one of the best examples of how a natural occurring fungal pathogen may be manipulated as a management component within an overall IPM program.

3.1.4. Entomophthorales Mycoses of Muscoid Flies. *Entomophthora muscae* causes widespread natural epizootics in several adult Diptera, including the common house fly *Musca domestica* (41, 45, 108–114). MacLeod et al. (115) have reviewed most of the early work on *E. muscae,* although taxonomic rather than epizootiological research was stressed. Summaries emphasizing the epizootiology of this pathogen are available only from a composite of sources (32, 41, 45, 112, 114).

Field level epizootiological studies of *E. muscae* mycosis have been conducted on the wheat bulb fly *Leptohylemyia coarctata* (41), the seed corn maggot *Delia (= Hylemya) platura* (45, 113), and the onion fly *D. antiqua* (45, 114, 116). Most of these studies are descriptive in nature, providing either a qualitative or quantitative record of the host population level, disease prevalence, and environmental conditions. Gross qualitative assessment of the dynamics of *E. muscae* with various host species have provided variable results as epizootics have been typified as occurring under cool-wet conditions (112, 113, 117, 118) or under warm-dry conditions (41, 116). Quantitative studies have shown no clear correlations between the prevalence of *E. muscae* mycosis and ambient environmental conditions (e.g., average daily temperature, moisture, wind, and/or sunlight hours), although positive correlations with host and/or pathogen density have been noted in several studies (41, 45, 116).

Although no clear relationship between ambient environmental conditions and disease prevalence was found using correlative techniques, Carruthers et al. (45) suggested that abiotic conditions influence the initiation and dynamics of *E. muscae* under field conditions. They speculated that more specific cause and effect associations between the environment and the host/pathogen life system were involved in the regulation of disease development (i.e., microenvironmental conditions affecting specific biological processes such as sporulation, germination, or infection), and that specific relationships of this type should be evaluated rather than correlating the average ambient conditions with over-all disease prevalence. A more mechanistic epizootiological approach has been suggested for assessing the influence of abiotic conditions on the development and spread of *E. muscae* under field conditions (32, 119). This approach is based on the synthesis of specific experimental results through the use of simulation modeling. In essence, this approach allows the development of a quantitative hypothesis that is testable by comparison to descriptive epizootiological data acquired from an independent source. This research approach will be discussed further in Section 4 with specific examples presented from the *D. antiqua/E. muscae* life system.

The effect of abiotic conditions on *E. muscae* sporulation, conidial germination, host infection, and incubation have been evaluated under laboratory conditions with a variety of different host species. Kramer (120) has shown that *E. muscae* primary conidia are produced at atmospheric moisture levels as low as 0% RH but that conidial production was significantly enhanced by RH levels greater than 55%. Similar results were found by Mullins and Rodriguez (121) for primary conidia also produced from *M. domestica* cadavers exposed to RH levels of 20, 50, and 80%, with sporulation being maximized at 50% RH. In the same study, conidia were found to be released over a 24 hour sporulation period. Under field conditions, *E. muscae* conidial production from *D. antiqua* adults was completed 12 hours post-mortem (32).

The overall number of conidia produced during the sporulation period is a function of host body size, moisture (121), and temperature (32). Conidial germination is also a function of temperature and moisture (121 – 123). Secondary conidia are produced over a wide range of conditions, while germ tubes, the invasive stage, are only produced under near-saturated conditions. Although germ tubes are only formed under near-saturated atmospheric conditions, Kramer (122) has shown infection to occur even at 0% RH. Kramer hypothesized that even under dry atmospheric conditions the microenvironment within the boundary layer surrounding the host insect was near saturation, thus allowing germination and infection.

Conidial germination and survival under field conditions have shown similar response patterns to those seen in the laboratory, that is, high moisture levels stimulate germination and prolong spore viability (123). Spore longevity and germination were further enhanced in dense crop canopies as foliage prolonged the duration of available moisture and protected the conidia from exposure to direct solar radiation.

The spatial configuration and seasonal dynamics of crop and border canopies also play an important role in the interactions between *D. antiqua* and *E. muscae* in the onion agroecosystem. Onion flies aggregate in areas of high canopy density such as field borders or in adjacent rows of other crops with dense canopies rather than being dispersed throughout the more open and exposed onion canopy. This spatial pattern results in high host densities in habitats that favor longevity and germination of *E. muscae* conidia and thus increases the observed rate of infection. Cultural manipulation and avoidance of pesticide applications in field border areas enhanced background disease levels allowing *E. muscae* to be used as one of several management strategies in onion fly control (32, 119).

Since *E. muscae* typically infects only adult flies, its primary role as a biological control agent is expressed as the reduction in fecundity of the host insect rather than as direct mortality. Thus the phenology of mycosis in respect to the ovipositional cycle of the host population is of key importance in determining population level impacts (32). Infections associated with the wheat bulb fly and the onion fly populations reduced oviposition substantially (32, 41) because the incubation period of the fungus is similar to or less than the preovipositional period required by the host fly (58). In hosts with shorter preovipositional periods, such as *M. domestica,* the impact on the population may be less, depending on other factors affecting adult longevity and fecundity. Laboratory experiments with *E. muscae* mycosis of *D. antiqua* (32) have also shown a reduction in the fecundity of gravid ovipositing flies beginning ca. 2 days (33 degreedays) postexposure with a total cessation of egg laying occurring ca. 4 days (66 degreedays) postexposure. *Delia antiqua* mortality due to *E. muscae* in the same experiments occurred ca. 7 days postexposure.

The importance of *E. muscae* as a naturally occurring biological control agent has been emphasized in several agriculturally oriented studies where moderate to large reductions have been noted in pest populations (41, 45, 114, 116). Although population level impacts clearly suggest that this pathogen has potential for incorporation into an overall pest management program, there has been only a limited amount of research of such possibilities. Compatibility with selected agrochemicals, including fungicides and insecticides, has been assessed under laboratory conditions (124, 125), but only observational data have been collected under actual field conditions (123). Methods of pathogen manipulation or augmentation for biological control purposes have been discussed for *D. antiqua* management in the onion agroecosystem (32, 119), but little actual implementation other than the reduction of unnecessary pesticide use (26) has been accomplished.

3.1.5. *NOMURAEA RILEYI* INFECTION OF LEPIDOPTERAN LARVAE.   *Nomuraea rileyi* is known primarily from its natural and man-induced infections of a variety of different lepidopteran pests (20). Although low level prevalence and epizootics have been recorded from several natural and agricultural situations (20), the majority of the epizootiological research associated with *N. rileyi* has centered around its development and spread in soybean populations in the central and southern United States (20, 33, 42, 67). Natural epizootics are consistently noted decimating susceptible host populations but usually too late in the growing season to prevent economic crop losses (33, 42, 67, 127, 128). Numerous factors, including the timing of the initial infection of the host population, the developmental lag associated with the disease incubation period, and the rate of spread of the pathogen are known to delay naturally occurring *N. rileyi* epizootics (20). The duration of the disease incubation period is positively correlated with the host instar, and any delay in initial host infection further retards the production of secondary inoculum and thus the development of disease epizootics (67, 128).

Initial infections of the velvet bean caterpillar *Anticarsia gemmatalis* in Florida were believed to be caused by inoculum carried into soybean fields by ovipositing moths rather than by local, overwintering populations of *N. rileyi* (129). Ignoffo et al. (130), in studying natural and induced epizootics of soybean caterpillars, outlined the natural disease cycle and environmental situations that led to epizootics in these hosts. They proposed that *N. rileyi* overwintered locally in soybean fields and that the level of overwintering inoculum was one of the key factors in the development of epizootics.

Secondary spread of *N. rileyi* in the field depends on both the movement of infected host larvae in the soybean canopy and the airborne dispersal of infective conidia (67). Fuxa (33) has shown that infected insects initially occur in limited loci and the disease can spread throughout a field very rapidly even before moderate levels of disease prevalence occur. Early in the season infected hosts are characterized by a more highly aggregated spatial pattern than surrounding uninfected larvae. This aggregation pattern suggests that higher inoculum levels in the vicinity of initial loci contribute to the expansion of these loci while also providing an inoculum source for hosts in other sections of the field.

Dry windy conditions enhance conidial disperal in the field but also limit conidial survival and germination and infection (54, 67, 131, 132). High moisture levels, either elevated RH or free water in the form of dew, are required for infections (67). Heavy rainfall or extended light rainfall are detrimental to the development of *N. rileyi* epizootics because infective conidia are washed from the foliage lowering the probability that they will come in contact with potential hosts (67). Therefore, alternation of wet and dry environmental conditions are believed to provide the best situation for the spread of this disease and the development of subsequent epizootics (20, 67, 130).

To aid in the understanding and manipulation of *N. rileyi* under field conditions, Kish and Allen (67) developed one of the first quantitative models in insect pathology/epizootiology. Their model was based on a series of stimulus response experiments conducted specifically for model construction and on data that they collected from the literature. They used a regression approach to estimate the number of *A. gemmatalis* larvae killed by *N. rileyi*, given cadaver size and density from the field, ambient environmental conditions, and the dose-response relationship that produced different levels of infection. Their model incorporated conidial dispersal, an adjustment for deposition on non-host sites, mortality of conidia caused by direct ultraviolet radiation and heavy rainfall, and the effect of environmental moisture conditions on subsequent infection. Model predictions were compared to independently collected field data and found to accurately predict field mortality in approximately 50% of the cases examined.

Several methods for the use of *N. rileyi* to control insect pests have been tested in a variety of different cropping systems (20). Most tests involved the application of *N. rileyi* conidia as a spray or dust on lepidopteran pests (20, 42, 54, 133, 134). Although moderate to high levels of disease prevalence typically followed these applictions, the lag period associated with later instar host mortality prevented them from substantially reducing damage levels in the test crops. An alternate approach evaluated by Ignoffo et al. (42) used early season prophylactic applications of *N. rileyi* to control young instars of soybean defoliators and to induce epizootics prior to the time that economic pest levels are reached. This strategy proved successful if adequate environmental conditions for the development of an epizootic were to follow. If adequate conditions for disease development did not occur, rescue treatments of *Bacillus thuringiensis* were suggested to circumvent damage (54).

Other investigators have tried to induce epizootics with a variety of methods. These include increasing early season inoculum levels by distributing pieces of cadavers containing *N. rileyi* conidia (135), irrigating the crop during dry periods (67), applying the pathogen through irrigation water (55), or manipulating various cultural practices (47). Of these methods, the manipulation of cultural practices seemed to be the most attractive approach for enhancing natural control by *N. rileyi.* Although no cause and effect relationships were established, Sprenkel et al. (47) showed that early plantings of soybeans in North Carolina exhibited higher natural levels of parasites and pathogens, particularly *N. rileyi,* and that yields increased, probably due to this effect.

3.2. Forest Ecosystem

3.2.1. FOREST ECOSYSTEM AS AN INFECTION COURT.

The forest has different ecological conditions for disease development than those in the agroecosystem. Even though some forests might be considered fairly homogeneous (e.g., spruce-balsam fir forests of Canada and northern Maine), they are more diverse in both species composition and in habitat than agricultural production systems. The hardwood forests of the northeastern United States and Europe exhibit even more diversity in habitat and species mix than is found in the spruce-fir forest. Although such diversity is thought to stabilize forest communities, forest pests still reach high densities and cause severe damage. Epizootics of insect pathogenic fungi have been noted in several forest defoliator species, particularly during times of high population density, and, in some cases, fungi have been cited as the controlling biological agents.

Environmental conditions in the forest canopy are more heterogeneous than in the agricultural situation since both terrain and canopy structure are highly diverse. Microclimatic differences may allow disease development in certain habitats while not in others, even though they may be in proximity. In addition, a much more complicated physical structure (soil, litter, understory growth, tree bole, canopy, etc.) makes the spatial interactions in the forest environment highly complex. This is not to say that intimate host pathogen relationships do not exist in forest situations, but that their population responses are expected to be dampened when compared to the agricultural situation.

Since forests are natural ecosystems that are less frequently affected by catastrophic disturbances compared with agricultural systems, and since host and pathogen have evolved together in these more stable heterogeneous environments, more complex associations between these organisms are expected. Understanding and manipulating these systems for pest management purposes provide an almost unlimited challenge.

3.2.2. MYCOSIS OF THE SPRUCE BUDWORM.

The spruce budworm *Choristoneura fumiferana* is a major forest insect pest causing defoliation and bud damage to balsam fir and spruce trees throughout northeastern North America (136). A wide range of insect pathogens attack *C. fumiferana,* but the pathogenic fungi *Erynia radicans* and *Entomophaga aulicae* produce the highest levels of mortality (43, 137).

Descriptive studies (43) suggest that fungus prevalence in *C. fumiferana* populations is somewhat dependent on host density, although significant statistical correlations were not obtained. A definite age-specific dependence was noted, since the first through fourth instars were never found infected, whereas fifth instars through pupae commonly were. This difference was thought to be primarily dependent on the cryptic bud feeding habits of the young instars and the wandering nature of the older instars. Differences in disease prevalence were noted also in different canopy strata, with larvae collected from upper branches showing lower levels of mycosis than those collected farther down the tree. Vandenberg and Soper (43) hypothesized that these increased prevalences were due to a greater degree of host exposure, more favorable physical conditions for fungal spores, and the potential for spore accumulation in the lower canopy areas.

The effects of abiotic conditions on fungus disease prevalence are not clear although Vandenberg and Soper (43) showed that cool, wet conditions typically preceded increases in mycoses. In conjunction with an evaluation of *E. radicans* as a mycoinsecticide, Soper (87) found that low temperature might be the limiting factor in disease development in the field since cool conditions significantly retarded sporulation and germination of *E. radicans.* Perry et al. (138) and Van Roemund et al. (139) experimented with the mode of both resting spore and conidial germination of *E. radicans* under laboratory conditions. Using these data and simulation techniques, Perry and Whitfield (137) examined the phenology of *C. fumiferana* linked with *E. radicans* mycosis with temperature as the single environmental stimulus variable. Their model provided a reasonable pattern of fungal occurrence and is a natural beginning point for further system analysis. They suggested incorporation of additional physical factors (e.g., leaf wetness, RH, solar radiation) to aid in the prediction of specific processes associated with fungal development.

3.2.3. *ERYNIA CRUSTOSA* MYCOSIS OF THE FOREST TENT CATERPIL-LAR.   The forest tent caterpillar *Malacosoma disstria* is a lepidopteran defoliator that is frequently found throughout eastern North American forests. Outbreaks of this pest occur in cyclic patterns with an approximate 10 year periodicity (140). Several natural biological control agents are found in association with *M. disstria* populations including the fungal pathogen *Entomophthora crustosa (= E. megasperma)* (141–143). It has been suggested that these biological control agents, including *E. crustosa,* are involved in the observed population oscillations (140–142). Studies documenting these cycles and the role of *E. crustosa* in the regulation of this insect have been conducted in western Ontario, Canada (Soper, unpublished data). After several years of low level populations, *M. disstria* populations rapidly expanded and began causing heavy defoliation. During the first year of an outbreak, only low levels of both parasites and pathogens were detected. In the second outbreak season, mycosis usually reached moderate levels late in the life cycle of the forest tent caterpillar. The third outbreak season typically ended with a collapse of the *M. disstria* population, caused by either *E. crustosa* or a nucleopolyhedrosis virus.

Of particular interest in this host/pathogen life system is the within-season phenology associated with the interactions of these two populations. Although all immature life stages of this insect are susceptible to *E. crustosa* in the laboratory, only fourth and fifth instars are infected under natural field conditions. Infection seems to correlate highly with changes in the behavior of *M. disstria* caterpillars as they reach later developmental stages. The first three instars are found only in the upper portions of trees, either on boles or on foliage. As the caterpillars mature to the fourth and fifth stages they become more active and disperse from their initial feeding sites. This dispersal results in substantial movement between trees over the forest floor, and it is presumed that initial infections of *M. disstria* occur in the litter zone during this period of movement. Secondary infection therefore can take place only late in the life cycle of the host. This limitation on secondary inoculum production in turn limits the rate at which an epizootic can develop. Instead of disease prevalence reaching high levels within a given season, the density-dependent response of *E. crustosa* is delayed.

Although our evaluation of the population level response of this system is somewhat speculative, the behavioral habits and spatial patterns associated with the interaction of this host/pathogen life system provide an example in which the complexity of the habitat structure and behavioral responses of the host are involved in the regulation of disease development.

3.3. Rangeland Ecosystem

3.3.1. RANGELAND AS AN INFECTION COURT. Rangelands fall into an ecological category somewhere between the agroecosystem on one side and forests on the other. Rangelands are composed of a wide variety of vegetation types and in terms of species numbers may actually be more diverse than either of the previously mentioned ecosystems. Although many species of plants are found in rangelands, grasses dominate because they are capable of thriving in areas where moisture is below levels capable of sustaining forest vegetation. Rangelands are also found intermixed with forested areas where edaphic factors limit the development of woody plants (144).

These grass-dominated ecosystems are more uniform in physical makeup and structure than forest ecosystems and thus tend toward the homogeneity seen in the agricultural situation. Canopy structure is more uniform and less complex than that found in the forest. This, in turn, yields higher host population densities and more uniform environmental conditions, which are believed to stimulate the development of disease epizootics (see Section 3.1.1). On the other hand, rangelands are more stable than agricultural systems. Catastrophic alterations of these ecosystems are rare thus giving them some commonality with the forest situation. As host and pathogen populations have been free to coevolve in this environment relatively undisturbed, intricate patterns of association are expected. While host/pathogen interactions in rangelands may be more complex and stable than those found in agroecosystems, they are expected to induce higher levels of disease than might be expected in the forest situation because of the more homogeneous nature of this habitat.

3.3.2. ENTOMOPHTHORALES MYCOSIS OF GRASSHOPPERS.   Grasshoppers are no doubt the dominant group of herbivorous insects associated with many rangelands around the world (145, 146). Their populations are known to have plagued man from the earliest days of recorded history, and they continue to do so today. *Entomophaga grylli* (= *Entomophthora)* is a fungal pathogen of many different grasshopper species and is known worldwide (147–150). Epizootics of this pathogen have been closely linked to major population reductions of several grasshopper species (149, 151). In western North America, *E. grylli* has caused high levels of mortality in two major grasshopper pests, *Camnula pellucida* and *Melanoplus bivittatus* (152). Epizootics are particularly spectacular in populations of *C. pellucida* (Fig. 12.1) and have been noted as the key factor in checking population outbreaks of this insect in Canada (152, 153). Such epizootics have been grossly correlated with warm, moist environmental conditions (152), although no empirical or statistical analyses were performed.

Field studies of this disease have suggested that two different strains of *E. grylli* are active in western North America, because contiguous populations of *C. pelucida* and *M. bivittatus* responded differently and cannot be cross-infected under controlled conditions (152). Using isoenzyme analysis, Soper et al. (154) have been able to differentiate two major pathotypes (I and II) of *E. grylli* from seven different grasshopper species. Pathotype I infects grasshoppers of several different genera while Pathotype II is only known from *Melanoplus* spp. These pathotypes differ not only in their host range but also in their basic patterns of life history (Fig. 12.4). Further evaluation has shown that these two pathotypes may be two separate species (155).

Grasshopper populations are highly mobile in open rangeland making it difficult to assess host and pathogen densities and disease prevalence in the same population between years. To avoid these problems, epizootiological investigation of *C. pellucida* and pathotype I of *E. grylli* were initiated in isolated mountain valleys of Arizona during the summer of 1981 (Soper, unpublished data). Several years of observational data from these locations suggested that a combination of host and pathogen density-dependent relationships, spatial dynamics, and varying environmental conditions regulated disease prevalence. Population cycles took from three to five years to change from low host and pathogen densities to outbreak levels, through a disease epizootic, and back to low population densities (Fig. 12.5). Although it was felt that weather conditions play an important role in the initiation and development of this disease, it was interesting to note that populations in similar and adjacent mountain valleys (with similar weather patterns) followed this basic cycle but were out of synchrony with each other by one or more seasons. Observational and experimental studies are continuing on this host/pathogen system.

3.4. Soil Stratum

3.4.1. SOIL AND LITTER AS A HABITAT.   Soil is a complex environment which is not easily studied or understood. It maintains large populations of

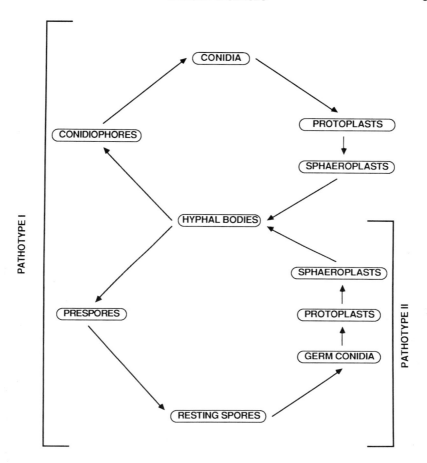

PATHOTYPE I

PATHOTYPE II

Figure 12.4. Life cycle variations for the two North American pathotypes of *Entomophaga grylli*. Pathotype I is found in areas where moisture is available on a regular basis whereas pathotype II is found in drier environments.

biotic fauna and flora that interact in intricate patterns of interdependence. These populations have complex relations with the physical environment, which acts as a dominant force in the population dynamics of many soil-dwelling organisms. Several insect pathogenic fungi, including, *Beauveria* spp. and *Metarhizium anisopliae,* infect numerous soil-dwelling insects, particularly immature Coleoptera. The capacity of these fungi to move through the soil, thus transmitting disease between subterranean larvae, and their ability to induce epizootics demonstrates their fitness for this environment (156). Several other fungi, not specifically adapted to infect soil-dwelling hosts, have also evolved at least a temporary association existing in the soil as resting spores.

Although the soil is often thought of as very stable, many environmental factors fluctuate to a great degree, even beyond normally acceptable ranges for

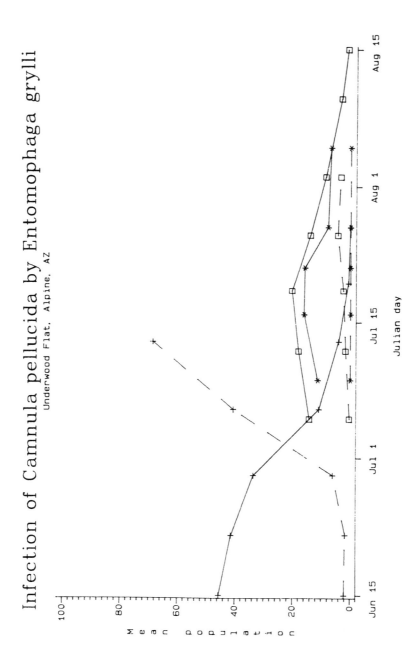

Figure 12.5. Three consecutive years of population data for the *Camnula pellucida* (hoppers per m² — solid lines) / *Entomophaga grylli* (percent infection — dashed lines) life system. Grasshopper populations in this area of Arizona cycle from low (1982) to moderate (1983) to high (beginning — 1984) and finally back to low (end — 1984) densities. This system oscillates in an approximate 3 to 5 year pattern and seems to be regulated by *E. grylli*.

many organisms. Temperature and moisture levels provide two important examples. Temperature in the top few centimeters of soil may range over 40°C between daylight and night hours. In many soils, the upper few centimeters of soil reach temperatures well above 50° C, a temperature lethal to the vegetative stages of most insect pathogenic fungi. In addition to their ability to withstand high and variable temperatures, soil-dwelling life stages of these fungi have developed adaptations to withstand both high moisture and drought stress (79).

3.4.2. *METARHIZIUM* INFECTION OF THE COCONUT PALM RHINOCEROS BEETLE.

The coconut palm rhinoceros beetle *Oryctes rhinoceros* is a major pest of Asian- and Pacific-grown coconut and oil palms. Although the damaging adult stage feeds on palm fronds, boring into the axils and destroying developing plant tissues, the immatures are found in soil, litter and rubbish piles (157, 158). Larvae of this pest are naturally infected with the fungus *Metarhizium anisopliae* and both larvae and adults are susceptible also to a baculovirus *Rhabdionvirus oryctes.* These pathogens have been cited as significant mortality factors reducing populations of *O. rhinoceros* (157, 159, 160). Simultaneous introductions and epizootiological evaluations of *B. oryctes* and *M. anisopliae* were made on the island of Tongatapu, Tonga, where the virus was found to induce disease epizootics while the fungus remained at low levels of prevalence (161, 162). Comparative epizootiological studies following these introductions revealed that both pathogens were active in the soil-litter release sites for several months but that the virus spread more rapidly and caused mortality in natural breeding sites 2.5 months after the original release. *Metarhizium anisopliae* was not found in natural breeding sites for 15 months and then only at extremely low levels (162). Two years postrelease, *M. anisopliae* prevalence was barely detectable in natural breeding sites.

Laboratory bioassays of *M. anisopliae* against *O. rhinoceros* have shown that two pathotypes of this fungus, a long- and a short-spored form, affect this host differentially. The long-spored form of *M. anisopliae* was found to be more pathogenic to the beetle causing ca. 100% mortality. The fungus infected all life stages in the laboratory, and the incubation period was increased with the stage of the host (163). These results suggest that the pathogen may than be limited in the field by environmental conditions and/or spatial and temporal interactions with its host rather than due to the lack of pathogenicity.

Young (162) suggests that both low moisture conditions and limited inoculum potential kept *M. anisopliae* at low levels in Tonga. Moisture has been cited as an abiotic factor limiting *M. anisopliae* mycosis in several studies (164–166). However, in laboratory studies, Latch (163) has found that larval mortality is not significantly reduced even under moisture conditions as low as 30% and he has concluded that moisture levels in soil and litter breeding sites on the Pacific Islands are normally sufficient to allow infection. Limited dispersal and longevity of inoculum within the environment provide the best explanation of the low prevalence levels of this pathogen following introduction.

Field studies on the use of *M. anisopliae* as a microbial insecticide for the control of *O. rhinoceros* have shown high levels of host mortality when the inoculum is distributed on an oat grain substrate. When formulated and distributed using these techniques, *M. anisopliae* has persisted in original introduction sites for up to 2 years (167). Again, minimal spread to adjacent natural breeding areas has been detected suggesting that this pathogen is limited in prevalence by its dispersal capabilities.

3.4.3. SURVIVAL OF SPORES IN SOIL.   The resistant stages of many insect pathogenic fungi end up in the soil even when the susceptible stages of the host occur aerially or on the surface of plants. The persistence of these stages in the soil has long-term significance for epizootics. It is likely that resistant spores can remain viable in the soil for very long periods. The cicada pathogen *Massospora cicadina* belongs to the Entomophthorales and, as is typical of that group, produces resting spores. This pathogen is restricted to the periodical cicadas *Magicicada* spp. Since adults of *M. cicadina* are present only once in 13 or 17 years in any large geographical area, and since this pathogen does not infect the subterranean nymphal stages, the resting spores must remain dormant at least for this same period of time (168). A similar longevity was shown for *Massospora levispora* in which resting spores survived at least 9 years in the soil (30). The inoculum potential of resting spores, as measured by the reduction of infection during 2 subsequent years with equal host population densities, showed a decrease of only 0.4 infected/100 adults/year. This limited reduction in primary infection, between years, suggested extremely stable resting spore levels in this research area. Although circumstantial, this is the best evidence for extended longevity of entomopathogenic fungal spores under natural field conditions.

Krejzova (169) has shown that resting spores of Entomophthorales can survive 6 1/2 years when stored in the laboratory at 4°C, and *Conidiobolus thromboides* resting spores have been shown to maintain high viability (90%) after laboratory storage for 12 years (Soper, unpublished). Resting spores of *Erynia radicans* in soil samples have caused infection in spotted alfalfa aphids several years after introduction into Moonbi, Australia (R. A. Milner, unpublished) even though the aphids had not been present in the field for at least 1 year. This is again circumstantial evidence, but it indicates that not all resting spores germinate the year after production.

Survival of spores is affected by different soil types. Oliveria et al. (170) compared survival of *Metarhizium anisopliae* conidia in various soils. They concluded that the presence of fertilizers and organic matter improved spore survival. This agrees with earlier work (171) which indicated that high organic matter favored survival of *Torrubiella* sp., *B. bassiana*, and *M. anisopliae*. Mikuni et al. (172) investigated the occurrence of *M. anisopliae* conidia in soil of a mulberry plantation for a 25 month period in connection with a study of the green muscardine disease of the silkworm *B. mori*. The fungus could be recovered throughout this period with little seasonal variability. They also showed that the fungus could multiply in the soil.

It is well known that pH has a significant effect on fungal development and it is not surprising that soil pH likewise influences fungal survival. In fact, certain pathogens are more likely to survive low pH whereas others do better at high pH. Beet weevil grubs *Cleonus punctiventris* are commonly attacked by *M. anisopliae* in acid soils and by *Sorosporella uvella* in alkaline soils (173, 174). When attempting to understand the epizootiology of an entomopathogenic fungus, the possibility of resistant stages in the soil and their survival must be considered. This is especially true if the objective is to obtain data on the persistence of the pathogen for evaluation as a long-term biological control agent.

3.5. Aquatic Ecosystems

3.5.1. EFFECTS OF AQUATIC ECOSYSTEMS ON EPIZOOTICS. Aquatic ecosystems are equally diverse in structure and ecological response as the terrestrial ecosystems. This diversity, whether associated with the physical movement of water (i.e., lentic or lotic habitats) or with its chemical quality, highly affects host and pathogen populations and their interactions (175, 176). This variability makes it difficult to generalize about the effects that the aquatic ecosystems will have on the epizootiology of fungal diseases but some aspects are noteworthy.

Aquatic pathogens have evolved a variety of morphological adaptions and/ or biological associations that allow effective interactions with their hosts. Some aquatic fungi have uniquely shaped conidia (177) or motile zoospores (178) that aid them in contacting hosts. For example, flagellated zoospores of the mosquito pathogen *Lagenidium giganteum* are thought to actively locate host larvae by chemotaxis (179), a characteristic unknown in terrestrial fungal pathogens. Other pathogens have specialized by infecting only the aerial adult stages of aquatic insects, thus avoiding many of the unique problems associated with aquatic habitats. Infection of aerial adults also provides an upstream dispersal mechanism allowing pathogens to compensate for aquatic drift. On the other hand, fungi from the genus *Coelomomyces,* have an intimate association with the aquatic environment and actually require a copepod intermediate host in the disease cycle with mosquito larvae (180).

Water as a surrounding medium presents different environmental hazards and benefits when compared to terrestrial habitats. Tanada (175) suggests that the effects of temperature and sunlight, two extremely important abiotic factors in terrestrial environments, are minimized in aquatic situations while factors such as pH, salinity, dissolved inorganic and organics solutes, current, and the ephemeral nature of many aquatic habitats highly affect pathogen persistence. Again, a variety of mechanisms have evolved to overcome these hazards. For example, *Coelomomyces* sp. are known to withstand moderate levels of salinity by increasing the thickness of sporangial walls (181) and are commonly found in association with salt marsh mosquitoes (182). Actually, fungal pathogens of many invertebrates are known from both marine and brackish waters (175, 183), thus the distribution of fungal pathogens of insects seems to be limited more by the lack of hosts in these habitats than by salinity. Less information is available on the effects of organic pollution, oxygen concentration, inorganic solutes, and other chemical factors.

3.5.2. *LAGENIDIUM GIGANTEUM* INFECTION OF MOSQUITO LARVAE. *Lagenidium giganteum* infects a wide variety of mosquito species in many parts of the world (178, 184, 185). It is fully dependent on the aquatic environment for interaction with its hosts although it may lay dormant for extended periods in the dry soil of ephemeral aquatic habitats in the form of thick-walled oospores (178, 186–189). Even when adequate moisture is available, oospore germination occurs over an extended period of time presumably to ensure some degree of synchrony with host larvae in these variable habitats (188, 189). This persistance has been particularly helpful in the use of *L. giganteum* as a microbial control agent for multivoltine mosquito pests (189).

Germtubes originating from oospores each form 15 to 20 biflagellated motile zoospores (188) that infect mosquitoes by direct penetration of the larval cuticle (190). Host death occurs approximately 2 days following infection (191) but both pathogenicity and the duration of the incubation period is dependent on the age and species of the host (192). *Lagenidium giganteum* may then cycle asexually through the development of zoospore forming sporangia or sexually by the development of oospores. Domnas et al. (193) determined that zoospore formation was dependent on exogenous sterols in the aquatic environment surrounding the host. This discovery subsequently led to the recognition that *L. giganteum* could be manipulated in the laboratory and the natural environment allowing increased potential in its use as a microbial control agent of mosquitoes (185, 194).

Field studies on the epizootiology of *L. giganteum* and mosquito larvae have been conducted almost exclusively in microbial control situations (189, 190, 194). These studies have produced varying results depending on the mosquito host and the specific habitats involved. Applications of laboratory-produced zoospores to California rice fields were found to cause between 70 and 100% infection in sentinal *Culex tarsalis* larvae placed at the time of application. Sentinal larvae placed in the same rice plots 2 weeks posttreatment still became infected at levels up to 24% (189). Persistence and recycling of *L. giganteum* following introductions was common (187, 189, 194).

As summarized by Lacey and Undeen (185), a variety of environmental conditions limits the persistence of *L. giganteum* under natural field conditions. Temperature extremes, organic pollution, salinity, and anaerobic conditions are the most frequently cited factors limiting zoosporogenesis and thus infectivity. In spite of these limiting factors, *L. giganteum* is an excellent candidate for microbial control due to its virulence and persistence in the aquatic environment, the development of *in vitro* production techniques, (188, 194) and its compatibility with a number of pesticides (195).

3.6. Summary of Epizootics by Disease Life System

The examples of epizootiological studies presented in this section were selected because of our familiarity with them and because they provided a reasonable cross-section of both the techniques and the current level of understanding of host and pathogen dynamics. These host/pathogen life systems have a substantial amount of commonality in the way fungal diseases develop in respect to their hosts and the environment. Host and pathogen population levels are certainly key factors in fungal dynamics, as are specific environmental conditions such as temperature, moisture, and solar radiation. Although this commonality exists generally, the actual mechanisms of interaction among hosts, pathogens, and their environment are very complex and system specific. For example, moisture seems to be necessary in almost every situation for fungal infections, but the mechanisms through which this requirement is met are extremely diverse from one life system to the next. The same could be said for virtually any of the factors involved in the regulation of dynamics of fungal diseases.

As an area of research, we have just begun to explore the universe of insect/fungal disease epizootiology. Periodically, we obtain brief glimpses of how these life systems truly function in nature, but to date, it is clear that we do not have a thorough understanding and/or a predictive ability for any single epizootiological system. If we are to gain such an understanding for basic or applied purposes, a systematic and cooperative approach will have to be developed among all those involved in this area of research.

4. A GENERALIZED CONCEPTUAL MODEL FOR RESEARCH AND EVALUATION OF INSECT/FUNGAL DISEASE DYNAMICS

The development of insect pathogens as biological control agents for IPM will necessitate a comprehensive understanding of host/pathogen population dynamics and the environmental factors affecting these dynamics. Gaining such an understanding will necessitate detailed experimentation and evaluation of natural and man-manipulated epizootiological systems. It will require a mixture of directed theoretical considerations, descriptive characterizations, experimental analysis, and system synthesis.

Such complexity dictates a systematic methodology to help guide, analyze, and evaluate detailed and diverse research of this type. The problem-solving process commonly referred to as the "systems approach" (196–198) offers a structured format within which complex problems of this type can be managed.

In this section, we show how the systems approach may be applied to fungal epizootiology.

4.1. The Systems Approach in Ecology

Ecological systems belong to a special set of real world problems where system structure (i.e., the biological components and their interrelationships) is usually preexisting but unknown. Scientific experimentation is the primary methodology used to identify the underlying structure and the functional processes linking that structure. Mathematical modeling, a component of the systems approach, has been successfully used to guide and link such experimentation, thus helping to develop a more comprehensive understanding of ecosystems and their operation (59, 61, 119, 197) (see Ch. 3).

The systems approach is more than just applying mathematical modeling to complex problems. It is a problem-solving methodology that begins with a set of objectives and, through a structured process, results in an operational system capable of satisfying original or slightly modified needs (198). The process is composed of several interconnected decision-making phases which aid the researcher in formulating, evaluating, and solving problems. Each phase of the systems approach is iterative in nature and composed of numerous steps (for an in-depth explanation see, 197, 198). We will be concentrating on system identification and conceptual modeling although an operational model will be presented as an example of fungal epizootiology.

4.2. A Generalized Host/Pathogen Model

To utilize the power of the systems approach we must first frame our problem from a biological perspective by identifing the populations of interest and our objectives in studying them (system identification). Once this has been accomplished, we may then move ahead toward meeting those objectives. Here, we hope to provide insight into how this approach can be used in a general way to aid in the understanding of specific host/pathogen interactions and how those interactions may lead to the regulation of host and pathogen populations.

In evaluating disease phenomena, both host and pathogen populations are of equal importance and must be considered not only independently but also as components which interact with one another and with their abiotic and biotic environments (68). Plant pathologists have long recognized these associations, and also that man occupies a special position in respect to these systems. We are highly affected both as members of the ecosystems in which these interactions occur and as potential managers of them. This four-way interaction has been depicted as the disease tetrahedron (27) and it is from this basic paradigm that we hope to gain further insight into the overall dynamics of disease life systems (Fig. 12.6).

Each of the vertices of the disease tetrahedron is extremely important in the initiation and development of disease under natural or man-manipulated con-

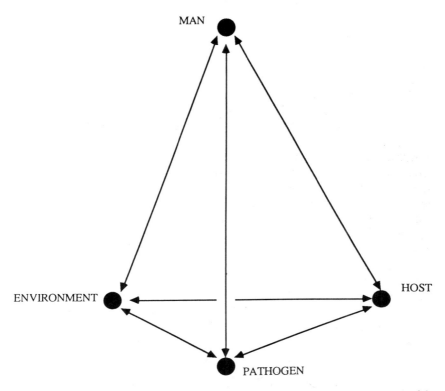

Figure 12.6. The disease tetrahedron representing interactions between the primary components of disease development including the effects induced by man.

ditions. Although interactions between each of these components typically occur, they do not always affect one another equally. Therefore, the links or interactions between these components themselves are dynamic and require special attention in the study of disease epizootiology. In Section 3 of this chapter, we discussed the interactions of several host and pathogen populations within the context of the ecosystem giving our impression of gross effects induced by major differences in the environment at the mesolevel. Now we attempt to be more specific in respect to the components of the disease tetrahedron and their interactions and to provide a perspective from which the dynamics of fungal diseases may be evaluated.

Both host and pathogen populations possess certain independent characteristics that allow them to function within their specific environments. We will emphasize only a few of these characteristics as entire chapters in this book have been devoted to the subject of the host (see Ch. 4), the pathogen (see Ch. 5), and the environment (see Ch. 7).

In developing a generalized conceptual model of insect host/fungal pathogen interactions, we feel it is necessary to expand upon the basic model developed by Anderson and May (37) (Fig. 12.2) to include additional details known to be important in the development of insect diseases caused by fungi. Although a generalized model cannot capture all of the essence of any specific epizootiological system, enough commonality exists between fungal disease systems to support this approach in developing a more comprehensive understanding of the processes and dynamics involved in regulating population numbers. For example, the inclusion of host age structure is of extreme importance in understanding disease dynamics as many of the biological processes that regulate disease development are differentially affected by the age or life stage of the host. Likewise, consideration of pathogen life stages (conidia, resting spores, etc.) and their interactions with the host population and the environment provide a more mechanistic understanding of disease epizootiology that is lacking in the analytical model of Section 2.1.1. In fact, the pathogen is not even considered as a separate component of most analytical models used in epidemiology with the entire infection process typically being condensed to a single static parameter. Although conceptually interesting, these models lack realism and quite possibly the essence of the biological systems they were constructed to represent. An alternative approach is to focus on specific biological mechanisms, their interactions, and then link them qualitatively and quantitatively to mimic real host/pathogen life systems. The systems approach and simulation modeling provide the tools through which this may be accomplished.

A conceptual model representing a singl-host single-pathogen life system is represented in Figure 12.7. This functional block diagram was structured to represent a holometabolous insect host, susceptible only thoughout the larval period, a fungal pathogen that produces both conidia and resting spores, and their interactions in an unspecified terrestrial environment. Although many other host/pathogen/environment combinations are possible, we will use this basic paradigm in developing the remainder of our discussion.

The life cycles of both the host and pathogen are represented separately where free-living or healthy portions of the host populations exist and in combination where the disease process is under development (infected hosts). Both the host and pathogen populations have been divided into subcomponents or life stages (titled rectangular boxes) that respond uniquely in the initiation and development of disease. These stages are connected in the model by arrowed lines which represent either the flow of energy or of individual members of a population from one stage to the next, through time. This movement is regulated by a variety of intrinsic and extrinsic factors that are represented as transfer functions and inputs to transfer functions (e.g., temperature and moisture) in the block diagram. These transfer functions thus represent specific linkages between the host, pathogen, and environment which follow the common form of stimulus-response reactions (27). That is, certain abiotic or biotic factors (stimuli) evoke specific and predictable reactions (responses) from the subcomponents on which they act. By isolating these stimulus-response reactions and quantifying them experimentally, a cause and effect understanding of their operation provides the first step in developing a predictive understanding of epizootiological dynamics.

4.3. Consideration of Specific Model Subcomponents

Several common biological processes as well as many specific to fungal pathogens are represented in the conceptual model presented in the previous section. These specific processes have been considered as discrete subcomponents (197) of many host/fungal pathogen life systems, the most important of which we will explore in further detail.

Physiological development is of one of the most commonly evaluated processes associated with population dynamics and epizootiology. In most poikilothermic animals, the rate of development is controlled not only by the genetic composition of the population but also by external environmental factors. Although several different variables may influence insect development, temperature is known to dominate the process, producing characteristic nonlinear stimulus-response functions (199, 200). Temperature-dependent development is also known to regulate the rate of fungal pathogen maturation during both free-living (49, 123) and parasitic (53, 201, 202) phases of the life cycle. Characterizing physiological development is important in understanding disease dynamics as individual host and pathogen phenologies and thus their combined synchrony are dependent on this process. Fungal development may also be affected by factors other than temperature. For example, the incubation period of *Beauveria bassiana* mycosis of the European corn borer is primarily regulated by temperature although both the density of pathogen inoculum and the stage of the host at the time of exposure can alter this relationship (Fig. 12.8). In addition to the effects of temperature, free-living stages of most fungal pathogens are dependent on moisture for development.

Sporulation and germination are two life processes of fungal pathogens that are commonly both temperature- and moisture-dependent (32, 50, 57, 123). In fact, many fungal pathogens, particularly the Entomophthorales, induce host mortality following specific diurnal patterns that synchronize sporulation and germination with high moisture periods in the environment (Figs. 12.7 and 12.9) (32, 203–205). Millstein et al. (57) have shown that the process of sporulation in *Erynia radicans* may be predicted based on accumulated relative humidity (humidity hours) independent of temperature, although in other host/pathogen systems temperature is positively correlated with spore production (Fig. 12.9) (32, 206–209). Environmental moisture on the other hand, may actually affect sporulation more as a switch, allowing the process to either halt or proceed rather than act as a continuous rate-controlling stimulus. Under fluctuating moisture conditions, many fungal pathogens begin sporulation given adequate moisture, dehydrate during dry periods, and then resporulate when moisture again reaches suitable levels (Fig. 12.7 sporulation delay) (50, 206, 210). This dehydration-rehydration phenomenon has allowed the recent

Single Host/ Single Pathogen Life System

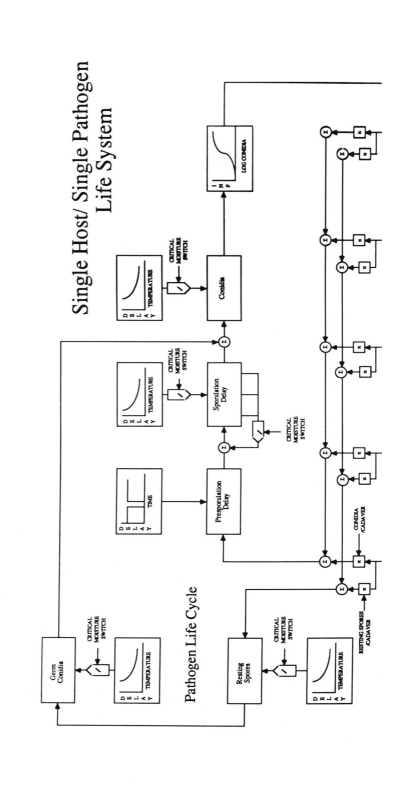

Pathogen Life Cycle

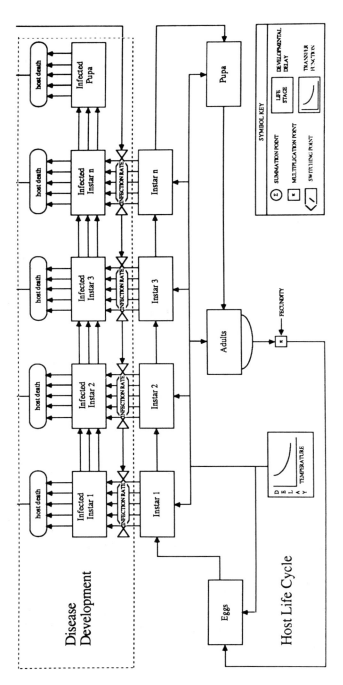

Figure 12.7. A conceptual host/fungal pathogen model outlining specific subcomponents and interactions from a mechanistic perspective.

391

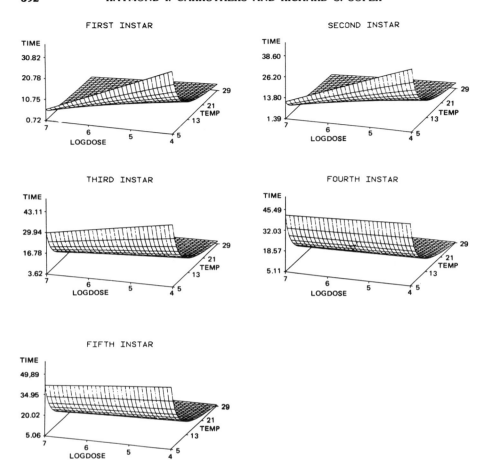

Figure 12.8. Age-specific response surfaces for the incubation period of *Beauveria bassiana* mycosis of the European corn borer *Ostrinia nubilalis*. Note the exponential decay pattern associated with the duration of the incubation period as temperature increases and the effect of the *B. bassiana* inoculation levels in the early instars.

development of marcesence technology for the production, formulation, and application of fungal pathogens (87).

Spore germination in many fungi is dependent on the availability of free water (27). Insect pathogenic fungi basically follow this pattern as high moisture levels are necessary for spore germination (123, 211, 212), although moisture must be assessed within the microhabitats where germination takes place rather than using gross ambient conditions. Again, as was seen in the sporulation process, moisture acts as an on or off switch, while temperature regulates the rate of germination (Fig. 12.10) (211).

Exposure and infection are fundamental components of host and pathogen interactions, yet little attention has been given to this area of study in natural or man-manipulated epizootiological systems (213). This is particularly true with respect to diseases caused by fungi. Even though a substantial amount of laboratory experimentation has been conducted on the dose-mortality response of fungal pathogens, these data provide little realism in the field situation. Pinnock and Brand (213) describe several time-dependent factors that affect these interactions and suggest that they are strongly influenced by environmental conditions. Current field studies being conducted on *Entomophaga grylli* infection of *Camnula pellucida* illustrate some of these effects. A constant range of cadaver densities produced widely varying numbers of conidia under different environmental conditions. After adjusting for differences in the rate of sporulation, a logistic infection model (214) provided reasonable fit to these data (Fig. 12.11). However, further analysis revealed significant variability in infection between experimental periods due to other differential responses of the host and pathogen to varying environmental conditions. To date, field experiments with *E.*

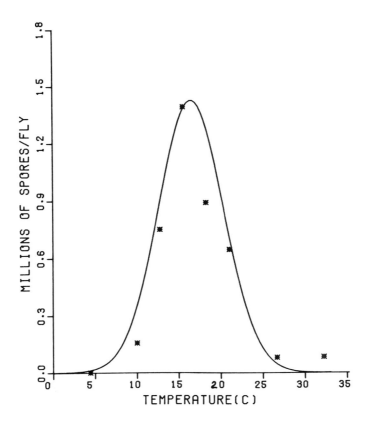

Figure 12.9. Temperature-dependent sporulation in the *Entomophthora muscae* mycosis of the onion fly *Delia antiqua.*

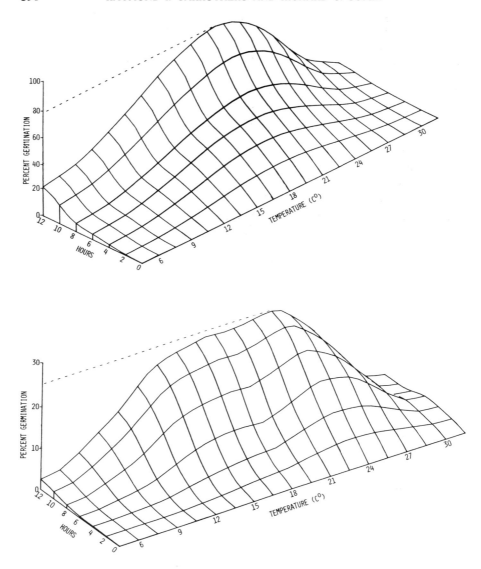

Figure 12.10. Response surfaces representing temperature, moisture, and time-dependent germination of *Entomophthora muscae* primary conidia. The upper graph is for germination in free water while the lower graph is for 100% R.H. No germination was indicated at 95% R.H. or lower in these studies.

grylli and *C. pellucida* have shown that habitat and microenvironmental conditions affect sporulation, spore germination, conidial survival, conidial dispersal, host/pathogen contact and thus ultimately the rate of infection.

Infection itself is composed of several subcomponents, each requiring some level of understanding before predictive capabilities can be developed. An area

FIELD DOSE—MORTALITY
LOGISTIC MODEL

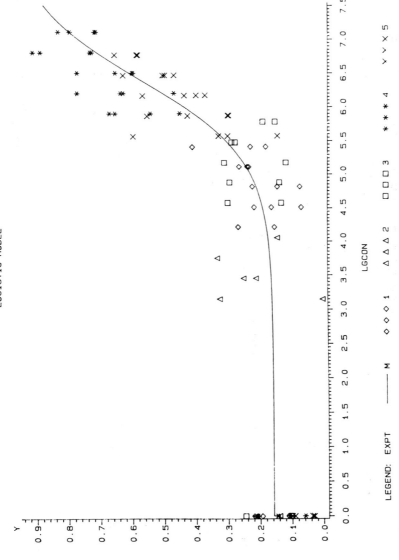

Figure 12.11. Field dose-response relationship for *Entomophaga grylli* infection of *Camnula pellucida* (y = proportion infection). The different symbols represent data from different experimental dates. Fixed numbers of cadavers (0, 1, 2, 4, 8, and 16/m²) produced different levels of conidia (lgcon = $\log_{10}$ conidia) depending on temperature and moisture conditions. Additional differences were also found (although not indicated in figure) in infection levels between dates while controlling for conidial dose.

particularly lacking in understanding is the spatial interaction of the host and pathogen populations at both the micro- and meso-levels of concern. Fungal pathogens add further complexity in this area as many species, particularly in the Entomophthorales, are active rather than passive dispersers. Several species of these fungi produce some behavioral control over their hosts at the time of death which positions the cadaver in locations beneficial for spore dispersal. Many of these pathogens also produce spores that are forcefully ejected into the environment, providing still further complexity in their spatial interactions with their hosts.

At the time of dispersal, fungal pathogens as thin-walled propagative spores (typically conidia), are particularly susceptible to desiccation, critical temperatures, and solar radiation (215–216). Environmentally induced mortality of fungal pathogens is particularly high during this process as spores are exposed to detrimental environmental conditions and are removed in large numbers from areas where viable hosts can be contacted (215). Age-specific survival of both the host and pathogen populations is extremely important in understanding the dynamics of disease systems. This is true not only for independent stages of the host and pathogen but also when differential mortality, other than that caused by the pathogen, occurs between healthy and infected hosts. This mortality not only affects the host population but also the pathogen. Likewise, mortality induced by mycosis may affect other mortality factors (i.e., parasites and predators) involved in regulating host populations.

Resting spore survival and germination under natural conditions are two areas still poorly understood in most insect pathogenic fungi. A variety of stimuli seem to be responsible for the germination of resting spores of different species, although moisture, temperature, and photoperiod most commonly produce effects in fungal pathogens (217–219). Most epizootiological studies to date, have evaluated resting spore germination either through assessing primary infection of newly emerging hosts (30, 32) or by the use of sentinel hosts in field bioassays. In either case, a more thorough understanding of this aspect of fungal life cycles is currently needed.

Although we have only lightly touched on a limited subset of the subcomponents and interactions of fungal pathogens, their insect host, and the environment, it should be clear that each of these processes is quiet complex and that an entire epizootiological system composed of these component pieces is even more complex. The response of a total host/pathogen life system to a change in a single environmental stimulus (e.g., temperature, moisture, etc.) or set of stimuli cannot intuitively be evaluated as each process may be affected differently either augmenting or hindering disease development. In addition, biological systems are not a simple sum of their parts, that is, synergism or interference between different subcomponents adds further complexity in the response of both the host and pathogen populations. The systems approach and simulation modeling provide a methodology within which complex biological systems of this type can be synthesized and evaluated as comprehensive units.

4.4. Synthesis and Application of Component Models

The biological subcomponents and interactions discussed in Section 4.3 are the primary elements and processes that are important in understanding fungal disease epizootiology. This is not to say that they are universally or exclusively important as each host/pathogen system is unique and warrants individual consideration, but these subcomponents are felt to represent the basic building blocks and linkages common to many host/fungal pathogen life systems. Research emphasizing identification and quantification of these subcomponents, their interactions with the environment and with other elements of disease life systems are essential for the development of a comprehensive understanding of fungal disease epizootiology. Synthesis of these elements into a more wholistic representation of a host/pathogen life system further aids in the understanding of disease dynamics by allowing quantitative hypothesis generation and testing.

Figure 12.7 represents a conceptual synthesis of several subcomponents important in the epizootiology of insect pathogenic fungi. Linkages between the host, pathogen, and the environment have been specified through interconnecting these subcomponents with arrowed lines representing the direction of influence in the system. This in itself is a gross simplification as the actual mechanisms of the interactions are more complex than represented in this functional block diagram. For actual model formulation, more detailed conceptual representations may be necessary to outline and meet specific research objectives but the process is essentially equivalent to the construction of this generalized model. Specific research models, however, may be developed in either greater detail to explore a subcomponent more precisely or in more generality to evaluate the response of an entire host/pathogen life system.

The synthesis process involves quantification of these biological components and linking them according to mechanistic interactions. Methods for quantifying these biological components and processes are beyond the scope of this chapter but a variety of mathematical techniques are available to simulate the stimulus-response patterns and interactions discussed in Section 4.3 (see Ch. 3) (39, 119, 198, 220).

Synthesis, however, is more than mathematically characterizing and linking these biological processes, it also involves analysis and comparison to descriptive data (model validation), hypothesis generation and testing, and reassessment of specific research priorities. Through this process, quantitative hypotheses representing our understanding of host/fungal pathogen dynamics can be generated and tested against independently collected data. Our ability or lack of ability to predict these data typically leads to further experimentation or analysis on specific system subcomponents or interactions. This evaluation-reevaluation process is one of the primary strengths of the systems approach and allows research to be directed iteratively toward meeting specific objectives.

A model of *Delia antiqua,* the onion fly, and its pathogen *Entomophthora muscae* in the onion agroecosystem provides an example for the application of the systems approach in evaluating epizootiological systems. This host/pathogen system will be discussed from its original conceptualization through early stages of evaluation. For more details on this analysis, see Haynes et al. (68), Carruthers (32), and Carruthers et al. (119).

A detailed conceptual representation of the Michigan onion agroecosystem, its biological components, and their interconnections with other system components, the environment, and man was given by Haynes et al. (68). This conceptualization was developed for the primary objectives of designing and managing a multi-pest agricultural production system (68). For the more restricted objectives of a biological control subproject (119), this overall conceptualization was reduced for more specific evaluation (Fig. 12.12). The conceptual model represents a specific expansion of the basic disease tetrahedron, identifying specific abiotic and biotic components that affect the initation and development of *E. muscae* mycosis of the onion fly.

Note that in the organization of the components of Figure 12.12, *E. muscae* is the reference point (the onion plant is used as the reference point in the overall onion agroecosystem project, Haynes et. al, (68)). In organizing the components in this way, the ecosystem may be viewed from the perspective of *E. muscae* rather than from that of any other organism. This perspective allows the ordering of the biological linkages between *E. muscae* and other components in the agroecosystem and aids in identifying key components and linkages important to the dynamics of this disease system.

Six biological components were found to be of major significance in regulating the levels of *E. muscae* mycosis of the onion fly: the primary host insect (*D. antiqua*), a secondary host fly (*D. platura*), the pathogen (*E. muscae*), the onion crop, adjacent crops, and border plants (see 32, 45, 119). A simulation model was developed and used over a 5 year period to guide and synthesize research efforts associated with insect pest dynamics in the onion agroecosystem. A functional block diagram representing the interactions of *E. muscae* with the onion fly (Fig. 12.13) provides an example of the relationships between two of the components from this model. The specific model subcomponents (see Fig. 12.13) are similar to those presented in Section 4.3 although in this host/pathogen system, infection only occurrs in the adult stage. Additional model components are necessary to describe the effect of *E. muscae* mycosis on the fecundity of the host (Fig. 12.14), the effects of spatial patterns on infection, and interactions with other host insects within the onion agroecosystem (see 32, 119).

Simulations were conducted to verify specific stimulus-response functions derived from both laboratory and field experiments and to identify areas where data were either lacking or incomplete for predicting descriptive data from the field. Lack of fit in the synthesis process typically led to reassessments. For example, early simulation results for the onion fly in the absence of *E. muscae* were found to differ substantially from observed field data. Estimates of adult emergence patterns predicted by the model accurately described the initiation of each generation, although as the season progressed, the results differed significantly from the observed data. Estimated adult density patterns also deviated

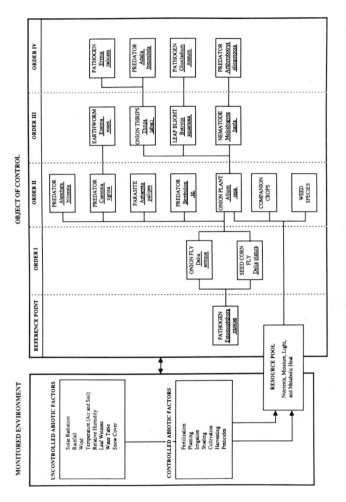

Figure 12.12. Conceptual representation of the onion agroecosystem showing the relationships between *Entomophthora muscae* and other abiotic and biotic factors affecting its development. Notice that each of the vertices of the disease tetrahedron (Fig. 12.6) are represented in this model.

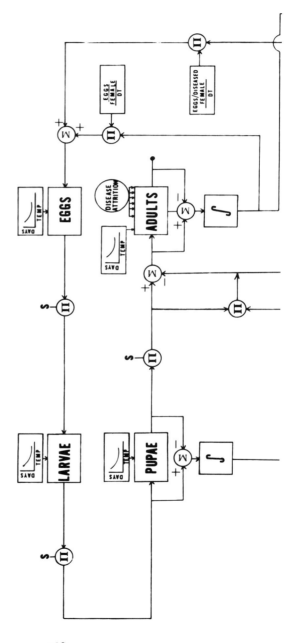

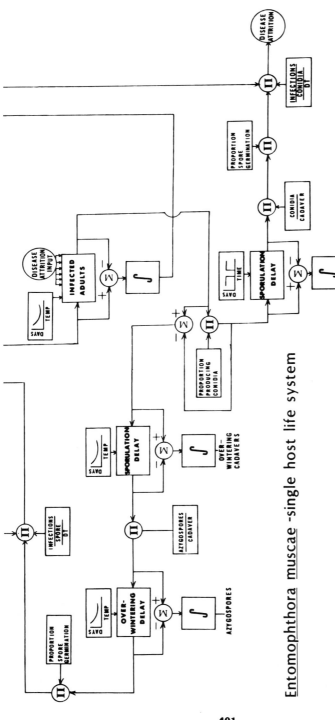

Figure 12.13. Functional block diagram representing a computer simulation model of the interactions between *Entomophthora muscae* and the onion fly *Delia antiqua*.

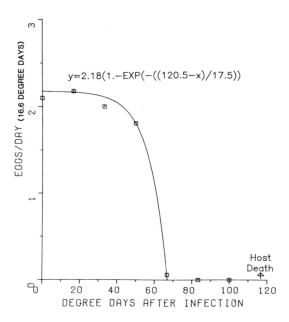

Figure 12.14. Premortatily reduction in fecundity of *Delia antiqua* females infected with *Entomophthora muscae*.

substantially from the observed field patterns. Incorporation of *E. muscae* and an alternate disease host into the simulation corrected the results and provided a much more realistic model response (see 119).

Final simulation results in combination with field and laboratory studies were used to alter the control recommendations for Michigan-grown onions. Indirect effects of both fungicides and insecticides interfered with the control of this pest species (124). This information led to a reduction of chemical use while maintaining or increasing the effectiveness of onion maggot control (32, 45, 68, 126). Examples of simulation output and a brief explanation of the results are given in Figures 12.15 to 12.18.

4.5. Alteration of the Basic Model

Organizing ideas and research results in the form of conceptual and/or mathematical models is useful when working toward very specific goals or towards an understanding of the dynamics of epizootiological systems in general. In either case, the specifics involved in using this process for other host/pathogen systems are certain to be different from those presented in Figure 12.7 and the previous sections, even though many of the underlying biological components may be similar. This model was presented as a paradigm for consideration of a method for approaching research and evaluation of insect/fungal disease life systems

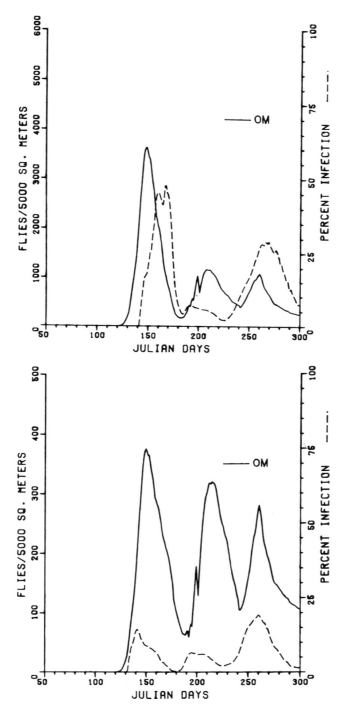

Figure 12.15–16. Basic model output for *Entomophthora muscae* infection of a single host fly (the onion maggot — OM). Note the density-dependent response of disease development that is indicated between these two model runs and the impact on the host population (the host density of Fig. 12.15 is 10 × that of Fig. 12.16; all other parameters were the same).

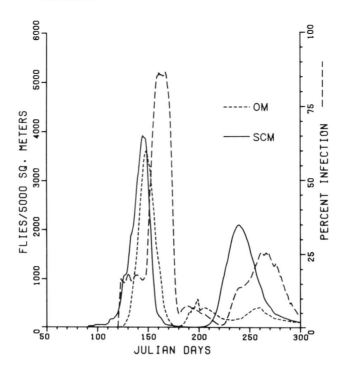

Figure 12.17. Model output showing the effects of adding an additional host (the seed corn maggot — SCM) in the simulation (all other paramenters set as in Fig. 12.15). Adding a second host significantly increased the prevalence of *Entomophthora muscae* and its impact on the primary pest (OM) population. Also note the earlier emergence time of the SCM allowing *E. muscae* to reach higher levels earlier in the season. These effects aid in the natural control of the onion maggot.

rather than outlining the model structure itself. To that end, alterations of this basic model may take numerous shapes and be used in a diversity of ways in exploring host/pathogen dynamics. Clearly, the biological dimensions of space and genetics were not explicitly taken into consideration in Sections 4.1 to 4.3 yet these factors may be considered implicitly in the components and transfer functions of any operational model.

5. SUMMARY

Research and evaluation of insect/fungal epizootiology is a relatively new field of study that is expected to lead toward a better understanding of disease dynamics and an ability to predict and manipulate these life systems for insect control purposes. Our current level of understanding falls short of these ideals although a diversity of research is providing a more concise image of fungal disease epizootiology and thus, the factors involved in both host and pathogen population regulation.

It has been stated that specific research on insect/fungal epizootiology has been hindered in the past by the lack of uniformity in research methods (21). Although we believe this to be true, we also feel that a more complete theoretical or conceptual base must be developed, discussed, challenged, and altered if our knowledge of these life systems is to increase substantially. To that end, we suggest that more emphasis be placed on the development and publication of conceptual models of insect/fungal disease life systems. Sharing of these ideas will not only further our theoretical understanding of these biological systems, but it will also provide a base from which more specific research can be conducted, evaluated, and synthesized, thus increasing our operational proficiency as well.

The development of a thorough understanding of disease epizootiology is a tremendous undertaking that will require extensive individual and cooperative research efforts covering many different aspects of host and pathogen dynamics. Evaluation of such complex and diverse life systems will require a variety of scientific expertise and ingenuity, but in addition, it will necessitate

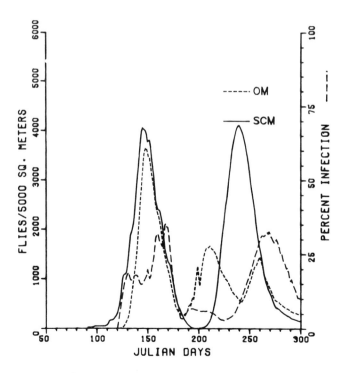

Figure 12.18. The effects of fungicide applications in reducing *Entomophthora muscae* prevalence and secondary effects on second and third generation insect pest levels were examined using computer simulation (Carruthers et al., 119). This output was generated from the same parameters as those used in Figure 12.17 with the addition of two fungicide applications (Maneb 80WP, applied on days 150 and 170 for downy mildew control). Note the sharp decline in *E. muscae* prevalence following day 150.

an integrating philosophy. We feel that the systems approach to research, analysis, and synthesis provides a unique format within which coordinated and cooperative research of this sort can be achieved. This chapter was designed to stimulate thought in this area and we hope that, if nothing more, it has accomplished that goal.

ACKNOWLEDGMENTS

We would like to thank A. E. Hajek and T. S. Larkin for critical review of early drafts of this manuscript, and R. A. Humber for his endless supply of biological details.

REFERENCES

1. A. Bassi, *Del mal del sengno, calcinaccio o moscardino, malattia che affligge i bachi de seta e sul modo di liberarne le bigattaie anche le piu infestate.* Part I, Teoria. Orcesi, Lodi, 1835.

2. D. W. Roberts and R. A. Humber, "Entomogenous Fungi," in G.T. Cole and B. Kendrick, Eds., *Biology of Conidial Fungi,* Vol. 2., Academic Press, New York, 1981, p. 201.

3. Anonymous, *Microbial Processes: Promising Technologies for Developing Countries,* Natl. Acad. Sci., Washington D. C., 1979.

4. G. C. Varley, G. R. Gradwell, and M. P. Hassell, *Insect Population Ecology: An Analytical Approach,* Univ. Calif. Press, Berkeley, 1973.

5. M. P. Hassell, *The Dynamics of Arthropod Predator-Prey Systems,* Princeton Univ. Press, Princeton, 1979.

6. P. W. Price, *Evolutionary Biology of Parasites,* Princeton Univ. Press, Princeton, 1980.

7. J. E. Vanderplank, *Plant Diseases: Epidemics and Control,* Academic Press, New York, 1963.

8. J. Kranz, "Epidemiology, Concepts and Scope," in S. P. Raychaudhuri and J. P. Verma, Eds., *Current Trends in Plant Pathology,* Lucknow Univ. Botany Dept., Lucknow, 1974, p. 26.

9. J. C. Zadoks and L. M. Koster, *A Historical Survey of Botanical Epidemiology,* Meded. Landbouwhogeschool, Wageningen, 1976.

10. G. Macdonald, *The Epidemiology and Control of Malaria,* Oxford Univ. Press, London, 1957.

11. G. Macdonald, *Dynamics of Tropical Disease,* Oxford Univ. Press, London, 1973.

12. K. Dietz, "The Incidence of Infectious Diseases Under the Influence of Seasonal Fluctuations," in J. Berger, Ed., *Mathematical Models in Medicine,* Springer-Verlag, Berlin, 1976, p. 1.

13. C. A. Mimms, *The Pathogenesis of Infectious Diseases,* Academic Press, London, 1977.

14. H. D. Burges, Ed., *Microbial Control of Pests and Plant Diseases 1970–80,* Academic Press, London, 1981.

15. A. W. Sweeney, Preliminary field tests of the fungus *Culicinomyces* against mosquito larvae in Australia, *Mosq. News,* **41,** 470 (1981).

16. W. A. Gardner and J. R. Fuxa, Pathogens for suppression of the fall armyworm, *Fla. Entomol.* **63,** 439 (1980).

17. P. Ferron, "Pest Control by the Fungi *Beauveria* and *Metarhizium*," in H. D. Burges, Ed., *Microbial Control of Pests and Plant Diseases 1970–80,* Academic Press, London, 1981, p. 465.

18. R. A. Hall, "The Fungus *Verticillium lecanii* as a Microbial Insecticide Against Aphids and Scales," in H. D. Burges, Ed., *Microbial Control of Pests and Plant Diseases 1970–80,* Academic Press, London, 1981, p. 483.

19. G. E. Allen, C. M. Ignoffo, and R. P. Jaques, Eds., *Microbial Control of Insect Pests: Future Strategies in Pest Management Systems,* Univ. Fla., Gainesville, 1978.

20. C. M. Ignoffo," The Fungus *Nomuraea rileyi* as a Microbial Insecticide," in H. D. Burges, Ed., *Microbial Control of Pests and Plant Diseases 1970–80,* Academic Press, London, 1981, p. 513.

21. N. Wilding, "Pest Control by Entomophthorales," in H. D. Burges, Ed., *Microbial Control of Pests and Plant Diseases 1970–80,* Academic Press, London, 1981, p. 539.

22. J. P. Latge, P. Silvie, B. Papierok, G. Remaudiere, C. A. Dedryver, and J. M. Rabasse. "Advantages and Disadvantages of *Conidiobolus obscurus* and *Erynia neoaphidis* in the Biological Control of Aphids," in R. Cavalloro, Ed., *Aphid Antagonists,* A. A. Balkema, Rotterdam, 1983, p. 20.

23. G. Latteur and J. Godefroid, "Trial of Field Treatments Against Cereal Aphids with Mycelium of *Erynia neoaphidis* Produced In Vitro," in R. Cavalloro, Ed., *Aphid antagonists,* A. A. Balkema, Rotterdam, 1983, p. 2.

24. A. J. Nicholson, The balance of animal populations, *J. Anim. Ecol.,* **2,** 132 (1933).

25. H. G. Andrewartha and L. C. Birch, *The Distribution and Abundance of Animals,* Univ. Chicago Press, Chicago, 1954.

26. L. R. Clark, P. W. Geier, R. D. Hughes, and R. F. Morris, *The Ecology of Insect Populations in Theory and Practice,* Methuen, London, 1967.

27. J. C. Zadoks and R. D. Schein, *Epidemiology and Plant Diseases Management,* Oxford Univ. Press, New York, 1979.

28. A. A. Berryman, *Population Systems, a General Introduction,* Plenum Press, New York, 1981.

29. H. Dingle and J. P. Hegmann, Eds., *Evolution and Genetics of Life Histories,* Springer-Verlag, New York, 1982.

30. R. S. Soper, L. F. R. Smith, and A. J. Delyzer, Epizootiology of *Massospora levispora* in an isolated population of *Okanagana rimosa, Ann. Entomol. Soc. Am.,* **69,** 275 (1976).

31. R. S. Soper and D. M. MacLeod, *Descriptive Epizootiology of an Aphid Mycosis,* USDA Tech. Bull., No. 1632, (1981).

32. R. I. Carruthers, *The Biology and Ecology of Entomophthora muscae in the Onion Agroecosystem,* Ph. D. Thesis, Michigan State Univ., East Lansing, 1981.

33. J. R. Fuxa, Dispersion and spread of the entomopathogenic fungus *Nomuraea rileyi* in a soybean field, *Environ. Entomol.,* **13,** 252 (1984).

34. R. M. Anderson, Parasite pathogenicity and the depression of host population equilibria, *Nature,* **279,** 150 (1979).

35. R. M. Anderson and R. M. May, Population biology of infectious diseases: Part I, *Nature,* **280,** 361 (1979).

36. R. M. May and R. M. Anderson. Population biology of infectious diseases: Part II, *Nature,* **280,** 455 (1979).

37. R. M. Anderson and R. M. May, The population dynamics of microparasites and their invertebrate hosts, *J. Anim. Ecol.,* **291,** 451 (1980).

38. G. C. Brown and G. L. Nordin, An epizootic model of an insect-fungal pathogen system. *Bull. Math. Biol.,* **44,** 731 (1982).

39. G. C. Brown, Stability in an epizootiological model with age-dependent immunity and seasonal host reproduction, *Bull. Math. Biol.,* **46,** 139 (1984).

40. K. S. S. Nair and F. L. McEwen, *Strongwellsea castrans* as a parasite of the adult cabbage maggot, *Hylemya brassicae,* in Canada, *J. Invertebr. Pathol.,* **22,** 442 (1973).

41. N. Wilding and F. B. Lauckner, *Entomophthora* infecting wheat bulbfly at Rothamsted, Hertfordshire, 1967–71, *Ann. Appl. Biol.,* **76,** 161 (1974).

42. C. M. Ignoffo, B. Puttler, D. L. Hostetter, and W. A. Dickerson, Susceptibility of the cabbage looper, *Trichoplusia ni,* and the velvet bean caterpillar, *Anticarsia gemmatalis,* to several isolates of the entomopathogenic fungus *Nomuraea rileyi, J. Invertebr. Pathol.,* **28,** 259 (1976).

43. J. D. Vandenberg and R. S. Soper, Prevalence of Entomophthorales mycosis in populations of spruce budworm, *Choristoneura fumiferana, Environ. Entomol.,* **7,** 847 (1978).

44. N. Wilding, *Entomophthora* species infecting pea aphids, *Trans. R. Entomol. Soc.,* **127,** 171 (1975).

45. R. I. Carruthers, D. L. Haynes, and D. M. MacLeod, *Entomophthora muscae* mycosis of the onion fly, *Delia antiqua, J. Invertebr. Pathol.,* **45,** 81 (1985).

46. D. G. Harcourt, J. C. Guppy, D. M. MacLeod, and D. Tyrrell, The fungus *Entomophthora phytonomi* pathogenic to the alfalfa weevil, *Hypera postica, Can. Entomol.,* **106,** 1295 (1974).

47. R. K. Sprenkel, W. M. Brooks, J. W. Van Duyn, and L. L. Deitz, The effects of three cultural variables on the incidence of *Nomuraea rileyi,* phytophagous Lepidoptera, and their predators on soybeans, *Environ. Entomol.* **8,** 334 (1979).

48. B. W. Taylor, J. A. Harlos, and R. A. Brust, *Coelomomyces* infections of the adult female mosquito *Aedes trivittatus* in Mamitoba, *Can. J. Zool.,* **58,** 1215, (1980).

49. B. Payandeh, D. M. MacLeod, and D. R. Wallace, An empirical regression function suitable for modelling spore germination subject to temperature threshold, *Can. J. Bot.,* **56,** 2328 (1978).

50. N. Wilding, Effect of humidity on the sporulation of *Entomophthora aphidis* and *E. thaxteriana, Trans. Br. Mycol. Soc.,* **53,** 126, (1969).

51. N. Wilding, The effect of temperature on the infectivity and incubation periods of the fungi *Entomophthora aphidis* and *E. thaxteriana* for the pea aphid *Acyrthosiphon pisum, Proc. IVth. Int. Coll. Insect Pathol.,* College Park, Maryland, p. 84 (1970).

52. J. L. Kerwin, Biological aspects of the interactions between *Coelomomyces psorophorae* zygotes amd the larvae of *Culiseta inornata:* environmental factors, *J. Invertebr. Pathol.,* **41,** 233 (1983).

53. M. W. Stimmann, Effect of temperature on infection of the garden symphylan by *Entomophthora coronata, J. Econ. Entomol.,* **61,** 1558 (1968).

54. C. M. Ignoffo, C. Garcia, D. L. Hostetter, and R. E. Pinnell, Stability of conidia of an entomopathogenic fungus, *Nomuraea rileyi,* in and on soils, *Environ. Entomol.,* **7,** 724 (1978).

55. J. J. Hamm and W. W. Hare, Applications of entomopathogens in irrigation water for control of fall armyworms and corn earworms on corn, *J. Econ. Entomol.,* **75,** 1074 (1982).

56. A. A. Callaghan, Light and spore discharge in Entomophthorales, *Trans. Br. Mycol. Soc.,* **53,** 87 (1969).

57. J. A. Millstein, G. C. Brown, and G. L. Nordin, Microclimatic humidity influences on conidial discharge in *Erynia* sp. (Entomophthorales: Entomophthoraceae), an entomopathogenic fungus of the alfalfa weevil (Coleoptera: Curculionidae), *Environ. Entomol.,* **11,** 1166 (1982).

58. R. I. Carruthers and D. L. Haynes, Laboratory transmission and in vivo incubation of *Entomophthora muscae* in the onion fly, *Delia antiqua J. Invertebr. Pathol.,* **45,** 282 (1985).

59. B. C. Patten, *Systems Analysis and Simulation in Ecology,* Academic Press, New York, 1981.

60. H. J. Gold, *Mathematical Modeling of Biological Systems: An Introductory Guidebook,* John Wiley & Sons, New York, 1977.

61. W. G. Ruesink, Status of the systems approach to pest management, *Annu. Rev. Entomol.,* **21,** 27, (1976).

62. W. G. Ruesink, "Analysis and Modeling in Pest Management," in R. L. Metcalf and W. H. Luckman, Eds., *Introduction to Insect Pest Management,* Wiley-Interscience, New York, 1982, p. 353.

63. R. L. Tummala, "Concepts of On-Line Pest Management," in R. L. Tummala, D. L. Haynes, and B. A. Croft, Eds., *Modeling for Pest Management: Concepts, Applications and Techniques,* Michigan State Univ., East Lansing, 1976, p. 28.

64. C. A. Shoemaker, "The Role of Systems Analysis in Integrated Pest Management," in C. B. Huffaker, Ed., *New Technology of Pest Control,* John Wiley & Sons, New York, 1980, p. 25.

65. W. M. Getz and A. P. Gutierrez, A perspective on systems analysis in crop protection and insect pest management, *Annu. Rev. Entomol.,* **27,** 447 (1982).

66. S. M. Welch, Developments in computer-based IPM extension delivery systems, *Annu. Rev. Entomol.,* **29,** 359 (1983).

67. L. P. Kish and G. E. Allen, The biology and ecology of *Nomuraea rileyi* and a program for predicting its incidence on *Anticarsia gemmatalis* in soybean, *Fla. Agric. Exp. Stn. Bull. 795,* (1978).

68. D. L. Haynes, R. L. Tummala, and T. L. Ellis, Ecosystem management for pest control, *BioScience,* **30,** 690 (1980).

69. J. Jacobs, "Diversity, Stability and Maturity in Ecosystems Influenced by Human Activities," in W. H. Dobben and R. H. Lowe-McConnell, Eds., *Unifying Concepts in Ecology,* Junk, Hague & PUDOC, Wageningen, 1975, p. 187.

70. R. G. Wiegert, *Benchmark Papers in Ecology: Energetics,* Dowden, Hutchingson and Ross, Stroudsburg, Pennsylvania., 1976.

71. R. Levins and M. Wilson, Ecological theory and pest management, *Annu. Rev. Entomol.,* **25,** 287 (1980).

72. A. P. Gutierrez and Y. H. Wang, "Models for Managing the Economic Impact of Pest Populations in Agricultural Crops," in C. B. Huffaker and R. L. Rabb, Eds., *Ecological Entomology,* John Wiley & Sons, New York, 1984, p. 729.

73. T. C. Edens and H. E. Koenig, Agroecosystem management in a resource-limited world, *BioScience,* **30,** 695 (1980).

74. R. L. Metcalf and W. H. Luckmann, Eds., *Introduction to Insect Pest Management,* Wiley-Interscience, New York, 1975.

75. R. L. Rabb, G. K. DeFoiart, and G. G. Kennedy, "An Ecological Approach to Managing Insect Populations," in C. B. Huffaker and R. L. Rabb, Eds., *Ecological Entomology,* John Wiley & Sons, New York, 1984, p. 697.

76. Y. Tanada, "Epizootiology of Infectious Diseases," in E. Steinhaus, Ed., *Insect Pathology: An Advanced Treatrise.* Vol. 2, Academic Press, New York, 1963, p. 423.

77. N. Wilding, *Entomophthora* species infecting pea aphids, *Trans. R. Entomol. Soc.,* **127,** 171 (1975).

78. N. Wilding and B. F. Lauckner, *Entomophthora* infecting wheat bulb fly at Rothamsted, Hertfordshire, 1967–71, *Ann. Appl. Biol.,* **76,** 161 (1974).

79. D. W. Roberts and A. S. Campbell, Stability of entomopathogenic fungi, *Misc. Publ. Entomol. Soc. Am.,* **10,** 19 (1977).

80. J. D. Ovington, "Strategies for Management of Natural and Man-made Ecosystems," in W. H. Dobben and R. H. Lowe-McConnell, eds. *Unifying Concepts in Ecology,* Junk, Hague & PUDOC, Wageningen, 1975 p. 239.

81. R. S. Soper, "Role of Entomophthoran Fungi in Aphid Control for Potato Integrated Pest Management," in J. Lashomb and R. Cassigrande, Eds., *Advances in Potato Pest Management,* Hutchinson Ross, Stroudsburg, 1981, p. 153.

82. R. S. Soper. "Pathogens of Leafhoppers and Planthoppers," in L. R. Nault and J. G. Rodriguez, Eds., *Leafhoppers and Planthoppers,* John Wiley & Sons, New York, 1985, p. 469.

83. J. Missonier, Y. Robert, and G. Thoizon, Epidemiological circumstances which seem to promote Entomophthorosis in three aphids, *Entomophaga*, **15**, 169 (1970).

84. E. G. Voronina, Entomophthorosis epizootics of the pea aphid *Acyrthosiphon pisum*, *Entomol. Rev.*, **4**, 444 (1971).

85. G. Z. Ventskevich, *Agrometerology*, Isr. Program. Sci. Transl., Jerusalem, 1961.

86. I. Ben-Ze'ev and R. G. Kenneth, *Zoophthora radicans* and *Zoophthora petchi*, two species of the "Sphaerosperma group" attacking leaf-hoppers and frog-hoppers, *Entomophaga*, **26**, 131 (1981).

87. R. S. Soper, "*Erynia radicans* as a Mycoinsecticide for Spruce Budworm Control," in *Proc. Symp. Microbial Control Spruce Budworm and Gypsy Moths*, Windsor Locks, Connecticut, April 10–12, 1984, p. 69 (1985).

88. R. J. Milner and R. S. Soper, Bioassay of *Entomophthora* against the spotted alfalfa aphid, *Therioaphis trifolii* f. *maculata*, *J. Invertebr. Pathol.*, **37**, 168 (1981).

89. R. J. Milner, R. S. Soper, and G. G. Lutton, Field release of an Israeli strain of the fungus *Zoophthora radicans* for biological control of *Therioaphis trifolii* f. *maculata*, *J. Austral. Entomol. Soc.* **21**, 113 (1982).

90. J. C. Arthur, Disease of clover-leaf weevil. *Entomophthora, phytonomi* Arthur. *4th Annu. Rep. N. Y. Agric. Exp. Stn. 1885*, 258 (1886).

91. J. C. Arthur, A new larval *Entomophthora*, *Bot. Gaz.*, **11**, 14 (1886).

92. U. S. Dep. Agric., The clover leaf weevil and its control, *Farmer's Bull. 1484* (1956).

93. I. Ben-Ze'ev and R. G. Kenneth, *Zoophthora phytonomi* and *Conidiobolus osmodes*, two pathogens of *Hypera* species coincidental in time and place, *Entomophaga* **25**, 171 (1980).

94. B. Puttler and L. W. Coles, Biology of *Biolysia tristis* and its role as a parasite of the clover leaf weevil, *J. Econ. Entomol.*, **55**, 831 (1962).

95. R. J. Dysart and B. Puttler, The alfalfa weevil parasite *Bathyplectes curculionis* in Illinios and notes on its dispersal, *J. Econ. Entomol.*, **58**, 1154 (1965).

96. B. Puttler, Interrelationship of *Hypera postica* and *Bathyplectes curculionis* in the eastern United States with particular reference to encapsulation of parasite eggs by weevil larvae, *Ann. Entomol. Soc. Am.*, **60**, 1031 (1967).

97. B. Puttler, *Hypera postica* and *Bathyplectes curculionis:* encapsulation of parasite eggs by host larvae in Missouri and Arkansas, *Environ. Entomol.*, **3**, 881 (1974).

98. A. A. Muka, A disease of the alfalfa weevil in New York, *Proc. 18th Forage Insect Res. Conf. 1976*, **28** (1976).

99. B. Puttler, D. L. Hostetter, S. H. Long, and R. E. Pinnell, *Entomophthora phytonomi*, a fungal pathogen of the alfalfa weevil in the Mid-Great Plains, *Environ. Entomol.*, **7**, 670 (1978).

100. G. C. Brown and G. L. Nordin, An epidemiological model of *Entomophthora phytonomi-Hypera postica* populations, *Soc. Invertebr. Pathol. Newsl.*, **7**, 9 (1980).

101. L. M. Los and W. A. Allen, Visual technique for determining presence and stage of *Zoophthora phytonomi* in dead alfalfa weevil larvae, *J. Econ. Entomol.*, **75**, 375 (1982).

102. W. A. Gardener, Occurrence of *Erynia* sp. in *Hypera postica* in central Georgia, *J. Invertebr. Pathol.*, **40**, 146 (1982).

103. R. L. Pienkowski and P. R. Mehring, Influence of Avermectrin B_1 and carbofuran on feeding by alfalfa weevil larvae, *J. Econ. Entomol.*, **76**, 1167 (1983).

104. D. G. Harcourt, J. C. Guppy, D. M. MacLeod, and D. Tyrrell, Two *Entomophthora* species associated with disease epizootics of the alfalfa weevil, *Hypera postica*, in Ontario, *Great Lakes Entomol.*, **14**, 55 (1981).

105. B. Puttler, D. L. Hostetter, S. H. Long, and A. A. Borski Jr., Seasonal incidence of the fungus *Entomophthora phytonomi* infecting *Hypera postica* larvae in central Missouri, *J. Invertebr. Pathol.*, **35**, 99 (1980).

106. P.L. Watson, R.J. Barney, J. V. Maddox, and E. J. Armbrust, Sporulation and mode of infection of *Entomophthora phytonomi,* a pathogen of the alfalfa weevil, *Environ. Entomol.,* **10,** 305 (1980).

107. G. C. Brown and G. L. Nordin, "Alfalfa Crop Management Augmented by Managing Diseases of Insect Pests," in R. E. Frisbie, Ed., *Consortium for Integrated Pest Management Success Stories, 1982, USDA-CSRS/EPA Project Rep.,* p. 23 (1982).

108. R. Thaxter, The Entomophthoraceae of the United States, *Mem. Boston Soc. Nat. Hist.,* **4,** 133 (1888).

109. C. C. Yeager, *Empusa* infection of the housefly in relation to moisture conditions of northern Idaho, *Mycologica,* **31,** 154 (1939).

110. R. Baird, Notes on a laboratory infection of Diptera caused by the fungus *Empusa muscae* Cohn, *Can. Entomol.,* **89,** 423 (1957).

111. L. A. Miller and R. H. McClanahan, Note on occurrence of the fungus *Empusa muscae* (Cohn) on adults of the onion maggot, *Hylemya antiqua* (Meig.) (Diptera: Anthomyiidae), *Can. Entomol.,* **91,** 525 (1959).

112. J. P. Kramer, Experimental studies on the phycomycosis of Muscoid flies caused by *Entomophthora muscae* (Cohn), *N. Y. Entomol. Soc.,* **79,** 42 (1971).

113. Y. C. Berisford and C. H. Tsao, Field and laboratory observations of an entomogenous infection of the adult seed corn maggot, *Hylemya platura, J. Georgia Entomol. Soc.,* **9,** 104 (1974).

114. M. Loosjes, Ecology and genetic control of the onion fly, *Hylemya antiqua* (Meig.), *Agric. Res. Rep. 857,* Pudoc, Wageningen, 1976.

115. D. M. MacLeod, E. Muller-Kogler, and N. Wilding, *Entomophthora* species with *E. muscae*-like conidia, *Mycologia,* **68,** 1 (1976).

116. J. P. Perron and R. Crete, Premieres observations sur le champignon, *Empusa muscae* Cohn. parasitant la mouche de l'oignon, *Hylemya antiqua* (Meig.) dans le Quebec, *Ann. Entomol. Soc. Quebec,* **5,** 25 (1960).

117. H. C. Gouch, Studies on the wheat bulb fly (*Leptohylemyia coarctata* Fall.) I. Biology. *Bull. Entomol. Res.,* **37,** 251 (1946).

118. T. Petch, Notes on entomogenous fungi, *Trans. Br. Mycol. Soc.,* **19,** 161 (1934).

119. R. I. Carruthers, G. H. Whifield, R. L. Tummala, and D. L. Haynes, A systems approach to research and simulation of insect pest dynamics in the onion agro-ecosystem, *Ecol. Modelling,* (in press).

120. J. P. Kramer, The house fly mycosis caused by *Entomophthora muscae:* influence of relative humidity on infectivity and conidial germination, *N. Y. Entomol. Soc.,* **88,** 236 (1980).

121. B. A. Mullins and J. L. Rodriguez, Dynamics of *Entomophthora muscae* conidial discharge from *Musca domestica* cadavers, *Environ. Entomol.,* **14,** 317 (1985).

122. J. P. Kramer, *Entomophthora muscae:* moisture as a factor affecting its transmission and conidial germination, *Acta Mycol.,* **16,** 133 (1980).

123. R. I. Carruthers and D. L. Haynes, Effects of temperature, moisture, and habitat on *Entomophthora muscae* conidial germination and survival in the onion agroecosystem, *Environ. Entomol.,* (in press).

124. R. I. Carruthers, G. H. Whitfield, and D. L. Haynes, Pesticide-induced mortality of natural enemies of the onion maggot, *Delia antiqua, Entomophaga,* **30,** 151 (1985).

125. B. A. Mullins and J. L. Rodriguez, Insecticide effects on *Entomophthora Muscae, Entomophaga,* (in press).

126. G. H. Whitfield, R. I. Carruthers, and D. L. Haynes, Phenology and control of the onion maggot in Michigan onion production, *Agric. Ecosyst. Environ.,* **12,** 189 (1985).

127. E. A. Heinrichs, H. A. Gastal, and M. H. Galileo, Incidence of natural control agents of the

velvetbean caterpillar and response of its predators to insecticide treatments in Brazilian soybean fields, *Brasilia,* **14,** 79 (1979).

128. D. G. Boucias, D. L. Bradford, and C. S. Barfield, Susceptibility of the velvetbean caterpillar and soybean looper to *Nomuraea rileyi:* effects of pathotype, dosage, temperature, and host age, *J. Econ. Entomol.,* **77,** 247 (1984).

129. L. P. Kish, *The Biology and Ecology of Nomuraea rileyi,* Ph. D. Diss., Univ. Fla., Gainesville, 1975.

130. C. A. Ignoffo, C. Garcia, D. L. Hostetter, and R. E. Pinnel, Laboratory studies of the entomopathogenic fungus *Nomuraea rileyi:* soil born contamination of soybean seedlings and dispersal of diseased larvae of *Trichoplusia ni, J. Invertebr. Pathol.,* **29,** 147 (1977).

131. C. M. Ignoffo and O. F. Batzer, Microencapsulation and ultraviolet protectants to increase sunlight stability of an insect virus, *J. Econ. Entomol.,* **64,** 850 (1977).

132. W. A. Gardner, R. M. Sutton, and R. Noblet, Persistence of *Beauveria bassiana, Nomuraea rileyi,* and *Nosema necatrix* on soybean foliage, *Environ. Entomol.,* **6,** 616 (1977).

133. L. W. Getzin, *Spicaria rileyi,* an entomogenous fungus of *Trichoplusia ni, J. Invertbr. Pathol.,* **3,** 2 (1961).

134. A. K. A. Mohamed, J. V. Bell, and P. P. Sikorowski, Field cage tests with *Nomuraea rileyi* against corn earworm larvae on sweet corn, *J. Econ. Entomol.,* **71,** 102 (1978).

135. R. K. Sprenkel and W. M. Brooks, Artificial dissemination and epizootic initiation of *Nomuraea rileyi,* an entomogenous fungus of lepidopterous pests on soybeans, *J. Econ. Entomol.,* **68,** 847 (1975).

136. R. F. Morris, The dynamics of epidemic spruce budworm populations, *Mem. Entomol. Soc. Can.,* **31,** 1 (1963).

137. D. F. Perry and G. H. Whitfield, "The Interrelationships Between Microbial Entomopathogens and Insect Hosts: A System Study Approach with Particular Reference to the Entomophthorales and the Eastern Spruce Budworm," in J. M. Anderson, A. D. M. Rayner, and D. Walton, Eds., *Animal-Microbial Interactions,* Cambridge Press, London, 1985, p. 307.

138. D. F. Perry, D. Tyrrell, and A. J. Delyzer, The mode of germination of *Zoophthora radicans* azygospores, *Trans. Br. Mycol. Soc.,* **78,** 221 (1982).

139. H. J. W. Van Roemund, D. F. Perry, and D. Tyrrell, The influence of temperature, light, nutrients and pH in the determination of the mode of conidial germination in *Zoophthora radicans, Trans. Br. Mycol. Soc.,* **82,** 31 (1983).

140. C. E. Brown, Habits and control of the forest tent caterpillar, *Dep. For., Ottawa, Canada, Cat. No. Fo-23-966,* (1966) p. 1.

141. L. P. Abrahamson and J. D. Harper, Microbial insecticides control forest tent caterpillar in southwestern Alabama, *U. S. For. Serv. Res. Note, 80-157,* (1972) p. 1.

142. R. D. Frye and D. A. Ramse, Natural control agents in forest tent caterpillar populations, *Farm Res.,* **14,** 1 (1975).

143. D. M. MacLeod and D. Tyrrell, *Entomophthora crustosa* as a pathogen of the forest tent caterpillar, *Can. Entomol.,* **111,** 1137 (1979).

144. E. P. Odum, *Fundamentals of Ecology,* 3rd ed., Saunders, Philadelphia, 1971.

145. B. P. Uvarov, *Locusts and Grasshoppers,* Imp. Bur. Entomol., London, 1928.

146. P. W. Riegert, A history of grasshopper abundance surveys and forecasts of outbreaks in Saskatchewan, *Mem. Entomol. Soc. Can.,* **52,** 1 (1968).

147. S. H. Skaife, The locust fungus *Empusa grylli* and its effect on its host, *S. Afr. J. Sci.,* **22,** 298 (1925).

148. J. Roffey, The occurrence of the fungus *Entomophthora grylli* Fres. on locusts and grasshoppers in Thailand, *J. Invertebr. Pathol.,* **11,** 237 (1968).

149. D. M. MacLeod and E. Muller-Kogler, Entomogenous fungi: *Entomophthora* species with pear-shaped to almost spherical conidia, *Mycologia,* **65,** 823 (1973).

150. R. J. Milner, On the occurrence of *Entomophthora grylli*, a fungal pathogen of grasshoppers in Australia, *J. Aust. Entomol. Soc.*, **17**, 293 (1978).

151. D. M. MacLeod, "Entomophthorales Infections, " in E. A. Steinhaus, Ed., *Insect Pathology: An Advanced Treatise*, Vol. 2, Academic Press, New York, 1963, p. 189.

152. R. Pickford and P. W. Riegert, The fungus disease caused by *Entomophthora grylli* Fres., and its effect on grasshopper populations in Saskatchewan in 1963, *Can. Entomol.*, **96**, 1158 (1964).

153. R. C. Treherne and E. R. Buckell, Grasshoppers of British Columbia, *Bull. Dom. Can. Dep. Agric.* **39**, 1 (1924).

154. R. S. Soper, B. May, and B. Martinell, *Entomophaga grylli* enzyme polymorphism as a technique for pathotype identification, *Environ. Entomol.*, **12**, 720 (1983).

155. R. A. Humber, *Conidiobolus* and *Entomophaga* (Entomophthorales): emendations and included species, *Mycotaxon* (in preparation).

156. T. R. Gottwald and W. L. Tedders, Suppression of pecan weevil populations with entomopathogenic fungi, *Environ. Entomol.*, **12**, 471 (1984).

157. A. Catley, The coconut rhinoceros beetle *Oryctes rhinoceros*, *Pans*, **15**, 18 (1969).

158. G. O. Bedford, Biology, ecology and control of palm rhinoceros beetles, *Annu. Rev. Entomol.*, **25**, 309 (1980).

159. K. Friederichs, Uber pleophafie des insektenpilzes, *Metarrhizium anisopliae*, *Zentralbl. Bakteriol. Parasitenkd. Infektionskr.*, **2**, 335 (1920).

160. A. M. Huger, A virus disease of the Indian rhinoceros beetle *Oryctes rhinoceros* caused by a new type of insect virus, *Rhabdionvirus oryctes*, *J. Invertebr. Pathol.*, **8**, 38 (1966).

161. B. Zelazny, Studies on *Rhabdionvirus oryctes*. II. Effect on adults of *Oryctes rhinoceros*, *J. Invertebr. Pathol.*, **22**, 122 (1973).

162. E. C. Young, The epizootiology of two pathogens of the coconut palm rhinoceros beetle, *J. Invertebr. Pathol.*, **24**, 82 (1974).

163. G. C. M. Latch, Studies on the susceptibility of *Oryctes rhinoceros* to some entomogenous fungi, *Entomophaga*, **21**, 31 (1976).

164. K. K. Nirula, K. Radha, and K. P. V. Menon, The green muscardine disease of *Oryctes rhinoceros* L. I. Symptomatology, epizootiology and economic importance. *Indian Coconut J.*, **9**, 3 (1955).

165. B. Schaerffenberg, Untersuchungen uber die Wirkung der insektentotenden Pilze *Beauveria bassiana* und *Metarrhizium anisopliae*, *Entomophaga* **13**, 175 (1968).

166. K. J. Marshall, Report of the insect pathologist UNDP/SPC project for research on the control of the coconut palm rhinoceros beetle, Semi-annual Report of the Project Manager for the period June 1969–November 1969 18 (1970).

167. G. C. M. Latch and R. E. Falloon, Studies on the use of *Metarhizium anisopliae* to control *Oryctes rhinoceros*, *Entomophaga*, **21**, 39 (1976).·

168. R. S. Soper, A. J. Delyzer, and L. F. R. Smith, The genus *Massospora* entomopathogenic for cicadas. Part II. Biology of *Massospora levispora* and its host *Okanagana rimosa*, *Ann. Entomol. Soc. Am.*, **69**, 275 (1976).

169. R. Krejzova, The resistance of cultures and dried resting spores of three species of the genus *Entomophthora* to ajatin and the viability of their resting spores after long-term storage in the refrigerator, *Cesk. Mykol.*, **27**, 107 (1973).

170. D. P. Oliveria, B. M. Chaves, and E. G. Loures, Estudo comparativo da sobrevivencia de *Metarhizium anisopliae* Sorokin em diferentes tipos de solo, *Rev. Theobroma*, **11**, 233 (1981).

171. S. R. Dutky, Insect microbiology, *Adv. Appl. Microbiol.*, **1**, 175 (1959).

172. T. Mikuni, K. Kawakami, and M. Nakayama, Survival of an entomogenous fungus, *Metar-*

hizium anisopliae, causing the muscardine disease of the silkworm, *Bombyx mori,* in the soil of mulberry plantations, *J. Seric. Sci. Japan,* **51,** 325 (1982).

173. V. P. Pospelov, "Biological Methods of Controlling the Beet Weevil," in N. M. Kulagin and G. K. Pyatnitzkii, Eds., *The Beet Weevil and its Control,* Demyovo, Moscow, (*Rev. Appl. Entomol.* **A30,** 66, (1940)).

174. G. K. Pyantnitzkii, "Agrotechnical Methods of Controlling the Beet Weevil," in N. M. Kulagin and G. K. Pyatnitzkii, Eds., *The Beet Weevil and its Control,* Demyovo, Moscow, (*Rev. Appl. Entomol.* **A30,** 66 (1940)).

175. Y. Tanada, "Persistence of Pathogens in the Aquatic Environment," in A. W. Bourquin, D. G. Ahearn, and S. P. Meyers, Eds., *Impact of the Use of Microorganisms on the Aquatic Environment,* Environ. Prot. Agency (EPA–660/3–75–001), Corvallis, Oregon, 1975.

176. J. L. Kerwin, Biological aspects of the interaction between *Coelomomyces psorophorae* zygotes and the larvae of *Culiseta inornata:* environmental factors, *J. Invertebr. Pathol.,* **41,** 233 (1983).

177. E. Descals, J. Webster, M. Ladle, and J. A. B. Bass, Variations in asexual reproduction in species of *Entomophthora* on aquatic insects, *Trans. Br. Mycol. Soc.,* **77,** 85 (1981).

178. J. N. Couch and S. V. Romney, Sexual reproduction in *Lagenidium giganteum, Mycologia,* **65,** 250 (1973).

179. A. J. Domnas, "Biochemistry of *Lagenidium giganteum* Infection of Mosquito Larvae," in E. W. Davidson, Ed. *Pathogenesis of Invertebrate Microbial Diseases,* Totowa, New Jersey, 1981.

180. H. C. Whistler, S. L. Zebold, and J. A. Shermanchuk, Alternate host for mosquito parasite *Coelomomyces. Nature,* **251,** 715 (1975).

181. J. S. Pillai and I. H. O'Loughlin, *Coelomomyces opifexi,* II. Experiments in sporangial germination, *Hydrobiologia,* **40,** 77 (1972).

182. T. G. Andreadis and L. A. Magnarelli, New variants of the *Coelomomyces psorophorae* "complex" from the salt-marsh mosquitoes *Aedes cantator* and *Aedes sollicitans, J. Med. Entomol.,* **21,** 379 (1984).

183. T. C. Cheng, "Use of Microorganisms to Control Aquatic Pests Other Than Insects," in A. W. Bourquin, D. G. Ahearn, and S. P. Meyers, Eds., *Impact of the Use of Microorganisms on the Aquatic Environment,* Environ. Prot. Agency (EPA–660/3–75–001), Corvallis, Oregon, 1975, p. 105.

184. B. A. Federici. "Mosquito Control by the Fungi *Culicinomyces, Lagenidium* and *Coelomomyces,"* in H. D. Burges, Ed., *Microbial Control of Pests and Plant Diseases 1970–80,* Academic Press, London, 1981, p. 555.

185. L. A. Lacey and A. H. Undeen, Microbial control of black flies and mosquitoes, *Annu. Rev. Entomol.,* **31,** 265 (1986).

186. R. K. Washino, J. L. Fetter-Lasko, C. K. Fukushima, and K. Gonot, The establishment of *Lagenidium giganteum,* an aquatic fungal parasite of mosquitoes, three years after introduction. *Proc. Calif. Vector Mosq. Control Assoc.,* **44,** 52 (1976).

187. J. L. Fetter-Lasko and R. K. Washino, A three year study of the ecology of *Lagenidium* infections of *Culex tarsalis* in California, *Proc. Calif. Mosq. Vector Control Assoc.,* **45,** 106 (1977).

188. J. L. Kerwin and R. K. Washino, Artificial culture of the sexual and asexual stages of *Lagenidium giganteum, Proc. Calif. Mosq. Vector Control Assoc.,* **50,** 43 (1982).

189. J. L. Kerwin and R. K. Washino, Efficacy of *Romanomermis culicivorax* and *Lagenidium giganteum* for mosquito control: strategies for use of biological control agents in rice fields of the Central Valley of California, *Proc. Calif. Mosq. Vector Control Assoc.,* **52,** 86 (1984).

190. E. M. MaCray, Jr., D. J. Womeldorf, R. C. Husbands, and D. A. Eliason, Laboratory observations and field tests with *Lagenidium* against California mosquitoes. *Proc. Calif. Mosq. Vector Control Assoc.* **41,** 123 (1973).

191. A. J. Domnas, E. Giebel, and T. M. McInnis Jr., Biochemistry of mosquito infection: preliminary studies of biochemical changes in *Culex pipiens quinquefasciatus* following infection with *Lagenidium giganteum. J. Invertebr. Pathol.,* **24,** 293 (1974).

192. M. S. Goettel, M. K. Toohey, and J. S. Pillai, Preliminary laboratory infection trials with a Fiji isolate of the mosquito pathogenic fungus *Lagenidium, J. Invertebr. Pathol.,* **41,** 1 (1983).

193. A. J. Domnas, J. P. Srebro, and B. F. Hicks, Sterol requirements for zoospore formation in the mosquito-parasitizing fungus *Lagenidium giganteum, Mycologia,* **69,** 875 (1977).

194. S. T. Jaronski and R. C. Axtell, Persistence of the mosquito fungal pathogen *Lagenidium giganteum* after introduction into natural habitats, *Mosq. News.* **43,** 332 (1983).

195. T. L. Merriam and R. C. Axtell, Relative toxicity of certain pesticides to *Lagenidium giganteum,* a fungal pathogen of mosquito larvae, *Environ. Entomol.,* **12,** 515 (1983).

196. C. W. Churchman. *The Systems Approach,* Dell Publ., New York, 1968.

197. R. L. Tummala, W. Ruesink, and D. L. Haynes, A discrete component approach to the management of the cereal leaf beetle ecosystem, *Environ. Entomol.,* **4,** 175 (1975).

198. T. J. Manetsch and G. L. Park, *Systems Analysis and Simulation with Application to Economic and Social Systems,* Parts I & II, Michigan State Univ., East Lansing, 1977.

199. P. J. H. Sharpe and D. W. DeMichele, Reaction kinetics of poikilotherm development. *J. Theor. Biol.,* **64,** 649 (1977).

200. T. Wagner, Hsin-I Wu, P. J. H. Sharpe, R. M. Schoolfield and R. N. Coulson, Modeling insect development rates: a literature review and application of a biophysical model, *Ann. Entomol. Soc. Am.,* **77,** 208 (1984).

201. I. M. Hall and J. V. Bell, Further studies on the effect of temperature on the growth of some Entomophthoraceous fungi, *J. Invertebr. Pathol.,* **3,** 289 (1960).

202. R. I. Carruthers, Z. Feng, D. S. Robson, and D. W. Roberts, In vivo temperature-dependent development of *Beauveria bassiana* mycosis of the European corn borer, *Ostrinia nubilalis, J. Invertebr. Pathol.,* **46,** 305 (1985).

203. G. G. Newman and G. R. Carner, Diel periodicity of *Entomophthora gammae* in soybean looper, *Environ. Entomol.,* **33,** 888 (1974).

204. D. G. Holdom, *Studies on the Biology, Nutrition and Physiology of Entomophthora planchoniana, a Pathogen of the Bluegreen Aphid, Acyrthosiphon kondoi,* Ph. D. Thesis, Univ. Queensland, Queensland, Australia, 1984.

205. R. G. Milner, D. G. Holdom, and T. R. Glare, Diurnal patterns of mortality in aphids infected by entomophthoran fungi, *Entomol. Exp. Appl.,* **36,** 37 (1984).

206. N. Wilding, Discharge of conidia of *Entomophthora thaxteriana* form the pea aphid *Acyrthosiphon pisum, J. Gen. Microbiol.,* **69,** 417 (1971).

207. R. G. Milner, Patterns of primary spore discharge of *Entomophthora* spp. from the blue green aphid, *Acyrthosiphon kondoi, J. Invertebr. Pathol.,* **38,** 419 (1981).

208. S. M. Pady, C. L. Kramer, D. L. Long, and T. D. McBride. Spore discharge in *Entomophthora grylli. Ann. Appl. Biol.,* **67,** 145 (1971).

209. R. G. Milner and G. G. Lutton, Effect of temperature on *Zoophthora radicans:* an introduced microbial control agent of the spotted alfalfa aphid *Therioaphis trifolii, J. Aust. Entomol. Soc.,* **22,** 267 (1983).

210. N. Wilding, The Survival of *Entomophthora* spp. in mummified aphids at different temperatures and humidities, *J. Invertebr. Pathol.,* **21,** 309 (1973).

211. W. G. Yendol, Factors affecting germination of *Entomophthora* conidia, *J. Invertebr. Pathol.,* **10,** 313 (1968).

212. M. Shimazu, Factors affecting conidial germination of *Entomophthora delphacis, Appl. Entomol. Zool.,* **12,** 260 (1977).

213. D. E. Pinnock and R. J. Brand, "A Quantitative Approach to the Ecology of the Use of Pathogens for Insect Control," in H. D. Burges, Ed., *Microbial Control of Pests and Plant Diseases 1970–80,* Academic Press, London, 1981, p. 655.

214. R. J. Brand and D. E. Pinnock, "Application of Biostatistical Modelling to Forecasting the Results of Microbial Control Trials," in H. D. Burges, Ed., *Microbial Control of Pests and Plant Diseases 1970–80,* Academic Press, London, 1981, p. 667.

215. C. T. Ingold, *Fungal Spores: Their Liberation and Dispersal,* Clarendon Press, Oxford, 1971.

216. S. Galaini, *The Efficacy of Foliar Applications of Beauveria bassiana Conidia Against Leptinotarsa decemlineata,* M. S. Thesis, Cornell Univ., Ithaca, New York, 1984.

217. N. Wilding, Resting spore formation and germination in *Entomophthora fresenii, Rep. Rothamsted Exp. Stn.* 1970, pt **1,** 207 (1971).

218. D. R. Wallace, D. M. MacLeod, C. R. Sullivan, D. Tyrrell, and A. J. DeLyzer, Induction of resting spore germination in *Entomophthora aphidis* by long-day conditions, *Can. J. Bot.,* **54,** 1410 (1976).

219. D. F. Perry and J. P. Latge, Dormancy and germination of *Conidiobolus obscurus* azygospores. *Trans. Br. Mycol. Soc.,* **78,** 221 (1982).

220. A. L. Pugh, *Dynamo User's Manual,* MIT Press, Cambridge, 1963.

13

PROTOZOAN DISEASES

JOSEPH V. MADDOX

Illinois Natural History Survey and
Illinois Agricultural Experiment Station
Champaign, Illinois

1. INTRODUCTION

The large group of organisms once combined in the phylum Protozoa have recently been divided into seven different phyla (1). Four of these phyla, all unicellular eukaryotes, contain species parasitic to insects (2). For the purposes of this chapter, representatives of the phyla Ciliophora, Sarcomastigophora, Apicomplexa, and Microspora will be considered.

There are thousands of references dealing with protozoan infections in insects. Many of these, while invaluable to other areas of pathology, are not directly involved with epizootiology. The focal point of this chapter is the occurrence of protozoan infections in insect populations. Included in this review are references dealing with characteristics of the parasite, characteristics of the host, and influence of the habitat as they relate to epizootiology.

The accounts of protozoan infections in insect populations may be divided into the following categories based on the type of collection data available: (a) anecdotal reports with no quantitative information about percentage infection or host population density, (b) percentage infection but no information on population host density, (c) percentage infection and host population density. Most reports of protozoan infections in insect populations fall into categories a or b. As isolated observations these are not very helpful, but trends can often be noticed for similar pathogens, pathogens infecting similar hosts, or pathogens occurring in similar habitats.

Observations in any of the above categories may consist of a single observation at a single location or a series of observations at the same location. The observational time frame may be once each season for a number of years or several observations per year over one or more seasons. Disease prevalence may be observed in all developmental stages of the host or only a single developmental stage. From the standpoint of epizootiological information, the most desirable type of information is obtained by monitoring the host population throughout the season for several seasons. This should ideally include information about disease prevalence for all developmental stages of the host, including the number of individuals examined, and the host population density.

2. CHARACTERISTICS OF THE GROUPS

2.1. Sarcomastigophora

Relatively few amoebic diseases of insects have been described, and most work has been done on amoebic diseases of grasshoppers and honeybees. *Malamoeba locustae* is a parasite of several species of grasshoppers and crickets. It is not transmitted vertically but is transmitted horizontally by oral ingestion of resistant cysts. These hatch and primary trophozoites invade caecal and midgut epithelium. Secondary trophozoites later invade gut muscularis, hemocoele, and Malpighian tubules (3). Within 9 to 18 days after ingestion of cysts, infected grasshoppers have newly produced cysts in their feces (3, 4). About 45 species of grasshoppers are susceptible to *M. locustae,* and infections are generally chronic (4-6). There are only two reports of high prevalence rates in field collections, both in African populations of the brown locust, *Locustana pardalina* (7).

Malpighamoeba mellificae infects the lumen of the Malpighian tubules of adult honeybees (8). The life cycle is identical to that of *M. locustae* except that cysts are passed in the feces 22 to 24 days after ingestion of spores. Honey production may be slightly reduced, but the effects on individual bees are minimal (9). There are a number of publications dealing with the epizootiology of this pathogen (8-10).

Thirty percent of bark beetles *Dryocoetes autographus* collected during 1980-81 in spruce stands in the Bavarian Alps were infected with *Malamoeba scolyti,* a recently described species (11). Their importance as mortality factors in bark beetles is unknown.

The order Kinetoplastida contains most of the flagellate species parasitic to insects. Eight genera have insect associations, and the six genera limited to insects contain approximately 130 species (12,13). Over 90% of the described species parasitize the orders Diptera and Hemiptera, and most infections are limited to the alimentary tract. Infection is usually by ingestion of flagellated stages or cysts present in fecal-contaminated food. Transovum transmission as well as transmission by parasitoids has been suggested but not conclusively demonstrated experimentally. Most entomophilic Kinetoplastida have a wide host range and cause chronic infections, although few specifics are known about their pathogenicity. There are numerous accounts of percentage infections in single collections, but data on host population density and seasonal trends are largely unavailable.

2.2. Ciliophora

Most ciliates associated with insects are mutualists in the digestive tracts of termites or cockroaches or occur externally as epibionts of aquatic insects (14). The ciliate genera *Tetrahymena* and *Lambornella* contain most if not all of the described species of entomophilic endoparasitic ciliates. *Tetrahymena* spp. are often free living and are probably facultative endoparasites. *Lambornella* spp. possess invasion cysts that attach to the mosquito cuticle and subsequently penetrate the cuticle and enter the hemocoele. Whether they are obligate or facultative parasites is unresolved. For all entomophilic ciliates only horizontal transmission is known, although ciliates have been recovered from adult mosquitoes. The extracorporeal ecology of this group of ciliates has not been investigated.

There are few reports of prevalence of ciliate infections and no information about host population density correlated with disease prevalence or sequential samples from the same location. Kellen et al. (15) found less than 1% infection in *Aedes sierrensis,* Chapman et al. (16) found "small numbers" of infected

Aedes vexans, Sanders (17) described 24% infection in one of 18 tree holes in California, and Clark and Brandl (18) reported 50% infection in *Aedes sierrensis.*

The prevalence of ciliate parasites in tree-hole habitats and the greater susceptibility of first and second instar mosquito larvae suggest that the epizootiology of ciliates is closely associated with their movement from one tree hole to another, persistence within a tree hole once introduced, and the age distribution of the larval mosquito population in the tree hole.

2.3. Apicomplexa

Only two groups of this phylum, the gregarines and the adeline coccidia, develop exclusively in invertebrates (2).

The eugregarines, occurring primarily in intestines of Coleoptera, Diptera, and Orthoptera, are among the more common parasites of insects. The life cycle of all species is similar (12). Spores are ingested by a susceptible host. Sporozoites exit ingested spores in the host gut and invade intestinal epithelial cells. After an initial intracellular phase, most trophozoites insert a structure called the epimerite into a host cell but obtain nutrients from the gut lumen. Gametocysts are formed by the encystment of two sporadins. There is no schizogonic phase in the life cycle. Gametocysts are usually passed out in the host feces after which isogamous or anisogamous gametes are formed. These combine to form a zygote which develops into a thick-walled oocyst or spore. The number of sporozoites in each spore varies with the species. Transovarial transmission does not occur. Eugregarines are generally considered to be of low pathogenicity to their hosts. Anecdotal reports about infection rates are available for many eugregarine species, and for the eugregarine, *Lankesteria culicis,* epizootiological information is abundant.

The neogregarines are common insect parasites infecting primarily members of the orders Lepidoptera, Coleoptera, Hemiptera, Diptera, and Orthoptera (19). They differ from the eugregarines in having one or more cycles of schizogony, an intracellular developmental cycle, and adipose tissue as a site of infection. The biology of two species, *Mattesia grandis* and *Mattesia trogodermae,* has been studied extensively. Only oral transmission is known. There are many reports of naturally occurring infections caused by these two species as well as by other species of neogregarines.

Most species of Coccidia are parasites of vertebrate digestive tracts, and the relatively few species that occur in insects are considered "accidental" (20). They differ from the Gregarines primarily in having small mature gamonts. Coccidian infections are most often found in Siphonaptera, Coleoptera, Diptera, and various stored product insects. Only oral transmission is known. The best known coccidian genus is *Adelina,* and there are many reports of *Adelina* infections in insect populations.

2.4. Microspora

The microsporidia are the most important group of protozoan pathogens of insects. Over 250 species have been described from insects, but microsporidiologists agree that of the total number in existence only a small fraction have been described. Most, if not all, species are transmitted orally by the ingestion of small spores. Upon ingestion by a suitable insect host these spores extrude a long hollow tube, the polar filament, through which the infective agent or sporoplasm travels. The polar filament extrudes with such force that it places the sporoplasm inside or in close proximity to a midgut epithelial cell. After this initial invasion, repeated binary and/or multiple fission is followed by sporogony which produces spores. There are many variations of this basic reproductive scheme in different groups of microsporidia. Several excellent reviews are available (20-24).

Microsporidian species have been described from most orders of insects and range from chronic to highly virulent. In addition to the oral route, transovarial transmission, transovum transmission, and transmission by parasitoids have been well documented (2, 25, 26). Although little is known about most of the described species of microsporidia, for a few species there is extensive information about host-pathogen relationships, extracorporeal ecology, and epizootiology.

3. INSECT ASSOCIATIONS

3.1. Forest Insects

The natural forest ecosystem is a relatively stable habitat in which insect-protozoan associations are expected to be cyclical and somewhat predictable. Most of our information about protozoan diseases of forest pests meets these expectations. An accumulating body of evidence strongly suggests that the protozoa are one of several groups of insect pathogens responsible for regulating cyclic changes of insect populations (27). Host collections from the same location for several successive seasons usually harbor the same protozoa with percentages of infection ranging from 0 to 90%, indicating a cyclic association. Many of these observations have not included data on host population density, making it difficult to infer a direct effect on host populations.

3.1.1. BARK BEETLES OR ENGRAVERS.   Several species of microsporidia have been described from beetles of the family Scolytidae. These are small beetles that as both adults and larvae live beneath tree bark. They construct many winding galleries in the wood of dead or dying trees just under the bark. The microsporidian infections reported in these beetles are summarized in Table 13.1.

TABLE 13.1
Prevalence of Microsporidian Infections in Bark Beetles Collected in Czechoslovakia, Poland, and the United States

Microsporidium Species	Host Species	Date Collected	Location Collected	Stage	Number Examined	% Infected	References
Nosema curvidentis	*Pityocteines curvidens*	Before 1961	Czechoslovakia	Adults	22	50	Weiser (28)
Nosema curvidentis	*Pityocteines curvidens*	Before 1961	Czechoslovakia	Pupae	12	35	Weiser (28)
Nosema curvidentis	*Pityocteines curvidens*	Before 1961	Czechoslovakia	Larvae 3–4 instar	74	0.2	Weiser (28)
Nosema curvidentis	*Scolytus*	Before 1968	Czechoslovakia (St. Maria)	Adults	13	30	Weiser (28)
Nosema curvidentis	*Scolytus*	Before 1968	Czechoslovakia (St. Maria)	Larvae	12	17	Weiser (28)
Nosema curvidentis	*Scolytus*	Before 1968	Czechoslovakia (Polarikovo)	Larvae	90	7.7	Weiser (28)
Nosema scolyti	*S. multistriatus*	1965	Poland	Larvae	239	41	Lipa (29)
Nosema scolyti	*S. pygmaeus*	1965	Poland	Larvae	126	21	Lipa (29)
Nosema scolyti	*S. scolytus*	1965	Poland	Larvae	126	12	Lipa (29)
Nosema scolyti	*S. multistriatus*	1965	Poland	Adults	44	50	Lipa (29)
Nosema scolyti	*S. pygmaeus*	1965	Poland	Adults	96	23	Lipa (29)
Nosema scolyti	*S. scolytus*	1965	Poland	Adults	49	14	Lipa (29)
Unikaryon minutum	*Dendroctonus frontalis*	Sept. 18, 1975	Elgin AFB Florida	Adults	395	71.1	Atkinson and Wilkinson (30)
Unikaryon minutum	*Dendroctonus frontalis*	Nov. 20, 1975	Elgin AFB Florida	Adults	309	66.9	Atkinson and Wilkinson (30)
Unikaryon minutum	*Dendroctonus frontalis*	Feb. 3, 1976	Elgin AFB Florida	Adults	29	79.3	Atkinson and Wilkinson (30)
Unikaryon minutum	*Dendroctonus frontalis*	Sept. 9, 1976	Athens, Georgia	Adults	50	2.0	Atkinson and Wilkinson (30)

Nosema curvidentis is primarily a parasite of the fat body, making horizontal transmission the most important, if not the only means of transmission. Weiser (28) suggested that differences in the disease prevalence between larvae and adults were caused by behavioral differences in these two stages. Larvae start galleries from the original egg packet and move into intact uninfected bark. Infection only occurs when the new gallery crosses an old gallery containing spores from the remains of dead insects. Since adults move about more freely and feed in several locations before laying eggs, they have a better chance of ingesting spores. Obviously as the larval population increases and/or the infested tree becomes more riddled with contaminated galleries, the larval infection rate would increase.

Microsporidium scolyti and *Nosema scolyti,* which often occur as mixed infections and could possibly be a single dimorphic species, cause massive infections of the gut epithelium of several species of bark beetles (29). Spores are passed in fecal material, making transovum transmission a possibility.

Differences in prevalence of these microsporidia in populations of the three species of *Scolytus* could be explained by differences in host feeding behavior or host population density.

Adults of the southern pine beetle, *Dendroctonus frontalis,* collected on three dates from one Florida location, had percentages of *Unikaryon minutum* infection ranging from 66.9 to 79.3%, while one collection from Athens, Georgia, had an infection rate of only 2.0% (30). All of these adults were obtained by holding samples of equal size from five infested trees at each collection date in emergence cages and capturing adult beetles. While the samples were not taken to determine host population density, they probably reflect the relative population densities of the different collections. The collection from Elgin AFB, Florida, may reflect an epizootic in decline while the collection from Georgia was an enzootic.

The white pine weevil, *Pissodes strobi,* while not a true bark beetle, lives under tree bark as a larva and is infected by a *Nosema* sp. Collections taken from Allegany County, Maryland, in the fall of 1972 had 30.4% infection in larvae (N = 23) and 68.1% in adults (N = 47). The *Nosema* heavily infected most body tissues including gut and reproductive organs suggesting both *per os* and transovarial transmission. The microsporidium was suspected of causing previously unexplained mortality in the southern pine beetle, although no supportive data were available (31).

The bark beetle-microsporidian associations pose some interesting research possibilities. The epizootiological characteristics of the bark beetle-microsporidian association probably resemble microsporidian associations of stored grain pests more than those of most other forest pests, such as phytophagous Lepidoptera and sawflies. Each infested tree is an isolated arena in which horizontal transmission occurs. Adult beetles moving from tree to tree transport the microsporidia, and horizontal transmission within the tree is largely dependent on host population density.

3.1.2. PHYTOPHAGOUS LEPIDOPTERA.   Most examples of protozoan infections in forest pests involve phytophagous Lepidoptera. There are numerous
reports of protozoa, primarily microsporidia, infecting forest Lepidoptera.

Nosema fumiferanae has been studied extensively and was reviewed thoroughly by Wilson (32). *N. fumiferana* will infect lepidoptera other than the
spruce budworm *Choristoneura fumiferana* but probably does not have a broad
host spectrum. The primary host, *C. fumiferana,* has one generation per year.
Adults emerge in July or August and lay masses of eggs on conifer needles. Eggs
hatch in about 10 days, and larvae spin a hibernaculum, molt to second instar,
and overwinter. The following spring larvae emerge from the hibernaculum,
mine into old needles, then bore into expanding buds. There are six larval
instars, and in heavy infestations there is often more than one larva per bud.
Horizontal transmission occurs when larvae eat food contaminated with fecal
material or larval regurgitation. Wind and rain have been implicated as factors
responsible for spreading spores from feces and remains of dead infected larvae.

Transovarial transmission occurs, but the infection has little effect on eggs or
first or second instar larvae. Similar observations of *Orthosoma* sp. in the winter
moth *Operophtera brumata* led Canning (2) to postulate that spores may not be
in larval tissue but rather in remnants of yolk taken into the midgut late in
embryogenesis. Such an adaptation would allow much more efficient dissemination of spores during the larval stages because few young larvae would die,
allowing older larvae to produce many more spores for horizontal transmission. Other studies of the effect of *N. fumiferanae* on spruce budworm survival
indicate that under laboratory conditions infections are chronic (33-35).

Nosema fumiferanae occurs naturally in spruce budworm populations, and
information relative to both seasonal and within season occurrence is abundant
(36-43). Some of these observations are summarized in Tables 13.2 and 13.3.
All observations except the seasonal observations for 1955 clearly indicate a
gradual increase in *N. fumiferanae* prevalence. After the years 1959 and 1973,
when infection levels exceeded 80%, population density of the spruce budworm
decreased impressively. During 1955 the percentage of larvae infected with *N.
fumiferanae* decreased, probably because spruce budworm population densities were below the population threshold density necessary for horizontal
spread of the pathogen. These data, while valuable in their present form, would
be much more valuable from an epizootiological point of view if they were
accompanied by estimates of host population density at the time samples for
determination of percent infection were taken.

The green tortrix *Tortrix viridana* a primary lepidopteran defoliator of European forests, is naturally infected by microsporidia. Franz and Huger (44)
described an epizootic caused by *Nosema tortricis* in central Germany during
1968-1969. A disease prevalence of 29% was observed in first instar larvae,
which were obtained by warming twigs collected early in the spring of
1968-1969. At this time, population density was very high, 1.2 to 3.2 larvae per
bud; heavy defoliation occurred when larval populations reached 0.1 larvae per
bud. Midway in the season the prevalence in larvae had risen to 59%, and the

TABLE 13.2

Prevalence of *Nosema fumiferanae* in Larval Spruce Budworm Populations in
Parkinson Township, Ontario, Canada, 1971–1973

Year	Date	% Infection with *Nosema fumiferanae*
1971	May 27	32.6
1971	June 3	22.4
1971	June 10	50.6
1971	June 17	40.0
1972	May 24	37.0
1972	May 31	33.0
1972	June 7	33.0
1972	June 14	40.7
1972	June 22	41.6
1973	May 10	67.0
1973	May 23	66.8
1973	June 20	82.0

Source: Modified from Wilson (40).

prevalence in adults was 60%. In 1969, first instar population density was 0.5
larvae per bud and the prevalence 60%. Late in the season on May 12, third
instar prevalence was 80%, and by May 27, when larvae were in the fifth instar,
the prevalence had dropped to 24%, and larval population density had decreased to the point where larvae were difficult to find.

Lipa and Madziara-Borusiewicz (45) observed a similar epizootic in the
green tortrix in Poland. They identified three species of microsporidia, *N.*

TABLE 13.3

Prevalence of *Nosema fumiferanae* in Larval Spruce Budworm Populations,
Uxbridge Forest, Ontario, Canada

Date	Stage	% Infection
1955	Overwintering larvae	36.4
May 10, 1955	4th instar larvae	28.0
May 26, 1955	5th instar larvae	27.2
June 7, 1955	6th instar larvae and pupae	21.8
June 20, 1955	Pupae–adults	22.4
1956	Overwintering larvae	45.7
1957	Overwintering larvae	56.1
1958	Overwintering larvae	69.1
1959	Overwintering larvae	81.3

Source: Modified from Thomson (37, 39).

tortricis, Octosporea viridanae, and *Thelohania weiseri,* as being responsible for this epizootic. Since all three microsporidia were reported to be parasites of the adipose tissue, it is possible that two or more of these species are conspecific. This information summarized in Table 13.4 is based on total number of microsporidian infections of all three species. During 1972, larvae collected early in the season, May 15, had 16% infection in the higher and 44% in the middle part of the tree crown. Midway in the season, May 25, the prevalence was 30 to 40% in both crown levels, and it reached 100% at the end of the feeding period, May 20. The microsporidia infecting the green tortrix were transovarially transmitted to a large percentage of the offspring of infected adult females. Although no data are available, these microspordia are probably more pathogenic than *N. fumiferanae,* since larval mortality caused by *N. tortricis* was more dramatic than that reported for *N. fumiferanae.* There are many additional examples of microsporidian infections in populations of phytophagous Lepidoptera and Hymenoptera from forests and orchards (46–55).

The microsporidia of forest Lepidoptera and Hymenoptera have a number of similarities. Most are transovarially transmitted, disease prevalence generally varies from season to season, and they are often important naturally occurring regulators of forest pests.

3.2. Pasture, Range, and Forage Insects

Pasture and range crop ecosystems are less stable than the forest ecosystems, but more stable than most other agroecosystems. Crops are harvested or grazed, but plants are usually perennial and fields are only occasionally disturbed by plowing. As expected, many of the protozoan agents infecting insects in these environments are well established and cause periodic epizootics. Epizootics have been reported in several insect species from this group, but information about disease prevalence and host population density is extensive for only two species of microsporidia: *Pleistophora oncoperae* from *Oncopera alboguttata,* a lepi-

TABLE 13.4

Microsporidian Infections in Larvae of the Green Tortrix, *Tortrix viridana,* in Poland

Year	% Infection Caused by Microsporidia	Total No. of Larvae Collected from 5 Trees
1970	12%	1,316
1971	14%	1,386
1972	65%	3,716
1973	—	54
1974	—	72

Source: Modified from Lipa and Madziare-Borusiewicz (45)

dopteran pest of Australian pastures (56–58), and *Nosema locustae,* a widely distributed pathogen of many species of grasshoppers (59–62).

The insect pest *O. alboguttata* is a small moth, the female of which scatters her eggs at random while flying about 1 m above the ground. Young larvae live and feed in litter on the soil surface, but as older larvae they form vertical tunnels in the soil and emerge only at night to feed. *Pleistophora oncoperae* infects primarily muscle tissue, but adipose and connective tissues also are infected. Spores are not present in larval fecal material, making larval death, cannibalism, or a parasitoid vector a prerequisite for horizontal transmission. The rate of transovum transmission was 92% in eggs laid by heavily infected females. Most of the larvae hatching from such eggs died in the early larval stadia, but the disease did not affect fecundity or egg hatch (56). The numbers of adults examined and infected over the period 1972 to 1979 are shown in Figure 13.1. Although the numbers of adults examined do not represent absolute population densities, they do reflect relative population levels. Visible damage to pastures occurred in 1972, '73, '77, and '78 while no adults could be found in '75, '76, and '79. The presence of this microsporidium was thought to be responsible for the population crashes in 1975 and 1979. Interestingly, samples taken throughout the year from three sites in 1972–73 and from one site in 1973–74 did not demonstrate striking increases in infection levels throughout the season nor could disease prevalence be significantly associated with larval population density (56). This was probably because progeny of infected females die as young larvae. The progeny of healthy females develop infections late in the season as older larvae, thus delaying the horizontal transmission effect. Since spores are not present in the feces of infected larvae, this delayed effect is not unexpected.

Microsporidian infections also have been reported from other species of insects forming webs or tunnels under grass sod (63, 64). Representatives of the insect genus *Crambus* have behavioral characteristics similar to *O. alboguttata.* Eggs are scattered over sod by the female, and larvae form silken tunnels under sod. Insects in the genus *Crambus* often have high microsporidian infection levels, and epizootics are frequent. Although most epizootics have not been well documented, the mechanisms influencing these epizootics are probably similar to those affecting microsporidian epizootics in *O. alboguttata.*

Nosema locustae is a pathogen of at least 58 species of Orthoptera and has a wide geographical distribution. Adipose tissue is the principal infection site, and spores are not passed in the feces. Extensive information is available about the epizootiology of naturally occurring infections as well as the production and application of *N. locustae* for the control of rangeland grasshoppers (19, 59–62, 65–67). Although infection levels in naturally occurring epizootics averaged only 2 to 8% from 1963 through 1967, studies of these epizootics were the basis for the development of spore application techniques used for grasshopper control (60). The occurrence of epizootics in grasshopper populations in isolated Idaho rangelands was dependent on three species of grasshoppers, *Oedaleonotus enigma, Melanoplus bivittatus,* and *Melanoplus sanguinipes.* All three

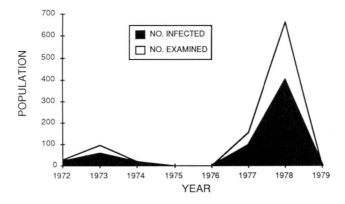

Figure 13.1. The prevalence of *Pleistophora oncoperae* infections in adult *Oncopera alboguttata* from 1970–79. (Adapted from Milner and Lutton, ref. 56.)

species of grasshoppers exhibit cannibalism, which is conducive to the horizontal transmission of *N. locustae* since infections are restricted to adipose tissue. The grasshoppers *O. enigma* and *M. bivittatus* are both early season species, while the third grasshopper, *M. sanguinipes,* occurs later in the summer. Epizootics occur when the early season grasshoppers appear unusually early and acquire *N. locustae* infections, thus producing an inoculum for the predominant summer season species, *M. sanguinipes.* Heavily diseased *M. sanguinipes* produce an inoculum for the next generation of early season grasshoppers. The presence of these three grasshopper species and the timing of their developmental sequences are the most important factors in the occurrence of epizootics. Epizootics tend to occur much more frequently in areas that supported development of all three host species (60).

3.3. Field Crops

Compared to other agroecosystems, there is little environmental stability in most row crop systems. Each year the soil is disturbed by plowing. Harvesting usually includes the partial destruction or modification of the crop residue, and agricultural chemicals are often applied to the crop or to the soil. Epizootics caused by protozoan pathogens frequently occur in populations of some insect pests of row crops but do not generally occur in predictable cycles, and disease agents are not so uniformly distributed as in the insects of more stable habitats. Information about infection levels of protozoan agents is abundant for insect pests of field crops, probably a result of research emphasis.

Most protozoan pathogens of this group of insects do not rely on the survival of extracorporeal stages during periods when the crop is not present, but rather survive in infected living hosts, often in diapausing or estivating stages. The disease is, therefore, present in the host population when the crop is colonized.

This makes horizontal transmission dependent on the initial infection level of the colonizing insect as well as on host behavior and population density.

Most Protozoa infecting this group of insects exhibit transovum or transovarial transmission. Since crops are often colonized by adults that have moved from other fields, vertical transmission may be the only method by which the disease can be introduced into the first generation larval population.

Nosema pyrausta, a pathogen of the European corn borer *Ostrinia nubilalis,* is one of the most important biological mortality agents present in corn borer populations. Compared with most other protozoan diseases, considerable information about host-pathogen relationships, mechanisms of transmission, and prevalence correlated with host population density is available for *N. pyrausta* (68–78).

N. pyrausta is well adapted for its association with the corn borer. Transmission is efficient by both vertical and horizontal means. Infected females lay eggs, which have a high percent infection, on the host plant, usually corn. Healthy females lay over 30 egg masses, each of which contains from 15 to 18 eggs. Infected females produce less than half the fertile eggs produced by a healthy female. The larval corn borers tunnel in the stalk of the cornplant, where horizontal transmission occurs. *N. pyrausta* infects most body tissues, but the Malpighian tubules are infected early in the course of the infection causing spores to be passed in the feces of infected larvae (68,70). Corn stalks inhabited by infected larvae quickly become contaminated with *N. pyrausta* spores, and subsequent larval inhabitants of that stalk also become infected. The frass from corn stalks infested with corn borer larvae falls from larval entrance holes and broken stalks. Some of this frass falls or is wind blown into leaf sheaths of adjacent plants. Young larvae often enter the plant at this point, thus coming in contact with the spores. After hatching, vertically infected larvae may move short distances to adjacent plants providing another means for the pathogen to spread from plant to plant (74). There is also evidence that the parasitoid, *Macrocentrus grandii,* is responsible for some horizontal spread of the disease (79, 80), but the epizootiological importance of this method of transmission is unknown. It is probably most important at low host population densities.

Since the horizontal spread of infection in corn borer populations depends on the probability of healthy larvae inhabiting a corn stalk with infected larvae, the initial infection level of vertically infected larvae and the larval population density are two of the most important variables affecting infection levels in corn borer populations.

Long-term observations of *N. pyrausta* prevalence and the population density of the European corn borer are characterized by three of the four features of population cycles driven by infectious diseases: (1) peak in percent infection occurs shortly after peak in host density, (2) the host population density falls below the threshold necessary for the maintenance of the epizootic, and (3) when the population is below this threshold, prevalence declines significantly (2). *N. pyrausta* is not extremely virulent but causes some mortality in all life stages of the borer. Its greatest effect, however, occurs in transovarially infected larvae, over 60% of which die before pupation. Older larvae, infected *per os* with moderate spore dosages, usually survive to become infected adults and produce infected offspring. All stages, when infected, are more susceptible to mortality caused by stress, such as temperature extremes and crowding (78).

There are other examples of protozoa having long-term associations with field crop insects. The bean leaf beetle *Cerotoma trifurcata* is infected by at least two different species of microsporidia, one of which causes disease in up to 33% of adult beetles and occurs yearly (81, 82). The seasonal prevalence of these microsporidia over a 3-year period gave no clear indication of the factors involved in the development of epizootics, but the differential prevalence of the two microsporidia remained the same. One microsporidian species infected 0 to 12.9% of the hosts while the other species infected 0 to 5.8%.

Nosema heliothidis, a pathogen of the corn earworm *Heliothis zea* occurs regularly in host field populations throughout much of the eastern USA (83– 85). During 1979, adults were collected in light traps at about weekly intervals from mid-May to late October in North Carolina. Prevalence of the disease ranged from 10 to 80% infection in these adults and averaged about 30% (85). The highest prevalence was at the end of the season. Larvae collected from the same location during 1971 had 84.9% infection (N = 106) when collected from corn, 62.3% (N = 61) when collected from tobacco, and 52.8% (N = 53) when collected from soybeans. The differences in prevalence of infection in *H. zea* collected from different crops may have resulted in part from host behavior and the biology of *N. heliothidis.* Larval corn earworms are very cannibalistic. When a larva infests an ear of corn it feeds in the tip of the ear just inside the husk. Adult corn earworms lay single eggs on corn silks, and the newly hatched larvae move into the husk. If the ear is already occupied by a corn earworm, the older earworm usually eats the new invader. If any of the new invaders are infected with *N. heliothidis,* the original invader becomes infected. Because *N. heliothidis* is transovarially transmitted (84, 86), infected females, by laying infected eggs, can spread *N. heliothidis* throughout the earworm population in a manner similar to the spread of *N. pyrausta.* Corn earworms have an increasingly aggregated distribution on soybeans, tobacco, and corn, respectively. This tendency to aggregate more on some crops than on others probably causes more efficient horizontal transmission and thus the differential prevalence. Although *N. heliothidis* is not a virulent microsporidium, it causes sublethal effects that probably have significant effects on the population dynamics of *H. zea* (87).

Most of the protozoa that occur regularly in populations of field crop insect pests are transmitted transovarially. One exception is the microsporidium *Vairimorpha necatrix* which infects larval fat body. It seldom causes epizootics in field crop insects though it will infect a wide range of lepidopteran pests. Epizootics have been reported in the true armyworm *Pseudaletia unipuncta* in Hawaii and Illinois (88–90). Both of these were isolated cases, and there was no pattern of yearly infections or cyclic prevalence. The Illinois epizootic was accompanied by high levels of insect parasitoids, which may have acted as vectors of the microsporidium (90).

3.4. Stored Product Insects

Stored product environments range from ecological islands, such as undisturbed sacks of grain and packages of cereals, to continually changing environments, such as grain storage facilities to which grain is frequently added and removed. In such storage facilities, residual food pockets in inaccessible areas are a source of insects and often a source of insect pathogens for reinfesting newly added grain supplies, and are ideal habitats for the development of epizootics. In areas where grain is constantly being added or removed, epizootics have much less chance of developing.

In contrast to insect pests of forest and row crops, most pest populations in stored grain products have no distinct generations. The infective stages of the pathogen are added to the food medium by infected individuals and are constantly available to all stages of the host. Overwintering and spatial distribution of the hosts are much less important for pathogens of insect pests of stored products.

Stored product pests, especially representatives of the insect orders Lepidoptera and Coleoptera, have many protozoan pathogens, primarily microsporidia, neogregarines, and coccidia. The flour beetles *Tribolium confusum* and *Tribolium castaneum* infected with the microsporidium *Nosema whitei*, the neogregarine *Mattesia dispora,* and the coccidian *Adelina tribolii* have received significant research attention.

The relationships between *N. whitei* and the flour beetles *T. confusum* and *T. castaneum* are similar and have been described in detail by Watson and Milner, respectively (91 – 94). *N. whitei* infects only fat body and is only transmitted *per os.* Transovarial and transovum transmissions do not occur. Spore concentrations greater than 10^7 spores/g of food medium cause 100% mortality from bacterial septicemia before microsporidian spores are produced in host tissue, whereas initial dosage between 10^5 and 10^7 spores/g of media results in 100% infection with spore production. If a first instar *T. confusum* larva is infected and dies as a heavily infected sixth instar, 10^{10} spores/g of larval body weight are produced.

Tribolium confusum and *T. castaneum* have been the subjects of many studies of population dynamics. The major factors affecting the dynamics of healthy populations are crowding and cannibalization of eggs and larvae. The introduction of *N. whitei* into a *T. confusum* population causes a slower growth rate and reduced fecundity of infected individuals. This, in turn, affects crowding and cannibalism, and thus the population dynamics of *T. confusum* (91).

In an undisturbed container of food, an infected *T. confusum* population differs from a healthy population as follows (91).

1. Relative numbers of each larval stadium differ from healthy populations.
2. Cannibalism increases because of this change in relative numbers of larval stadia.

3. There are fewer individuals at each larval stadium because of increased mortality.
4. The population grows more slowly due to reduced fecundity of infected adults.
5. The population peaks at levels well below the carrying capacity of the food supply.
6. The maximum number of spores in the medium is reached at approximately the same time that the medium is inoculated with any spore concentration large enough to produce initial infections.
7. After the spore concentration in the medium exceeds 10^7 spores/g, the host population crashes.

This means that regardless of the initial spore inoculum, if this inoculum is large enough to cause infections, epizootics in confined *T. confusum* populations develop in a similar and predictable manner.

Surveys of *N. whitei* in natural flour beetle populations illustrate its widespread occurrence (95–97). However, the percentage infection at a specific sample site is not particularly valuable epizootiological information unless the detailed history of the site is known; the infection level may be greatly affected by man's manipulation of the stored grain. Naturally occurring epizootics usually do not control grain pest populations until after the economic threshold of the pest has been exceeded.

The studies of Park and Frank (98–100) are regularly cited for their contribution to intraspecific and interspecific competition theory developed through studies of the flour beetles *T. confusum* and *T. castaneum*. Several of these studies included the effect of the coccidian *Adelina tribolii* (99). This pathogen infects both *T. confusum* and *T. castaneum* and is similar to *N. whitei* in the way it affects its hosts. Not unexpectedly, densities of uninfected *T. castaneum* populations stabilized at a level three times higher than densities of infected counterpart cultures. This was determined in a four-year study during which a population census of all stages was taken and fresh food given at monthly intervals. Densities of infected *T. confusum* populations, on the other hand, did not differ significantly from those of their uninfected counterparts. This difference may result from the stronger cannibalistic tendency of *T. castaneum*. The studies of Park and Frank differed from Watson's (91) in that the culture medium was replaced completely every 30 days; thus the spore concentration in the medium never reached the level at which 100% mortality occurred.

The neogregarines *Farinocystis tribolii, Mattesia dispora*, and *Mattessia trogodermae*; the coccidia *Adelina trogoderma* and *Adelina tribolii*; and the microsporidium *Nosema oryzaephili* have all been reported as having both high and low prevalence in populations of stored product insects (95–97, 101–103). These pathogens are similar in being primarily, if not exclusively, parasites of adipose tissue, having only *per os* transmission, and causing high rates of mortality when spore concentrations reach high levels in the medium.

Epizootics caused by these parasites are probably similar to those described for *N. whitei* and *A. tribolii*.

The microsporidia *Vairimorpha plodiae, V. heterosporum,* and *V. invadens* were described from lepidopteran pests of stored products, but no long-term data on infection levels are available (104–106). These microsporidia have transovarial as well as *per os* transmission and infect most body tissues. It is likely that in closed containers epizootics caused by these pathogens would be similar to those caused by the protozoan pathogens of adipose tissue; experiments similar to the 4-year study of Park (99), where the medium is changed monthly, would be very interesting.

Prevalence of *V. plodiae* varied from 0 to 80% infection in the Indian meal moth *Plodia interpunctella* collected two times per year in 1971–72 from grain storage facilities at 14 locations in central Illinois, USA (107). A total of 34 samples were taken from the 14 locations. Only four samples had no infected insects, indicating that *V. plodiae* is widely distributed and causes periodic epizootics in Indian meal moth populations in grain storage facilities in Illinois.

3.5. Parasitoids and Predators

Parasitoids and predators may play important roles as uninfected or infected vectors of protozoan agents, but in this section only those pathogens causing actual infections in parasitoids or predators will be considered. There are few detailed studies of protozoan infections of parasitoids and predators. This is understandable because of the often complex relationships. When infected with a pathogen of the host, the epizootiology of the disease in the parasitoid or predator is closely tied to the epizootiology of the disease in the host. When the disease is exclusively a pathogen of the parasitoid or predator, horizontal transmission is often difficult to evaluate.

Microsporidia have been described from many insect predators, but prevalence rates were given for only a few species. *Pleistophora eretesi,* a pathogen of the dytiscid beetle *Eretes sticticus,* and *Toxoglugea tillargi*, a pathogen of the odonate *Tholymis tillarga,* caused less than 3% infection in India (108, 109). *Microsporidium calopterygis* was found infecting a single nymph of the odonate *Calopteryx virgo* in Czechoslovakia, while near Sarajevo, Yugoslavia, about 50% of nymphal *Calopteryx* sp. were infected (28). Infections were confined to fat body and there was no evidence that infected prey were the source of the infections. The microsporidium *Nosema nepae* infected 12.5% (N = 16) of the hemipteran predator *Nepa cinerea* in Poland (110). *Nosema coccinellae* infected three of 15 species of coccinellids collected in Poland and the Soviet Union. *Coccinella septempunctata,* the most abundant host species, was collected from 18 locations from 1961 through 1965, and diseased insects were found at four locations, ranging from 4 to 30% infection (111). *Nosema hippodamiae* caused a localized epizootic in the convergent ladybird beetle *Hippodamia convergens* in California (112). Adults and pupae had prevalence rates of about 18% infection from two different collection sites, but no larvae were found infected. This microsporidium infected adipose and midgut tissues. The predaceous mecopteran *Hylobittacus apicalis* is commonly infected throughout Illinois with the microsporidium *Nosema apicalis* (113). Infection levels varied from 10 to 100% from 1971 through 1976. In 1981 prevalence was monitored throughout the season. The overall infection level was 29.6%, but interestingly, 41.1% of females were infected but only 14.3% of males. Prevalence tended to increase from 5% on June 15 to 100% on August 4. Since transovarial transmission did not occur and infection was restricted to the fat body, the authors suggested that the infections resulted either from eating infected prey or from an undetermined cannibalistic behavior of the mecopteran predators. The infected prey theory seemed the least likely because small insects collected from the mecopteran habitat were not infected and because infected Mecoptera were consistently found at widely separated collection sites. In addition, the same microsporidium was found in other nonpredatory species of Mecoptera suggesting that epizootics in Mecoptera are the result of behavioral characteristics of the group. This example illustrates the difficulty of determining mechanisms of horizontal transmission in protozoan pathogens of predators. No detailed studies of the protozoan pathogens of predators have fully described how horizontal transmission takes place and how epizootics develop.

Several detailed studies of the interactions of parasitoids, protozoan agents, and insect hosts have been published (79, 80, 114–123). Most of these studies deal with the susceptibility of the parasitoids to the microsporidia infecting the host or with the role of the parasitoid in transmitting microsporidia from host to host. Most of these associations were reviewed by Brooks (122).

There is very little information about infection levels in parasitoids. The meager information available suggests, as one would expect, that disease prevalence in parasitoids is closely tied to that in their hosts. This generalization is, however, affected by the transovarial transmission of some of these microsporidia from infected parasitoid adults to their offspring.

Microsporidia that are pathogens exclusively of the parasitoids and do not infect the parasitoid's host have an entirely different epizootiological picture. These associations are throughly discussed by Brooks et. al. (121–123) but may be summarized as follows. *Campoletis sonorensis,* a parasitoid of the corn earworm *Heliothis zea*, is infected by the microsporidium *Nosema campoletidis* which does not infect *H. zea*. Transovarial transmission in the parasitoid averages about 72%. Horizontal transmission occurs when both infected and healthy larvae parasitize the same corn earworm host. *Nosema heliothidis* is primarily a pathogen of the corn earworm, but also infects the parasitoid, in which it is transovarially transmitted, although prevalence of the disease rapidly decreases with subsequent parasitoid generations. The prevalence of both microsporidia in parasitoid collections made in Clayton, North Carolina during 1970–1971 are shown in Table 13.5. The infection levels of *N. heliothidis* are probably caused by a low level of transovarial transmission, about 5% after the fourth parasite generation, as compared to an infectioni level of 72% for *N.*

TABLE 13.5

Prevalence of *Nosema heliothidis* and *Nosema campoletidis* Infections in Field Collections of Adult *Campoletis sonorensis*, Clayton, North Carolina

Date Collected	No. Collected	No. Infected With	
		N. heliothidis	*N. campoletidis*
June 1, 1970	127	1	10
June 4, 1970	11	0	7
June 11, 1970	37	0	31
June 18, 1970	138	1	5
June 26, 1970	13	0	4
Sept. 2, 1971	19	0	1
Sept. 16, 1971	12	0	3
Sept. 23, 1971	30	1	2

Source: Modified from Brooks and Cranford (121)

campoletidis under identical conditions. Another possible factor is the greater virulence of *N. heliothidis* to parasitoids. Both microsporidia *N. heliothidis* and *N. campoletidis* depend on their host, the corn earworm, for hosizontal transmission. In the case of *N. heliothidis,* the corn earworm host is an active participant since the earworm is infected by *N. heliothidis,* which is transmitted in the earworm population both *per os* and transovarially. Conversely, the corn earworm is a passive participant for *N. campoletidis,* since the earworm is not infected by this microsporidium. In this case the earworm host is analogous to any other food source having the chance of being occupied by healthy and/or infected insects.

Protozoan pathogens of both predators and parasitoids pose some fascinating epizootiological questions, not only because they have such complex interrelationships, but also because as vectors of disease agents they often play an important role in the epizootiology of the pathogens of the host.

3.6. Social Insects

Our concepts of horizontal transmission, density dependence, and threshold host density as major factors in the development of epizootics must be modified when evaluating diseases of social insects. Characteristics of the insect society itself, such as division of labor, rearing the young, and cleaning the colony, greatly influence infection levels. Since the colony is in many ways like an individual organism, percentage infection and the development of epizootics may be considered either within or among colonies.

Not unexpectedly, *Nosema apis,* a disease of the honeybee *Apis mellifera* has received more research attention than all other protozoan diseases of social insects combined. *Nosema apis,* described in 1909, infects primarily the midgut epithelium of adult honeybees (9,124). Larval honeybees are not infected, nor do infected adults develop from bees fed spores as larvae. Few, if any, other insects are infected by *N. apis.* Infected worker bees live only half as long as healthy bees, and queens die within a few weeks after becoming infected. Honey production and hive vigor are significantly reduced. Spores are present in the feces of infected adults, and only horizontal transmission is known. *Nosema apis* is found worldwide, although infection levels differ geographically (9, 28, 124, 125).

Numerous observations throughout the world have shown that prevalence, whether expressed both as percent colonies infected or percent infected bees per infected colony, has a consistent seasonal variation, with the number of infections being highest in the spring and decreasing during the summer (9, 124–129). Bailey's (125) explanation for this seasonal variation is that bees defecate on the combs more frequently in late winter because of the accumulation of rectal contents during their winter confinement in the hive. The rise in percentage infections occurs in the spring when brood-rearing begins and bees are mainly confined to the cluster. Infected bees continue to contaminate the combs with feces, promoting horizontal transmission when healthy worker bees clean contaminated combs. This usually occurs as the brood expands. Queens do not clean combs and, therefore, seldom acquire infections (9, 125).

In summer, bees defecate outside the hive, and infected bees die without transmitting the disease to other bees. Bees continue to clean the hive, and newly emerged bees are not infected, thus the number of infections decreases and occasionally declines to zero in vigorous colonies. Vigorous bee colonies acquiring *N. apis* infections naturally or experimentally during spring or summer months often exhibit minimal detrimental effects. But similar colonies experimentally infected in the fall often fail to survive the winter.

Any beekeeping operation that results in bees defecating inside the hive or that interrupts cleaning operations of the hive increases the probability of *N. apis* infections and promotes the development of epizootics. Colonies that are transported, especially long distances, have a greater chance of developing *N. apis* epizootics, probably because confined bees become dysenteric. Colonies in urban areas generally have higher rates of *N. apis* infections, presumably because they have poorer foraging and are moved more often than their rural counterparts. Queens do become infected, although much less frequently than workers. When this happens the queen dies and the colony usually does not survive.

The amoeba *Malpighamoeba mellificae* causes "amoeba disease" in adult bees (9). Only adult bees are infected and oral ingestion of cysts is the only method of transmission. The amoebae live in Malpighian tubules and are passed in the feces of infected bees. *Malpighamoeba mellificae* infections are largely chronic and have a minimal effect on both individual bees and colonies. The seasonal infection levels of *M. mellificae* are very similar to those of *N. apis*; prevalence is highest in the spring and declines during summer months.

Protozoan pathogens also cause epizootics in ants and termites. Prevalence rates were determined monthly (1967–1969) for the microsporidium *Gurleya spraguei*, a parasite of the termite *Macrotermes estherae* (130). The infection levels for termites collected in Andhra Pradesh, India, are given in Figure 13.2. *Gurleya spraguei*, unlike *N. apis* and *M. mellificae*, infects only adipose tissue and thus is not passed in fecal material. The pathology of the microsporidium in individual termites or termite colonies was not reported. The social activities of the termite colonies probably account for the seasonal fluctuations in prevalence. The seasonal infection levels were similar to those reported for *N. apis* and *M. mellificae* in honeybee adults except that the highest levels occurred in winter rather than spring. Prevalence declined sharply during the summer and began to increase in late summer.

At least seven different species of microsporidia and several neogregarines have been described from ants. Although prevalence rates have not been recorded for any protozoan pathogen over several seasons, numerous examinations of ants have produced interesting information on intracolonial and intercolonial infection levels, mostly for the fire ants *(Solenopsis)* (131–134).

The microsporidium *Thelohania solenopsae* has been recovered from at least 12 species of *Solenopsis* from Uruguay and Argentina (134). It infects

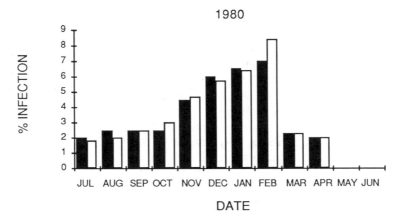

Figure 13.2. The prevalence of the microsporidium *Gurleya spraguei* in the termite *Macrotermes estherae*, Andhra Pradesh, India, 1967–1969. (Adapted from Kalavati, ref. 130.)

ovaries of queens and adipose tissue of workers and sexuals (135). It is presumably transmitted transovarially but cannot be transmitted by feeding spores to ants. It produces chronic pathogenic effects, but infected colonies do not live as long as healthy colonies under laboratory conditions. Sixty-seven of 865 colonies of fire ants (7.8%) collected around Matto Grosso, Brazil were infected with this microsporidium. Two distinct types of spores were produced, but their significance is unknown (134).

Another microsporidium, *Burenella dimorpha,* from *Solenopsis* spp. also produces two types of spores (134, 136). The binucleate spores are infective *per os* but the uninucleate octospores are not. Infections are fatal in the pupal stage. Infected pupae often are cannibalized by adult ants, but the adults do not ingest spores because of a filter mechanism. The spores, contained in an infrabuccal pellet, are fed to fourth instar larvae which is the only stage that can be infected because younger larvae receive only liquid food (134).

Infected ants were found in only 3.9% of 307 colonies of *S. geminata* in Florida, although a small sample from a different location in Florida indicated infection levels of over 40% (133). Intracolonial prevalence was usually below 5% infection but in some colonies occasionally reached 100%.

Another interesting feature of this microsporidium is its persistence in colonies of other ant species. *Burenella dimorpha* is easily transmitted *per os* to the fire ant species *S. invicta, S. richteri,* and *S. xyloni* (136), but once established, infections do not persist in laboratory colonies of these species (134).

The neogregarine *Mattesia geminata,* a pathogen of *S. geminata* in Florida, usually has intercolonial prevalence of less than 2% infection, but greater than 90% was found in three of 25 infected colonies (134). About 20% of the colonies collected from one site over a 2-year period were infected, whereas only one of 307 colonies from 74 other sites in Florida and Georgia was infected.

The epizootiology of the protozoan diseases of social insects is perhaps the most complex of all protozoan-insect associations. The social behavior of the insect colony, probably more than any other factor, influences infection levels and the development of epizootics. *Nosema apis*, one of the most thoroughly investigated protozoan pathogens, infects insects in distinct seasonal patterns of prevalence as a result of the social behavior of the bee colony. When this seasonal pattern changes it is usually because of behavioral changes in the bee colony. The other protozoan pathogens of social insects have not been studied as thoroughly as *N. apis*. While the mechanisms responsible for seasonal fluctuations of microsporidian infections in Indian termites are unknown, they are probably related to behavioral characteristics of the termite colony.

Although infection levels throughout a single season are not available for protozoan diseases of ants, the variable infection levels both within and among colonies strongly suggest that both intracolonial and intercolonial infection levels are dynamic.

3.7. Aquatic Insects

There is a voluminous literature on protozoan pathogens of aquatic insects. Many species of protozoan pathogens have been described from aquatic insects, and information about infection levels and geographical distribution is available for many of these pathogens. However, the environmental ingredients necessary for the development of epizootics are unknown for most of them, especially the microsporidia. This is because many of the microsporidia of aquatic insects have a complex life cycle with two or more developmental sequences. In one sequence, diploid, binucleate spores are formed in the oenocytes of adult females, while in another sequence haploid, membrane-bound octospores are formed usually in adipose tissue of larvae. Transovarial transmission is known for all species of microsporidia in this group, and the larval adipose infections resulting from transovarial transmission are usually fatal, producing large numbers of spores. However, ingestion of haploid larval spores by other larvae normally does not produce larval infections.

Recent studies have demonstrated the presence of an alternate copepod host for some species of *Amblyospora* (137, 138). Spores produced by a third developmental cycle in the alternate host were infectious when fed directly to mosquito larvae. If this phenomenon is widespread among the microsporidia of aquatic insects it could explain their confusing epizootiological nature.

For the purposes of this review, the protozoan pathogens of aquatic insects will be divided into three groups: (1) microsporidia with two developmental cycles and transovarial transmission, (2) microsporidia that are readily transmitted *per os* and usually have transovum transmission, and (3) gregarines having only *per os* transmission.

The first group of microsporidia contains at least eight genera and 40 described species plus many reported, but undescribed, species (24, 139, 140). Representatives of two larger genera, *Parathelohania* and *Amblyospora*, have been reported from at least 60 mosquito species, and several excellent references deal with classification, life cycles, and biology of this group (139–145). There are many reports of these microsporidia infecting field-collected larvae, probably because heavily infected larvae are milky-white and easily distinguished from healthy larvae. Field-collected larvae, representing many mosquito species in most areas of the world, seldom have infection levels exceeding 5%; most have 1% or less (146–150). True epizootics are extremely rare, as emphasized by the observation that while thousands of mosquito collections have been examined for the presence of this group of microsporidia there are fewer than 10 reports of infection levels of 50% or higher. Although *per os* transmission has been experimentally proven for only a few species, it has been demonstrated that without horizontal transmission, infections would soon disappear from mosquito populations, because vertical transmission alone cannot sustain them (151). Andreadis (152) monitored an epizootic of *Amblyospora* sp. infecting *Aedes cantator* in pools of coastal salt-marshes in Connecticut, USA during 1982. Of 2203 larvae collected from pools scattered throughout this area, 98.1% were infected. This infection level was constant throughout larval development from the first through the fourth larval stadia, indicating that horizontal transmission did not occur, and suggesting that the high percentage of infection was the result of transovarial infection from infected females. Horizontal transmission must have occurred extensively in some preceding generation in order to produce such high prevalence. The alternate copepod host demonstrated for some species of *Amblyospora* could explain these results since horizontal transmission could occur only when the alternate host was present. Epizootics, although rare, obviously occur with this group of microsporidia, but all of the factors responsible for producing them are not known.

The discovery of an alternate host (137, 138) for some species of *Amblyospora* offers some exciting possibilities for research on the epizootiology of this group of microsporidia.

Microsporidia such as *Nosema algerae* and *Vavraia culicis* have both *per os* and transovum transmission (153–155). Transovarial transmission is not known to occur. *Nosema algerae* and *V. culicis* frequently cause epizootics in laboratory colonies of mosquitoes. Most laboratory larval-rearing facilities are ideal for horizontal transmission, and unless eggs are washed to prevent transovum transmission, infections quickly spread throughout the colony (156).

Nosema algerae infects a wide range of mosquito hosts and probably has a wide geographical distribution, but epizootics in field populations are rare (2, 143, 147, 157). Results of field tests with *N. algerae* in Panama and in Pakistan may partially explain why epizootics are uncommon. In both Pakistan (Maddox unpublished) and Panama (158), open pool tests with *N. algerae* spores resulted in moderate prevalence of 27% infection in *Anopheles stephensi* larvae at one week after treatment. However, the infection levels dropped steadily each week thereafter and in Panama fell to less than 1% at 67 days after treatment. In Pakistan, treated areas had no infected mosquitoes the season following treatment. Results of a laboratory bioassay conducted in Petri dishes containing water only or water plus bottom soil showed median infection concentrations 40 times lower in dishes containing water only (159). *Nosema algerae* spores have a high sedimentation rate and quickly settle to the bottom where they become inaccessible in the soil. *Nosema algerae* and similar mosquito pathogens, therefore, must have low horizontal transmission coefficients in most mosquito habitats. In laboratory colonies of mosquitoes, larval rearing trays are normally shallow with smooth bottoms, making horizontal transmission much more efficient. In nature, *N. algerae,* with its extensive host range and transovum transmission, is probably introduced into many habitats by infected females but becomes established with low levels of infection in only a few of them. Microsporidian epizootics have been reported from other groups of aquatic insects (24, 28): Ephemeroptera (159), Trichoptera, and the dipteran families Simuliidae (160–163) and Chironomidae (164, 165). Most of the microsporidia from these hosts have not been extensively studied, but from the information available it appears that many of these microsporidia are similar to the dimorphic or polymorphic microsporidia from mosquitoes. Infection levels are usually low and horizontal transmission often cannot be demonstrated in the laboratory.

The gregarine *Lankesteria culicis* is a widely distributed pathogen of the intestinal tracts of mosquito larvae (166, 167). Only *per os* transmission is known to occur. It causes chronic infections with minimal pathogenic effects on its mosquito hosts. Most surveys for occurrence and infection levels of *L. culicis* have involved *Aedes aegypti* in small containers such as tires and tree holes. Both the percentage of sites with infected hosts and the percentage of infected mosquito larvae in such sites were high. In an area of western Samoa 80% of 45 tree holes contained infected larvae, and 40% of the individual mosquito larvae in gregarine-infested holes were infected (168). In the southeastern US 54% of all sites examined contained infected larvae and 60% of larvae from gregarine-infested sites were infected (169). The percentage of sites with gregarines increased throughout the season in containers monitored three times. An epizootic probably occurs whenever a sufficient number of spores is introduced into a breeding container, but because *L. culicis* is not very virulent, the mosquito population does not decline significantly. The most important factor in the epizootiology of this pathogen is the degree of transmission between breeding containers.

4. CONCLUSIONS

From an epizootiological point of view the protozoa are one of the most interesting and diverse groups of insect pathogens. There are many unresolved problems, some of which pose serious difficulties. There are also many unexplored areas of research ranging from insect control to basic epizootiological theory.

4.1. Problem Areas

One of the most serious problems in protozoan epizootiology is the proper identification of the pathogen. This is especially true for the Microsporida because of the large number of undescribed and poorly described species. Electron microscopy is often necessary for identification and it is impossible to make ultrastructural preparations for each insect examined in the course of an epizootiological study. Thus several species of microsporidia may actually be present in what appears to be an epizootic caused by a single species. There are also many misidentifications of the protozoa causing epizootics in insect populations.

The life cycles of protozoa as they relate to epizootiology, host-pathogen relationships, and transmission mechanisms have been studied for a very small percentage of protozoan insect pathogens. This is especially evident in many of the protozoa of aquatic insects for which portions of the life cycles and the factors responsible for horizontal transmission are unknown. Since the horizontal transmission method often cannot be identified, it is frequently impossible to develop an epizootiological scenario for this group of pathogens or to evaluate all of the factors contributing to the development of an epizootic.

Unlike many of the fungal, viral, and nematode diseases for which infection often means death, many of the protozoan diseases are chronic or benign. Therefore, determining just what impact a protozoan infection will have on an insect population is difficult. We use percentages of the population infected, but we seldom know how these infections ultimately affect the insect population. The sublethal effects of Protozoa are often overlooked as important factors in the dynamics of insect populations (87). A growing body of evidence now suggests that pathogens that are very efficiently transmitted and have a moderate degree of virulence can be effective biological control agents (27).

Most observations of protozoan infections in insect populations are not accompanied by data on the population density of the host, a problem shared with many other groups of insect diseases.

4.2. Suggested Research Areas

The tremendously wide range of associations of protozoan diseases in insects affords many opportunities for epizootiological studies. Many such studies have implications extending far beyond insect-disease associations and into areas of basic epidemiological theory.

Chronic infections of insects are common among the protozoa and have not been extensively studied because they were previously thought to have little effect on the dynamics of insect populations. A thorough study of some of these chronic protozoan diseases, especially in combination with stress mechanisms such as crowding, nutritional deficiencies, or temperature extremes, would be valuable. Chronic diseases as predisposing factors for the development of more virulent pathogens also should be investigated.

There are many associations in which parasitoids, hyperparasitoids, and/or predators are involved with the epizootiology of protozoan diseases, especially the microsporidia. Only a few of these associations have been examined, and they have revealed some fascinating interactions. Not only are protozoan diseases often vectored in these associations, but in addition, the population dynamics and the reproductive strategies of many parasitoids and hyperparasitoids are probably greatly influenced by the presence of these protozoan diseases. This could be a very fertile area of research.

The epizootiology of the diseases of social insects offers the possibility of integrating studies of insect pathology and insect behavior. The long-term associations of protozoan diseases with social insects have probably acted as selection pressures on social behavior. Studies on social behavior of insects with and without long-term associations with specific protozoan pathogens could lend interesting insights into how these pathogens may have affected the evolution of social behavior.

There are depressingly few long-term studies of protozoan epizootiology. It would be invaluable to have more studies conducted over several years including infection levels, host population densities, and age-specific life tables. Most of the available data about infection levels and host population densities were collected during outbreaks of the host when prevalence was high. These studies should be continued during times of low host density and low prevalence. The host population threshold density necessary for the spread of the disease is unknown for most protozoan pathogens, and how protozoan diseases are maintained in host populations at low host densities is a topic for which virtually no information is available. There are opportunities for comparative epizootiological studies of protozoan pathogens of an insect host occupying natural ecosystems as opposed to different types of agroecosystems. The same protozoan pathogen in different hosts and different pathogens in the same host often exhibit markedly different epizootiological characteristics. Studies addressing the reasons for these differences would be invaluable in interpreting epizootiological data as well as in revealing cause and effect relationships between the biological characteristics of pathogen and host and the dynamic characteristics of the epizootic.

Finally, we should consider the evolutionary strategies of the protozoan parasites themselves. It is tempting to list "good" and "bad" characteristics of protozoan pathogens relative to effectiveness in regulating insect populations or to the magnitude and regularity of epizootics. Methods of transmission, number of infective agents produced, generation time, virulence, and extracorporeal longevity of infective agents are characteristics whose relative importance is dictated by the characteristics of the host insect and the ecosystem in which the host lives. One possible approach is to consider the attributes of protozoan pathogens as they relate to the r and K-selection strategies proposed by Dobzhansky (170) and MacArthur and Wilson (171). Most studies of r and K strategies have concerned macro rather than microorganisms, but several workers have applied this concept to parasites and to protozoa. An excellent example of this type of approach is that of Andrews and Rouse (172) who developed a conceptional model showing how plant pathogens could be positioned along an r-K continuum based on relative commitment of resources to maintenance, growth, and reproduction. Similar evaluations of protozoan entomopathogens could give us additional insights about the epizootiology of this very successful group of organisms.

REFERENCES

1. N. D. Levine, J. O. Corliss, F. E. G. Cox, G. Deroux, J. Grain, B. M. Honigberg, G. F. Leedale, A. R. Loeblich, III, J. Lom, J. L. D. Lynn, E. G. Merinfeld, F. C. Page, G. Poljansky, V. Sprague, J. Vavra, F. G. Wallace, A newly revised classification of the Protozoa, *J. Protozool.*, **27**, 37 (1980).

2. E. U. Canning, An evaluation of protozoal characteristics in relation to biological control of insects, *Parasitol.*, **84**, 119 (1982).

3. W. A. Evans and R. G. Elias, The life cycle of *Malamoeba locustae* (King et Taylor) in *Locusta migratoria* (R. et F.), *Acta Protozool.*, **7**, 229 (1970).

4. J. E. Henry, *Malamoeba locustae* and its antibiotic control in grasshoppers, *J. Invertebr. Pathol.*, **11**, 224 (1968).

5. S. A. Hanrahan, Ultrastructure of *Malamoeba locustae* (K & T), a protozoan parasite of locusts, *Acrida*, **4**, 235 (1975).

6. H. E. Prinsloo, Parsitiese mikro-organismes by die bruinsprinkaan *Locustana pardalina* (Walk.), *S.-Afr. Tydskr. Landbouwalinsk*, **3**, 551 (1960).

7. A. Lea, Recent outbreaks of the brown locust, *Locustana pardalina* (Walk.) with special references to the influence of rainfall, *J. Entomol. Soc. S. Africa*, **21**, 18 (1958).

8. H. Prell, Beitrage zur Kenntnis der Amobenseuche der erwachsenen Honigbiene, *Arch. Bienenkd.*, **7**, 113 (1926).

9. L. Bailey, *Infectious Diseases of the Honey-Bee*, Land Books, London, 1963.

10. M. H. Hassanein, Some studies on amoeba disease, *Bee World*, **33**, 109 (1952).

11. K. Purrini and Z. Žižka, More on the life cycle of *Malamoeba scolyti* (Amoebidae:Sacromastigophora) parasitizing the bark beetle *Dryocoetes autographus* (Scolytidae, Coleoptera), *J. Invertebr. Pathol.*, **42**, 96 (1983).

12. W. M. Brooks, "Protozoan Infections," in G. Cantwell, Ed., *Insect Diseases*, Vol. 1, Marcel Dekker, New York, 1974, p. 237.

13. F. G. Wallace, "Biology of the Kinetoplastida of Arthropods," in W. H. R. Lumsden and D. A. Evans, Eds., *Biology of Kinetoplastida*, Vol. 2, 1979, p. 213.

14. J. O. Corliss and D. W. Coats, A new cuticular cyst-producing tetrahymenid ciliate, *Lambornella clarki* n. sp. and the current status of ciliatosis in culicine mosquitoes, *Trans. Am. Microscop. Soc.*, **95**, 725 (1976).

15. W. R. Kellen, W. Wills, and J. E. Lindegren, Ciliatosis in *Aedes sierrensis* (Ludlow), *J. Insect Pathol.*, **3**, 335 (1961).

16. H. C. Chapman, D. B. Woodward, and J. J. Petersen, Pathogens and parasites in Louisiana Culicidae and Chaoboridae, *Proc. N. J. Mosq. Exterm. Assoc.*, **54**, 54 (1967).

17. R. D. Sanders, Microbial mortality factors in *Aedes sierrensis* populations, *Proc. Calif. Mosq. Control Assoc.*, **40**, 66 (1972).

18. T. B. Clark and D. G. Brandl, Observations on the infection of *Aedes sierrensis* by a tetrahymenine ciliate, *J. Invertebr. Pathol.*, **28**, 341 (1976).

19. J. E. Henry, Natural and applied control of insects by Protozoa, *Annu. Rev. Entomol.,* **26,** 49 (1981).

20. J. Weiser, "Sporozoan Infections," in E. A. Steinhaus, Ed., *Insect Pathology: An Advanced Treatise,* Vol. 2, Academic Press, New York, 1963, p. 291.

21. E. U. Canning, "Microsporida," in J. P. Kreier, Ed., *Parasitic Protozoa,* Vol. 4, Academic Press, New York, 1977, p. 155.

22. J. Vavra, "Development of the Microsporidia," in L. A. Bulla and T. C. Cheng, Eds., *Comparative Pathobiology,* Vol. 1, *Biology of the Microsporidia,* Plenum Press, New York, 1976, p. 87.

23. E. I. Hazard, E. A. Ellis, and D. J. Joslyn, "Identification of Microsporidia," in H. D. Burges, Ed., *Microbial Control of Pests and Plant Diseases 1970–1980,* Academic Press, London, 1981, p. 163.

24. V. Sprague, "Classification and Phylogeny of the Microsporidia," in L. A. Bulla and T. C. Cheng, Eds., *Comparative Pathobiology, Vol. 2, Systematics of the Microsporidia,* Plenum Press, New York, 1977, p. 1.

25. H. Blunck, R. Krieg, and E. Scholtyseck, Weitere untersuchungen uber die Mikrosporidien von Pieriden und deren Parasiten und Hyperparasiten, *Z. Pflantzenkranh.,* **66,** 129 (1959).

26. A. M. Tanabe and M. Tamashiro, The biology and pathogenicity of a microsporidian *(Nosema trichoplusiae* sp. n.) of the cabbage looper, *Trichoplusia ni* (Hubner) (Lepidoptera:Noctuidae), *J. Invertebr. Pathol.,* **9,** 188 (1967).

27. R. M. Anderson, Theoretical basis for the use of pathogens as biological control agents of pest species, *Parasitology,* **84,** 3 (1982).

28. J. Weiser, Die Mikrosporidien als Parasiten der Insekten, *Monogr. Angew. Entomol.,* **17,** 149 (1961).

29. J. J. Lipa, *Stempellia scolyti* (Weiser) comb. nov. and *Nosema scolyti* sp. n. microsporidian parasites of four species of *Scolytus* (Coleoptera), *Acta Protozool.,* **6,** 70 (1968).

30. T. H. Atkinson and R. C. Wilkinson, Microsporidian and nematode incidence in live-trapped and reared southern pine beetle adults, *Fla. Entomol.,* **62,** 169 (1979).

31. D. A. Streett, V. Sprague, and D. M. Harman, Brief study of microsporidian pathogens in the white pine weevil *Pissodes strobi, Chesapeake Sci.,* **16,** 32 (1975).

32. G. G. Wilson, "*Nosema fumiferanae,* a Natural Pathogen of a Forest Pest: Potential for Pest Management," in H. D. Burges, Ed., *Microbial Control of Pests and Plant Diseases 1970–1980,* Academic Press, London, 1981, p. 595.

33. M. M. Neilson, "Disease and the Spruce Budworm," in R. F. Morris, Ed., *The Dynamics of Epidemic Spruce Budworm Populations,* Mem. Entomol. Soc. Can., 1963, p. 272.

34. G. G. Wilson, The effects of feeding microsporidian, *Nosema fumiferanae,* spores to naturally infected spruce budworm *Choristoneura fumiferana, Can. J. Zool.,* **55,** 249 (1977).

35. H. M. Thomson, The effect of a microsporidian parasite on the development, reproduction, and mortality of the spruce budworm *Choristoneura fumiferana* (Clem.), *Can. J. Zool.,* **36,** 499 (1958).

36. H. M. Thomson, Some aspects of the epidemiology of a microsporidian parasite of the spruce budworm *Choristoneura fumiferana* Clem., *Can. J. Zool.* **36,** 309 (1958).

37. H. M. Thomson, Microsporidian disease of spruce budworm, *Can. Dep. Agric. Bi-mon. Prog. Rep.,* **11,** 2 (1955).

38. M. M. Neilson, Disease in spruce budworm adults, *Can. Dep. Agric. Bi-mon. Prog. Rep.,* **12,** 1 (1956).

39. H. M. Thomson, The possible control of a budworm infestation by a microsporidian disease, *Can. Dep. Agric. Bi-mon. Prog. Rep.,* **16,** 1 (1960).

40. G. G. Wilson, Incidence of microsporidia in a field population of a spruce budworm, *Can. For. Serv. Bi-mon. Res. Notes,* **29,** 35 (1973).

41. G. G. Wilson and W. J. Kaupp, Incidence of *Nosema fumiferanae* in spruce budworm *Choristoneura fumiferana* in the year following application, *Can. For. Serv. Bi-mon. Res. Notes,* **32,** 32 (1976).

42. G. G. Wilson, Microsporidian infection in spruce budworm *Choristoneura fumiferana* 1 and 2 years after application, *Can. For. Serv. Bi-mon. Res. Notes,* **34,** 16 (1978).

43. G. G. Wilson, Observations on the incidence rates of *Nosema fumiferanae* (Microsporidia) in a spruce budworm *Choristoneura fumiferana* (Lepidoptera:Tortricidae) population, *Proc. Entomol. Soc. Ontario,* **108,** 144 (1977).

44. J. M. Franz and A. M. Huger, Microsporidia causing the collapse of an outbreak of the green tortrix *Tortrix viridana* L. in Germany, *Proc. Int. Colloq. Insect Pathol., 4th,* College Park, MD, p. 48 (1971).

45. J. J. Lipa and K. Madziara-Borusiewicz, Microsporidians parasitizing the green tortrix *Tortrix viridana* L. in Poland and their role in the collapse of the tortrix outbreak in Puszcza Niepolomicka during 1970–1974, *Acta Protozool.,* **15,** 529 (1976).

46. J. J. Lipa and M. E. Martignoni, *Nosema phryganidiae* n. sp., a microsporidian parasite of *Phryganidia californica* Packard, *J. Insect Pathol.,* **2,** 396 (1960).

47. J. Weiser, Protozoan diseases of the gypsy moth, Prog. in Protozool., *Proc. 1st Int. Congr. Protozool,* Prague, Academia, 497 (1961).

48. W. A. Smirnoff, *Thelohania pristiphorae* sp. n., a microsporidian parasite of the larch sawfly *Pristiphora erichsonii* (Hymenoptera:Tenthredinidae), *J. Invertebr. Pathol.,* **8,** 360 (1966).

49. G. L. Nordin, R. G. Rennels, and J. V. Maddox, Parasites and pathogens of the fall webworm in Illinois, *Environ. Entomol.,* **1,** 351 (1972).

50. C. Sidor, Diseases provoked with microorganisms by some Lymantriidae in Yugoslavia and their importance for entomofauna, *Arch. Biol. Nauka,* Beograd, **28,** 127 (1976).

51. M. T. E. C. Cabral, Papel des doencas na limitacao natural das populacoes de *Lymantria dispar* L. (Lepidoptera:Lymantriidae), *An. Inst. Super. Agron., Lisboa,* **37,** 153 (1977).

52. P. A. Simchuk and A. I. Sikura, The microsporidian *Nosema carpocapsae* Paillot and its importance as a regulator of the codling moth, *Laspeyresia pomonella* (Lepidoptera, Tortricidae), *Entomol. Obozr.,* **57,** 495 (1978).

53. L. M. Zelinskaya, Role of microsporidia in the abundance dynamics of the gypsy moth *Porthetria dispar* in forest plantings along the lower Dnepr River (Ukrainian Republic, USSR), *Vestn. Zool.,* **1,** 57 (1980).

54. G. G. Wilson, The potential of *Pleistophora schubergi* in microbial control of forest insects, Can. For. Serv. Sault Ste. Marie, *For. Pest Manag. Inst. Rep. FPM-X-49,* 7 pp (1981).

55. J. J. Lipa, Epizootics observed in laboratory rearing and in field population of *Lobesia botrana* Den et Schiff. (Lepidoptera, Torticidae) caused by *Plistophora legeri* (Paillot) (Microsporidia), *Bull. Acad. Pol. Sci.,* **29,** 311 (1982).

56. R. J. Milner and G. G. Lutton, Interactions between *Oncopera alboguttata* (Lepidoptera:Hepialidae) and its microsporidian pathogen *Pleistophora oncoperae* (Protozoa:Microsporida), *J. Invertebr. Pathol.,* **36,** 198 (1980).

57. R. J. Milner, The role of disease during an outbreak of *Oncopera alboguttata* Tindale and *O. rufobrunnea* Tindale, (Lepidoptera: Hepialidae) in the Ebor/Dorrigo region of N.S.W., *J. Aust. Entomol. Soc.,* **16,** 21 (1977).

58. R. J. Milner and C. D. Beaton, *Pleistophora oncoperae* sp. n. (Protozoa:Microsporida) from *Oncopera alboguttata* (Lepidoptera:Hepialidae) in Australia, *J. Invertebr. Pathol.,* **29,** 133 (1977).

59. E. U. Canning, The pathogenicity of *Nosema locustae* Canning, *J. Insect Pathol.,* **4,** 248 (1962).

60. J. E. Henry, Epizootiology of infections by *Nosema locustae* Canning (Microsporidia:Nosematidae) in grasshoppers, *Acrida,* **1,** 111 (1972).

61. A. B. Ewen, Extension of the geographic range of *Nosema locustae* (Microsporidia) in grass-hoppers (Orthoptera:Acrididae), *Can. Entomol.,* **115,** 1049 (1983).

62. J. E. Henry and E. A. Oma, "Pest Control by *Nosema Locustae,* a Pathogen of Grasshoppers and Crickets," in H. D. Burges, Ed., *Microbial Control of Pests and Plant Diseases 1970– 1980,* Academic Press, London, 1981, p. 571.

63. I. M. Hall, Studies of microorganisms pathogenic to the sod webworm, *Hilgardia,* **22,** 535 (1954).

64. A. C. Banarjee, Microsporidian diseases of sod webworms in bluegrass lawns, *Ann. Entomol. Soc. Am.,* **61,** 544 (1968).

65. J. E. Henry, K. Tiahrt, and E. A. Oma, Importance of timing, spore concentrations, and levels of spore carrier in applications of *Nosema locustae* (Microsporida:Nosematidae) for control of grasshoppers, *J. Invertebr. Pathol.,* **21,** 263 (1973).

66. J. E. Henry, Experimental application of *Nosema locustae* for control of grasshoppers, *J. Invertebr. Pathol.,* **18,** 389 (1971).

67. J. E. Henry and E. A. Oma, Effects of infections by *Nosema locustae* Canning, *Nosema acridophagus* Henry, and *Nosema cuneatum* Henry (Microsporidia:Nosematidae) in *Melanoplus bivittatus* (Say) (Orthoptera:Acrididae), *Acrida,* **3,** 223 (1974).

68. H. L. Zimmack and T. A. Brindley, The effect of the protozoan parasite *Perezia pyraustae* Paillot on the European corn borer, *J. Econ. Entomol.,* **50,** 637 (1957).

69. J. P. Kramer, Observations on the seasonal incidence of microsporidiosis in European corn borer populations in Illinois, *Entomophaga,* **9,** 37 (1959).

70. J. P. Kramer, Some relationships between *Perezia pyraustae* Paillot (Sporozoa, Nosematidae), and *Pyrausta nubilalis* (Hubner) (Lepidoptera, Pyralidae), *J. Insect Pathol.,* **1,** 25 (1959).

71. R. S. VanDenburgh and P. P. Burbutis, The host-parasite relationship of the European corn borer *Ostrinia nubilalis* and the protozoan *Perezia pyrausta* in Delaware, *J. Econ. Entomol.,* **55,** 65 (1962).

72. F. B. Peairs and J. H. Lilly, *Nosema pyrausta* in populations of the European corn borer, *Ostrinia nubilalis* in Massachusetts, *Environ. Entomol.,* **3,** 878 (1974).

73. M. B. Windels, H. C. Chiang, and B. Furgala, Effects of *Nosema pyrausta* on pupa and adult stages of the European corn borer, *Ostrinia nubilalis, J. Invertebr. Pathol.,* **27,** 239 (1976).

74. L. C. Lewis, Migration of larvae of *Ostrinia nubilalis* (Lepidoptera:Pyralidae) infected with *Nosema pyrausta* (Microsporida:Nosematidae) and subsequent dissemination of this microsporidium, *Can. Entomol.,* **110,** 897 (1978).

75. R. E. Hill and W. J. Gary, Effects of the microsporidium *Nosema pyrausta* on field populations of European corn borers in Nebraska, *Environ. Entomol.,* **8,** 91 (1979).

76. L. C. Lewis, Persistance of *Nosema pyrausta* and *Vairimorpha necatrix* measured by microsporidiosis in the European corn borer, *J. Econ. Entomol.,* **75,** 670 (1982).

77. T. G. Andreadis, Epizootiology of *Nosema pyrausta* in field populations of the European corn borer (Lepidoptera:Pyralidae), *Environ. Entomol.,* **13,** 882 (1984).

78. J. P. Siegel, *The Epizootiology of Nosema pyrausta (Paillot) in the European Corn Borer Ostrinia nubilalis,* Ph.D. Thesis, Univ. Illinois, 1985.

79. T. G. Andreadis, Impact of *Nosema pyrausta* on field populations of *Macrocentrus grandii,* an introduced parasite of the European corn borer *Ostrinia nubilalis, J. Invertebr. Pathol.,* **39,** 298 (1982).

80. J. P. Siegel, J. V. Maddox, and W. G. Ruesink, Impact of *Nosema pyrausta* on the brachonid *Macrocentrus grandii* in central Illinois, *J. Invertebr. Pathol.,* **47,** 271 (1986).

81. W. M. Brooks, D. B. Montross, R. K. Sprenkel, and G. Carner, Microsporidioses of coleopterous pests of soybeans, *J. Invertebr. Pathol.,* **35,** 93 (1980).

82. P. G. Marrone, W. M. Brooks, and R. E. Stinner, The incidence of tachinid parasites and

pathogens in adult populations of the bean leaf beetle *Cerotoma trifurcata* (Forster) (Coleoptera:Chrysomelidae) in North Carolina, *J. Georgia Entomol. Soc.,* **18**, 363 (1983).

83. J. P. Kramer, On *Nosema heliothidis* Lutz and Splendor, a microsporidian parasite of *Heliothis zea* (Boddie) and *Heliothis virescens* (Fabricius) (Lepidoptera, Phalaenidae), *J. Insect Pathol.,* **1**, 297 (1959).

84. J. J. Lipa, Some observations on *Nosema heliothidis* Lutz and Splendor, a microsporidian parasite of *Heliothis zea* (Boddie) (Lepidoptera, Noctuidae), *Acta Protozool,* **6**, 273 (1968).

85. W. M. Brooks and J. D. Cranford, Host-parasite relationships of *Nosema heliothidis* Lutz and Splendor, *Misc. Publ. Entomol. Soc. Am.,* **11**, 51 (1978).

86. W. M. Brooks, Transovarial transmission of *Nosema heliothidis* in the corn earworm, *Heliothis zea, J. Invertebr. Pathol.,* **11**, 510 (1968).

87. R. R. Gaugler and W. M. Brooks, Sublethal effects of infection by *Nosema heliothidis* in the corn earworm, *Heliothis zea, J. Invertebr. Pathol.,* **26**, 57 (1975).

88. Y. Tanada and G. Y. Chang, An epizootic resulting from a microsporidian and two virus infections in the armyworm, *Pseudaletia unipuncta* (Haworth), *J. Invertebr. Pathol.,* **4**, 128 (1962).

89. Y. Tanada, Incidence of microsporidiosis in field populations of the armyworm *Pseudaletia unipuncta* (Haworth), *Proc. Hawaii. Entomol. Soc.,* **18**, 435 (1964).

90. J. V. Maddox, *Studies on a Microsporidiosis of the Armyworm, Pseudaletia unipuncta (Haworth),* Ph.D. Thesis, Univ. Illinois (1966).

91. P. L. Watson, *The Biology and Computer Simulation of the Population Dynamics of Tribolium confusum (Order Coleoptera, Family Tenebrionidae) with an Introduced Pathogen, Nosema whitei (Order Microsporidia, Family Nosematidae),* Ph.D. Thesis, Univ. Illinois (1979).

92. R. J. Milner, *Nosema whitei* Weiser, a microsporidian pathogen of some species of *Tribolium.* I. Morphology, life cycle, and generation time, *J. Invertebr. Pathol.,* **19**, 231 (1972).

93. R. J. Milner, *Nosema whitei,* a microsporidian pathogen of some species of *Tribolium.* III. Effect on *T. castaneum, J. Invertebr. Pathol.,* **19**, 248 (1972).

94. R. J. Milner, *Nosema whitei,* a microsporidian pathogen of some species of *Tribolium.* IV. The effect of temperature, humidity and larval age on pathogenicity for *T. castaneum, Entomophaga,* **18**, 305 (1973).

95. H. D. Burges, Enzootic diseases of insects, *Ann. N.Y. Acad. Sci.,* **217**, 31 (1973).

96. H. D. Burges and J. Weiser, Occurence of pathogens of the flour beetle, *Tribolium castaneum, J. Invertebr. Pathol.,* **22**, 464 (1977).

97. K. Purrini, Studies on microsporidian pathogens of insects living in mills in Yugoslavia, *Anz. Schadlingskd. Pflanz. Umweltschutz,* **48**, 104 (1976).

98. T. Park, Observations on the general biology of the flour beetle, *Tribolium confusum* Duval, *Q. Rev. Biol.,* **9**, 36 (1934).

99. T. Park, Experimental studies of interspecies competition. 1. Competition between populations of the flour beetles, *Tribolium confusum* Duval and *Tribolium castaneum* Herbst, *Ecol. Monogr.,* **18**, 267 (1984).

100. T. Park and M. B. Frank, The fecundity and development of the flour beetles *Tribolium confusum* and *Tribolium castaneum* at 3 constant temperatures, *Ecology,* **29**, 368 (1948).

101. I. M. Hall, F. D. Stewart, K. Y. Arakawa, and R. G. Strong, Protozoan parasites of species of *Trogoderma* in California, *J. Invertebr. Pathol.,* **18**, 252 (1971).

102. K. Purrini, *Adelina tribolii* Bhatia and *A. mesnili* Perez (Sporozoa, Coccidia) infecting insect pests of stores in Kosova region, Yugoslavia, *Anz. Schadlingskd. Pflanz. Umweltschutz,* **49**, 51 (1976).

103. F. O. Marzke and R. J. Dicke, Disease-producing protozoa in species of *Trogoderma, J. Econ. Entomol.,* **51**, 916 (1958).

104. W. R. Kellen and J. E. Lindegren, Biology of *Nosema plodiae* sp. n., a microsporidian pathogen of the Indian-meal moth *Plodia interpunctella* (Hubner) (Lepidoptera:Phycitidae), *J. Invertebr. Pathol.,* **11,** 104 (1968).

105. W. R. Kellen and J. E. Lindegren, Host-pathogen relationships of two previously undescribed midrosporidia from the Indian-meal moth, *J. Invertebr. Pathol.,* **14,** 328 (1969).

106. W. R. Kellen and J. E. Lindegren, *Nosema invadens* sp. n. (Microsporidia:Nosematidae), a pathogen causing inflammatory response in Lepidoptera, *J. Invertebr. Pathol.,* **21,** 293 (1971).

107. R. K. Sprenkel, *Studies on Nosema plodiae (Sporozoa, Microsporida) and Mattesia dispora (Sporozoa, Gregarina) Pathogens of the Indian Meal Moth, Plodia interpunctella (Lepidoptera, Pyralidae) in Illinois,* Ph.D. Thesis, Univ. Illinois, Urbana-Champaign (1973).

108. C. Kalavati and C. C. Narasimhamurti, A new microsporidian *Pleistophora eretesi* n. sp. from *Eretes sticticus* (L.) (Dytiscidae, Coleoptera), *Acta Protozool.,* **15,** 139 (1976).

109. C. Kalavati and C. C. Narasimhamurti, A new microsporidian parasite *Toxoglugea tillargi* sp. n. from an odonate *Tholymis tillarga, Acta Protozool.,* **17,** 279 (1978).

110. J. J. Lipa, Miscellaneous observations on protozoan infections of *Nepa cinerea* Linnaeus including descriptions of two previously unknown species of microsporidia, *Nosema bialoviesianae* sp. n. and *Thelohania nepae* sp. n., *J. Invertebr. Pathol.,* **8,** 158 (1966).

111. J. J. Lipa, *Nosema coccinellae* sp. n., a new microsporidian parasite of *Coccinella septempunctata, Hippodamia tredecimpunctata,* and *Myrrha octodecimguttata, Acta Protozool.,* **5,** 369 (1968).

112. J. J. Lipa and E. A. Steinhaus, *Nosema hippodamiae* n. sp., a microsporidian parasite of *Hippodamia convergens* Guerin (Coleoptera, Coccinellidae), *J. Insect Pathol.,* **1,** 304 (1959).

113. J. V. Maddox and D. W. Webb, A new species of *Nosema* from *Hylobittacus apicalis* (Insecta:Mecoptera:Bittacidae), *J. Invertebr. Pathol.,* **42,** 207 (1983).

114. H. Blunck, Mikrosporidien bei *Pieris brassicae* L., ihren Parasiten und Hyperparasiten, *Z. Angew. Entomol.,* **36,** 316 (1954).

115. H. W. Allen, *Nosema* disease of *Gnorimoschema operculella* (Zeller) and *Macrocentrus ancylivorus* Rohwer, *Ann. Entomol. Soc. Am.,* **47,** 407 (1954).

116. Y. Tanada, Field observations on a microsporidian parasite of *Pieris rapae* (L.), *Proc. Hawaii. Entomol. Soc.,* **15,** 609 (1955).

117. F. M. Laigo and M. Tamashiro, Interactions between a microsporidian pathogen of the lawn-armyworm and the hymenopterous parasite *Apanteles marginiventris, J. Invertebr. Pathol.,* **9,** 546 (1967).

118. J. J. Lipa, Observations on development and pathogenicity of the parasite of *Aporia crataegi* L. (Lepidoptera) *Nosema aporiae* n. sp., *Acta Parasitol. Pol.,* **5,** 559 (1957).

119. I. V. Issi and V. A. Masdennikova, The role of the parasite *Apanteles glomeratus* L. (Hymenoptera, Braconidae) in transmission of *Nosema polyvora* Blunck (Protozoa, Microsporidia), *Entomol. Obozr.,* **45,** 494 (1966).

120. Z. Hostounsky, *Nosema mesnili* (Paill.), a microsporidian of the cabbageworm *Pieris brassicae* (L.) in the parasites *Apanteles glomeratus (L.), Hyposoter ebenius* (Grav.) and *Pimpla instigator* (F.), *Acta Entomol. Bohemoslov.,* **67,** 1 (1970).

121. W. M. Brooks and J. D. Cranford, Microsporidoses of the hymenopterous parasites *Campoletis sonorensis* and *Cardiochiles nigriceps,* larval parasites of *Heliothis* species, *J. Invertebr. Pathol.,* **20,** 77 (1972).

122. W. M. Brooks, Protozoa: host-parasite-pathogen interrelationships, *Misc. Publ. Entomol. Soc. Am.,* **9,** 105 (1973).

123. J. N. McNeil and W. M. Brooks, Interactions of the hyperparasitoids *Catolaccus aeneoviridis* (Hym:Pteromalidae) and *Spilochalcis side* (Hym:Chalcididae) with the microsporidans *Nosema heliothidis* and *N. campoletidis, Entomophaga,* **19,** 195 (1974).

124. G. F. White, Nosmea disease, *U.S. Dep. Agric. Bull. 780*, 59 pp (1919).

125. L. Bailey, The epidemiology and control of nosema disease of the honey bee, *Ann. Appl. Biol.*, **32**, 379 (1955).

126. K. M. Doull, A theory of the causes of development of epizootics of nosema disease of the honey bee, *J. Insect Pathol.*, **3**, 297 (1961).

127. K. M. Doull and K. M. Cellier, A survey of the incidence of nosema disease *(Nosema apis* Zander) of the honey bee in South Australia, *J. Insect Pathol.*, **3**, 280 (1961).

128. F. E. Moeller, *Nosema Disease—Its Control in Honey Bee Colonies*, U.S. Dep. Agric. Tech. Bull. No. 1569, 16 pp (1978).

129. E. R. Jaycox, Surveys for nosema disease of honey bees in California, *J. Econ. Entomol.*, **53**, 95 (1960).

130. C. Kalavati, Four new species of microsporidians from termites, *Acta Protozool.*, **15**, 411 (1976).

131. G. E. Allen and W. F. Buren, Microsporidian and fungal diseases of *Solenopsis invicta* Buren in Brazil, *J. N. Y. Entomol. Soc.*, **82**, 125 (1974).

132. G. E. Allen and A. Silviera-Guido, Occurrence of microsporidia in *Solenopsis ricteri* and *Solenopsis* sp. in Uruguay and Argentina, *Fla. Entomol.*, **57**, 327 (1974).

133. D. P. Jouvenaz, G. E. Allen, W. A. Banks, and D. P. Wojcik, A survey for pathogens of fire ants *Solenopsis* spp. in the southeastern United States, *Fla. Entomol.*, **60**, 275 (1977).

134. D. P. Jouvenaz, Natural enemies of fire ants, *Fla. Entomol.*, **66**, 111 (1983).

135. J. D. Knell, G. E. Allen, and E. I. Hazard, Light and electron microscope study of *Thelohania solenopsae* n. sp. (Microsporida:Protozoa) in the red imported fire ant, *Solenopsis invicta*, *J. Intertebr. Pathol.*, **29**, 192 (1977).

136. D. P. Jouvenaz and E. I. Hazard, New family, genus and species of Microsporida (Protozoa:Microsporida) from the tropical fire ant, *Solenopsis geminata* (Fabricius) (Insecta:Formicidae), *J. Protozool.*, **25**, 24 (1978).

137. T. G. Andreadis, Experimental transmission of a microsporidian pathogen from mosquitoes to an alternate copepod host (*Amblyospora/Aedes cantator/Acanthocyclops vernalis*/ultrastructure), *Proc. Natl. Acad. Sci. Biol. Sci. Population Biol.* (in press).

138. A. W. Sweeny, E. I. Hazard, and M. F. Graham, Intermediate host for an *Amblyospora* sp. (Microspora) infecting the mosquito *Culex annulirostris*, *J. Invertebr. Pathol.*, **46**, 98 (1985).

139. E. I. Hazard and S. W. Oldacre, *Revision of Microsporida (Protozoa) Close to Thelohania with Descriptions of One New Family, Eight New Genera and Thirteen New Species*, U.S. Dep. Agric. Tech. Bull. 1530, 104 pp (1975).

140. J. Weiser, Contribution to the classification of Microsporidia, *Vestn. Cesk. Spol. Zool.*, **61**, 308 (1977).

141. E. I. Hazard and D. W. Anthony, *A Redescription of the Genus Parathelohania Codreanu 1966 (Microsporida:Protozoa) with a Reexamination of Previously Described Species of Thelohania Henneguy 1892 and Descriptions of Two New Species of Parathelohania from Anopheline Mosquitoes*, U.S. Dep. Agric. Tech. Bull. 1505, 26 pp (1974).

142. T. G. Andreadis, Life cycle and epizootiology of *Amblyospora* sp. (Microspora:Amblyosporidae) in the mosquito, *Aedes cantator*, *J. Protozool.*, **30**, 509 (1983).

143. T. Tsai, A. W. Grundmann, and D. M. Rees, Parasites of mosquitoes in southwestern Wyoming and northern Utah, *Mosq. News*, **29**, 102 (1969).

144. E. I. Hazard and J. W. Brookbank, Karyogamy and meiosis in an *Amblyospora* sp. (Microspora) in the mosquito *Culex salinarius*, *J. Invertebr. Pathol.*, **44**, 3 (1984).

145. T. G. Andreadis and D. W. Hall, Development, ultrastructure, and mode of transmission of *Amblyospora* sp. (Microspora) in the mosquito, *J. Protozool.*, **26**, 444 (1979).

146. H. C. Chapman, Biological control of mosquito larvae, *Annu. Rev. Entomol.*, **9**, 33 (1974).

147. H. C. Chapman, D. B. Woodard, and J. J. Peterson, Pathogens and parasites in Louisiana Culicidae and Chaoboridae, *Proc. N.J. Mosq. Exterm. Assoc.*, **54**, 54 (1967).

148. D. W. Jenkins, Pathogens, parasites and predators of medically important arthropods, Bull. WHO, 30, 150 pp (1964).

149. J. F. Anderson, Microsporidia parasitizing mosquitoes collected in Connecticut, *J. Invertebr. Pathol.*, **11**, 440 (1968).

150. J. M. Castillo, "Microsporidian Pathogens of Culicidae (mosquitoes)," in D. W. Roberts and J. M. Castillo, Eds., *Bibliography on Pathogens of Medically Important Arthropods: 1980*, Bull. WHO, 58, 1980, p. 33.

151. T. G. Andreadis and D. W. Hall, Significance of transovarial infections of *Amblyospora* sp. (Microspora: Thelohaniidae) in relation to parasite maintenance in the mosquito *Culex salinarius, J. Invertebr. Pathol.*, **34**, 152 (1979).

152. T. G. Andreadis, An epizootic *Amblyospora* sp. (Microspora: Amblyosporidae) in field populations of the mosquito, *Aedes cantator, J. Invertebr. Pathol.*, **42**, 427 (1983).

153. E. U. Canning, *Plistophora culicis* Weiser: its development in *Anopheles gambiae, Trans. Roy. Soc. Trop. Med. Hyg.*, **51**, 8 (1957).

154. E. I. Hazard and C. S. Lofgren, Tissue specificity and systematics of a *Nosema* in some species of *Aedes, Anopheles,* and *Culex, J. Invertebr. Pathol.*, **18**, 16 (1971).

155. A. H. Undeen and N. E. Alger, The effect of the microsporidian *Nosema algerae* on *Anopheles stephensi, J. Invertebr. Pathol.*, **25**, 19 (1975).

156. N. E. Alger and A. H. Undeen, The control of a microsporidian, *Nosema* sp., in an anophelene colony by an egg-rinsing technique, *J. Invertebr. Pathol.*, **15**, 321 (1970).

157. J. V. Maddox, N. E. Alger, A. Ahmad, and M. Aslamkhan, The susceptibility of some Pakistan mosquitoes to *Nosema algerae* (Microsporidia), *Pak. J. Zool.*, **9**, 19 (1977).

158. D. W. Anthony, K. E. Savage, E. I. Hazard, S. W. Avery, M.D. Boston, and S. W. Oldacre, Field tests with *Nosema algerae* Vavra and Undeen (Microsporida, Nosematidae) against *Anopheles albimanus* Weidemann in Panama, *Misc. Publ. Entomol. Soc. Am.*, **11**, 17 (1975).

159. E. Fahy, A recurrent epizootic of a microsporidian parasite in a mayfly population, *Entomol. Gaz.*, **26**, 146 (1975).

160. J. Maurand, Les microsporidies des larves de Simulies: systematique, donnies cytochimiques, pathologiques et ecologiques, *Ann. Parasitol. Hum. Comp.*, **50**, 371 (1975).

161. M. Laird, M. Colbo, J. Finney, J. Makry, and A. Undeen, "Pathogens of Simuliidae (Blackflies)," in D. W. Roberts and J. M. Castillo, Eds., *Bibliography on Pathogens of Medically Important Arthropods: 1980*, Bull. WHO, 58, 1980, p. 105.

162. J. Vavra and A. H. Undeen, Microsporidia (Microspora: Microsporida) from Newfoundland blackflies (Diptera:Simuliidae), *Can. J. Zool.*, **59**, 1431 (1981).

163. H. A. Jammback, *Caudospora* and *Weiseria*, two genera of microsporidia parasitic in blackflies, *J. Invertebr. Pathol.*, **16**, 3 (1970).

164. W. Hilsenhoff and O. L. Lovett, Infection of *Chironomus plumosis* (Diptera:Chironomidae) by a microsporidian *(Thelohania* sp.) in Lake Winnebago, Wisconsin, *J. Invertebr. Pathol.*, **8**, 512 (1966).

165. D. K. Hunter, Response of populations of *Chironomus californicus* to a microsporidian *(Gurleya* sp.), *J. Invertebr. Pathol.*, **10**, 387 (1968).

166. E. M. McCray Jr., R. W. Fay, and H. F. Schoof, The bionomics of *Lankesteria culicis* and *Aedes aegypti, J. Invertebr. Pathol.*, **16**, 42 (1970).

167. R. D. Walsh and J. K. Olson, Observations on the susceptibility of certain culicine mosquito species to infection by *Lankesteria culicis* (Ross), *Mosq. News*, **36**, 154 (1976).

168. J. S. Pillai, H. J. C. Neill, and P. F. Stone, *Lankesteria culicis,* a gregarine parasite of *Aedes polynesiensis* in western Samoa, *Mosq. News*, **36**, 150 (1976).

169. G. R. Hayes Jr. and L. E. Haverfield, Distribution and density of *Aedes aegypti* (L.) and *Lankesteria culicis* (Ross) in Louisiana and adjoining areas, *Mosq. News,* **31,** 28 (1971).

170. T. Dobzhansky, Evolution in the Tropics, *Am. Sci.,* **38,** 209 (1950).

171. R. H. MacArthur and E. O. Wilson, *The Theory of Island Biogeography,* Princeton Univ. Press, Princeton, N.J., 1967.

172. J. H. Andrews and D. I. Rouse, Plant pathogens and the theory of r- and K-selection, *Am. Nat.,* **120,** 283 (1982).

14

DISEASES CAUSED
BY NEMATODES

HARRY K. KAYA

Division of Nematology
Department of Entomology
University of California
Davis, California

---

1. INTRODUCTION

Certain nematodes are facultative and obligate endoparasites of many insect species. They were first recorded from insects in the seventeenth century (1), and by 1975, more than 3000 nematode-insect relationships had been recorded (2). During the past decade, interest in entomogenous nematodes has increased, particularly in their use as biological control agents. This interest has been demonstrated by numerous review articles (3–10), books (11–13), and comprehensive bibliographies (14, 15).

Nematode parasitism of insects may result in sterility, reduced fecundity, delayed development, aberrant behavior, or host death. These effects may play a significant role in regulating insect populations. Although numerous reports on the prevalence of nematode parasitism in a given insect population have been published, long-term epizootiological studies are generally lacking. This chapter will focus on the current knowledge of epizootiology of nematode diseases.

2. LIFE CYCLES OF ENTOMOGENOUS NEMATODES

Inasmuch as most entomologists and many insect pathologists are not familiar with the diverse life cycles of the various types of nematodes, the life cycles of the families Mermithidae, Steinernematidae, Heterorhabditidae, Allantonematidae, and Sphaerulariidae are briefly outlined. This information is important in developing and understanding epizootiological principles in relation to these organisms.

Most entomogenous nematodes have a life cycle that includes the egg, four immature larval (often referred to as juveniles in entomological literature) stages, and the dioecious adults. Often molt occurs within the egg and the second stage emerges. Life cycles vary from simple to complex.

2.1. Mermithidae

Mermithid nematodes are obligate invertebrate parasites, mostly of aquatic and terrestrial insects. The newly emerged infective second stage nematode directly penetrates the cuticle of its host, develops rapidly, and exits as a fourth stage postparasite, killing its host in the process. The postparasite molts to the adult stage and mates. Most mermithids have this typical life cycle including *Romanomermis culicivorax (Reesimermis nielseni),* a parasite of mosquitoes (16), and *Hexamermis arvalis,* a parasite of alfalfa weevil (17).

Variations in the general mermithid life cycle occur with *Mermis nigrescens* (18) and *Pheromermis pachysoma* (19). Eggs of *M. nigrescens* are ingested by its orthopteran host, and the second stage nematodes emerge in the gut and penetrate into the host's hemocoel. The remainder of the life cycle in the host follows the typical mermithid pattern. *P. pachysoma* infects paratenic hosts such as mosquitoes, caddisflies, and crane flies, and encysts as a second stage nematode in various tissues. Vespids prey on adults of these insects and feed them to their larvae. The mermithids penetrate into the larval hemocoel and resume development, eventually emerging as postparasites from adult vespids seeking water. The postparasites molt, mate, and lay eggs which are ingested by the aquatic larval stages of the paratenic hosts.

Parthenogenesis is common in mermithids. Although males occur, *M. nigrescens* can reproduce parthenogenetically (20). *Filipjevimermis leipsandra,* a parasite of *Diabrotica* spp., reproduces parthenogenetically (11).

2.2. Steinernematidae and Heterorhabditidae

Steinernematids and heterorhabditids are mutualistically associated with bacteria in the genus *Xenorhabdus* (11). The bacteria occur in the intestinal lumen of the infective third stage nematode. This nonfeeding stage is ensheathed in the second stage cuticle and is often called the "dauer" stage.

Steinernematids have a simple life cycle. Host finding by dauers can be an active process in response to physical and chemical cues (21–24). Dauers enter suitable hosts via natural body openings (mouth, anus, or spiracles), exsheath, and enter into the hemocoel (11). The nematodes release the bacteria which kill the host by septicemia within 48 hours. Feeding upon bacterial cells and host tissues, the nematodes develop rapidly. As resources are depleted, the progeny of the second and third generations develops into infectives which exit from the cadaver and seek new hosts.

Heterorhabditids have a life cycle similar to steinernematids. Infectives enter their hosts as described above (11) or penetrate directly through the cuticle (25). Unlike steinernematids, which are dioecious, heterorhabditid infectives are hermaphroditic (11), and one nematode per host is sufficient for reproduction. The progeny of the hermaphrodite is dioecious and produces the new hermaphroditic dauers.

Steinernematids and heterorhabditids have broad host ranges and kill their hosts rapidly (10, 11). Thus, interest in them as biological control agents is great, and in vitro methods of mass production have been developed (26).

2.3. Neotylenchidae

Few species in the family Neotylenchidae parasitize insects. *Deladenus siricidicola* parasitizes the woodwasp, *Sirex noctilio,* an important pest of *Pinus radiata* in Australia and New Zealand. It is a facultative parasite and can establish successive generations on the fungus *Amylostereum areolatum,* which is symbiotically associated with the woodwasp.

The infective stage of *Deladenus* is the mated adult female which penetrates directly through the cuticle of the woodwasp larva (27). The nematode develops in the host's hemocoel. Toward the end of pupation, the female nematode produces several thousand progenies which invade the ovaries, and the female *Sirex* becomes sterile. The infected-female *Sirex* adult oviposits eggs, each containing 50 to 100 nematodes, along with spores of the symbiotic fungus into the pine tree. The pathogenic fungus kills the tree; the nematodes feed on the fungal hyphae, molt, and reproduce. Because healthy *Sirex* females oviposit in the same tree, hosts are available for the nematode to infect. When the juvenile nematodes are developing near a *Sirex* larva, instead of becoming mycetophagous, egg-laying adults, they transform into the infective female form, mate, and infect their host. Infected male woodwasps are "dead-end" hosts because there is no natural mechanism for releasing the nematodes into the environment.

2.4. Sphaerulariidae and Allantonematidae

Nematodes in these families are obligate parasites of insects and have similar life cycles. The mated female penetrates the cuticle of its larval host (2, 11, 28). In most cases, the nematode does not produce progeny until the host is an adult. The adult insects have reduced fecundity or are sterilized by the nematodes. The progeny exits through the host's reproductive tract or intestine. The nematodes molt, mate in the host's environment, and seek new hosts. A representative of the Sphaerulariidae is *Sphaerularia bombi,* a parasite of bumble bees (*Bombus* spp.). Representative nematodes in the Allantonematidae include *Contortylenchus* spp., parasites of bark beetles, and *Howardula* spp., parasites of certain Diptera, Coleoptera, and Thysanoptera.

Allantonematids in the genus *Heterotylenchus* have a complex life cycle (29). The life cycle of *H. autumnalis,* a parasite of the face fly, *Musca autumnalis,* is typical for this genus. Except for a brief free-living stage, this nematode occurs entirely in its host, alternating between gamogenetic and parthenogenetic stages. Infected face fly females "nemaposit" male and female nematodes on fresh cow dung. The mated female (gamogenetic) nematodes infect healthy face-fly larvae by penetrating directly through the integument into the hemocoel. After the infected face flies become adults, the gamogenetic female lays eggs which develop into parthenogenetic females. These females lay eggs producing the second generation gamogenetic male and female progenies which invade the host's ovaries resulting in sterility. Rather than laying eggs, the female flies deposit nematodes. Infected male flies are "dead-end" hosts.

3. EPIZOOTIOLOGY

Epizootiological principles of infectious diseases in insects have been discussed in detail previously (see Chs. 4–7). These principles also apply to nematode diseases. Because epizootiology has not been adequately studied, much of the information in this chapter is derived from laboratory or controlled field studies. Therefore, some caution is necessary because conclusions or statements about epizootiology may not always be applicable to natural populations.

Key factors involved in the cause, initiation, and development of epizootics are the host population (see Ch. 4), the pathogen population (see Ch. 5), transmission (see Ch. 6), and the abiotic and biotic environments (see Ch. 7). Although they are considered separately, these factors interact and overlap considerably.

3.1. Host Population

The susceptibility of insects to nematode infection varies with different larval instars and developmental stages. Mosquitoes in the first and second instars are more susceptible to *Romanomermis culicivorax* infection than third or fourth instars (30). The thicker cuticle of older instars apparently serves as a barrier to infection. The infective stage of some mermithids must reach the ganglion or become melanized and encapsulated by the host's defenses (31, 32). When *Filipjevimermis leipsandra* infects first or second instar *Diabrotica* spp., it is usually successful in reaching the ganglion (31), but infection of the third instar results in encapsulation by the host's hemocytes, and the nematode is killed (33).

Older larvae of blackflies (34) and mosquitoes (35) are more susceptible to *Steinernema feltiae (Neoaplectana carpocapsae)* than younger ones. The resistance to infection by these insects is the result of physical exclusion or injury to the nematode caused by the larval mouthparts.

Lepidopterous prepupae and adults are more susceptible to *S. feltiae* than pupae (36, 37). Certain dipterous larvae and adults are more susceptible to steinernematids and heterorhabditids than pupae that occur in the puparia (38, 39). In *Scolytus multistriatus*, mature larvae, pupae, and adults are susceptible to steinernematids and heterorhabditids although infection time of pupae takes longer than that of larvae or adults (40).

Insect behavior affects the susceptibility of the host. Active mosquito larvae are less susceptible than sluggish insects to *R. culicivorax,* apparently because the nematode has greater difficulty in attaching to its host (41, 42). High survival of first instar artichoke plume moths exposed to *S. feltiae* is believed to be in part due to their feeding on peripheral leaves where the microclimate is less favorable for nematode survival (43).

The ability of *R. culicivorax* to infect mosquitoes varies with the host species (44). By 1979, 87 species from 13 genera were tested for their susceptibility to this mermithid. Most *Anopheles* spp. are susceptible, but one species is highly refractory and another is intermediate in susceptibility. Similarly, the genera *Aedes, Culex,* and *Psorophora* have at least one species that is refractory to *R. culicivorax.* These mosquitoes resist infection by encapsulation and melanization of the nematode.

Mosquito larvae are susceptible to infection by steinernematids (45 – 47) and heterorhabditids (48), but the initial nematodes to penetrate into the hemocoel are usually melanized and encapsulated. Mosquito larvae with only a few encapsulated nematodes can survive to adulthood (45).

Heterotylenchus spp. infect a number of muscid flies, but larvae of the housefly, *Musca domestica,* are refractory to *H. autumnalis* (49). The nematode penetrates into the larval hemocoel but is encapsulated by hemocytes.

3.2. Pathogen Population

The properties of the nematode population most significant in epizootiology are (1) virulence and infectivity, (2) capacity to survive (persistence), and (3) capacity to disperse (dispersal).

3.2.1. VIRULENCE AND INFECTIVITY.   Virulence and infectivity have been studied primarily with facultative parasites. Species and strains of steinernematids, heterorhabditids (50 to 53), and neotylenchids (27) with differences in virulence and infectivity have been isolated from various insect species and geographical locations. For example, the New Zealand strain of *H. heliothidis* is more virulent to the sheep blowfly, *Lucilia cuprina,* than the Tasmanian or North Carolina strain (50). However, an undescribed *Heterorhabditis* sp. is the most virulent against the blowfly while *S. bibionis* is the least effective. In another study, the Breton strain of *S. feltiae* is more infective to larvae of the rat flea, *Ctenocephalides felis,* than the DD-136 strain (52).

Bedding et al. (51) examined interspecific and intraspecific differences in infectivity of heterorhabditids and steinernematids to insects. All species/strains of these nematodes killed the tested insect species, but no one species/strain of nematode was the most infective for all insects. *S. bibionis* field-isolated from *L. cuprina* or from *Otiorhynchus sulcatus* was the least infective for these insects. *H. bacteriophora* isolated from *Heliothis punctiger* was least infective to its host. Just because a nematode strain is isolated from an insect species does not mean that it will be the most infective to that species. Thus, continued association with the same insect species may reduce virulence rather than enhance it (see Ch. 5).

Extensive testing determined the best strain of *D. siricidicola* for biological control of *S. noctilio.* Strains were collected from Europe, Japan, and New Zealand. Some strains did not sterilize the host and were eliminated from consideration. The strain selected to initiate an epizootic in a *Sirex* population was one that did not significantly affect the size of the emerging woodwasp and gave the highest levels of parasitism.

Although *R. culicivorax* has been studied extensively, there have been very few studies of its virulence or infectivity. The infectivity to mosquito larvae decreases 24 hours after nematode hatching and is virtually nonexistent after 72 hours even though 50% of the infective nematodes are still swimming (54).

3.2.2. PERSISTENCE. Entomogenous nematodes have various methods of surviving in the environment. Heterorhabditids and steinernematids have the dauer stage. Dauers of steinernematids can survive for long periods (up to several years) in water storage at 4 to 10°C (55, 56), but dauers of heterorhabditids do not survive as long (56). The free-living stages of *Sphaerularia bombi* can survive for 9 months on agar plates or in the refrigerator for several years (57).

Dauers of *S. feltiae* can persist for at least 6 months in sterilized soil in the laboratory and can survive the winter in soil at the latitude of St. Paul, Minnesota (58). They can persist for 3 months in crevices on apple tree trunks in Virginia (59) but die rapidly on foliage without adequate moisture (10, 11).

Mermithids persist in several ways such as in paratenic hosts (19), as long-lived adults (18), or as eggs (60, 61). For example, *M. nigrescens* adults are long-lived and the eggs can be stored for 15 months at 5°C without loss of viability (60).

The free-living mycetophagous stage of the neotylenchid, *D. siricidicola*, persists in the environment (27). This nematode has been maintained in the free-living stage for more than 12 years and nearly 200 generations with no loss of parasitic capability.

The infected host can serve as a means for nematode persistence. Thus, *H. autumnalis* overwinters in its diapausing face fly host (62), and *Howardula dominicki* overwinters in the tobacco flea beetle, *Epitrix hirtipennis* (63). These obligate parasites have synchronized their life cycles with their hosts allowing them to persist in their respective host populations.

3.2.3. DISPERSAL. Nematodes are unique among insect pathogens because they are motile and have sensory organs (amphids) which assist them in host finding. Consequently, they can actively seek their hosts. In addition, they are dispersed by physical factors such as wind and water or by their insect hosts.

Steinernematids and heterorhabditids can disperse in soil. *S. feltiae* moves by gliding, bridging, and leaping (64). It can leap up to 10 mm across a substrate. In moist sandy soil, the majority of *S. feltiae* infectives may remain at the point of placement although a few infectives disperse up to 14 cm laterally and 12 cm vertically (65, 66). *S. glaseri* behaves similarly (67). Heterorhabditids have a greater tendency to disperse downwards in sandy loam soil than steinernematids (68).

Soil type affects nematode dispersal (66, 67). Steinernematids move further in sand than in soil containing clay and silt. The presence of a host in the soil increases the number of dispersing infectives indicating that they can orient to kairomones in the soil.

Infectives of some terrestrial mermithids actively disperse from the soil onto plants where their hosts occur. *Agamermis decaudata* (69) and *Hexamermis arvalis* (17) disperse onto plants to contact their host. *M. nigrescens* adults also move from the soil onto plants to deposit eggs which are consumed by the grasshopper host (18).

Infectives of aquatic mermithids actively disperse in water to contact their hosts. Mosquitoes parasitized by *R. culicivorax* have been found up to 12 m from the point of application of 1500 postparasites (70). Moreover, other nematodes can be dispersed passively by water or by wind (71).

Host dispersal is important for many entomogenous nematodes. Because many entomogenous nematodes develop in their hosts without killing them, the adult insects serve as the primary means of dispersal. Allantonematids, sphaerulariids, and neotylenchids infecting bark beetles (28), woodwasps (27), bumble bees (72), face flies and phorids (29), and flea beetles (73) are dispersed in this fashion. *S. glaseri* has been found in the alimentary canal of Japanese beetle adults which are believed to disperse this nematode (74). Parasitoids of *Sirex* spp. infected with *Deladenus* spp. also can disperse the nematode into the hosts' environment (27). In essence, the infected insect disseminates the nematode into the environment where healthy hosts occur or may potentially occur.

3.3. Transmission

The major route of infection for most entomogenous nematodes into their hosts is by direct penetration of the cuticle. Mermithids use the odontostyle, and allantonematids, neotylenchids, and sphaerulariids use stylets to penetrate through the host's cuticle. Steinernematids and heterorhabditids enter through natural body openings and then penetrate by mechanical pressure into the hemocoel of the insect (11). Heterorhabditids also can penetrate directly through the cuticle (25).

As indicated earlier, the infected host is very important in dispersal, which is another aspect of transmission. Altered behavior of the infected host can disseminate the nematode into the appropriate environment. *Bombus* spp. queens infected with *S. bombi* do not make nests but disseminate nematodes by digging at sites used by healthy hibernating queens (72). Face fly females infected with *H. autumnalis* become permanent dung seekers and "nemaposit" nematodes (75). Healthy face flies oviposit in the same dung to provide hosts for the nematodes. Mermithid parasitism modifies the behavior of adult male blackflies which attach themselves to oviposition sites and simulate oviposition (76). These types of behavior ensure the dissemination of nematodes into the proper environment for the perpetuation of the species.

3.4. Environmental Factors

The effects of the abiotic environment on entomogenous nematodes, particularly *R. culicivorax* (6, 7, 11, 16) and steinernematids (4, 10, 11), have been reviewed. Some of these factors have been discussed in previous sections. Among the physical factors, moisture is most important for nematode survival.

Unless nematodes have mechanisms to undergo anhydrobiosis (77), they desiccate and die. Some may have this ability but the conditions to enhance this survival mechanism have not yet been determined. However, entomogenous nematodes can survive under less than optimal humidity or without standing water. For example, *S. feltiae* can survive in the soil after 12 days at 79.5% RH (78), *R. culicivorax* can overwinter in California rice fields that have been drained and disced (70), and *Octomyomermis muspratti* eggs remain viable in moist sand for several years until proper hatching conditions occur (61).

Temperature extremes can be an important limiting factor for infection. *R. culicivorax* infects mosquito larvae at a temperature as low as 10°C (79), but the optimum is 21 to 33°C (80). Different species have different temperature requirements. *R. culicivorax* has higher temperature requirements for postparasite development than does another species of *Romanomermis* (81). *S. feltiae* can infect its host at 9°C (55), but the limit for development is between 15 and 30°C (82, 83). Heterorhabditids have similar developmental temperatures (84). Temperatures above 35°C are lethal to *S. feltiae* (58). Interestingly, *S. feltiae* reared at a given temperature will tend to aggregate at or near their prior cultured temperature (85).

Salts inhibit the infectivity of *R. culicivorax* (30, 86), and epizootics in mosquito larvae in brackish or polluted waters (wastewater, feedlot runoff) are unlikely. In contrast, *O. muspratti* can tolerate diluted seawater containing organic matter and thus has potential for epizootics in brackish and polluted waters (61).

Low oxygen tension increases the survival of the infective stage of *R. culicivorax* but reduces infectivity (87). Anoxia makes the infectives more quiescent which enhances survival. The postparasitic stage of this mermithid is a facultative anaerobe which can continue development on the bottom of mosquito ponds (86). Infectives of *S. feltiae* also can survive at low oxygen tension (89).

Solar radiation and ultraviolet light are detrimental to *S. feltiae* (90). From an epizootiological perspective, these radiations will not significantly affect entomogenous nematodes because most occur in soil or other protected habitats. However, light is important for *M. nigrescens* because eggs are deposited on foliage only during daylight after a rainstorm (18).

In most instances, entomogenous nematodes have been compatible with parasitoids and predators. Although parasitoids and predators are susceptible to *S. feltiae* (10), they are often separated by time and space. Parasitoids of *Sirex* spp. are infected by *Deladenus* spp. but are not killed and may assist in the dispersal of the nematode (27). *S. feltiae* appears to be compatible with baculoviruses and certain entomogenous fungi (10).

Nematodes have a number of natural enemies (91). Ostracods, gammarids, and copepods prey upon the infectives of *R. culicivorax* (6). Additionally, the postparasites are attacked by a number of aquatic invertebrates including diving beetles, gammarids, dragonfly and damselfly naiads, and crayfish (6). A mite, *Dendrolaelaps neodisetus*, feeds upon *Contortylenchus brevicomi* and reduces the prevalence of nematode infection in bark beetles (92).

Pathogens have been reported from entomogenous nematodes. A fungal pathogen infected laboratory cultures of *R. culicivorax* (93). An iridescent virus infecting a pillbug also infects its mermithid parasite (94), and an insect microsporidium infects *S. feltiae* (95).

The internal environment of the host is important for sex determination in certain dioecious mermithids. The ratio of male to female is influenced by the number of nematodes within the same host. Hosts with three or fewer *M. nigrescens* produce females; hosts with more than 24 mermithids produce only males; and hosts with 4 to 23 mermithids produce both males and females (20). A mosquito host with more than one *R. culicivorax* produces a preponderance of males (96). Moreover, inadequately fed or overcrowded mosquitoes will produce more males than females.

The occurrence of more than one infective nematode of the same species in a given host is common. For example, the average number of infective *Contortylenchus* in the bark beetle *Dendroctonus rufipennis* was 5.5 (97). In face flies, the number of *H. autumnalis* averaged 1.6, 1.9, and 2.4 in three successive years (98). The number of infective nematodes per host differed considerably from the Poisson distribution indicating that superinfections were not random. These data suggest that the insect may be differentially susceptible to the nematode because of age or genotype, or the nematode, insect, or both may be distributed in clumps. In blackflies, the mermithid *Mesomermis flumenalis* temporarily paralyzes its host during the infection process (76). Such paralyzed blackflies are easily attacked and are penetrated by other infectives. This observation also may explain multiple attacks by nematodes in other insects.

4. SELECTED EXAMPLES OF NEMATODE EPIZOOTIOLOGY IN INSECT POPULATIONS

Prevalences of nematode infection have been recently summarized for grasshoppers (18), bark beetles (28), hymenopterans (26), blackflies (99), mosquitoes (16), and other dipterans (29). Prevalences of infection are affected by a number of factors including host habitat, host density, host stage, sampling methods, and seasonal variation.

4.1. Mermithidae

Mermithids are commonly found in grasshopper populations (18). The prevalence of infection varies from year to year and from site to site. In Vermont, mermithid infection ranged from 2 to 75% (100), while in Illinois and Indiana mermithids were found in 14% of the adults and 23% of the grasshopper nymphs (101). Decline and disappearance of a natural population of the grasshopper, *Hesperotettix viridis pratensis,* were attributed to severe infection by *M. nigrescens* in North Dakota (102). This grasshopper species fed primarily in low, moist habitats, favorable for the survival of the mermithid. In contrast, other grasshopper species that were collected from drier habitats had a low prevalence of disease.

A 10 year study of the impact of parasites on grasshopper populations in Ontario revealed no evidence that parasites including mermithids affected the trend in grasshopper abundance (103). Mermithids accounted for the highest percentage of parasitism (3–21%) in 7 of the 10 years. Infections correlated positively with the amount of rainfall in July in wet years and negatively with the hours of sunshine in July in dry years. Similarly, *F. leipsandra* infection in the banded cucumber beetle was greatest during periods of high rainfall and high temperatures (104).

R. culicivorax has been the most studied mermithid because of its biological control potential (16). Originally found in the southeastern United States, it could recycle in more northern habitats and in cultivated fields. When this mermithid was inoculated into semipermanent and permanent water sites containing *Anopheles crucians* in Louisiana, epizootics occurred in subsequent years (105). The first infections were detected in mid-April (about 5%) and increased to 94 to 100% during June. This site contained no standing water in July and August. Mermithid prevalence in the mosquito population averaged 31% from September until late October when it increased significantly to 100% in the first week of November.

Recently, Hominick and Tingley (9) critically assessed the potential of mermithids as agents for vector control. They analyzed data from the mermithid literature according to theoretical aspects of population behavior. They examined six biological attributes that determined the degree of depression of mosquito populations. The six attributes were (1) density-dependent constraints on parasite reproduction, (2) distribution of parasite numbers per host, (3) density of host population, (4) pathogenicity of mermithids, (5) reproduction of mermithids and their hosts, and (6) predation.

Under the first attribute, factors such as sex determination, crowding, fecundity, intensity of infection, and density of postparasites were considered. For the second, patterns of mermithid dispersion within host populations in relation to the dynamics of host-parasite population interactions were examined. However, little information was available about the distribution of mermithids in aquatic insects in natural habitats. Generally, mermithids tended to show low degrees of overdispersion in their host population. Moreover, insect populations were not uniformly susceptible to mermithids.

The third attribute, density of the host population, determined whether the mermithid could recycle in the population. Mermithids with their limited reproductive capabilities required a dense host population for establishment and recycling. The fourth attribute dealt with pathogenicity. Mermithids were highly pathogenic with low transmission efficiency and were characterized by low prevalence in the host population and by low numbers of mermithids per host.

The fifth attribute was the reproductive potential of mermithids and hosts. Mermithids required a reproductive potential several orders of magnitude greater than that of their hosts in order to significantly affect the host population. Although females of *R. culicivorax* had a greater reproductive potential than female mosquitoes when generation times were considered, the nematode reproduced at levels far less than its host. Finally, predation might reduce the infective and postparasitic stages of mermithids as well as the infected host.

Based on these analyses the authors concluded that mermithid populations are controlled by tight density-dependent constraints, and these constraints can cause only moderate long-term depressions in host populations.

4.2. Steinernematidae and Heterorhabditidae

Steinernematids and heterorhabditids occur naturally in soils throughout much of the world (106). Detection of these nematodes in soil with trap insects has been successful (107). The distribution of *S. bibionis* in forest soils in Czechoslovakia (108) and of other steinernematids and heterorhabditids in soils in North Carolina has been determined with this method (106). Placement of host larvae in soil under citrus trees in Florida has shown that prevalence of *S. feltiae* and *Heterorhabditis* sp. is highest during summer months (70%) and lowest during winter months (10%) (109). In Czechoslovakia, monthly samples indicate that *S. kraussei* infects about 3% of the diapausing sawflies in a spruce stand (110).

Steinernematids and heterorhabditids have great potential as biological control agents of insects. They are highly pathogenic and have high survivability and high transmission efficiency under laboratory conditions. In nature, they appear to have high transmission efficiency based on trap insects placed in the soil. However, effects of predation and soil microorganisms on these nematodes remain unknown.

4.3. Neotylenchidae

The introduction of *D. siricidicola* into *Sirex noctilio* populations in Australia has resulted in classical biological control of this insect (27). The success of this program can be attributed to several factors. The nematode is a facultative parasite which can maintain itself in the absence of its host. When the host is in the area, the nematode becomes parasitic. The strain that is released into the field does not affect the size of its host, and infected hosts disperse as far as healthy hosts. The strain is also virulent (i.e., effective in sterilizing its host) and highly infectious. It has high infectivity and a high transmission efficiency. In 1970, 50 infected *Sirex* females emerged from nematode-infested logs placed in a northern Tasmanian forest with low host density. By 1972, the nematodes had been found in 37% of *Sirex*-infested trees; by 1974, they had been in 70% of such trees with over 90% of the *Sirex* emerging from the trees infected. Moreover, the number of *Sirex*-killed trees has been reduced dramatically.

4.4. Allantonematidae

Heterotylenchus autumnalis persists in face fly populations from year to year, but the general consensus at this time is that the nematode does not regulate the host population (29, 111). The prevalence of infection varies from locality to locality and from year to year. The sampling method also can affect results (75). Although *Heterotylenchus* spp. have many of the attributes of *Deladenus* spp., they are not facultative parasites and cannot persist in the absence of their host. Furthermore, *H. autumnalis* is not as virulent as *Deladenus* spp. and attempts to increase infection rates in the laboratory have failed (112). *Heterotylenchus autumnalis* and other allantonematids have low infectivity and low transmission efficiency.

5. CONCLUSIONS

There have been few epizootiological studies of nematode diseases in insect populations. The case of *Deladenus siricidicola* is the exception rather than the rule. Nematodes are present in many insect populations. Every combination of high versus low virulence or transmission efficiency has been observed. However, most nematodes have not been evaluated systematically. Nematodes are density-dependent regulating agents, but some are density-independent agents and may not regulate the host population or cause epizootics. The theoretical assessment by Hominick and Tingley (9) should stimulate research to obtain more information about host/nematode relationships.

REFERENCES

1. W. R. Nickle and H. E. Welch, "History, Development, and Importance of Insect Nematology," in W. R. Nickle Ed., *Plant and Insect Nematodes*, Marcel Dekker, New York, 1984, p. 627.

2. G. O. Poinar, Jr., *Entomogenous Nematodes*, E. J. Brill, Leiden, Netherlands, 1975.

3. J. R. Finney, "Potential of Nematodes for Pest Control," in H. D. Burges, Ed., *Microbial Control of Pests and Plant Diseases 1970–1980*, Academic Press, New York, 1981, p. 603.

4. R. Gaugler, Biological control potential of neoaplectanid nematodes, *J. Nematol.,* **13,** 241 (1981).

5. W. R. Nickle, Mermithid parasites of agricultural pests, *J. Nematol.,* **13,** 262 (1981).

6. E. G. Platzer, Biological control of mosquitoes with mermithids, *J. Nematol.,* **13,** 257 (1981).

7. J. J. Petersen, Current status of nematodes for the biological control of insects, *Parasitology,* **84,** 177 (1982).

8. J. M. Webster, Biocontrol: the potential entomophilic nematodes in insect management, *J. Nematol.,* **12,** 270 (1980).

9. W. M. Hominick and G. A. Tingley, Mermithid nematodes and the control of insect vectors of human disease, *Biocontrol News Inf.,* **5,** 7 (1984).

10. H. K. Kaya, "Entomogenous Nematodes for Insect Control in IPM Systems," in M. A. Hoy

and D. C. Herzog, Eds., *Biological Control in Agricultural IPM Systems,* Academic Press, New York, 1985, p. 283.

11. G. O. Poinar, Jr., *Nematodes for Biological Control of Insects,* CRC Press, Boca Raton, Fla., 1979.

12. W. R. Nickle, *Plant and Insect Nematodes,* Marcel Dekker, New York, 1984.

13. I. A. Rubzov, *Aquatic Mermithidae of the Fauna of the USSR,* Vol. II, Oxonian Press Pvt. Ltd., New Delhi, 1981.

14. M. R. N. Shephard, *Arthropods as final hosts of nematodes and nematomorphs. An Annotated Bibliography 1900–1972,* Farnham Royal: Commonwealth Agric. Bur., Tech. Commun., **45,** 1974.

15. R. Gaugler and H. K. Kaya, Bibliography of the entomogenous nematode family Steinerne-matidae, *Bibliogr. Entomol. Soc. Am.,* **1,** 43 (1983).

16. J. J. Petersen, "Nematode Parasites of Mosquitoes," in W. R. Nickle, Ed., *Plant and Insect Nematodes,* Marcel Dekker, New York, 1984, p. 797.

17. G. O. Poinar, Jr. and G. G. Gyrisco, Studies on the bionomics of *Hexamermis arvalis* Poinar and Gyrisco, a mermithid parasite of the alfalfa weevil, *Hypera postica* (Gyllenhal), *J. Insect Pathol.,* **4,** 469 (1962).

18. J. M. Webster and C. H. S. Thong, "Nematode Parasites of Orthopterans," in W. R. Nickle, Ed., *Plant and Insect Nematodes,* Marcel Dekker, New York, 1984, p. 697.

19. G. O. Poinar, Jr., R. S. Lane, and G. M. Thomas, Biology and redescription of *Pheromermis pachysoma* (V. Linstow) n. gen., n. comb. (Nematoda: Mermithidae), a parasite of yellow-jackets (Hymenoptera: Vespidae), *Nematologica,* **22,** 360 (1976).

20. J. R. Christie, Some observations on sex in the Mermithidae, *J. Exp. Zool.,* **53,** 59 (1929).

21. J. Schmidt and J. N. All, Chemical attraction of *Neoaplectana carpocapsae* (Nematode: Steinernematidae) to insect larvae, *Environ. Entomol.,* **7,** 605 (1978).

22. J. Schmidt and J. N. All, Attraction of *Neoaplectana carpocapsae* (Nematode: Steinernemati-dae) to common excretory products of insects, *Environ. Entomol.,* **8,** 55 (1979).

23. J. A. Byers and G. O. Poinar, Jr., Location of insect hosts by a nematode, *Neoaplectana carpocapsae,* in response to temperature. *Behaviour,* **79,** 1 (1982).

24. R. Gaugler, L. LeBeck, B. Nakagaki, and G. M. Boush, Orientation of the entomogenous nematode *Neoaplectana carpocapsae* to carbon dioxide, *Environ. Entomol.,* **9,** 649 (1980).

25. R. A. Bedding and A. S. Molyneux, Penetration of insect cuticle by infective juveniles of *Heterorhabditis* spp. (Heterorhabditidae: Nematoda), *Nematologica,* **28,** 354 (1982).

26. R. A. Bedding, Large scale production, storage and transport of the insect-parasitic nema-todes *Neoaplectana* spp. and *Heterorhabditis* spp., *Ann. Appl. Biol.,* **104,** 117 (1984).

27. R. A. Bedding, "Nematode Parasites of Hymenoptera," in W. R. Nickle, Ed., *Plant and Insect Nematodes,* Marcel Dekker, New York, 1984, p. 755.

28. H. K. Kaya, "Nematode Parasites of Bark Beetles," in W. R. Nickle, Ed., *Plant and Insect Nematodes,* Marcel Dekker, New York 1984, p. 727.

29. C. J. Geden and J. G. Stoffolano, Jr., "Nematode Parasites of Other Dipterans," W. R. Nickle, Ed., *Plant and Insect Nematodes,* Marcel Dekker, New York, 1984, p. 849.

30. J. J. Petersen and O. R. Willis, Some factors affecting parasitism by mermithid nematodes in southern house mosquito larvae, *J. Econ. Entomol.,* **63,** 175 (1970).

31. G. O. Poinar, Jr., Parasitic development of *Filipjevimermis leipsandra* Poinar and Welch (Mermithidae) in *Diabrotica v. undecimpunctata* (Chrysomelidae), *Proc. Helminthol. Soc. Wash.,* **35,** 161 (1968).

32. R. Gaugler, S. Wraight, and D. Molloy, The bionomics of a mermithid parasitizing snowpool *Aedes* spp. mosquitoes, *Can. J. Zool.,* **62,** 670 (1984).

33. G. O. Poinar, Jr., R. Leutenegger, and P. Gotz, Ultrastructure of the formation of a melanotic

capsule in *Diabrotica* (Coleoptera) in response to a parasitic nematode (Mermithidae), *J. Ultrastruct. Res.,* **25,** 293 (1968).

34. R. Gaugler and D. Molloy, Instar susceptibility of *Simulium vittatum* (Diptera: Simuliidae) to the entomogenous nematode, *Neoaplectana carpocapsae, J. Nematol.,* **13,** 1 (1981).

35. R. H. Dadd, Size limitations on the infectibility of mosquito larvae by nematodes during filter-feeding, *J. Invertebr. Pathol.,* **18,** 246 (1971).

36. H. K. Kaya and A. H. Hara, Differential susceptibility of lepidopterous pupae to infection by the nematode *Neoaplectana carpocapsae, J. Invertebr. Pathol.* **36,** 389 (1980).

37. H. K. Kaya and B. J. Grieve, The nematode *Neoaplectana carpocapsae* and the beet army-worm *Spodoptera exigua:* infectivity of prepupae and pupae in soil and of adults during emergence from soil, *J. Invertebr. Pathol.,* **39,** 192 (1982).

38. J. B. Beavers and C. O. Calkins, Susceptibility of *Anastrepha suspensa* (Diptera: Tephritidae) to steinernematid and heterorhabditid nematodes in laboratory studies, *Environ. Entomol.,* **13,** 137 (1984).

39. H. K. Kaya, Effect of the entomogenous nematode *Neoaplectana carpocapsae* on the tachinid parasite *Compsilura concinnata* (Diptera: Tachinidae), *J. Nematol.,* **16,** 9 (1984).

40. G. O. Poinar, Jr. and N. Deschamps, Susceptibility of *Scolytus multistriatus* to neoaplectanid and heterorhabditid nematodes, *Environ. Entomol.,* **10,** 85 (1981).

41. D. B. Woodward and T. Fukuda, Laboratory resistance of the mosquito *Anopheles quadri-maculatus* to the mermithid nematode *Diximermis peterseni, Mosq. News,* **37,** 192 (1977).

42. J. J. Petersen, Development of resistance by the southern house mosquito to the parasitic nematode *Romanomermis culicivorax, Environ. Entomol.,* **7,** 518 (1978).

43. M. A. Bari and H. K. Kaya, Evaluation of the entomogenous nematode *Neoaplectana carpo-capsae* (=*Steinernema feltiae*) Weiser (Rhabditida: Steinernematidae) and the bacterium *Bacillus thuringiensis* Berliner var. *kurstaki* for suppression of the artichoke plume moth (Lepidoptera: Pterophoridae), *J. Econ. Entomol.,* **77,** 225 (1984).

44. J. J. Petersen and H. C. Chapman, Checklist of mosquito species tested against the nematode parasite *Romanomermis culicivorax, J. Med. Entomol.,* **15,** 486 (1979).

45. J. F. Bronskill, Encapsulation of rhabditoid nematodes in mosquitoes, *Can. J. Zool.,* **40,** 1269 (1962).

46. G. O. Poinar, Jr. and R. Leutenegger, Ultrastructural investigation of the melanization process in *Culex pipiens* (Culicidae) in response to a nematode, *J. Ultrastruct. Res.,* **36,** 149 (1971).

47. T. G. Andreadis and D. W. Hall, *Neoaplectana carpocapsae:* encapsulation in *Aedes aegypti* and changes in host hemocytes and hemolymph proteins, *Exp. Parasitol.,* **39,** 252 (1976).

48. G. O. Poinar, Jr. and H. N. Kaul, Parasitism of the mosquito *Culex pipiens* by the nematode *Heterorhabditis bacteriophora, J. Invertebr. Pathol.,* **39,** 382 (1982).

49. A. J. Nappi and J. G. Stoffolano, Jr., *Heterotylenchus autumnalis:* hemocytic reactions and capsule formation in the host, *Musca domestica, Exp. Parasitol.,* **29,** 116 (1971).

50. A. S. Molyneux, R. A. Bedding, and R. J. Akhurst, Susceptibility of larvae of the sheep blowfly *Lucilia cuprina* to various *Heterorhabditis* spp., *Neoaplectana* spp., and an undescribed steinernematid (Nematoda), *J. Invertebr. Pathol.,* **42,** 1 (1983).

51. R. A. Bedding, A. S. Molyneux, and R. J. Akhurst, *Heterorhabditis* spp., *Neoaplectana* spp., and *Steinernema kraussei:* interspecific and intraspecific differences in the infectivity for insect, *Exp. Parasitol.,* **55,** 249 (1983).

52. J. Silverman, E. G. Platzer, and M. K. Rust, Infection of the rat flea *Ctenocephalides felis* (Bouche) by *Neoaplectana carpocapsae* Weiser, *J. Nematol.,* **14,** 394 (1982).

53. R. A. Bedding and L. A. Miller, Use of a nematode, *Heterorhabditis heliothidis,* to control black vine weevil, *Otiorhynchus sulcatus,* in potted plants, *Ann. Appl. Biol.,* **99,** 211 (1981).

54. J. J. Petersen, Development and fecundity of *Reesimermis nielseni*, a nematode parasite of mosquitoes, *J. Nematol.,* **7,** 211 (1975).

55. S. R. Dutky, J. V. Thompson, and G. E. Cantwell, A technique for the mass propagation of the DD-136 nematode, *J. Insect Pathol.,* **6,** 417 (1964).

56. R. A. Bedding, Low cost in vitro mass production of *Neoaplectana* and *Heterorhabditis* species (Nematoda) for field control of insect pests, *Nematologica,* **27,** 109 (1981).

57. G. Stein, Weitere Beitrage zür Biologie von *Sphaerularia bombi* Leon Dufour 1837, *Z. Parasitenk.,* **17,** 383 (1956).

58. D. C. Schmiege, The feasibility of using a neoaplectanid nematode for control of some forest insect pests, *J. Econ. Entomol.,* **56,** 427 (1963).

59. S. R. Dutky, Insect microbiology, *Adv. Appl. Microbiol.,* **1,** 175 (1959).

60. S. M. Craig and J. M. Webster, Viability and hatching of *Mermis nigrescens* eggs and subsequent larval penetration of the desert locust *Schistocerca gregaria, Nematologica,* **24,** 472 (1978).

61. J. J. Petersen, Observations on the biology of *Octomyomermis muspratti*, a nematode parasite of mosquitoes, *J. Invertebr. Pathol.,* **37,** 290 (1981).

62. J. G. Stoffolano, Jr., The synchronization of the life cycle of diapausing face flies, *Musca autumnalis,* and of the nematode, *Heterotylenchus autumnalis, J. Invertebr. Pathol.,* **9,** 395 (1967).

63. K. D. Elsey, Bionomic comparison of two parasites of the tobacco flea beetle, a nematode and a wasp, *Tob. Sci.,* **20,** 177 (1976).

64. E. M. Reed and H. R. Wallace, Leaping locomotion by an insect-parasitic nematode, *Nature,* **206,** 210 (1965).

65. P. L. Moyle and H. K. Kaya, Dispersal and infectivity of the entomogenous nematode, *Neoaplectana carpocapsae* Weiser (Rhabditida: Steinernematidae), in sand, *J. Nematol.,* **13,** 295 (1981).

66. R. Georgis and G. O. Poinar, Jr., Effect of soil texture on the distribution and infectivity of *Neoaplectana carpocapsae* (Nematoda: Steinernematidae), *J. Nematol.,* **15,** 308 (1983).

67. R. Georgis and G. O. Poinar, Jr., Effect of soil texture on the distribution and infectivity of *Neoaplectana glaseri* (Nematoda: Steinernematidae), *J. Nematol.,* **15,** 329 (1983).

68. R. Georgis and G. O. Poinar, Jr., Vertical migration of *Heterorhabditis bacteriophora* and *H. heliothidis* (Nematoda: Heterorhabditidae) in sandy loam soil, *J. Nematol.,* **15,** 652 (1983).

69. J. R. Christie, Life history of *Agamermis decaudata*, a nematode parasite of grasshoppers and other insects, *J. Agric. Res.,* **52,** 161 (1936).

70. B. B. Westerdahl, R. K. Washino, and E. G. Platzer, Successful establishment and subsequent recycling of *Romanomermis culicivorax* (Mermithidae: Nematoda) in a California rice field following postparasite application, *J. Med. Entomol.,* **19,** 34 (1982).

71. H. B. Hungerford, Biological notes on *Tetradonema plicans*, Cobb, a nematode parasite of *Sciara coprophila* Lintner, *J. Parasitol.,* **5,** 186 (1919).

72. G. O. Poinar, Jr. and P. A. Van der Laan, Morphology and life history of *Sphaerularia bombi*, *Nematologica,* **18,** 239 (1972).

73. K. D. Elsey, Dissemination of *Howardula* sp. nematodes by adult tobacco flea beetles (Coleoptera: Chrysomelidae), *Can. Entomol.,* **109,** 1283 (1977).

74. R. W. Glaser, *Studies on* Neoaplectana glaseri, *a nematode parasite of the Japanese beetle* (Popillia japonica), N.J. Dep. Agric. Circ. No. 211, 34 pp. (1932).

75. H. K. Kaya, R. D. Moon, and P. L. Witt, Influence of the nematode, *Heterotylenchus autumnalis,* on the behavior of face fly, *Musca autumnalis, Environ. Entomol.,* **8,** 537 (1979).

76. D. P. Molloy, Mermithid parasitism of black flies (Diptera: Simuliidae), *J. Nematol.,* **13,** 250 (1981).

77. J. H. Crowe and K. A. C. Madin, Anhydrobiosis in nematodes: evaporative water loss and survival, *J. Exp. Zool.*, **193**, 323 (1975).

78. W. R. Simons and G. O. Poinar, Jr., The ability of *Neoaplectana carpocapsae* (Steinernematidae: Nematodea) to survive extended periods of desiccation, *J. Invertebr. Pathol.*, **22**, 228 (1973).

79. T. D. Galloway and R. A. Brust, Effects of temperature and photoperiod on the infection of two mosquito species by the mermithid, *Romanomermis culicivorax*, *J. Nematol.*, **9**, 218 (1977).

80. B. J. Brown and E. G. Platzer, The effects of temperature on the infectivity of *Romanomermis culicivorax*. *J. Nematol.*, **9**, 166 (1977).

81. J. J. Petersen, Comparative biology of the Wyoming and Louisiana populations of *Reesimermis nielseni*, parasitic nematode of mosquitoes, *J. Nematol.*, **8**, 273 (1976).

82. H. K. Kaya, Development of the DD-136 strain of *Neoaplectana carpocapsae* at constant temperatures, *J. Nematol.*, **9**, 346 (1977).

83. A. E. Pye and M. Burman, *Neoaplectana carpocapsae*: infection and reproduction in large pine weevil larvae, *Hylobius abietis*, *Exp. Parasitol.*, **46**, 1 (1978).

84. J. E. Milstead, Influence of temperature and dosage on mortality of seventh instar larvae of *Galleria mellonella* (Insecta: Lepidoptera) caused by *Heterorhabditis bacteriophora* (Nematoda: Rhabditoidea) and its bacterial associate *Xenorhabdus luminescens*, *Nematologica*, **27**, 167 (1981).

85. M. Burman and A. E. Pye, *Neoaplectana carpocapsae*: movements of nematode populations on a thermal gradient, *Exp. Parasitol.* **49**, 258 (1980).

86. B. J. Brown and E. G. Platzer, Salts and the infectivity of *Romanomermis culicivorax*, *J. Nematol.*, **10**, 53 (1978).

87. B. J. Brown, and E. G. Platzer, Oxygen and the infectivity of *Romanomermis culicivorax*, *J. Nematol.*, **10**, 110 (1978).

88. J. L. Imbriani and E. G. Platzer, Gaseous requirements of postparasitic development of *Romanomermis culicivorax*, *J. Nematol.*, **13**, 470 (1981).

89. M. Burman and A. E. Pye, *Neoaplectana carpocapsae*: respiration of infective juveniles, *Nematologica*, **26**, 214 (1980).

90. R. Gaugler, and G. M. Boush, Effects of ultraviolet radiation and sunlight on the entomogenous nematode, *Neoaplectana carpocapsae*, *J. Invertebr. Pathol.*, **32**, 291 (1978).

91. R. Mankau, Biological control of nematode pests by natural enemies, *Annu. Rev. Phytopathol.*, **18**, 415 (1980).

92. D. N. Kinn, Life cycle of *Dendrolaelaps neodisetus* (Mesostigmata: Digamasellidae), a nematophagous mite associated with pine bark beetles (Coleoptera: Scolytidae), *Environ. Entomol.*, **13**, 1141 (1984).

93. A. M. Stirling and E. G. Plazter, *Catenaria anguillulae* in the mermithid nematode *Romanomermis culicivorax*, *J. Invertebr. Pathol.*, **32**, 348 (1978).

94. G. O. Poinar, Jr., R. T. Hess, and A. Cole, Replication of an iridovirus in a nematode (Mermithidae), *Intervirology*, **14**, 316 (1980).

95. G. V. Veremtchuk and I. V. Issi, On the development of the microsporidian of insects in the entomopathogenic nematode *Neoaplectana aqriotois* Veremtchuk (Nematoda: Steinernematidae), *Parazitologiya*, **4**, 3 (1970). (In Russian).

96. J. J. Petersen, Factors affecting sex determination in a mermithid parasite of mosquitoes, *J. Nematol.*, **4**, 83 (1972).

97. C. L. Massey, *Biology and Taxonomy of Nematode Parasites and Associates of Bark Beetles in the United States*, USDA Agric. Handb. No. 446, U. S. Printing Office, Washington, D. C. (1974).

98. E. S. Krafsur, C. J. Church, M. K. Elvin, and C. M. Ernst, Epizootiology of *Heterotylenchus autumnalis* (Nematoda) among face flies (Diptera: Muscidae) in Central Iowa, USA, *J. Med. Entomol.*, **20**, 318 (1983).

99. R. Gordon, "Nematode Parasites of Blackflies," in W. R. Nickle, Ed., *Plant and Insect Nematodes*, Marcel Dekker, New York, 1984, p. 821.

100. R. W. Glaser and A. M. Wilcox, On the occurrence of a *Mermis* epidemic amongst grasshoppers, *Psyche*, **25**, 12 (1918).

101. W. P. Hayes and J. D. DeCoursey, Observations of grasshopper parasitism in 1937, *J. Econ. Entomol.*, **31**, 519 (1938).

102. S. Mongkolkiti and R. M. Hosford, Jr., Biological control of the grasshopper *Hesperotettix viridis pratensis* by the nematode *Mermis nigrescens*, *J. Nematol.*, **3**, 356 (1971).

103. R. W. Smith, A field population of *Melanoplus sanguinipes* (Fab.) (Orthoptera: Acrididae) and its parasites, *Can. J. Zool.*, **43**, 179 (1965).

104. C. S. Creighton and G. Fassuliotis, Seasonal population fluctuations of *Filipjevimermis leipsandra* and infectivity of juveniles on the banded cucumber beetle, *J. Econ. Entomol.*, **73**, 296 (1980).

105. J. J. Petersen and O. R. Willis, Establishment and recycling of a mermithid nematode for the control of mosquito larvae, *Mosq. News*, **35**, 526 (1975).

106. R. J. Akhurst and W. M. Brooks, The distribution of entomophilic nematodes (Heterorhabditidae and Steinernematidae) in North Carolina, *J. Invertebr. Pathol.*, **44**, 140 (1984).

107. R. A. Bedding and R. J. Akhurst, A simple technique for the detection of insect parasitic rhabditid nematodes in soil, *Nematologica*, **21**, 109 (1975).

108. Z. Mracek, J. Gut, and S. Gerdin, *Neoaplectana bibionis* Bovien, 1937, an obligate parasite of insects isolated from forest soil in Czechoslovakia, *Folia Parasitol. (Praha)*, **29**, 139 (1982).

109. J. B. Beavers, C. W. McCoy, and D. T. Kaplan, Natural enemies of subterranean *Diaprepes abbreviatus* (Coleoptera: Curculionidae) larvae in Florida, *Environ. Entomol.*, **12**, 840 (1983).

110. Z. Mracek, Horizontal distribution in soil, and seasonal dynamics of the nematode *Steinernema kraussei*, a parasite of *Cephalcia abietis*, *Z. Angew. Entomol.*, **94**, 110 (1982).

111. H. K. Kaya and R. D. Moon, The nematode *Heterotylenchus autumnalis* and face fly, *Musca autumnalis*: a field study in northern California, *J. Nematol.*, **10**, 333 (1978).

112. C. M. Jones and J. M. Perdue, *Heterotylenchus autumnalis*, a parasite of the face fly, *J. Econ. Entomol.*, **60**, 1393 (1967).

PART

IV

PRACTICAL ASPECTS

15

APPLIED EPIZOOTIOLOGY: MICROBIAL CONTROL OF INSECTS

JAMES D. HARPER

Department of Zoology and Entomology
Alabama Agricultural Experiment Station
Auburn University, Auburn, Alabama

1. INTRODUCTION

One of the driving forces in the rapid development of insect pathology as a separate discipline was the concept that the microorganisms involved could be utilized for the regulation of insect populations. As the body of basic knowledge regarding entomopathogens increased, so did our interest in utilizing that knowledge for practical purposes. From this interest has grown a subdiscipline within insect pathology — that of microbial control.

Microbial control of insects, simply stated, is the use of entomopathogenic microorganisms to regulate insect population levels. Normally, microbial control involves the induction of high prevalence of disease in an insect population. Such prevalence conforms to our definition of an epizootic, that is, an unusually high number of cases of disease in an animal population (1). The procedures used to achieve microbial control include the intentional creation of epizootics or the utilization of naturally occurring epizootics. Microbial control, then, can be referred to as applied epizootiology.

Much basic knowledge has been generated, mostly within the past 40 years, on the major groups of entomopathogens of pest insects throughout the world. This information has been generated at all levels of biological organization, from molecular to ecosystem, and all is critical to our ability to utilize these organisms effectively as biological control agents.

Previous chapters have thoroughly discussed the importance of the components of an epizootic — pathogen, pathogen transmission, host insect, and habitat or environment. Research emphasis has focused on the microorganisms themselves because, in most cases, this component is the most readily manipulated. In some cases, pathogens have been mass cultured for subsequent release. In others, culture has not been economically feasible, and other methods of inducing epizootics such as habitat or host population manipulation have been developed. I shall first discuss microbial control from a general perspective, considering ecological, economic, and practical principles. A number of successful or partially successful control programs will then be examined from an epizootiological point of view to show how interactions between host, pathogen, pathogen transmission, and environmental factors operate.

2. GOALS OF APPLIED EPIZOOTIOLOGY

The purpose of any insect pest management program is the reduction of economic loss to a crop or other system caused by an insect pest or pest complex.

Chemical insecticides are used to achieve this goal by reducing pest populations on a temporary basis. This is often accomplished rapidly and effectively. Microbial control may be used in the same way, but the unique action of entomopathogens provides other mechanisms of pest population suppression which may be more cost effective and environmentally desirable.

Goals of microbial control programs are varied and are based on the many different interactions between specific entomopathogens and their hosts. One desirable goal is permanent reduction of the general equilibrium level of pest population density. This is accomplished by introducing entomopathogens into pest habitats where they are able to cycle within the pest populations on an indefinite basis and effect continuous and significant levels of control. A second goal is the initiation of short-term epizootics in which the pathogen may cycle and cause mortality only in one or several host generations. A third goal is the increase of pathogen numbers over those already present in a habitat. This increases disease prevalence above natural levels and can result in a temporary lowering of pest population density. A fourth goal is the temporary suppression of pest populations through applications of rapidly acting entomopathogens which do not recycle. This is accomplished in much the same manner as with chemical pesticides. A fifth goal is management of naturally occurring entomopathogens in pest populations or their habitats in a way that is economically beneficial.

3. ECOLOGICAL, BIOLOGICAL, TECHNOLOGICAL, AND ECONOMIC CONSIDERATIONS

3.1. Transmission

Transmission is one of the key ecological factors that must be understood before entomopathogens can be manipulated. In some programs transmission will not be critical, as in the use of *Bacillus thuringiensis* against agricultural and forest pests. Since this pathogen has little horizontal or vertical transmissibility in nature (2), it is simply used in an inundative manner whenever suitable pest populations reach economic action levels (see Section 4.1.2). For most entomopathogens, however, the user should expect some level of natural recycling with transmission to uninfected hosts. Ideally, prevalence of disease in subsequent pest generations becomes greater than in the treated pest generation. Transmission is essential if this is to occur. Understanding the mechanisms of transmission of a given entomopathogen will allow prediction of its ability to spread within a host population and, in some cases, of its potential as an applied microbial control agent.

Knowledge of transmission is also important in the selection of pathogens for use in control programs as well as in program design. Organisms that are normally transmitted vertically are likely to be poor candidates for inundative spray release programs. Normally they will lack effective mechanisms of survival outside of their hosts for extended periods. On the other hand, organisms with resistant stages (for example, endospores, resting spores, and occlusion bodies) that persist in the environment in the absence of host insects would, in general, be better candidates for spray-disseminated microbial insecticides. Even for these organisms, however, transmission during both enzootic and epizootic phases may rely to a great degree on highly specific mechanisms rather than on the general contamination of host-food substrate.

Natural dispersal of pathogen units from the host occurs for many types of entomopathogenic microorganisms and does not appear to be a completely random process. Certain entomogenous fungi produce infectious conidia or spores only when environmental conditions are suitable for their survival and for the infection of new hosts (3–5). Insects infected by baculoviruses release concentrated masses of occlusion bodies onto host substrates when host tissues disintegrate after death. This material can be dispersed by rain and other mechanisms (6), but much of it remains protected from inactivation by ultraviolet irradiation for some time in the decomposed host tissues. Several pest species, such as *Heliothis zea* and *Spodoptera frugiperda* on soybean and *Trichoplusia ni* on cruciferous crops, appear to prefer to feed on foliage heavily contaminated with disintegrated cadavers. This results in a high level of horizontal transmission (Harper, unpublished data). Most infected insects die in the specific microhabitat of their cohorts, further increasing opportunities for horizontal transmission. Additional references to the importance of transmission will be presented later in this chapter.

3.2. Application Technology

Epizootics can be initiated in many cases simply by spraying entomopathogenic inoculum in a field or host habitat using standard technology developed for application of chemical pesticides. This technology, however, has some serious drawbacks for dispersal of entomopathogens. Unfortunately, as pointed out by Yearian and Young (7), because of the extensive spray technology developed for chemical pesticides, we are likely to continue to spend much energy trying to adapt these systems to entomopathogens. For some pathogens, spray dispersal may result in a distribution pattern of pathogen units in the host environment which approximates natural patterns. Generally, however, this will not be the case. Natural distribution of pathogens tends to be discontinuous with clumps around sites where hosts die. Spray dispersal can be utilized as a mechanism of short cutting or speeding the processes of natural transmission. However, serious problems are associated with the technique. Because entomopathogens are so labile with respect to ultraviolet irradiation, temperature, and desiccation (6, 8–11), it is somewhat amazing that dispersal results in any appreciable degree of control.

In most spray situations, waste is built into every application. Pathogen propagules deposited into direct sun exposure are rapidly inactivated. Host plant substrates must be covered with sufficient propagules per unit area to insure that enough insect pests are infected for population regulation. Thus orders of magnitude of more inoculum must be introduced into the environment than are needed for infections simply to compensate for waste due to application inefficiency. Thus, spray application may be only rarely desirable as a mechanism for artificially disseminating microbial insecticides, despite the large amount of effort previously and currently being expended on it.

A more reasonable approach to microbial control would be to emphasize natural transmission (12, 13). Insects themselves are best able to distribute inoculum of an entomopathogen to locations where it will most efficiently come in contact with other members of their species. The concept of autodissemination, or utilizing insects to distribute their own entomopathogens, has been tested and proven feasible for several groups of insect hosts and pathogen groups. Most such research has been on transmission of entomopathogenic viruses (14). The use of the baculovirus of the palm rhinoceros beetle *Oryctes rhinoceros,* is a particularly interesting example which will be discussed in Section 4.1.1. Autodissemination has been attempted by attracting moths to light or pheromone traps containing viruses. Contaminated moths carry the viruses to oviposition sites and contaminate the plants, which their offspring will consume (14). Organisms that are generally transmitted vertically might best be introduced into pest populations via living, infected hosts. This should be particularly valuable for introduction of microsporidia or other pathogen groups for which transovarial and transovum transmissions are known (15). Autodissemination techniques certainly deserve more attention based on these and other examples to be discussed. When compared with conventional spray application techniques, autodissemination requires much smaller quantities of infectious material because less is wasted.

Other innovative mechanisms of artificial introduction or transmission of entomopathogens are needed. Ignoffo's (16) appeal for innovative research on application technology made in 1972 is no less critical now for the use of microbial agents for insect population management than it was then.

3.3. Habitat Stability

Many examples of successful biological control with entomopathogens as well as parasitoids and predators are found in relatively stable or permanent habitats such as forests, orchards, turf, and pasture. Pests in these habitats tend to be K-selected species (17). They are characterized as having relatively stable but low population levels over time. Natural enemies tend to play an important role in their regulation because a stable interrelationship can develop between the ever-present pest and the natural control agents.

One cause of failure of permanent establishment of biological control agents in temporary habitats such as row crops is that the agents, including entomopathogens, have difficulty persisting. The continuously changing pattern of host plant and pest complex makes survival of the pathogen very tenuous. Pest populations in such habitats tend to increase rapidly in order to be able to utilize host food plants while they are available. They are characteristically r-selected species (17). As discussed by Price (18), natural enemies rarely play a significant role in the population regulation of such species because they do not have time to become established in the host population before it or its habitat is gone. Some entomopathogens appear to share this characteristic with parasitoids and predators.

A possible example of this phenomenon is the entomopathogenic fungus *Nomuraea rileyi.* This pathogen often causes nearly complete host population mortality of lepidopterous pests in soybean fields in the southeastern United States, but frequently not until severe defoliation has occurred (19, 20). Whether this delay represents the failure of a natural mortality factor to respond to the population buildup of an r-selected host, however, is not certain. It may simply represent failure of all the necessary conditions for epizootic development to be present until late in the growing season.

On the other hand, entomopathogens often are able to function permanently and effectively as biological control agents in temporary habitats. This is particularly evident in those pathogens largely dependent on vertical transmission, especially those with true transovarial or transovum transmission. In such cases, the entomopathogen is always present in the pest population, and enzootic or epizootic mortality occurs regularly. For example, the microsporidium *Nosema pyraustae* is transovarially transmitted within European corn borer *Ostrinia nubilalis,* host populations and exerts significant natural control (21).

The short generation time of pathogens and their ability to rapidly generate numbers of infectious units that are greater than needed to insure high levels of host infection make them capable of functioning as effective regulators of r-selected pest species. In this respect, many entomopathogens are superior to parasitic and predatory insects which have much lower intrinsic rates of reproduction.

3.4. Density Dependence

Most entomopathogens are density-dependent mortality factors (22). As their host populations increase, the percentage of infected hosts and the amount of infectious disease propagules in the habitat also increase. As host numbers decline, the percentage of infected individuals likewise declines as does production of the pathogen. Depending on how such pathogens are used, this density dependence may be advantageous, disadvantageous, or of no importance. Density dependence is important to the outcome of introductions of pathogens with the capability of recycling or multiplying in, being transmitted to, and invading new hosts. Those that do not recycle do not function as density-dependent mortality factors; they are used in much the same fashion as chemical insecticides, albeit of a much more selective nature.

Entomopathogens often exhibit a rapid numerical response to changes in their host densities. Because of their short generation times and high reproductive rates, obvious and dramatic epizootics seem to occur almost overnight. Because of the high pathogen numbers that develop as an epizootic progresses, these pathogens can overwhelm their hosts at rates that few parasitoid or predator species can achieve. Many entomopathogens have short generation times relative to those of their hosts. During a single generation of the Douglas fir tussock moth *Orgyia pseudotsugata,* for example, three replicatory cycles of baculovirus can occur (23). The potential buildup of pathogen numbers during three successive generations in a heavy host population is so high that probability of the escape from contact with the pathogen for a given larva becomes almost negligible. Such rapidly developing epizootics can be induced, but more frequently occur naturally.

Theoretically, the density-dependent organism, because of its self-perpetuating nature, is the more desirable type for manipulation. Once introduced into the host population, it continues to exert some level of regulation on the pest population, in some situations for years. The costs of control, which may be high initially, are justified because reduced expenditures, sometimes none, are needed for control of future pest generations (see Section 4.1.1 for examples). From a practical viewpoint, density-independent organisms may also serve as effective microbial control agents. The insecticidal approach to the use of entomopathogens will be discussed with respect to *Bacillus thuringiensis* (see Section 4.1.2).

3.5. Other Pathogen Characteristics

Entomopathogens have certain other characteristics that make them valuable as mortality agents. Most are sufficiently host specific that they pose no significant threat to nontarget insects or other life forms. As living organisms they are capable of being cultured or produced in the laboratory. For many pathogens, this technology is quite advanced. For others, it is yet to be perfected.

Entomopathogens generally are capable of infecting several of their hosts' life stages, unlike many parasitoids and predators. Most larval pathogens, for example, infect most or all larval instars. Parasitoids and many important predator groups select their prey on the basis of size and are thus much more restricted in their utilization of a given host species, though there are important exceptions.

3.6. Pest Population Parameters

A second level of consideration for biological control is a sound understanding of what constitutes a pest population. The early definitions of pest population parameters, such as the economic injury level, economic threshold, and general equilibrium position, were an important step toward pest management because these concepts led to a clearer understanding of problems associated with the overuse of chemical pesticides during the 1940's and 1950's (24).

These concepts are presented in most current texts on entomology and will not be repeated here. It should be stated, however, that they were responsible for a revolution in attitudes and use patterns for chemical pesticides. Moreover, they demonstrated why biological control agents, including entomopathogens, could be potentially valuable as management tools despite the fact that, in many cases, they could not provide the same degree of population reduction as chemical insecticides.

By merging the concepts of ecology and economics to arrive at the threshold concept or action level, economically acceptable control could be achieved without the near total reduction in population that had come to be expected with chemical insecticides. Thus, in some situations, suppression of only a small percentage of the pest population to below the economic threshold using a specific biological agent could be as cost effective as suppression with a chemical control agent. The latter might simply overkill the pest, destroy all or part of the complex of beneficial organisms, and result in the need for additional control inputs because of the pest population rebound effect. For a given point in the development of any crop, for any stored product, or for any human or animal health situation, an acceptable level of damage by an insect pest can be defined based on the costs of control versus the increased benefits of yield, product quality, comfort, health, etc. Successful control only requires that the population be held below that level.

4. GENERAL MECHANISMS FOR ACHIEVING MICROBIAL CONTROL

The methodologies for regulation of insect pest populations with entomopathogens are diverse. This is not surprising, given the spectrum of entomopathogenic bacteria, viruses, fungi, protozoa, and nematodes that are now under investigation. Each organism has its own mode of action, has its own mechanisms of interacting with its host, fills a unique niche, and is influenced by unique combinations of physical and biological conditions. If these organisms are to be used successfully and to maximum advantage, these factors must be thoroughly understood.

Approaches to microbial control can be categorized in a number of ways. Kaya (25), for example, considered use patterns to be short-term, long-term, or integrated in nature. Longworth and Kalmakoff (13) divided use patterns into insecticidal release, limited release, and manipulation of natural enzootics. Hamm (26) discussed insecticidal epizootic induction, colonization, and auto-dissemination for achieving microbial control of insects. I shall categorize microbial control according to the general principles established by researchers in biological control with predators and parasites. The three approaches to microbial control are (1) introduction, (2) augmentation, and (3) conservation. Each of these methods and their subcategories will be discussed and illustrated by examples of sucessful microbial control programs, with an analysis of the key pathogen, host, and environmental factors. With few exceptions, successful utilization of entomopathogens as regulatory agents requires that epizootiological concepts or knowledge be addressed.

4.1. Introductions

The basic concept of introduction of entomopathogens is that of placing them in an ecosystem where they did not previously occur, or in some cases, as stated by Harper (27), where they were present but not functional as entomopathogens. In the classical sense of biological control, such introductions are inoculative and result in the permanent establishment of the control agent with subsequent reduction of the general equilibrium position of the host population. In another sense, pathogens can be introduced, usually by inundation, and provide short-term reductions in pest populations, either for brief periods with no amplification or for longer periods with only temporary amplification. The latter may be for a single season, multiple seasons, a single host generation, or multiple host generations; ultimately, however, disease prevalence will be reduced to zero.

4.1.1. INOCULATIVE INTRODUCTIONS.   There are several examples of situations in which the introduction of insect pathogens has been successful in more or less permanently regulating a pest population. The best known of these is the use of *Bacillus popilliae* for control of the Japanese beetle *Popillia japonica*. As early as the 1930s, *B. popilliae* and the closely related *B. lentimorbus* were introduced into turf grasses in the northeastern United States (28 – 31). Several studies have demonstrated that these early introductions at many sites were still controlling *P. japonica* larvae three decades after treatment (32, 33).

The success of this program revolves around a very stable system of habitat, host, and pathogen. An epizootic is initiated by inoculating lawns at intervals with specific quantities of dried spores of the two bacteria. Each application spot serves as a focus of infection for beetle larvae. Larval *P. japonica* become infected by consuming spores as they feed on grass roots. The larvae continue to feed, gradually moving away from the infection sites. Days to weeks later, depending on temperature and other factors, they die and release millions of infectious spores at their sites of death. These foci of spores function just like the original artificially applied foci. As this process repeats itself over several years, the soil becomes extensively contaminated with spores (28).

Once the soil environment becomes well contaminated, a balance of host numbers and pathogen density becomes established through an effective feedback mechanism. When high numbers of larvae occur, spore numbers increase after larval deaths and the probability of survival for any one grub decreases. A reduced host population results in less replenishment of the pathogen population and a reduction in the probability of infection for a given grub. If larval numbers begin to increase, incidence of disease will rise as will the pathogen population density. The system does not fluctuate rapidly because spores are relatively long-lived in the stable soil environment, and new larvae continually enter the environment from eggs laid by adults originating in untreated turf. Within the soil, spores are not subjected to ultraviolet irradiation or to the temperature extremes which adversely affect most entomopathogens. The turf remains healthy because the numbers of larvae that are able to survive in the pathogen-contaminated soil are below the economic threshold (28).

Another commonly cited example of a successful and permanent introduction of an entomopathogen into a new pest-habitat complex is that of the baculovirus of the European spruce sawfly *Gilpinia hercyniae*. This case is typical of a pest species that was introduced into a new habitat, in this case from Europe into eastern Canada, without its natural biological control agents. In the absence of these agents, it established itself as a defoliator of young spruce trees. A baculovirus was found in Canadian populations some 8 years after the pest was first discoveied (34). The source of this virus is unknown, but Balch and Bird (35) theorized that it was inadvertently introduced into the population in cocoons containing contaminated parasitoids which were imported into Canada.

Regardless of how it was introduced, this virus began infecting sawfly colonies in a small area and rapidly spread over a very large portion of the range of the pest (34). Later, new epicenters of the disease were established in areas where the spruce sawfly spread by artificially introducing the virus (36, 37). Treatment of only a few trees with a virus suspension initiated an epizootic which spread within 2 years over an infestation of several square kilometers.

In this example, both host and habitat are relatively stable components of the system. The host occurs on a fairly continuous basis from year to year and has two generations per year. The virus population can grow during the season, with epizootics typically increasing in intensity toward late summer (36). The virus is virulent and is efficiently transmitted by a number of means, both within and between generations (38). All of these factors provide a continuity between virus and host which is important for maintenance of the virus in the environment.

A third example of intentional introduction of an entomopathogen into a host population is that of the baculovirus of the coconut beetle *Oryctes rhinoceros* on several South Sea islands (39). In this case, the virus, which infects both the immature and adult stages of the host, could not be applied effectively to the growing tip of the tree where the adult beetles feed, to the larval feeding sites in the tops of beetle-killed coconut trees, or in felled palm trunks and compost. As with many internal plant feeders, the insect is relatively immune to directed sprays of pesticides, whether chemical or microbial, once it is within the host plant substrate. However, it was demonstrated that healthy adults became infected when they consumed the virus-contaminated frass of infected adults. Infected females exhibited reduced fecundity and feeding and contaminated their own eggs. Researchers took advantage of these factors for coconut beetle control.

Large numbers of adult beetles were hand collected and submersed or allowed to swim in a water suspension of the virus for several hours. A high percentage of infection was achieved in this manner. Males were then released. These beetles then flew to feeding sites which they contaminated during their normal feeding activities. Females became infected through their normal gregarious feeding activities with these males and subsequently oviposited contaminated eggs. Larvae emerging from these eggs in turn became infected and died or developed into transmitting adults. The net result of the introductions on many islands was a decrease in both beetle population and damage to coconut trees.

In this example, the epizootic was enhanced by the constantly available larval and host habitats, vertical transmission involving adult tissue infection, and host behavior resulting in horizontal transmission of the virus. As in previous examples, continued activity of the virus depends on a population of hosts that is not completely destroyed by the virus. Rather, a continuous movement of virus into larval or adult feeding sites is needed to insure that the virus is maintained in the environment.

Other examples of pathogen introductions that have resulted in long-term control are referenceable but are not as well documented as the above. Several baculoviruses of sawflies, in addition to that of *G. hercyniae,* have been successfully introduced into new areas (37). Notable among these are introductions of the baculoviruses of the European pine sawfly *Neodiprion sertifer* and the redheaded pine sawfly *N. lecontei* (38). Both viruses have been introduced in several areas of the United States and Canada, with subsequent development of epizootics. The *N. lecontei* virus has now been registered by the Canadian government and the *N. sertifer* virus by the United States Environmental Protection Agency as microbial insecticides.

A nuclear polyhedrosis virus (NPV) of the soybean looper *Pseudoplusia includens* was introduced into field plots in Louisiana and Arkansas in 1972 after its importation from Guatemala (40, 41). Large-area spray applications were made in Louisiana in 1975. Since that time the virus has persisted in the soil and has caused a low prevalence of disease. Generally, prevalence has not exceeded 25% infection and is usually less (41). This example illustrates an introduction that did not result in suppression of the host population below economic injury levels. However, the fact that persistence has been demonstrated is important. Generally, it has been concluded that permanent establishment of biocontrol agents in row-corp agroecosystems has relatively little chance of success because of the temporary nature of the pest habitat through annual destruction, crop rotation, use of annual crops, soil tillage, and other factors.

There are several factors that probably limit the epizootic potential of the *P. includens* NPV. First, the field environment is not stable. Good farming practices dictate that the soybean crop be rotated between years to other crops. This causes a break in continuity between years of pest populations that can support epizootics and support replenishment of virus in the soil reservoir. In its country of origin, *P. includens* develops on other host plants if the soybean is not available. Further, in tropical areas of Central and South America, host plants are available year round. This provides for potentially continuous maintenance of the pest-virus interaction. In the more temperate climate of the southeastern United States, there is less abundance and variety of alternate host plants as well as the length of time they are available.

Secondly, even if soybean is planted in the same fields year after year, soybean looper population levels vary considerably from one year to the next. The soybean looper is an annual immigrant into the United States and does not overwinter in soybean production areas. This limits both the seasonal and geographic distribution of the pest. In the same way, it limits the potential for continuity of virus-infected insects to maintain the soil reservoir titer.

The baculovirus of the velvetbean caterpillar *Anticarsia gemmatalis* has a history similar to that of the soybean looper baculovirus. The *A. gemmatalis* NPV was imported into the United States in the mid-1970's from southern Brazil and Argentina where it was a common mortality factor in populations of the velvetbean caterpillar on soybeans (42, 43). It was released in the United States and can overwinter (44, 45), but annual host immigration without overwintering and instability of the soybean agroecosystem again slow or prevent the NPV from reaching its epizootic potential. Nevertheless, field trials in several states have demonstrated that this virus has potential for host population regulation because of its ability to cycle and spread rapidly. The low disease prevalence in epizootics of both *P. includens* and *A. gemmatalis* NPVs may hold promise for more significant levels of control of their respective hosts in future years. In the meantime, annual recycling of the NPVs is insuring renewal and perhaps slow increases of the pathogen populations, as well as providing some pest population suppression at no additional input costs.

Laird (46) demonstrated that a fungus, *Coelomomyces* sp., could be successfully moved from southeast Asia to a South Pacific island where it did not occur and could be permanently established for regulation of mosquito species. In this example, the pest population was reduced but not below economically important levels.

Other examples of entomopathogen introductions resulting in long-term pest suppression could be given, but the ones cited illustrate the importance of understanding the system components and why the epizootics occurred in each case. Each host-pest-habitat combination cited represents a different combination of characteristics. The host species in these cases included Diptera, Lepidoptera, Hymenoptera, and Coleoptera, and the entomopathogens included viruses, bacteria, and fungi. In each case, the combination of components has characteristics that promote permanent pest control. The host habitat in most of these examples was relatively stable. However, the suppression of *P. includens* with NPV illustrates the potential for permanent or prolonged microbial control of pests of annual row crops.

4.1.2. INUNDATIVE INTRODUCTIONS.   A second type of introduction is illustrated by the inundative release of microorganisms to achieve relatively rapid suppression of a pest population. By definition, this must involve the release of a pathogen that does not already occur in that habitat. Harper (27) has referred to the use of *Bacillus thuringiensis* as an example of such introductions. While *B. thuringiensis* is a more or less cosmopolitan soil-inhabiting microorganism (47), naturally occurring infections are rare (44). Further, the strains of *B. thuringiensis* that are isolated from soils are quite varied and are not generally those found in commercial products unless they had a previous spray history (47). Thus, for practical purposes of epizootiology or microbial control, it almost always is "absent," and it is appropriate to consider the use of products based on *B. thuringiensis* as true introductions.

When *B. thuringiensis* is introduced into a pest's habitat, the number of cases of disease rapidly increases. This increase conforms to the definition of an epizootic. However, because of the unique mode of action of *B. thuringiensis* and the nature of commercial products based on it, there is no recycling of the organism within the host environment (48). For some insects, the activity of *B. thuringiensis* is dependent primarily, if not entirely, on the delta endotoxin. This is true for mosquito and simuliid larval response to *B. thuringiensis* var. *israelensis* (49) as well as for certain lepidopterous larvae to commercial varieties of *B. thuringiensis.* Other insects, however, are more susceptible to a mixture of both the exotoxin and endospores, both of which are found in commercial products. The bacterium is known to produce a variety of different toxins as it grows and metabolizes (48, 49).

Regardless of the mode of action, the infected insect ultimately dies with typical symptoms of bacterially caused mortality. These symptoms include a relatively tough, leathery integument, particularly in lepidopterous hosts. Even if death is caused by the delta-endotoxin, spores normally germinate and grow vegetatively in the cadaver. After several days, the dead insect contains spores and parasporal crystals. However, these do not cause further mortality in the host population for several reasons. First, because the insect integument remains leathery, the body of the dead insect does not rupture and release spores and crystals to contaminate the food source of other hosts. Second, the dead insects generally fall from their substrate or, in the case of mosquito larvae, sink before putrefactive processes cause disintegration of the integument and release of the internal fluids.

Once in or on the soil or substrate, the bacterium begins to be deactivated by normal environmental factors. Endospores and endotoxin crystals become separated after bacterial cell lysis. Thus, the probability of a sufficient dosage of both endotoxin and endospores ever reaching a new host becomes smaller with time. The mode of action of the pathogen coupled with the behavioral response of the dying host does not result in an efficient mechanism for spread and for the development of a natural epizootic as is the case with so many other insect pathogens.

In summary, the action of *B. thuringiensis* is primarily the same as that of a chemical insecticide (50). Its activity period is confined to that of its persistence after application. Its advantage over nonselective insecticides, however, is its specificity. By not causing direct mortality in populations of beneficial species in a treated habitat, these organisms are free to feed on new pest insects entering the habitat, to shift their activities temporarily to other species which were not killed by the treatment, or to leave the habitat in search of other food sources.

Other inundative introductions of entomopathogens also have been directed at suppression of pest populations on an immediate basis without the likelihood of recycling. A typical example is *Heliothis zea* and *H. virescens* management on cotton with *Baculovirus heliothidis* (51, 52). In this case, a baculovirus was produced commercially to be applied in an insecticidal fashion.

Heliothis spp. eggs generally are deposited by female moths on the leaves, petioles, new squares, or flowers of the cotton plant. Upon hatching, the larvae feed briefly on the exposed portions of the plant and then move and feed inside the young squares, where they are unlikely to consume sprayed virus. As the larvae grow, they simply move from one square or boll to another, boring into each new fruit.

In order to be effective, the NPV must be present at the time of egg hatch so that the newly emerged larvae eat a lethal dosage during their brief open-feeding period. Once inside the boll, the larva behaviorally avoids the virus until it leaves the protection of the square or boll in search of a new one. Even this does not render them particularly susceptible because they consume only a small area of potentially contaminated fruit surface and are larger larvae, requiring higher dosages of virus for infection.

A diseased larva usually dies after it enters a square or boll. It develops the typical symptoms of nuclear polyhedrosis, ultimately disintegrating to release the newly replicated infectious virus into its environment. However, the new virus is confined to the inner portions of the fruit and does not contaminate the general plant surfaces. Thus the majority of new virions are not available to the next group of *Heliothis* larvae. This interruption of transmission must be countered by repeated applications of the virus so that inoculum is present when the young larvae are present and feeding outside of the fruit. Despite this drawback to the use of the virus, it has been used successfully to control *Heliothis* spp. on cotton. As with the use of *B. thuringiensis* against a variety of pests, this pathogen spares beneficial insects on cotton. The importance of predator and parasite activity on cotton pests has long been recognized, and techniques that are compatible with them, if cost effective, should be implemented.

Interestingly, the same NPV causes natural epizootics in *Heliothis zea* populations on other crops such as soybean and peanut (53), where the insect feeds on more exposed plant parts. This indicates that the cryptic feeding habit of *Heliothis* spp. on cotton and possibly antimicrobial characteristics of the cotton plant probably interfere with epizootics.

Baculoviruses have been introduced into soybean production systems in the southeastern United States (41) on several occasions. The NPVs of *P. includens* and *A. gemmatalis* have already been mentioned. In addition to their potential for colonization, they also can be applied for short-term population suppression (54). The use of these viruses may ultimately be based on periodic, perhaps even annual releases, on an as needed basis when population management dictates, with any subsequent natural infection occurring as a bonus.

The success of NPVs following inundative release in agricultural systems is related, in part, to their ability to be transmitted within those environments. Boucias et al. (55) in Florida demonstrated that after placement of *A. gemmatalis* NPV at specific locations in a field, insects in other areas of the field began dying of the same viral disease, indicating that the virus was spreading. Many predaceous insects in these treated fields contained infectious virus, suggesting that they were eating virus-infected insects. One pentatomid predator produced fecal pellets containing infectious virus after eating infected velvetbean caterpillar larvae (56).

The transmission of insect baculoviruses by birds and other predators has been suggested, hypothesized, and demonstrated by numerous authors (see 57). The importance of predator transmission of baculovirus and other pathogens has not been fully appreciated and deserves further study. It may be important in the uniform dispersion of inoculum over a pest habitat during the increase phase of an epizootic and in the spread of inoculum from an epicenter. This information lends further credence to the concept of developing pest management programs that utilize all possible methods for conservation of predaceous and parasitic arthropods.

4.2. Augmentation

If an entomopathogen is present in a pest-habitat system but does not prevent economic damage by the pest, it is possible to place additional pathogen units into the environment to increase disease prevalence. This procedure is termed augmentation. It is useful when the timing of natural epizootics is not favorable or when the prevalence of disease in the pest population is too low to be of economic value.

Theoretically, the probability of success of augmentation is high because the natural presence of the organism in the pest population indicates that it is well adapted to its environment. Increasing pathogen numbers should increase disease prevalence by overcoming or bypassing a weak link in the transmission of the disease or by speeding the process of epizootic development. If transmission, for example, is mediated by predaceous insects, and their populations are insufficient to effectively transmit the pathogen to other parts of the habitat, it should be possible to supplement their activity by artificially disseminating the pathogen. If the pathogen demonstrates a typical numerical response to its host population, the host population may reach economic injury levels before the entomopathogen can replicate and infect sufficient numbers of the pests to keep their population below the economic threshold. The augmentation of that pathogen at the appropriate time can result in effectively bypassing the natural processes that generate the numerical response.

Many of the entomopathogens currently registered for insect control are actually utilized in an augmentative fashion. In some cases, the classification of a specific use pattern depends on interpretation of the facts and definitions. *Baculovirus heliothidis* was given as an example of an introduced entomopathogen (see Section 4.1.2). It is possible that *B. heliothidis* may occur naturally in the cotton agroecosystem yet never cause detectable disease. If so, in the strict sense, the application of more virus is an augmentation. But from a practical standpoint, application of the virus is an introduction. For other entomopathogens, the question is less clouded.

Baculoviruses have been augmented successfully in larval populations of several lepidopterous forest defoliators. Baculoviruses of both the Douglas fir tussock moth *Orgyia pseudotsugata* and the gypsy moth *Porthetria dispar* have been developed and registered for use against these pests (58, 59). In both cases, the viruses cause natural epizootics in the pest populations. However, both host species alternate over periods of years between endemic and epidemic populations. Tussock moth outbreaks generally occur at intervals of perhaps 8 to 10 years and last for 3 to 4 years (60). Gypsy moths apparently fluctuate with similar periodicities in Europe, but have somewhat different patterns in North America, with longer outbreak phases (61). During these outbreak phases the prevalence of viral disease in the populations increases following a numerical response curve. By the time the populations crash due to high prevalence of virus and other factors, considerable damage has been caused by the pest populations. This may include the death of large areas of spruce, fir, and Douglas fir forests in the case of the Douglas fir tussock moth and the growth reduction and even death of various eastern deciduous hardwood species in the case of the gypsy moth.

In general, the strategy for use of both baculoviruses is to introduce them into their respective host populations when larvae are first hatched and before the naturally occurring virus has provided effective natural control (23, 62). These larvae are easiest to infect with an aerial application of virus because the dosage required is smallest for the young larvae and is thus most economical. Both of these insects are univoltine, but both have a total larval maturation period that is longer than the time required for the virus to infect, kill, and be released back into the environment. Within-generation recycling has been demonstrated to be important in the success of the tussock moth virus, with up to three cycles occurring if application is properly timed (23).

The production of virus in larvae infected by the initial application results in a biological magnification of virus in the environment. Both species demonstrate typical baculovirus infection symptomatology. Larvae normally die on the foliage, hang by their prolegs, rupture due to the fragile integumental tissues, and release their liquified, virus laden, body contents on the plant substrate below. Healthy larvae then feed on the contaminated foliage. Although not well quantified, there is evidence suggesting that healthy insects are not repelled and in fact may be attracted to the putrified tissue remnants (see Section 3.1). The augmentation procedure can cause initiation of epizootics at a somewhat earlier date than would occur naturally, resulting in damage prevention due to a rapid decline in the pest population.

Two fungi, *Hirsutella thompsonii* and *Nomuraea rileyi*, provide examples of the augmentative use of entomopathogenic fungi for population regulation. The deuteromycete *H. thompsonii* has been successfully produced on a commercial scale in the United States and has been approved for sale and use by the Environmental Protection Agency. It is important as a natural regulator of populations of several mite species, but it was initially developed for control of the citrus rust mite *Phyllocoptruta oleivora,* a serious pest of citrus in Florida (63).

Citrus rust mite numbers increase in Florida as the citrus fruits are developing. They cause damage by blemishing the fruit surface through feeding. Fungal prevalence increases with host numbers but does not peak until several weeks to a month or more after the mite numbers. In most situations, the fungus will ultimately reduce the mite populations to subeconomic levels, but not before economic damage has occurred. The standard use of fungicides to control plant pathogenic fungi on citrus also destroys *H. thompsonii* fungus and negatively impacts its ability to control the mite (63).

McCoy and Couch (64) have augmented the fungus in host populations by starting early epizootics and preventing damage. However, climatic conditions must be favorable, including free water on leaf surfaces, humidity of nearly 100%, and relatively high temperatures. These conditions normally occur in summer in Florida (63).

Successful use of *H. thompsonii* depends on the acceleration of the epizootic by 2 to 4 weeks. Careful attention must be given to a group of factors if the augmentation is to be successful — the mite population, rainfall, temperature, humidity, and natural disease prevalence, among others. Accurate weather predictions are important to predict whether the fungus will become established or desiccate after application.

H. thompsonii has been augmented in populations of this and other mites in other parts of the world with various degrees of success (63). In most of these cases, the approach appears to have been one of a general insecticidal application rather than its application being dictated by specific conditions of host and environment. Nevertheless, the material appears to show promise under a variety of conditions.

A second fungal pathogen recently investigated from an augmentative approach is *Nomuraea rileyi*, a pathogen of lepidopterous pests of soybean (65). This pathogen has been isolated from insect hosts from many parts of the world and can be found on a variety of crops. It is currently under investigation in the United States as a potential microbial pesticide. Ignoffo et al. (19) augmented *N. rileyi* in soybean plots and induced an epizootic approximately 2 weeks earlier than in untreated plots. Timing of the augmentative releases was such that significant protection was provided at a critical point of soybean growth. The natural epizootic, however, was too late to prevent damage.

Since the habitat was suitable for *N. rileyi* infection of host insects at the time of the augmentations, it was concluded that the added conidia speeded development of an epizootic due to the earlier expression of disease in the pest population. The infected individuals in turn produced more conidia which caused higher incidence in the pest population. A possible, though less likely, alternative is that the introduced conidia were of a strain of *N. rileyi* which was better adapted to the specific conditions of the test than was the naturally occurring local strain.

Henry and Oma (66) reviewed information about the microsporidium *Nosema locustae* as a microbial control agent for grasshoppers. Natural prevalence of the disease in most grasshopper populations in the western United States was low, generally less than 1% of hosts infected, but Henry (67) demonstrated that augmentation could cause epizootics in these grasshopper complexes. Aerial application of spore-contaminated bran caused significant mortality in grasshopper populations. Further data demonstrated that *N. locustae* cycled several times in the year of application, and that it overwintered at least once after release (66). Several factors were important for the epizootics. Once introduced, the pathogen was eaten by the insects. There were numerous routes of transmission including fecal, transovarial, and passage through cannibalism. More data are required before the full potential of this pathogen is understood. For now, it appears best categorized as an augmentation in which the microsporidium is capable of at least short-term colonization.

4.3. Conservation

Many examples of natural epizootics in pest insect populations have been recorded in the literature. These epizootics attract attention because they often occur rapidly and involve, in many cases, high percentages of the pest population. Even in populations of pests that do not have acutely infected individuals with obvious symptoms of disease, careful examination often reveals mortality due to inapparent disease. Effective pest management programs recognize the contributions of natural mortality factors in the dynamics of pest populations and seek to take advantage of them.

Many cases have been cited where predators and parasitoids have been killed by broad spectrum pesticides. Certain entomopathogens are similarly vulnerable, particularly fungi, after applications of fungicides for control of plant pathogens (see 68). Pesticide effects on other entomopathogens may be less direct. For example, pesticide reduction of host populations may prevent development of natural epizootics by limiting the host population density necessary to sustain an epizootic. Several studies in which pathogens were substituted for chemicals as the method of insect control over a multi-year period demonstrated that the amount of the pathogen needed for effective control decreased annually. This may have been due at least in part to an increase in pathogen numbers in the environment. Further research is needed in this area.

4.3.1. USE OF SELECTIVE CHEMICALS AND MINIMUM RATES.   Since pathogens may be inactivated by certain pesticides, selection of a pesticide for control of one pest problem should be carefully weighed in terms of its potential effects on other pests and biotic control agents in the same habitat. While entomo-pathogenic fungi are susceptible to many agricultural fungicides, they vary in their degree of susceptibility (69, 70). If several fungicides are available for a specific plant pathogenic fungus, care should be taken to select one that has the least impact on beneficial fungi in the same habitat. Compatibility of entomo-pathogens with chemical pesticides has been studied and reviewed by Benz (71) and Jaques and Morris (68). For the most part, few negative interactions have been noted with baculoviruses and *B. thuringiensis*, even when the pathogen and chemical are mixed. However, in those few cases where antagonism has been noted, a different pesticide should be used.

Antagonism between chemical pesticides and entomopathogens also can be minimized by using no more than the amount required for damage control. This should reduce both direct and indirect deleterious effects as well as costs to the producer.

4.3.2. Environmental Manipulation. Any agronomic practice that favors disease in pest populations can reduce control inputs. In row-crop agri-culture, practices that protect the pathogen from environmental degradation, for example, by ultraviolet (UV) light deactivation, can significantly enhance disease prevalence (see 72). This can be achieved by planting crops in such a way as to insure protection of the pathogen at a point in time when both host insect and pathogen are present. A closed crop canopy between rows generally protects exposed pathogens from UV radiation better than an open canopy. The higher humidity under a closed versus an open canopy similarly is more conducive to epizootics of entomopathogenic fungi (73). Timely irrigation also can improve conditions for a fungal epizootic by providing the humidity needed (74).

The soil serves as a reservoir for many pathogens. These can be moved by splashing or wind to the plant parts where they can be consumed by crown-feeding pests. Deep plowing can remove large numbers of pathogen units from the upper soil layer (75, 76). Minimum tillage cultivation systems may prove to be advantageous from this standpoint in the future. Minimum disturbance of the soil may also prove beneficial to the development of epizootics in soil inhabiting insect pests. Success of the milky disease in the Japanese beetle is due in part to the constancy of the host habitat, specifically to the nondisturbance of the upper soil horizons (see Section 4.1.1).

Practices that promote entomopathogen transmissibility increase the proba-bility of epizootics. Agronomic practices that favor conservation of predaceous and parasitic insect populations also favor the potential for epizootics of ento-mopathogens that are vectored by them.

5. CONCLUSION

Examples have been presented to illustrate principles that should be considered in order to achieve a successful microbial control program. The samples are not exhaustive; others from many parts of the world could have been cited. These were selected in part because they are representative of the spectrum of applied work in this area, in part because of the depth of knowledge relating to them, and in part because of this author's familiarity with them.

For an epizootic to be successful, the components of an epizootic—the pathogen, pathogen transmission, the host or pest, and their environment—must interact in such a way that a high number of cases of disease occur. This is true for natural epizootics and for the most part for applied epizootics. The major difference is that weak links in natural systems can be overcome artificially. Thus, if a pathogen does not occur in a given habitat, even though that habitat, including a host population, is suitable for the pathogen, the pathogen can be introduced. Similarly, habitat or host characteristics can be manipulated to provide conditions that will allow interactions leading to epizootics.

Considerable research effort is still needed in most areas of microbial control before we will be able to use entomopathogens to their greatest advantage. Of particular importance is an in-depth understanding of transmission. We generally understand the mechanisms for transmission of many pathogens, but the fine details often are lacking. For example, we know that parasitoids and predators are capable of moving pathogens within a host habitat, but we do not know how important this mechanism of transmission is to epizootics. Similarly, we know that wind and rain can move pathogen units, but again we have few quantitative data on the importance of these parameters. Until such quantitative data are generated about these and other parameters, many applied microbial control projects will continue on a trial and error basis.

The potential for continued development of microbial control as a component of integrated pest management systems is great. Applied epizootiology, including microbial control of insects, requires an in-depth knowledge of many separate disciplines. Among these are all aspects of entomology, as well as microbiology, pathology, epizootiology, ecology, meteorology, engineering, and many more. Use of any microbe for insect control will be successful only if each of these subjects and the interactions between them are thoroughly understood and put into use.

REFERENCES

1. E. A. Steinhaus and M. E. Martignoni, *An Abridged Glossary of Terms Used in Invertebrate Pathology,* 2nd ed., U. S. For. Serv., Pac. N.W. For. Range Exp. Stn., 1970.

2. H. T. Dulmage and K. Aizawa, "Distribution of *Bacillus thuringiensis* in Nature," in E. Kurstak, Ed., *Microbial and Viral Pesticides,* Marcel Dekker, New York, 1982, p. 209.

3. L. P. Kish and G. E. Allen, *The Biology and Ecology of Nomuraea rileyi and a Program for Predicting its Incidence on* Anticarsia gemmatalis *in Soybean,* Fla. Agric. Exp. Stn. Bull. 795, 1978.

4. J. A. Millsten, G. C. Brown, and G. L. Nordin, Microclimatic moisture and conidial production in *Erynia* sp. (Entomophthorales: Entomophthoraceae): in vitro production rate and duration under constant and fluctuating moisture regimes, *Environ. Entomol.,* **12,** 1344 (1983).

5. J. D. Harper, D. A. Herbert, and R. E. Moore, Trapping patterns of *Entomophthora gammae* (Weiser) (Entomophthorales: Entomophthoraceae) conidia in a soybean field infested with the soybean looper, *Pseudoplusia includens* (Walker) (Lepidoptera: Noctuidae), *Environ. Entomol.*, **13**, 1186 (1984).

6. Y. Tanada, "Persistance of Entomogenous Viruses in the Insect Ecosystem," in S. Asahina, J. L. Gressitt, Z. Hidaka, T. Nishida, and K. Nomura, Eds., *Entomological Essays to Commemorate the Retirement of Professor K. Yasumatsu,* Hokuryukan, Tokyo, 1971, p. 367.

7. W. C. Yearian and S. Y. Young, Application of microbial insecticides on field crops, *Misc. Publ. Entomol. Soc. Am.,* **10**, 21 (1978).

8. Y. Tanada, "Environmental Factors External to the Host," in L. A. Bulla, Ed., *Regulation of Insect Populations by Microorganisms,* Ann. N.Y. Acad. Sci. Vol. 217, 1973, p. 120.

9. R. P. Jaques, Stability of entomopathogenic viruses, *Misc. Publ. Entomol. Soc. Am.,* **10**, 99 (1977).

10. J. V. Maddox, Stability of entomopathogenic protozoa, *Misc. Publ. Entomol. Soc. Am.,* **10**, 3 (1977).

11. D. E. Pinnock, J. E. Milstead, M. E. Kirby, and B. J. Nelson, Stability of entomopathogenic bacteria, *Misc. Publ. Entomol. Soc. Am.,* **10**, 77 (1977).

12. G. E. Allen, C. M. Ignoffo, and R. P. Jaques, Eds., *Microbial Control of Insect Pests: Future Strategies in Pest Management Systems,* NSF—USDA—Univ. Fla., Gainesville, 1978.

13. J. F. Longworth and J. Kalmakoff, "An Ecological Approach to the use of Insect Pathogens for Pest Control," in E. Kurstak, Ed., *Microbial and Viral Pesticides,* Marcel Dekker, New York, 1982, p. 425.

14. I. E. Gard and L. A. Falcon, "Autodissemination of Entomopathogens: Virus," in G. E. Allen, C. M. Ignoffo, and R. P. Jaques, Eds., *Microbial Control of Insect Pests: Future Strategies in Pest Management Systems,* NSF—USDA—Univ. Fla., Gainesville, 1978, p. 46.

15. W. M. Brooks, "Protozoan Infections," in G. E. Cantwell, Ed., *Insect Diseases,* Vol. 1, Marcel Dekker, New York, 1974, p. 237.

16. C. M. Ignoffo, An appeal for research, *J. Invertebr. Pathol.,* **19**, i (1972).

17. R. H. MacArthur and E. O. Wilson, *The Theory of Island Biogeography,* Princeton Univ., Princeton, 1967.

18. P. W. Price, "Relevance of Ecological Concepts to Practical Biological Control," in G. C. Papavizas, Ed., *Biological Control in Crop Production,* BARC Symp. No. 5, Allanheld, Osmun, London, 1981, p. 4.

19. C. M. Ignoffo, N. L. Marston, D. L. Hostetter, and B. Puttler, Natural and induced epizootics of *Nomuraea rileyi* in soybean caterpillars, *J. Invertebr. Pathol.,* **27**, 191 (1976).

20. C. M. Ignoffo, B. Puttler, N. L. Marston, D. L. Hostetter, and W. A. Dickerson, Seasonal incidence of the entomopathogenic fungus *Spicaria rileyi* associated with noctuid pests of soybeans, *J. Invertebr. Pathol.,* **25**, 135 (1975).

21. T. G. Andreadis, Epizootiology of *Nosema pyrausta* in field populations of the European corn borer (Lepidoptera: Pyralidae), *Environ. Entomol.,* **13**, 882 (1984).

22. E. A. Steinhaus, The effects of disease on insect populations, *Hilgardia,* **23**, 197 (1954).

23. C. G. Thompson and B. Maksymiuk, "Laboratory and Simulated Field Tests," in M. H. Brookes, R. W. Stark, and R. W. Campbell, Eds., *The Douglas-Fir Tussock Moth: A Synthesis,* U.S.D.A. Tech. Bull. 1585, 1978, p. 147.

24. V. M. Stern, R. F. Smith, R. van den Bosch, and K. S. Hagen, The integration of chemical and biological control of the spotted alfalfa aphid. I. The integrated control concept, *Hilgardia,* **29**, 81 (1959).

25. H. K. Kaya, "Insect Pathogens in Natural and Microbial Control of Forest Defoliators," in J. F. Anderson and H. K. Kaya, Eds., *Perspectives in Forest Entomology,* Academic Press, New York, 1976, p. 251.

26. J. J. Hamm, Invertebrate pathology and biological control, *J. Ga. Entomol. Soc.*, **19**, 2nd suppl., 6 (1984).

27. J. D. Harper, "Introduction and Colonization of Entomopathogens," in G. E. Allen, C. M. Ignoffo, and R. P. Jaques, Eds., *Microbial Control of Insect Pests: Future Strategies in Pest Management Systems*, NSF—USDA—Univ. Fla., Gainesville, 1978, p. 3.

28. R. T. White and S. R. Dutky, Cooperative distribution of organisms causing milky disease of Japanese beetle grubs, *J. Econ. Entomol.*, **35**, 679 (1942).

29. W. E. Fleming, *Milky Disease for Control of Japanese Beetle Grubs*, U.S.D.A. Leaflet No. 500, 1961.

30. S. R. Dutky, "The Milky Diseases," in E. A. Steinhaus, Ed., *Insect Pathology: An Advanced Treatise* Vol. II, Academic Press, New York, 1963, p. 75.

31. M. G. Klein, "Advances in the Use of *Bacillus popilliae* for Pest Control," in H. D. Burges, Ed., *Microbial Control of Pests and Plant Diseases*, Academic Press, New York, 1981, p. 183.

32. T. L. Ladd, Jr. and P. L. McCabe, Persistence of spores of *Bacillus popilliae*, the causal organism of type A milky disease of Japanese beetle larvae, in New Jersey soils, *J. Econ. Entomol.*, **60**, 493 (1967).

33. P. O. Hutton, Jr. and P. P. Burbutis, Milky disease and Japanese beetle in Delaware, *J. Econ. Entomol.*, **67**, 247 (1974).

34. F. T. Bird and D. E. Elgee, A virus disease and introduced parasites as factors controlling the European spruce sawfly, *Diprion hercyniae* (Htg.), in central New Brunswick, *Can. Entomol.*, **89**, 371 (1957).

35. R. E. Balch and F. T. Bird, A disease of the European spruce sawfly, *Gilpinia hercyniae* (Htg.), and its place in natural control, *Sci. Agric.*, **25**, 65 (1944).

36. F. T. Bird and J. M. Burk, Artificially disseminated virus as a factor controlling the European spruce sawfly, *Diprion hercyniae* (Htg.), in the absence of introduced parasites, *Can. Entomol.*, **93**, 228 (1961).

37. J. C. Cunningham and P. F. Entwistle, "Control of Sawflies by Baculovirus," in H. D. Burges, Ed., *Microbial Control of Pests and Plant Diseases*, Academic Press, New York, 1981, p. 379.

38. F. T. Bird, Transmission of some insect viruses with particular reference to ovarial transmission and its importance in the development of epizootics, *J. Insect Pathol.*, **3**, 352 (1961).

39. G. O. Bedford, "Control of the Rhinoceros Beetle by Baculovirus," in H. D. Burges, Ed., *Microbial Control of Pests and Plant Diseases*, Academic Press, New York, 1981, p. 409.

40. J. M. Livingston and W. C. Yearian, A nuclear polyhedrosis virus of *Pseudoplusia includens* (Lepidoptera: Noctuidae), *J. Invertebr. Pathol.*, **19**, 107 (1972).

41. W. A. Jones, Jr., S. Y. Young, M. Shepard, and W. H. Whitcomb, "Use of Imported Natural Enemies Against Insect Pests of Soybean," in H. N. Pitre, Ed., *Natural Enemies of Arthropod Pests in Soybean*, South. Coop. Ser. Bull. 285, 1983, p. 63.

42. G. E. Allen and J. D. Knell, A nuclear polyhedrosis virus of *Anticarsia gemmatalis*: I. Ultrastructure, replication and pathogenicity, *Fla. Entomol.*, **60**, 233 (1977).

43. G. R. Carner and S. G. Turnipseed, Potential of a nuclear polyhedrosis virus for control of the velvetbean caterpillar in soybean, *J. Econ. Entomol.*, **70**, 608 (1977).

44. A. R. Richter and J. R. Fuxa, Timing, formulation, and persistence of a nuclear polyhedrosis virus and a microsporidium for control of the velvetbean caterpillar (Lepidoptera: Noctuidae) in soybeans, *J. Econ. Entomol.*, **77**, 1299 (1984).

45. F. Moscardi, G. E. Allen, and G. L. Greene, Control of the velvetbean caterpillar by nuclear polyhedrosis virus and insecticides and impact of treatments on the natural incidence of the entomopathogenic fungus *Nomuraea rileyi*, *J. Econ. Entomol.*, **74**, 480 (1981).

46. M. Laird, A coral island experiment: a new approach to mosquito control, *WHO Chron.*, **21**, 18 (1967).

47. A. J. DeLucca II, J. G. Simonson, and A. D. Larson, *Bacillus thuringiensis* distribution in soils of the United States, *Can. J. Microbiol.,* **27,** 865 (1981).

48. A. Burgerjon and D. Martouret, "Determination and Significance of the Host Spectrum of *Bacillus thuringiensis,"* in H. D. Burges and N. W. Hussey, Eds., *Microbial Control of Insects and Mites,* Academic Press, New York, 1971, p. 305.

49. P. Luthy and H. R. Ebersold, *"Bacillus thuringiensis* Delta-Endotoxin: Histopathology and Molecular Mode of Action," in E. W. Davidson, Ed., *Pathogenesis of Invertebrate Microbial Diseases,* Allanheld, Osmun, Totowa, 1981, p. 235.

50. E. F. Knipling, *The Basic Principles of Insect Population Suppression and Management,* U.S.D.A. Agric. Handb. No. 512, 1979.

51. R. G. Luttrell and W. C. Yearian, "Elcar® — Its Role in Midsouth Cotton Pest Management Programs," in *Summary Proc. Cotton Biol. Control Conf.,* Sandoz, Inc., 1981, p. 33.

52. J. R. Phillips, W. F. Nicholson, T. Teague, J. Bernhardt, and T. F. Mueller, "Community Insect Management Programs," in *Summary Proc. Cotton Biol. Control Conf.,* Sandoz, Inc., 1981, p. 43.

53. J. D. Harper, R. M. McPherson, and M. Shepard, "Geographical and Seasonal Occurence of Parasites, Predators and Entomopathogens," in H. N. Pitre, Ed., *Natural Enemies of Arthropod Pests in Soybean,* South. Coop. Ser. Bull. 285, 1983, p. 7.

54. J. M. Livingston, P. J. McLeod, W. C. Yearian, and S. Y. Young III, Laboratory and field evaluation of a nuclear polyhedrosis virus of the soybean looper, *Pseudoplusia includens, J. Ga. Entomol. Soc.,* **15,** 194 (1980).

55. D. Boucias, L. Rathbone, S. A. Abbas, and N. Hostettler, Predators as potential dispersal agents of the nuclear polyhedrosis virus of *Anticarsia gemmatalis* (Lep.: Noctuidae) in soybean, *Entomophaga* (in press).

56. M. S. T. Abbas and D. G. Boucias, Interaction between nuclear polyhedrosis virus-infested *Anticarsia gemmatalis* (Lepidoptera: Noctuidae) larvae and predator *Podisus maculiventris* (Say) (Hemiptera: Pentatomidae), *Environ. Entomol.,* **13,** 599 (1984).

57. Y. Tanada, "Ecology of Insect Viruses," in J. F. Anderson and H. K. Kaya, Eds., *Perspectives in Forest Entomology,* Academic Press, New York, 1976, p. 265.

58. M. E. Martignoni, "Production, Activity, and Safety," in M. H. Brookes, R. W. Stark, and R. W. Campbell, Eds., *The Douglas-Fir Tussock Moth: A Synthesis,* U.S.D.A. Tech. Bull. 1585, 1978, p. 140.

59. F. B. Lewis, "Registration," in C. C. Doane and M. L. McManus, Eds., *The Gypsy Moth: Research Toward Integrated Pest Management,* U.S.D.A. Tech. Bull. 1584, 1981, p. 514.

60. R. R. Mason and R. F. Luck, "Population Growth and Regulation," in M. H. Brookes, R. W. Stark, and R. W. Campbell, Eds., *The Douglas-Fir Tussock Moth: A Synthesis,* U.S.D.A. Tech. Bull. 1585, 1978, p. 41.

61. R. W. Campbell, "Historical Review," in C. C. Doane and M. L. McManus, Eds., *The Gypsy Moth: Research Toward Integrated Pest Management,* U.S.D.A. Tech. Bull. 1584, 1981, p. 65.

62. F. B. Lewis and W. G. Yendol, "Efficacy," in C. C. Doane and M. L. McManus, Eds., *The Gypsy Moth: Research Toward Integrated Pest Management,* U.S.D.A. Tech. Bull. 1584, 1981, p. 503.

63. C. W. McCoy, "Pest Control by the Fungus *Hirsutella thompsonii,"* in H. D. Burges, Ed., *Microbial Control of Pests and Plant Diseases,* Academic Press, New York, 1981, p. 499.

64. C. W. McCoy and T. L. Couch, Microbial control of the citrus rust mite with the mycoacaricide, Mycar®, *Fla. Entomol.,* **65,** 116 (1982).

65. C. M. Ignoffo, "The Fungus *Nomuraea rileyi* as a Microbial Insecticide," in H. D. Burges, Ed., *Microbial Control of Pests and Plant Diseases,* Academic Press, New York, 1981, p. 513.

66. J. E. Henry and E. A. Oma, "Pest Control by *Nosema locustae,* a Pathogen of Grasshoppers

and Crickets," in H. D. Burges, Ed., *Microbial Control of Pests and Plant Diseases,* Academic Press, New York, 1981, p. 573.

67. J. E. Henry, Experimental application of *Nosema locustae* for control of grasshoppers, *J. Invertebr. Pathol.,* **18,** 389 (1971).

68. R. P. Jaques and O. M. Morris, "Compatibility of Pathogens with Other Methods of Pest Control and with Different Crops," in H. D. Burges, Ed., *Microbial Control of Pests and Plant Diseases,* Academic Press, New York, 1981, p. 695.

69. R. S. Soper, F. R. Holbrook, and C. C. Gordon, Comparative pesticide effects on *Entomophthora* and the phytopathogen *Alternaria solani, Environ. Entomol.,* **3,** 560 (1974).

70. N. Wilding, The effect of systemic fungicides on the aphid pathogen, *Cephalosporium aphidicola, Plant Pathol.,* **21,** 137 (1972).

71. G. Benz, "Synergism of Micro-organisms and Chemical Insecticides," in H. D. Burges and N. W. Hussey, Eds., *Microbial Control of Insects and Mites,* Academic Press, New York, 1971, p. 327.

72. R. P. Jaques, "Manipulation of the Environment to Increase Effectiveness of Microbial Agents," in G. E. Allen, C. M. Ignoffo, and R. P. Jaques, Eds., *Microbial Control of Insect Pests: Future Strategies in Pest Management Systems,* NSF—USDA—Univ. Fla., Gainesville, 1978, p. 72.

73. J. G. Burleigh, Comparison of *Heliothis* spp. larval parasitism and *Spicaria* infection in closed and open canopy cotton varieties, *Environ. Entomol.,* **4,** 574 (1975).

74. L. P. Kish, "Manipulation of the Environment: Fungi," in G. E. Allen, C. M. Ignoffo, and R. P. Jaques, Eds., *Microbial Control of Insect Pests: Future Strategies in Pest Management Systems,* NSF—USDA—Univ. Fla., Gainesville, 1978, p. 85.

75. R. P. Jaques, Application of viruses to soil and foliage for control of the cabbage looper and imported cabbageworm, *J. Invertebr. Pathol.,* **15,** 328 (1970).

76. J. Kalmakoff and A. M. Crawford, "Enzootic Virus Control of *Wiseana* spp. in the Pasture Environment," in E. Kurstak, Ed., *Microbial and Viral Pesticides,* Marcel Dekker, New York, 1982, p. 435.

16

EPIZOOTIOLOGY: PREVENTION OF INSECT DISEASES

TOSIHIKO HUKUHARA

Faculty of Agriculture
Tokyo University of Agriculture and Technology
Fuchu, Tokyo, Japan

1. **Introduction**

2. **Disease-Free Stock**
   2.1. Historical Perspective
   2.2. Inspection and Selection of Disease-Free Stock
   2.3. Elimination of Disease from Infected Stock

3. **Sanitation**
   3.1. Rearing Facilities and Programs
   3.2. Disinfection of Rearing Area and Equipment
   3.3. Uncontaminated Diet
   3.4. Application of Chemicals to Insects
   3.5. Cleaning or Removal of Contaminated Rearing Unit
   3.6. Effects of Sanitary Regime

4. **Resistant Insect Varieties**

5. **Modifications of Adverse Rearing Conditions to Correct Predisposing Causes**

6. **Conclusions and Recommendations**

   References

1. INTRODUCTION

Since the dawn of history, man has been engaged in colonizing silkworm and honeybee. It is not surprising, therefore, that diseases of these insects were the first to be observed and recorded. Ancient Chinese such as Guan (?–645 B.C.) advocated that farmers who reared silkworms without any disease should be rewarded with gold and food (1). Various nostrums were recommended long before the cause of disease was known. Ancient Romans such as Virgil (70–19 B.C.) advised that various herbs and wine should be fed to ailing honeybee colonies (2), but there is no evidence that such treatment was beneficial. It was not until the nineteenth century that the microbial etiology of economically important disease was realized and effective control measures were developed (3).

At present, there is extensive rearing of many insects not only for their useful agricultural products but also for other practical purposes, for example, the release of sterile males; the application of pheromones; the testing of insecticides; the production of insect pathogens, parasites and predators; and many other experimental investigations and laboratory work.

When insect colonizations were first attempted, noninfectious diseases often occurred because of inadequate nutritional and environmental conditions. Once the rearing practices had been established, however, such problems were seldom important. On the other hand, the presence of infectious diseases is frequently a potential hazard in the established cultures of certain insects. If permitted to get out of hand, it can result in a complete loss of the culture.

The colonization of insects by man involves several significant departures from the situation in a natural ecosystem. One of the most important is the temporal and spatial accumulations of insects that are highly uniform in development and genetic constitution. Most rearing practices are designed to meet the market demands for uniformity, to increase yields, or to reduce labor and other costs of production. Accordingly, the potential for catastrophic losses due to infectious diseases is greater in the more intensively managed colonization systems. Inasmuch as disease is an inherent component of the system, insect rearing must be carefully managed on a continuous, knowledgeable basis. The management of insect disease is more population than individual oriented because of the economics involved. The principal strategy is the prevention of disease. Therapy is not usually undertaken except for cases where it is the best method of preventing the spread of infection.

The purpose of this chapter will be to explore the relationship between epizootiology of insect disease and its actual management in insect colonization. No attempt will be made to collate specific control recommendations, since such information is available in other publications (4–7). However, specific diseases and management tactics will be cited to illustrate basic principles or strategies.

2. DISEASE-FREE STOCK

Experiences gained during the past hundred years of sericulture have high-lighted the importance of disease-free stock in the success of a rearing program. This is usually the first factor that should be taken into consideration to initiate an insect culture (4). Frequently, "disease-free" is a misnomer because a low incidence of initial disease is the goal instead of absolute elimination. For some diseases, especially chronic ones, this approach alone may be adequate. A notable example is pebrine, a microsporidian disease of the silkworm.

2.1. Historical Perspective

In the mid-nineteenth century, France was one of the major sericultural coun-tries. Annual production of cocoons amounted to 26,000 tons in 1853, about one-tenth of the world's production (8). A devastating epizootic of pebrine was first noticed in 1849, and it increased in severity year after year to such an extent that the annual production was reduced to 4000 tons in 1865, equivalent to an annual loss of 100 million francs. Even before the causal nature of disease was understood, French farmers discovered empirically that the use of eggs from disease-free countries would ensure good yields of cocoons. Thus, a total of 800 tons of silkworm eggs were imported during a period of 20 years, from 1854 to 1873, at a cost of 300 million francs. The French obtained eggs from Spain, Italy, Turkey, China, and Japan, in chronological order, based on the absence of pebrine in each country.

Louis Pasteur (9) began investigating this disease in 1865 and, after 5 years, completed a practical method for selecting disease-free stocks. He first showed that the pathogen was transmitted through the eggs of a heavily infected moth. On the basis of this information, he selected eggs that gave rise to healthy larvae by microscopically examining the moths for the presence of "corpuscles," which we now know to be spores of *Nosema bombycis.* He recommended using eggs laid only by moths with no spores in their tissues. If a moth was found to harbor spores, its eggs were destroyed. This technique assured farmers against extensive loss in cocoon yield, because larvae from disease-free stocks rarely died before cocooning even if they were infected during the rearing. Based on this technique, French sericulture gradually recovered to such an extent that cocoon production reached the level of 11,400 tons in 1877. It declined thereaf-ter due to several economic causes, such as increase of labor costs and competi-tion with oriental silk and artificial fibers.

Pasteur's method has been applied more extensively by legal regulation in Japan, which has become the world's largest silk producer in this century. The first regulations of silkworm egg inspection were established by the Ministry of Agriculture and Commerce in 1886. The Sericulture Act was promulgated in 1911, and later amended in 1917, 1929, 1945, and 1956 (10). Through legal constraints, egg production has been restricted to licensed industrialists who have the expertise and appropriate facilities. The law has imposed upon the industry the responsibility for rearing silkworms under special sanitary conditions and for obtaining certificate of pebrine inspection for each lot of produced eggs. Certification has also been required for importing and exporting silkworm eggs. The statistics of the certification program attest that the law has been effective in suppressing pebrine (Fig. 16.1). The rejection rate of moths used for the maintenance of parental stock lines declined from 24% in 1898 to 3% in 1951. It is no more than 0.3% in recent years. In the production, moth lines from which the eggs are provided for the farmers, the rate of rejection of egg-production lots has been less than 4% since 1949.

Another example of legal regulation to obtain disease-free stocks is the quarantine of honeybees in the United States, as imposed by the Honeybee Act (12). It reads in part: "In order to prevent the introduction and spread of diseases . . . harmful to honeybees, . . . the importation into the United States of all honeybees is prohibited, except that honeybees may be imported . . . from countries determined by the Secretary of Agriculture . . . to be free of diseases. . . . "

2.2. Inspection and Selection of Disease-Free Stock

Inspection of individual insects and the selection of uninfected ones may be possible in small-scale rearings. In mass rearings, however, this is not practical and economical, and the method of inspection is to use a sampling plan based on the concept of statistical quality control. In the Japanese program of pebrine inspection, moths were examined individually until 1968, when a group in-

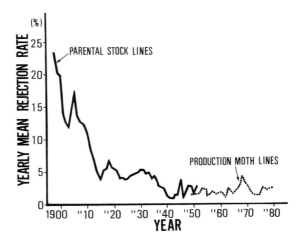

Figure 16.1. Prevalence of pebrine in silkworm egg producers in Japan (11). Rejection rates of individual moths and of egg-production lots are shown for parental stock lines and production moth lines, respectively.

spection method was adopted to reduce labor cost. The certification standard varies depending on the use of the eggs. If they are meant for the maintenance of parental lines, all mother moths must be free of spores. In the group inspection method, 14 or 28 moths are squashed together with a specifically designed instrument to obtain a microscopic preparation (13). If spores are found, all of the eggs laid by the moths are destroyed. The certification standard is less strict if the produced eggs are to be used in farmers' rearings. Mother moths are inspected according to a sampling inspection plan (Table 16.1). Each inspection unit is composed of 30 moths. The presence of a certain fraction (0.5%) of diseased moths can be tolerated by the farmers (15). The probability of receiving a lot worse than 0.5% is 0.0092 to 0.0149. A lot may be accepted at once if all units in the first sample are free from spores, or rejected at once if the number of contaminated units equals or exceeds the rejection number. If the first sample is inconclusive, the decision is based on the combined first and second samples.

Two protozoan pathogens were successfully eliminated from laboratory colonies of the boll weevil *Anthonomus grandis* by a system of "Unit Integrity" not unlike the approach of the above program of pebrine inspection (16). A "Unit" was composed of the most homogeneous group of weevils obtainable. Each unit was maintained separately until it was declared disease free, and as soon as disease was detected in a unit, the entire unit was destroyed. The system produced larger numbers of healthy insects more quickly than the selection of progeny from single pairs of weevils. It had the disadvantage of being more susceptible to failure when the disease prevalence was high.

2.3. Elimination of Disease from Infected Stock

Certain infectious diseases can be eliminated from infected stocks by artificially interrupting the transmission cycle. The egg is usually the best stage for applying the preventive measures. One of the most commonly used techniques is the surface disinfection of eggs with chemicals (17). The importance of egg disinfection to successful colonization has been demonstrated in cabbage looper rearing (18).

Surface disinfection interferes with transovum transmission by inactivating the pathogens on the egg surface but has no effect on those within the egg. High temperature is often less harmful to the embryo than to the pathogens within the ovum (19, 20). For example, corn borer eggs are freed from a microsporidium if immersed in a water bath for 30 minutes at 43.3°C (21). However, tolerated conditions for insect and pathogen have not been fully exploited to meet the ultimate aim of disease management.

3. SANITATION

Sanitation means the act or process of establishing the absence of any agents that can be injurious to health. The primary techniques are cleanliness and

TABLE 16.1

Sampling Inspection Plan for Pebrine (14)

Number of Mother Moths in the Lot	Number of Mother Moths		Rejection Number	Consumer's Risk
	First Sample	Second Sample		
390 or fewer	All moths	—	1	—
391–500	390	—	1	0.0107
501–600	450	—	1	0.0092
601–800	510	—	1	0.0117
801–1,000	570	—	1	0.0107
1,001–2,000	750	210	2	0.0138
2,001–3,000	870	480	3	0.0149
3,001–4,000	900	840	4	0.0135
4,001–6,000	960	1,140	5	0.0137
6,001–10,000	990	1,530	6	0.0146
10,001–30,000	1,020	1,680	6	0.0146
30,001 or more	1,050	1,740	6	0.0146

sterilization. Sterilization means the complete elimination of microbial viability regardless of the nature of the microorganisms. Sometimes the term is used loosely in referring to the destruction of only those microorganisms capable of causing disease. A better term for this latter meaning is "disinfection," that is, the killing of pathogenic agents by directly applied chemical or physical means (22). Rearing facilities and programs should be designed in such a way as to permit the full application of these techniques. Some techniques reduce the amount or efficacy of the initial pathogen population; others suppress the rate of epizootic development. Sanitation is such an important component of disease management that it is essential for the continued successful production of certain insects.

3.1. Rearing Facilities and Programs

Most modern insectaries are designed so that each step of the rearing operation is conducted in isolation to prevent the transfer of microorganisms from contaminated materials, workers, or air (23, 24). Regardless of how well the insectaries are arranged, contamination can still occur if traffic of personnel is not carefully regulated. Ideally, the personnel should be specialized so that those working in the rearing area would not enter the diet preparation area.

The basic step in inaugurating an insect-rearing project is the development of a rearing unit, such as a vial, tray, cage, or room. The use of small units greatly facilitates the sanitation of certain contaminated units without threatening the remainder of the culture. If insects are reared individually in separate units, the chance of horizontal transmission is held to a minimum. From an economic standpoint, however, it is usually more desirable to rear insects collectively in large units. Increased insect densities in a production unit or in a rearing space can result in greater mortality due to increased opportunity for the transmission of infectious diseases. For example, when honeybee colonies are crowded together in apiaries, this increases the possibility of the spread of disease to other colonies by robbing or drifting bees (12).

Crowding can be achieved also in "time." Pathogens are favored by continuous insect rearings because they are spared the rigors of surviving long periods without a suitable host. The intervention of insect-free periods between successive rearings provides the chance for sterilizing the rearing area and thus aids in preventing the accumulation of pathogens.

3.2. Disinfection of Rearing Area and Equipment

The entire rearing area and equipment should be thoroughly disinfected to reduce initial pathogen populations before the rearing is started. Use of disposable equipment such as plastic and waxed-card containers saves the labor of disinfection. Heat sterilization is the most reliable means of disinfection, but convenience and circumstance may demand the use of chemical sterilization (22). When chemical sterilization is used, the room and equipment should first be thoroughly cleaned of all foreign matter that might harbor microorganisms, because pathogens embedded within debris or dust are protected from disinfection. After the equipment has been cleaned and sterilized, it should be stored in a place free from air currents carrying dusts that might be laden with pathogens. Ultraviolet light is capable of providing continuous disinfection of contaminated atmospheres. It is useful in combination with air filtration which removes the larger dust particles, while the ultraviolet light is more effective against bacteria suspended in air.

3.3. Uncontaminated Diet

Of primary importance is the necessity of providing the insects with clean food, free from insect pathogens. Use of artificial diet for insect rearing ensures safety because it is more amenable to heat sterilization than natural diet (25). Natural plant food should be obtained from an area where the host plant is growing in the absence of diseased insects. Although it is usually possible to find such areas, certain diseases are so common in natural insect populations that the host plant nearly always is contaminated with pathogens. For example, naturally occurring residues of 4.6×10^3 – 3.3×10^7 polyhedra of cabbage looper nuclear polyhedrosis virus were found on cabbages per gram of heads harvested in Maryland (26). Pathogen residues can be reduced by controlling the natural populations of the insect being reared and/or the alternate insect hosts of the pathogen (27). Unfortunately, the residual activity of insecticides on the host plant severely restricts the practicability of this approach. One can disinfect fresh plant food with chemicals (18) or grow the food plant in a greenhouse free from the reared insect and the alternate insect hosts. These management tactics are, however, impractical for mass rearings.

3.4. Application of Chemicals to Insects

Certain chemicals applied to insects inhibit pathogenesis by suppressing the pathogen growth before or after infection. Routine chemical applications rarely prevent pathogenesis by 100% of the treated insects. However, adequate disease suppression in insect colonies can be achieved if incidence is maintained at a very low level.

There are two practical methods of chemical application: incorporation into the food and application to the insect body surface. Artificial diets usually contain several antimicrobial agents used as preservatives. Some of them inhibit the multiplication of certain insect pathogens. Some chemicals are incorporated into insect diets specifically to inhibit the growth of insect pathogens (28). For example, sucrose syrup containing oxytetracycline (Terramycin) is administered to honeybee larvae prior to the seasonal appearance of European foulbrood and as early in the season as practicable. Such a practice forestalls the multiplication and accumulation of the bacterial pathogen, and the disease outbreaks are decreased in severity (6). Based on this principle, a system of antibiotic treatment of honeybee colonies has been approved and administered officially in Britain since 1967. When diet incorporation is difficult, the chemicals can be sprayed onto the insect body. A dust formulation of a disinfectant (Pafusol) is applied in this manner in Japanese sericulture (Table 16.2).

3.5. Cleaning or Removal of Contaminated Rearing Unit

Not only diseased and dead insects but also insect excreta can be an important source of infection. Diseased insects emit certain pathogens in the feces. Unconsumed diet and feces can support the growth of saprophytic pathogens. Cleaning and removal of contaminated rearing beds suppress the rate of occurrence of new infections. Perhaps the best approach is to inspect and remove the source of infection, as is done routinely by adult nurse honeybees, which detect and remove from the comb larvae suffering from foulbrood before their cells are sealed (30). If inspections are impractical, the contaminated rearing beds can be isolated from healthy insects by spraying inert material, such as slaked lime, on the bed prior to the addition of a fresh supply of food.

In the case of certain viral and bacterial infections, the discovery of even a single diseased insect is reason enough to proceed rather ruthlessly and destroy the involved rearing unit. Destruction of honeybee colonies infected with foulbrood has been required by law in some countries.

3.6. Effects of Sanitary Regime

Measurement of losses due to disease is an important first step in evaluating the potential of disease management practices. The legal enforcement of the destruction of honeybee colonies with foulbrood in Switzerland from 1908 to 1933 and in Britain from 1942 to 1967 did not reduce the prevalence of European foulbrood. On the other hand, the prevalence of American foulbrood declined from 13% of inspected colonies to 2% after 25 years in Switzerland (31), and from 7 to 1% in Britain (32, 33).

Similar data have been collected for an insurance system for Japanese sericultural farmers. This system was established in 1947 to compensate for losses through natural disasters and unexpected calamities including disease outbreaks. The statistics are based on the official reports and are incomplete because the certification methods depend on their practicability and uniformity rather than on complete accuracy. Yet the general mortality trend over an extended period is informative (Table 16.2). The estimated disease loss in cocoon output was as high as 10.3% of the total cocoon production in 1956 (109,169 tons). It remained at a high level (7 to 10%) until 1964, but thereafter declined rapidly, and reached a low level (2 to 3%) in 1970s. Economic losses in recent years are estimated to be between 10 and 20 million dollars per year. The rapid decline is generally ascribed to strict sanitary regime in farmers' rearings as shown by the increased use of chemicals during this period (5, 34). Of the various diseases causing economic damage, the largest loss is caused by fla-

TABLE 16.2

Loss of Cocoon Yields due to Various Silkworm Diseases and the Amount of Chemicals Used in Japanese Silkworm Rearing (29)

Disease or Chemical	Year	
	1956	1979
Decrease in Cocoon Yield Expressed as Percentage of Total Yield		
Flacherie	7.8	1.2
Muscardine	1.2	0.1
Nuclear polyhedrosis	1.0	0.3
Aspergillosis	0.06	0.02
Parasitization by fly	0.01	0.001
Parasitization by mite	0.03	0.03
Tobacco poisoning	0.02	0.006
Intoxication by chemicals	0.02	0.08
Others	0.10	0.07
Amount of Chemicals Used per 100 kg Cocoon Obtained		
Formalin	175.4 g	2159.8 g
Chlorinated lime	2.4 g	—
Pafusol[a]	72.3 g	523.4 g
Slaked lime	2.2 kg	15.0 kg

[a]Mixture of paraformaldehyde, 1-oxy-4-methyl-4-isopropylbenzene, and Japanese acid clay.

cherie, a general term used for silkworm diseases with the common syndrome of flaccidity resulting mostly from viral and bacterial infections.

The application of formalin for disinfection over large areas and for many years accelerated evolutionary changes in pathogen populations. Formaldehyde-resistant strains of *Aspergillus* fungi became so prevalent in sericultural rearing facilities in the 1960s that the recommended concentration (2 to 3%) of formalin was no longer capable of ensuring adequate control (34).

Experimentally, resistant strains were selected rapidly after repeated exposures to increasing doses of formaldehyde (35). Both the naturally and experimentally induced resistant strains were more virulent to silkworm larvae than the sensitive isolates. Some strategies for managing the resistance problem include the use of mixtures of antifungal agents and the alternate use of disinfectant. However, we do not have appropriate pairs of chemicals to use alternatively. The development of new disinfectants by industry is becoming an increasingly more difficult problem. It is important, therefore, to prolong the useful life of existing disinfectants as much as possible.

4. RESISTANT INSECT VARIETIES

The use of resistant varieties is attractive from the managerial and economic standpoints because it requires no action during rearing. The magnitude of resistance ranges from very low to very high. When genetic resistance is so high as to completely prevent pathogenesis, there will be no initial disease to start an epizootic. When the pathogen growth in individual insects is partially suppressed or when hygienic behavior is effective, as in the case of honeybee strains resistant to American foulbrood (36), the disease prevalence increases more slowly in resistant insect populations than in susceptible populations.

The improvement of insect varieties is an activity involving breeding and selection for desired characters. The first task of a breeder, therefore, is to describe in rather specific terms what characters are desired in the variety to be produced. He certainly will list a number of characters. Disease resistance is only one of them. Thus, it is important to integrate all of these characters into one breeding program. The simplest conceptual and practical approach is to breed for "total ecosystem adaptation." This means, operationally, to breed and select for yield and quality at a few representative rearing sites under variable environmental conditions that affect disease development. When many undefined diseases influence yield and quality, this is the most direct route to a new desirable variety.

The breeding procedure is much easier if it involves a single prominent resistance mechanism whose inheritance is controlled by a single major gene. The gene can be transferred to a suitable variety by backcrossing. However, limiting factors can be anticipated on the basis of the experience of plant pathologists, who are far ahead of other protection sciences in utilizing genetic resistance for disease management (37). Some plant pathogen populations have been found to be remarkably adaptable and, therefore, have overcome the effects of host resistance conferred by a single gene. This adaptability has been confirmed in insect pathology by the occurrence, in farmers' rearings of silkworms, of a new isolate of *Bombyx* densonucleosis virus, which is virulent to silkworm varieties previously reported to be refractile to this virus (38, see Ch. 4).

Polygenically inherited resistance is likely to be stable. The greater the number of genes involved, the less likelihood that the resistance can be overcome by the pathogen. Use of polygenic resistance will require population breeding methods and changes in gene frequencies rather than pedigree breeding and gene transfers. The use of F_1 hybrids lends itself to the synthesis of multiply resistant hybrids by combining different dominant genes for resistance in the parental lines. For example, comparative tests of many silkworm varieties of their F_1 hybrids have shown that mortality during the rearing is reduced by 35 to 48% over the average of parental inbred lines (39).

5. MODIFICATIONS OF ADVERSE REARING CONDITIONS TO CORRECT PREDISPOSING CAUSES

Insects encounter inadequate environmental conditions less often in the insectary than in the field. Most rearing practices, which are designed primarily for good yield and quality, are usually adequate for maintaining a good physiological state of the insects. This is not always the case, however, and sometimes extensive losses occur as the result of deleterious effects of adverse noninfectious conditions.

By the beginning of the 19th century, sericulturists recognized the influence of several environmental factors on the course of silkworm diseases and considered them to be the actual causes. As it became known that silkworm diseases were caused primarily by microorganisms, the various environmental influences were considered as factors that predisposed the insect to disease. Pasteur (9), who demonstrated the contagiousness of flacherie, believed that certain kinds of bacteria in the gut of the diseased insect were responsible for the dysfunction of the gut and larval death. The bacterial multiplication was conditioned by certain predisposing causes, such as the excessive accumulation of larvae of different ages, excessive humidity, excessively high temperature during molting, prolonged lack of aeration, change in the quality of mulberry leaves, and the use of heated and unaerated leaves. The preventive measure recommended by Pasteur was to increase the vigor of the silkworms by all possible means.

Pasteur's theory on the etiology of flacherie was not universally accepted, because the disease was not always reproduced in healthy larvae by the peroral

administration of the bacteria that were suspected of being particularly important in inducing flacherie. Recent workers found that food condition greatly
influenced the pathogenicity of one of these bacteria. Some strains of the *Streptococcus faecalis-faecium* group were highly virulent to silkworm larvae reared
on an artificial diet but were not pathogenic to larvae reared on mulberry leaves
(40). Moreover, the abnormal proliferation of the streptococci in the gut prepared an environment favorable for the invasion of a potential pathogen, *Serratia piscatorum,* into the hemocoel (41). The question arises as to what factors
are involved in bringing about the increased virulence of the streptococci. Is it
due to the lack of antibiotics derived from the mulberry leaves? Do artificial
diets contain any nutritional elements favorable for the bacteria? Is the lower
oxygen level or the reduced intestinal microflora responsible? We still do not
have definitive answers to these questions.

Most beekeepers recognize that the control of Nosema disease is largely a
question of rearing practices. If brood temperature can be maintained within
the optimum range (92 to 95°F) by the colony, there will be less trouble from
the disease (12). A good apiary site should, therefore, be selected where the
colonies are protected from winds and have the full benefit of available sunlight.
Chilling of the brood may lead to physiological weakening that helps the microsporidia to spread in the colony (42). Some investigators propose a different
explanation for the effect of environmental temperatures, as revealed by the
seasonal fluctuation of the disease incidence (6). The decline of disease during
the summer may be ascribed to the reduced chance of contamination of combs
with the spore-containing excreta of infected worker bees which spend more
time outside in summer. Others believe that the combination of more flight
time and exposure to high temperatures is responsible for the apparent low
prevalence of Nosema disease in the southwestern United States, which has a
sunnier and more arid climate than the southeastern United States, where there
is relatively high prevalence (43).

Although predisposing causes are sometimes vague and ill-defined as exemplified above, it is apparent that they can play an important role in the inducement, provocation, or activation of an infectious process or disease. The predisposing causes give rise to disorders in the functioning of organs, lesions of
tissues, or alterations in cell metabolism. Physiologically disturbed insects are
more susceptible to infection by a pathogen of a given disease-producing power
than are normal insects. There is also the possibility, especially in the case of
virus infections, that a pathogen is maintained in the host tissues for a prolonged period without pathological manifestation until the disease process is
activated by some chemical or physical factors.

If the virulence of certain pathogens is so low that only weakened insects are
infected, the modification of rearing practices to correct the predisposing causes
can be effective in reducing the initial disease and/or epizootic. If the pathogens
are highly virulent, the modification must be combined with other disease
management techniques to achieve adequate disease suppression.

6. CONCLUSIONS AND RECOMMENDATIONS

Disease management is the selection and use of appropriate techniques to suppress disease to a tolerable level. Knowledge of epizootiology is the principal basis for selecting and deploying existing techniques and for the development of new ones. The concept of a tolerable level reflects many complex and interacting factors that are technical, economical, epizootiological, and managerial in nature. Disease management will be most successful if it is integrated into the insect production system and if it employs diverse approaches. Changes in insect production techniques will affect disease management, as exemplified by the development of artificial diets. In order to avoid unnecessary cost to insect operators and detriment to the environment, disease management should be utilized only when actual or predicted disease loss exceeds the tolerable level. A constraint to this approach is that it requires an insectary operator to have accurate prognoses about disease severity.

Disease predictions require a thorough understanding of epizootiology. It is a challenge to professional insect pathologists to develop a practical method of disease forecasting. Available methods generally have been based on an assessment of the initial pathogen population (5) and are of limited usefulness for diseases induced by pathogens that typically have low numbers initially but have the potential for many secondary cycles. Disease monitoring would be the basis for determining whether or when control measures will be required for such diseases. This has not been utilized extensively because of technical difficulty. Detection of pathogens in insect excreta may be useful for aiding disease forecasts because a low level of disease can be recognized before the appearance of macroscopic symptoms and without sacrificing the insects being reared (44).

Appropriate use of disease forecasts assumes the availability of effective techniques of disease suppression. Application of chemicals is the principal tactic used against bacterial, fungous, or protozoan infections that have already appeared in insect colonies. In the future, chemicals that effectively prevent or inhibit virus infections and those that are not toxic to the pathogens but enhance host resistance may become available (45). Chemicals may also be used to influence the intestinal microflora which may in turn inhibit pathogen activities. It is becoming increasingly apparent that those who wish to exploit such novel use of chemicals should have a greater understanding of predisposing causes and their manipulation. The amount of research being accomplished on this aspect is sadly out of balance with that being conducted on the pathogens.

REFERENCES

1. Z. Guan, *Guanzu San-chuang-shu-pjan,* Vol. 22, part 75, ?–645 B.C.; through Z. Wang, Knowledge on the control of silkworm disease in ancient China, *Symp. Sci. Hist.,* **8,** 15 (1965).

2. Virgil, *The Georgics of Virgil, Book IV,* 37–29 B.C.; transl. by C. D. Lewis, Oxford Univ. Press, Oxford, 1941.

3. E. A. Steinhaus, *Disease in a Minor Chord,* Ohio State Univ. Press, Columbus, 1975.

4. C. N. Smith, Ed., *Insect Colonization and Mass Production,* Academic Press, New York, 1966.

5. Ministry of Agriculture and Forestry, *Guide to the Extension of Preventive Methods of Silkworm Disease,* Nihon Sanshi-Shinbun-Sha, Tokyo, 1972.

6. L. Bailey, *Honey Bee Pathology,* Academic Press, New York, 1981.

7. T. J. Helms and E. S. Raun, "Perennial Laboratory Culture of Disease-Free Insects," in H. D. Burges and N. W. Hussey, Eds., *Microbial Control of Insects and Mites,* Academic Press, New York, 1971, p. 639.

8. A. Mozziconacci, *Le Ver à Soie du Mûrier,* Librairie Hachette, Paris, 1921.

9. L. Pasteur, *Études sur la Maladie des Vers à Soie,* Gauthier-Villars, Paris, 1870.

10. Anonymous, *Laws and Regulations Related to Sericulture,* Sanshi-Shinbun-Sha, Tokyo, 1968.

11. Ministry of Agriculture and Forestry, *Statistics and Data on Silkworm Eggs,* Minist. Agric. For., Tokyo, 1898, 1905, 1912, 1918, 1950–1984.

12. A. I. Root, *The ABC and XYZ of Bee Culture,* A. I. Root Co., Ohio, 1980.

13. T. Fujiwara, A technique for pebrine inspection of moths in group, *Tech. Bull. Seric. Exp. Stn.,* **120,** 113 (1984).

14. Ministry of Agriculture, Forestry and Fisheries, *Guide to the Practical Pebrine Inspection,* Nationwide Assoc. Silkworm Seed, Tokyo, 1979.

15. K. Ohshima, Studies on the rational simplification of mother moth inspection method for the exclusion of pébrine disease from the silkworm, *Bull. Seric. Exp. Stn.,* **13,** 1 (1949).

16. E. McLaughlin, Laboratory techniques for rearing disease-free insect colonies: elimination of *Mattesia grandis* McLaughlin, and *Nosema* sp. from colonies of boll weevils, *J. Econ. Entomol.,* **59,** 401 (1966).

17. M. Shapiro, "In Vivo Mass Production of Insect Viruses," in E. Kurstak, Ed., *Microbial and Viral Pesticides,* Marcel Dekker, New York, 1982, p. 463.

18. T. J. Henneberry and A. N. Kishaba, "Cabbage Loopers," in C. N. Smith, Ed., *Insect Colonization and Mass Production,* Academic Press, New York, 1966, p. 461.

19. J. Weiser, "Immunity of Insects to Protozoa," in G. J. Jackson, R. Herman, and I. Singer, Eds., *Immunnity to Parasitic Animals,* Appleton-Century-Crofts, Meredith Corp., New York, 1969, p. 129.

20. G. Brun, *Étude d'une Association du virus σ et de Son Hôte La Drosophile: L'État Stabilisé, Thèse,* Lab. Genetique, Fac. Sci. Orsay, France, 1963.

21. E. S. Raun, Elimination of microsporidiosis in laboratory-reared European corn borers by the use of heat, *J. Insect Pathol.,* **3,** 446 (1961).

22. J. J. Perkins, *Principles and Method of Sterilization in Health Sciences,* 2nd ed., Thomas, Springfield, 1970.

23. N. C. Leppla and T. R. Ashley, Eds., *Facilities for Insect Research and Production,* USDA Tech. Bull. 1576, 1978.

24. C. C. Doane and M. L. McManus, Eds., *The Gypsy Moth: Research Toward Integrated Pest Management,* USDA Tech. Bull. 1584, 1981.

25. B. Greenberg, Sterilizing procedures in mass rearing of insects, *Bull. Entomol. Soc. Am.,* **16,** 31 (1970).

26. E. D. Thomas, "Normal Virus Levels and Virus Levels Added for Control," in M. Summers, R. Engler, L. A. Falcon, P. V. Vail, Eds., *Baculoviruses for Insect Pest Control: Safety Considerations,* Am. Soc. Microbiol., Washington, D.C., 1975, p. 87.

27. K. Ono and T. Yamaguchi, Sericultural loss due to white muscardine derived from broad-winged planthoppers, *Rep. Gunma Seric. Exp. Stn.,* **51,** 27 (1978).

28. E. G. King, J. V. Bell, and D. F. Martin, Control of the bacterium *Serratia marcescens* in an insect host-parasite rearing program, *J. Invertebr. Pathol.,* **26,** 35 (1975).

29. Ministry of Agriculture and Forestry, *Annual Report of Sericultural Statistics,* Minist. Agric. For., Tokyo, 1956–1984.

30. A. W. Woodrow and J. J. States, Removal of diseased brood in colonies infected with A.F.B., *Am. Bee J.,* **81,** 22 (1943).

31. F. Leuenberger, Jahresbericht der Faulbrutversicherung, *Schweiz. Bienen-Zeit.,* **56,** 134 (1933).

32. Ministry of Agriculture, Fisheries and Food, *Survey of Bee Health in England and Wales,* Minist. Agric., Fish. Food, London, 1956–1959, 1969.

33. Ministry of Agriculture, Fisheries and Food, *Beekeeping Statistics,* Minist. Agric., Fish. Food, London, 1974.

34. K. Kawakami and T. Mikuni, Studies on the causative fungi of *Aspergillus* disease of the silkworm larvae, *Bull. Seric. Exp. Stn.,* **23,** 327 (1969).

35. S. Saijo, Studies on the chemical resistance of the parasitic fungi of *Aspergillus* disease in the silkworm, *Bombyx mori* L., *J. Seric. Sci. Japan,* **39,** 43 (1970).

36. W. C. Rothenbuhler, Genetics and breeding of the honey bee, *Annu. Rev. Entomol.,* **3,** 161 (1958).

37. R. A. Robinson, New concepts in breeding for disease resistance, *Annu. Rev. Phytopathol.,* **18,** 189 (1980).

38. H. Seki, Interstrain differences in the resistance to the infection with densonucleosis virus (Yamanashi isolate) in the silkworm *Bombyx mori, J. Seric. Sci. Japan,* **54,** 445 (1985).

39. K. Osawa and C. Harada, Studies on the F_1-hybrids of the silkworm, *Bull. Seric. Exp. Stn.,* **12,** 183 (1944).

40. T. Nagae, The pathogenicity of *Streptococcus* bacteria isolated from the silkworm reared on an artificial diet, *J. Seric. Sci. Japan,* **43,** 471, (1974).

41. R. Kodama and Y. Nakasuji, Further studies on the pathogenic mechanism of bacterial diseases in gnotobiotic silkworm larvae, *Inst. Fermentation, Osaka, Res. Comm.,* **5,** 1 (1971).

42. C. Toumanoff, *Les Maladies des Abeilles,* Rev. Fr. Apic., Paris, 1951.

43. G. E. Cantwell, "Honey Bee Diseases, Parasites, and Pests," in G. E. Cantwell, Ed., *Insect Diseases,* Vol. II, Marcel Dekker, New York, 1974, p. 501.

44. T. Hukuhara, Detection of a virus from silkworm feces, *Seric. Sci. Technol.,* **21**(10), 44, (1982).

45. T. Hukuhara, M. Iso, and H. Abe, The effects of serotonin on the midgut motility and virus resistance in the silkworm *Bombyx mori, J. Seric. Sci. Japan,* **55,** 158 (1986).

AUTHOR INDEX

SUBJECT INDEX